PELAGIC FISH

The Resource and its Exploitation

EDITED BY

J.R. BURT

R. HARDY

AND

K.J. WHITTLE

Torry Research Station
Aberdeen, Scotland

FISHING NEWS BOOKS

Fishing News Books
A division of Blackwell Scientific
Publications Ltd
Editorial offices:
Osney Mead, Oxford OX2 0EL
25 John Street, London WC1N 2BL
23 Ainslie Place, Edinburgh EH3 6AJ
3 Cambridge Center, Cambridge,
MA 02142, USA
54 University Street, Carlton,
Victoria 3053, Australia

First published 1992

Set by Best-set Typesetter Ltd., Hong Kong
Printed and bound in Great Britain by
The University Press, Cambridge

DISTRIBUTORS

Marston Book Services Ltd
PO Box 87
Oxford OX2 0DT
(*Orders*: Tel: 0865 240201
Fax: 0865 721205
Telex: 83355 MEDBOK G)

USA
Blackwell Scientific Publications Inc
3 Cambridge Center
Cambridge, MA 02142
(*Orders*: Tel: (800) 759 6102)

Canada
Oxford University Press
70 Wynford Drive
Don Mills
Ontario M3C 1J9
(*Orders*: Tel: (416) 441-2941)

Australia
Blackwell Scientific Publications
(Australia) Pty Ltd
54 University Street
Carlton, Victoria 3053
(*Orders*: Tel: (03) 347-0300)

British Library
Cataloguing in Publication Data

Pelagic fish: the resource and its exploitation.
I. Burt, J.R. II. Hardy, R. III. Whittle, K.J.
597.05248

ISBN 0-85238-180-8

Dedication

This volume is dedicated to the memory of James Burt, who died 30 June 1991, our leading co-editor, and the main organiser of Torry Research Station's Diamond Jubilee Conference, the contributions to which are the basis for this book.

James was born in July 1933 in Colombo, Ceylon, and came to Scotland to St Andrews, Fife, in 1946, where he completed his secondary education and subsequently received a PhD in Biochemistry at the University of St Andrews in 1959.

He joined Torry Research Station in Aberdeen, in 1958, as a Research Fellow, and spent some 33 years devoted to research in fish and food science and technology. He was appointed Head of the Research Station's Laboratory in Hull in 1974, returned to Aberdeen 10 years later as Assistant Director, and subsequently took on the new post of Commercial Director.

James enjoyed a position of high regard and respect within the national and international scientific community for his work. He played a leading role in a number of professional bodies and organisations, particularly the Institute of Food Science and Technology, the International Association of Fish Meal Manufacturers, and the UK Association of Fish Meal Manufacturers. He maintained close academic and research contacts with many Universities both at home and abroad. He will be sorely missed by his family, friends, and colleagues.

R. Hardy
K.J. Whittle
15 July 1991

Contents

Contributors, ix
Preface, xi
Keynote address, xiii

REVIEWS

1 The global pelagic fish resource and its biological potential, 1
R.S. Bailey

2 Catching methods for pelagic fish, 21
H.R. English and J.E. Tumilty

3 Preservation of pelagic fish quality for further processing on board and ashore, 38
N.K. Sørensen and A. Mjelde

4 Shipboard handling of pelagic fish with special emphasis on fast handling, rapid chilling and working environment, 55
K.B. Olsen

5 Pelagic fish in the Netherlands, 70
H. Houwing

RESEARCH PAPERS

6 RSW treatment of herring and mackerel for human consumption, 73
L. Kraus

7 Surimi from blue whiting (*Micromesistius poutassou*) produced on board m/s Uksnøy: process and quality, 82
R. Ofstad, E. Grahl-Madsen and C. Solberg

8 A techno-economic model for designing large factory vessels and its application to the Antarctic krill-finfish fishery, 94
J.K. McElroy

REVIEW

9 Composition and deterioration of pelagic fish, 115
J.R. Burt and R. Hardy

RESEARCH PAPERS

10 Total lipids and omega-3 fatty acids from seven species of pelagic fish, 142
R. Pozo, B. Pérez-Villarreal and E. Saitua

11 Oxidation of frozen sardine mince: effect of washing and antioxidants addition, 148
F. Mestiri, V. Verrez, L. Han-Ching, M. Cardinal and M. Jerome

12 Use of a portable microwave fat meter with herring (*Clupea harengus*), 157
M. Kent, A. Lees and R.H. Christie

13 Effect of frying on the nutritional quality of dried salted Indian mackerel (*Rastrelliger kanagurta*), 165
F.W. Ma'ruf, D.A. Ledward, R.J. Neal and R.G. Poulter

14 Fishing gear influence on the spoilage of albacore (*Thunnus alalunga*) in ice, 170
B. Pérez-Villarreal and R. Pozo

15 Quantitative and qualitative determination of biogenic amines in fish, 178
J.P. Gouygou, C. Sinquin, M. Etienne, A. Landrein and P. Durand

16 Colour changes in the skin in relation to the formation of histamine in dorado (Mahi-mahi, *Coryphaena hippurus*), 187
C.D. Wood, N. Camba and M. Grijalva

17 Further volunteer studies on scombrotoxicosis, 194
P. Ijomah, M.N. Clifford, R. Walker, J. Wright, R. Hardy and C.K. Murray

18 The application of electrophoretic techniques to the identification of species of canned tuna and bonito, 200
I.M. Mackie, M. Chalmers, P. Reece, A.E. Scobbie and A.H. Ritchie

19 Influence of iced storage and washing on the protein composition of sardine (*Sardina pilchardus*) minces, 208
I. Filipi, M.L. Nunes, N. Vrucinic-Filipi and D. Roseg

20 Influence of chilled and frozen storage of sardines on the gelling properties of surimi, 214
A. Leinot, C. Ozenne and J.C. Cheftel

REVIEW

21 By-products from pelagic fish, 222
N. Urdahl

RESEARCH PAPER

22 Fish meal quality, 232
G.P. Savage

REVIEW

23 Utilisation of pelagic fish for human consumption, 238
K.J. Whittle and C.D. Wood

RESEARCH PAPERS

24 Some factors affecting the viability of utilisation of small pelagics, 254
J.A. Young, K. Crean and D.A. Palfreman

25 Consumer attitudes towards pelagic fish, 260
D.W. Marshall and J. Currall

26 Deconstructing sensory preferences; social factors influencing the demand for dark fish, 267
L.R. Gofton and D.W. Marshall

27 Taking new product evaluation beyond the 'sensorial', 272
D.W. Marshall, N.J. Boyd and L.R. Gofton

28 Development of a new fishery in Denmark. Catching, handling and utilisation of roundnose grenadier and greater argentine, 278
E.P. Larsen

29 Utilisation of sand eel (*Ammodytes tobianus*), greater sand eel (*Hyperoplus lanceolatus*), sprat (*Sprattus sprattus*) and Norway pout (*Trisopterus esmarkii*) for human consumption, 285
J. Nielsen, B. Rønsholdt, N.C. Jensen and N. Alsted

30 The drying of small pelagic fish: the Icelandic experience, 291
S. Arason, G. Thoroddson and G. Valdimarsson

31 Utilisation of roe and milt from herring, 299
R. Richardsen

32 Utilisation of deboned retextured mackerel flesh, 307
A. Mills and H. Teepsoo

33 Response surface methodology in product development, 311
J.E.P. Currall

34 Properties of surimi and derived gels from pelagic fish, 317
R.J. Hastings, J.N. Keay and K.W. Young

CASE STUDY

35 The Java Sea purse seine fishery: resources, production methods and product markets – a case study, 325
J.K. McElroy

Index 344

The following paper was presented at the Diamond Jubilee Conference but, because it has already been published, it was not included in this volume:

Markets for UK caught pelagic fish
D. Cleghorn and J. Moore
(Previously published as 'Pelagic Species Review – Resources and Markets'. Sea Fish Industry Authority, Market Development Department. Report No. 2002 165 pp, January 1989, Edinburgh, Scotland.)

A number of posters covering interesting related topics were presented at the conference, and are listed below so that readers may contact the appropriate research group if they wish to explore the subject matter more closely.

POSTERS

Cephalopod identification by PAGE-SDS: preliminary results
I. Batista and R. Mendes
Instituto Nacional de Investigaçao das Pescas, Lisboa, Portugal.

Gel forming ability and protein-protein interaction in mackerel (*Scomber scombrus*)
A. Elahee Doomun and I.M. Mackie
Torry Research Station, Aberdeen, Scotland.

Dietary monounsaturated fatty acids in the diets of Greenland eskimos
E.O. Elvevoll
Institute of Fishery Technology Research, Tromsø, Norway.

Protein functionality changes in sardine stored in ice with and without salt added
A. Huidobro, P. Montero, M. Tejada, F. Jimenez Colmenero and A.J. Borderias
Instituto del Frio, Ciudad Universitaria, Madrid, Spain.

Freshness grading of herring
C.M. Jespersen
Technological Laboratory, Ministry of Fisheries, Lyngby, Denmark.

Grouping of species of fish by simple sensory methods
P. Howgate and A.W. MacRae
Torry Research Station, Aberdeen, Scotland.

Trimethylamine-N-oxide demethylase of saithe's kidney (*Pollachius virens*): Multi-enzymatic system?
A. Joly, P. Cottin, L. Han-Ching and A. Ducastaing
Laboratoire de Biochimie et Technologie des Aliments, Université de Bordeaux I, Talence Cedex, France.

Contributors

ALSTED N. *Danish Aqua Feed International, Brande, Denmark*

ARASON S. *Icelandic Fisheries Laboratories, Reykjavik, Iceland*

BAILEY R. *Marine Laboratory, Aberdeen, Scotland*

BOYD N.J. *Torry Research Station, Aberdeen, Scotland*

BURT J.R. *Torry Research Station, Aberdeen, Scotland*

CAMBA N. *Instituto Nacional de Pesca, Guayaquil, Ecuador*

CARDINAL M. *IFREMER, Nantes, France*

CHALMERS M. *Torry Research Station, Aberdeen, Scotland*

CHEFTEL J.C. *Université de Montpellier II (Sciences et Techniques de Languedoc), Montpellier, France*

CHRISTIE R. *Torry Research Station, Aberdeen, Scotland*

CLIFFORD M.N. *University of Surrey, Guildford, England*

CREAN K. *Humberside International Fisheries Institute, University of Hull, Hull, England*

CURRALL J. *Torry Research Station, Aberdeen, Scotland*

DURAND P. *IFREMER, Nantes, France*

ENGLISH H.R. *Sea Fish Industry Authority, Hull, England*

ETIENNE M. *IFREMER, Nantes, France*

FILIPI I. *University of Zagreb, Zadar, Yugoslavia*

GOFTON L.R. *University of Newcastle upon Tyne, Newcastle upon Tyne, England*

GOUYGOU J.P. *IFREMER, Nantes, France*

GRAHL-MADSEN E. *Institute of Fisheries Technology Research, Tromsø, Norway*

GRIJALVA M. *Instituto Nacional de Pesca, Guayaquil, Ecuador*

HAN-CHING L. *IFREMER, Nantes, France*

HARDY R. *Torry Research Station, Aberdeen, Scotland*

HASTINGS R.J. *Torry Research Station, Aberdeen, Scotland*

HOBBS G. *Torry Research Station, Aberdeen, Scotland*

HOUWING H. *Netherlands Organisation for Applied Scientific Research, IJmuiden, The Netherlands*

IJOMAH P. *University of Surrey, Guildford, England*

JENSEN N.C. *Andelssildeoliefabrikken, Esbjerg, Denmark*

JEROME M. *IFREMER, Nantes, France*

KEAY J.N. *Torry Research Station, Aberdeen, Scotland*

KENT M. *Torry Research Station, Aberdeen, Scotland*

KRAUS L. *Danish Institute for Fisheries Technology and Aquaculture, Hirtshals, Denmark*

LANDREIN A. *IFREMER, Nantes, France*

LARSEN E.P. *Danish Ministry of Fisheries, Technical University, Lyngby, Denmark*

LEDWARD D.A. *The Queen's University of Belfast, Belfast, Northern Ireland*

LEES A. *Torry Research Station, Aberdeen, Scotland*

LEINOT A. *Université de Montpellier II (Sciences et Techniques du Languedoc), Montpellier, France*

McELROY J.K. *Forum Fisheries Agency, Honiara, Solomon Islands*

MACKIE I.M. *Torry Research Station, Aberdeen, Scotland*

MA'RUF F.W. *University of Nottingham, Sutton Bonington, England*

MARSHALL D.W. *University of Edinburgh, Edinburgh, Scotland*

MESTIRI F. *IFREMER, Nantes, France*

MILLS A. *Sea Fish Industry Authority, Hull, England*

MJELDE A. *Norwegian Herring Oil and Meal Industry Research Institute, Bergen, Norway*

MURRAY C.K. *Torry Research Station, Aberdeen, Scotland*

NEALE R.J. *University of Nottingham, Sutton Bonington, England*

NIELSEN J. *Technological Laboratory, Danish Ministry of Fisheries, Copenhagen, Denmark*

NUNES M.L. *Instituto Nacional de Investigaçao das Pescas, Lisboa, Portugal*

OFSTAD R. *Institute of Fishery Technology Research, Tromsø, Norway*

OLSEN K.B. *Technological Laboratory, Danish Ministry of Fisheries, Copenhagen, Denmark*

OZENNE C. *Université de Montpellier II (Sciences et Techniques du Languedoc), Montpellier, France*

PALFREMAN D.A. *Humberside International Fisheries Institute, University of Hull, Hull, England*

PÉREZ-VILLARREAL B. *Research Institute for Fish Science and Technology, Sukarrieta (Bizkaia), Spain*

POULTER R.G. *Overseas Development Natural Resources Institute, Chatham, England*

POZO R. *Research Institute for Fish Science and Technology, Sukarrieta (Bizkaia), Spain*

REECE P. *Torry Research Station, Aberdeen, Scotland*

RICHARDSEN R. *Institute of Fisheries Technology Research, Tromsø, Norway*

RITCHIE A.H. *Torry Research Station, Aberdeen, Scotland*

ROSEG D. *University of Zagreb, Zagreb, Yugoslavia*

RØNSHOLDT B. *University of Aalborg, North Sea Centre, Hirtshals, Denmark*

SAITUA E. *Research Institute for Fish Science and Technology, Sukarrieta (Bizkaia), Spain*

SAVAGE G.P. *Lincoln University, Canterbury, New Zealand*

SCOBBIE A.E. *Torry Research Station, Aberdeen, Scotland*

SINQUIN C. *IFREMER, Nantes, France*

SOLBERG C. *Institute of Fisheries Technology Research, Tromsø, Norway*

SØRENSEN N.K. *Institute of Fisheries Technology Research, Tromsø, Norway*

TEEPSOO H. *Sea Fish Industry Authority, Hull, England*

THORODDSSON G. *Icelandic Fisheries Laboratories, Reykjavik, Iceland*

TUMILTY J.E. *Sea Fish Industry Authority, Hull, England*

URDAHL N. *Norwegian Herring Oil and Meal Industry Research Institute, Bergen, Norway*

VALDIMARSSON G. *Icelandic Fisheries Laboratories, Reykjavik, Iceland*

VERREZ V. *IFREMER, Nantes, France*

VRUCINIC-FILIPI N. *University of Zagreb, Zagreb, Yugoslavia*

WALKER R. *University of Surrey, Guildford, England*

WHITTLE K.J. *Torry Research Station, Aberdeen, Scotland*

WOOD C.D. *Natural Resources Institute, Chatham, England*

WRIGHT J. *University of Surrey, Guildford, England*

YOUNG J.A. *Humberside International Fisheries Institute, University of Hull, Hull, England*

YOUNG K.W. *Torry Research Station, Aberdeen, Scotland*

Preface

In 1979, Torry Research Station celebrated 50 years of research activities into fish handling and processing. To mark the occasion, the staff organised an international conference which set out to review the whole area of fish as food. The outcome of this Golden Jubilee Conference was an extensive compilation of review papers and original research reports which still stands as the definitive account of the current state of knowledge of the subject.

Ten years on, the Director and staff decided to mark the Institute's Diamond Jubilee with, among other activities, another international conference. There was little point in duplicating the wide subject matter of the previous conference since 10 years have not changed the overall picture dramatically. The emphasis and interests of research and development activities and of industry around the world do change more quickly however, and a subject of topical international interest was sought. The subject of pelagic fish came to mind immediately. The application of science and technology to optimise the utilisation of pelagic fish has proved difficult in most of the world's fishing countries. The problems are, however, becoming more and more important in view of the expert forecasts that demand will exceed supply by the turn of the century. It was appropriate and timely, therefore, to focus the attention of a group of eminent scientists from around the world on these problems.

The conference organisers succeeded in attracting speakers with established international reputations and this helped to ensure the participation of fish scientists and technologists from more than 20 countries. Fish is an important food resource throughout the world, and it is necessary to take account of all aspects of its eating quality, wholesomeness and safety. There is a need to provide a basis which will enable the handling, processing and distribution industry to deliver fish products to the consumer which meet the highest possible standards. The papers presented at the meeting indeed cover all these aspects and to complete the picture, papers were presented which examine the resource, what raw material is available, and the methods of harvesting it. Both areas can have a considerable influence on the quality of the product reaching the consumer.

The proceedings of the conference in this book are a compilation of the refereed papers presented, and represent an authoritative statement of where we are with the use of pelagic fish and what can be done to optimise their use. They are a fitting tribute to the activities of many research workers throughout the world and especially to 60 years of work at the Torry Research Station.

It is always difficult to predict what will happen in the future, as it will be influenced by political and social considerations as much as by science and technology. The proceedings of this conference present clear statements of the potential and the value of the pelagic fishery resources, what can be done to make better use of them and where research and development work are needed to make further progress.

G. Hobbs,
Director, Torry Research Station

Keynote address

Ing. H. Houwing,
Director of the Institute of Fishery Products TNO, IJmuiden, Netherlands

Address delivered at the opening of the conference

I felt highly flattered some weeks ago when I received the invitation to open this Torry Jubilee Meeting. I was also surprised. I considered it sort of unusual to grant such honour to a foreigner. However, this approach might be rewarding, now that European boundaries are fading away. At any rate it is quite original.

Please do not think that what I say is meant only for this country. Part of what I am going to say could have been said in another country, for instance Holland, as well.

What I am going to say is not on behalf of my organisation and has not been discussed with anyone else. It is purely a personal view. Even if it is my own wisdom, that does not necessarily mean it is the absolute truth. I guess it was the Ancient Greeks who said something like: 'Nothing is true, and even that is not true'. And if the Ancient Greeks did not actually say that, they could quite easily have done so.

Sixty years of research in fish technology: that is a respectable period of time. Most people arriving at that age are thinking of retirement. Their job is coming to an end and their future is characterised by immense amounts of leisure time. It is quite different for research institutes. From time to time the laboratories and accommodation are rebuilt, new directors and new scientists are appointed, and so on. However, if these changes are too radical and take place in too rapid succession, the building may come down. I shall return to this point later on in a way.

What never should happen in my view is what took place across the ocean some years ago, when fish technology institutes were bluntly liquidated. Such a measure carries with it an enormous waste of capital and know-how. Now and then we are visited by people from industry from that large bilingual country who complain about not getting proper answers to their questions in their own country. That is not really funny. But the funny, or rather sad, thing is that in Newfoundland for instance they have just started a new centre, on fish technology among other things. It is unbelievable. Obviously they know the proverb overseas: a fault confessed is half redressed.

I remember that some people at my institute during discussions some years ago asserted, 'If you want to know something, go and look it up in the library or the data bank'. But, even if you receive a proper answer, do you think it works that way? In my opinion it does not! It is like following a written course in skiing. Even if you have done your utmost, I can assure you that your initial upright position, on your skis, will turn into a supine position pretty soon. What I mean to say with this simile is that reading only is not enough. To know how things work in practice is at least as important as reading. And this practical knowledge can be acquired only by having things done by researchers either working in institutes or in the industry. In conclusion, my message to present and future decision-makers with respect to this matter can be brief: do not dismantle institutes as your colleagues overseas have done unless you are sure that there are good alternatives. An institute without capable people who have knowledge in their grey matter instead of in their data banks will never be able to solve problems properly.

You might ask, 'Why do you say this, now that we have gathered to celebrate and commemorate?' I think that this is an excellent opportunity to say what I want to say because so many distinguished guests from this country and from abroad are attending this celebration. It is quite plausible that some of these guests will be involved in matters of this kind in the years to come. Looking around in the past few years I have seem some black clouds gathering over Europe on fish technology institutes. However, in contrast in some countries, policy-makers seem to realise how important fish technology institutes are. Institutes in these countries are even allowed to expand to some extent. I am referring now to our French and Danish counterparts up until recently. They must have some wise guys on their boards.

That people think the way people across the ocean have thought recently can be understood (but by no means justified). The reason for this possible understanding is that, strangely enough, fishermen and those who depend on them or support them differ markedly in approach from their colleagues in other branches of food industry. I have often wondered why. One explanation might be that there is a shortage of fish rather than a surplus. To put it bluntly, it is no big problem to get rid of all fish caught or processed. So, the idea that there is no need for further research in technology, apart from shipbuilding etc., is quite tempting. These people in the fisheries industry seldom seem to wonder whether they could not have earned more money if they had taken more care of the way fish is handled onboard or in the auction hall or during further processing. I think they are absolutely wrong. Not all fish processed is of good quality. I have arrived at this inescapable conclusion after occasional claims from unsatisfied customers brought before our Institute. The incidents we have been faced with are presumably just the tip of the iceberg. In these cases of discord I have often concluded: 'This would not have happened if good manufacturing practices had been adhered to'. We have been discussing matters of this kind for many years in Codex Alimentarius meetings organized by FAO/WHO. I sincerely hope that these discussions will soon produce some effect.

Now I will switch to another subject. I would like to give one or two words of advice to all fish processors: be attentive! Processors of other foodstuffs, in the agricultural market, are deeply involved in quality matters. That is a necessity to them because, in their branch, they are faced with surpluses. Be aware that there is a limit to the elasticity of food markets from an economic point of view. When consumers eat chicken they do not eat fish at the same time. These competitors of the fisheries industry will do their utmost to create new markets, for example for small households. The key words for this market segment are: easy to prepare, wholesome, tasty etc., but also cheap. Foodstuffs for this target group must be relatively cheap because most youngsters do not always like to spend their last penny on food.

That is why I think research in fish processing is very important, because lost markets are very hard to recover.

Let us return now to the reason why we have gathered here: to celebrate in a proper way and with due respect the anniversary of the sexagenarian. In the period I have known Torry, which is much more than two decades, I have seen that Torry has a wide variety of expertise and a broad know-how in a diversity of disciplines such as sensory assessment, chemistry, biochemistry, microbiology, fish processing, physics and engineering. Torry staff are frequently called upon by foreigners for assistance. People looking for support make tours of established institutes. Among the institutes visited Torry is frequently the chosen one, for at least two reasons, I guess. One reason is undeniably the language spoken at Torry. English has increasingly become the *lingua franca* for science, technology, trade and transport. The other reason is undoubtedly Torry's vast expertise.

Let me give an example. If I were asked, 'Because I consider the sensory characteristics of fish products to be a paramount factor determining fish quality, where could I be trained in organoleptic assessment?' I would answer promptly 'Go and visit the specialists at Torry. They have thought things through so thoroughly that you will not easily find an encyclopaedia that can defeat them. I guess that most of it will be in their brains, not in a file.' Another example: staff of my Institute have been trained by Torry in facets of making surimi. I also remember that, many years ago, much know-how was gathered at Torry with regard to minced fish, its problems and its opportunities.

Another characteristic of Torry staff has always been that they are most helpful in finding answers to any questions coming from colleague institutes. This could be attributable partly to their membership of the West European Fish Technologists' Association (WEFTA) which is made up of fish technologists and workers in related disciplines, who have met annually since 1970. WEFTA meetings are of a non-public and highly informal nature. That is why we know each other so well. Torry has always advocated continuation of WEFTA meetings. WEFTA contacts are paying off now, for instance within the context of Codex Alimentarius where mutual understanding has greatly improved. Another example are the consultations in Brussels where, thanks to WEFTA, we could form a front, not against but with Community representatives.

The former Directors of Torry I have known, Dr Burgess and Dr Connell, and the present Director, Dr Hobbs, have always been firm promoters of WEFTA. I wish to thank them for this attitude, which may explain why the organisers of this conference have found participants and speakers so willing to come to this Mecca of fish science and technology. This is also why I am confident that all of us will be pleased to cooperate in making this conference succeed. Torry has always had a good nose for hot items and pelagic fish is one of them.

Britain can be proud of this successful offspring, which is so hale and hearty after sixty years of hard labour. I hope that this country will cherish it and keep it sound and well preserved, not as a museum but as a most lively centre of expertise on fish processing.

Before returning the floor to the Chairman I would like to wish the organisers a fine conference. I hope that it will yield what they have had in mind. Whether this comes true depends on the quality of both the speakers and the audience. I have no serious doubt on either aspect. If time permits, I should welcome lively discussions with many interesting and relevant questions.

It is an honour and pleasure, Mr Chairman, ladies and gentlemen, to open herewith this sexagenary Jubilee Conference of the Aberdonian Torry Research Station.

CHAPTER 1

The global pelagic fish resource and its biological potential

R.S. Bailey

INTRODUCTION

Of the total world catch of over 80 million tonnes of fish, roughly 36 million tonnes are accounted for by groups of typically pelagic fish (FAO 1989a). Of the total fish catch, approximately 25 million tonnes are used for reduction purposes and it is likely that a very high proportion of this comes from pelagic fish landed in bulk. The production of fish meal from oily fish is around 6 million tonnes (FAO 1989b) with an associated 1.4 million tonnes of oil, implying a conversion ratio of about 3.4:1. Allowing for approximations, these figures imply that about 70% of pelagic fish caught are used for purposes other than direct human consumption. In addition to meal and oil, over 1.7 million tonnes of canned pelagic fish products are produced.

The aim of this chapter is to summarise recent trends in the landings of the main groups of pelagic fish, to evaluate the likely future potential and to comment on the part that fishery management can play in the maintenance of the resource. Of necessity, the approach taken is broad-brush, but to highlight the problems involved in the management of pelagic fish resources, a number of examples are treated in greater detail.

Scope

Pelagic fish are those that typically live in midwater independently of the sea bed. This is not to say that they do not associate with the sea bed at times, or that demersal fish do not sometimes take up a pelagic existence. In practice, however, pelagic fish as normally recognised include a number of well known groups of fish. Most of them are characterised by their shoaling habits and many of them by the fact that they have a high oil content. The main species of pelagic fish are listed in Table 1.1. There are very few major groups, and they include: herrings and sprats; pilchards and sardines; menhaden; anchovies; capelin; sauries; horse mackerel; mackerel; and tuna. Also included is the blue whiting which, although not an oily fish, is largely pelagic in its habit. To these might be added other non-fish pelagic resources, such as the pelagic squids and Antarctic krill. Scientific names are given in the Table.

Of the main groups listed above, one or two are found in polar or boreal waters (capelin and herring), but for the most part the really large pelagic resources are found in temperate or subtropical waters. Moreover, the most important groups (sardines, anchovies, mackerel and horse mackerel) are represented by their own species in appropriate areas all round the world. The largest concentrations are found in the highly productive coastal upwelling areas, usually along western continental coasts (Fig. 1.1). The tunas are perhaps exceptional in being trans-oceanic migrants. The important point, however, is that a very high proportion of the total catches of pelagic fish is taken over a tiny proportion of the earth's surface, much of it close to land. This is of course why they can be exploited so profitably.

Besides these concentrated stocks of pelagic fish, the open oceans also harbour immense quantities of fish but they are so dispersed that they are normally not economic to exploit. In certain circumstances, however, they may concentrate as does the saury in Japanese seas, often with the assistance of lights. An estimate of the global annual production of pelagic fish, including the so-called mesopelagic species, is in the hundreds of millions of tonnes (Gulland 1971).

REVIEW OF RECENT LANDINGS

This section reviews changes in landings of the main pelagic fish resources over the past 50 years. This is based mainly on figures published in the FAO year

Table 1.1 The main species of pelagic fish.

Order	Family	Important species*		World catch in 1987 (000 tonnes)
Clupeoidei	Clupeidae	Atlantic herring	*Clupea harengus*	1561
		Pacific herring	*Clupea pallasi*	348
		Goldstripe sardinella	*Sardinella gibbosa*	122
		Indian oil sardine	*Sardinella longiceps*	263
		Round sardinella	*Sardinella aurita*	261
		Japanese pilchard	*Sardinops melanostictus*	5321
		California pilchard	*Sardinops caeruleus*	477
		South American pilchard	*Sardinops sagax*	4686
		Southern African pilchard	*Sardinops ocellatus*	101
		Atlantic menhaden	*Brevoortia tyrannus*	323
		Gulf menhaden	*Brevoortia patronus*	907
		Bonga shad	*Ethmalosa fimbriata*	116
	Engraulidae	Azov tyulka	*Clupeonella cultriventris*	404
		European pilchard	*Sardina pilchardus*	1086
		European sprat	*Sprattus sprattus*	318
		European anchovy	*Engraulis encrasicolus*	642
		Japanese anchovy	*Engraulis japonicus*	308
		Californian anchoveta	*Engraulis mordax*	167
		Peruvian anchoveta	*Engraulis ringens*	2101
		Southern African anchovy	*Engraulis capensis*	972
		Pacific anchoveta	*Cetengraulis mysticetus*	191
Salmonoidei	Osmeridae	Capelin	*Mallotus mallotus*	1105
Beloniformes	Scomberesocidae	Pacific saury	*Cololabis saira*	227
Gadiformes	Gadidae	Blue whiting	*Micromesistius poutassou*	708
		Southern blue whiting	*Micromesistius australis*	104
Percoidei	Carangidae	Atlantic horse mackerel	*Trachurus trachurus*	268
		Japanese jack mackerel	*Trachurus japonicus*	201
		Chilean jack mackerel	*Trachurus murphyi*	2682
		Mediterranean horse mackerel	*Trachurus mediterraneus*	108
		Cape horse mackerel	*Trachurus capensis*	556
		Cunene horse mackerel	*Trachurus trecae*	106
		Greenback horse mackerel	*Trachurus declivis*	107
		Japanese amberjack	*Seriola quinqueradiata*	159
Scombroidei	Scombridae	Chum mackerel	*Scomber japonicus*	1602
		Atlantic mackerel	*Scomber scombrus*	683
		Indian mackerel	*Rastrelliger kanagurta*	161
		Narrow-barred Spanish mackerel	*Scomberomorus commerson*	131
		Japanese Spanish mackerel	*Scomberomorus niphonius*	140
		Skipjack tuna	*Katsuwonus pelamis*	988
		Albacore	*Thunnus alalunga*	175
		Yellowfin tuna	*Thunnus albacares*	823
		Bigeye tuna	*Thunnus obesus*	238

* Species with landings of >100 000 tonnes in 1987.

book. The issues consulted were those for 1963 (vol. 16), 1970 (vol. 30), 1974 (vol. 38), 1976 (vol. 42), 1982 (vol. 54) and 1987 (vol. 64). These volumes contain a wealth of detailed information on fish landings and only broad summaries are given here to provide an indication of the stability of the resource. Table 1.2 shows national landings of the main groups of pelagic fish in 1987.

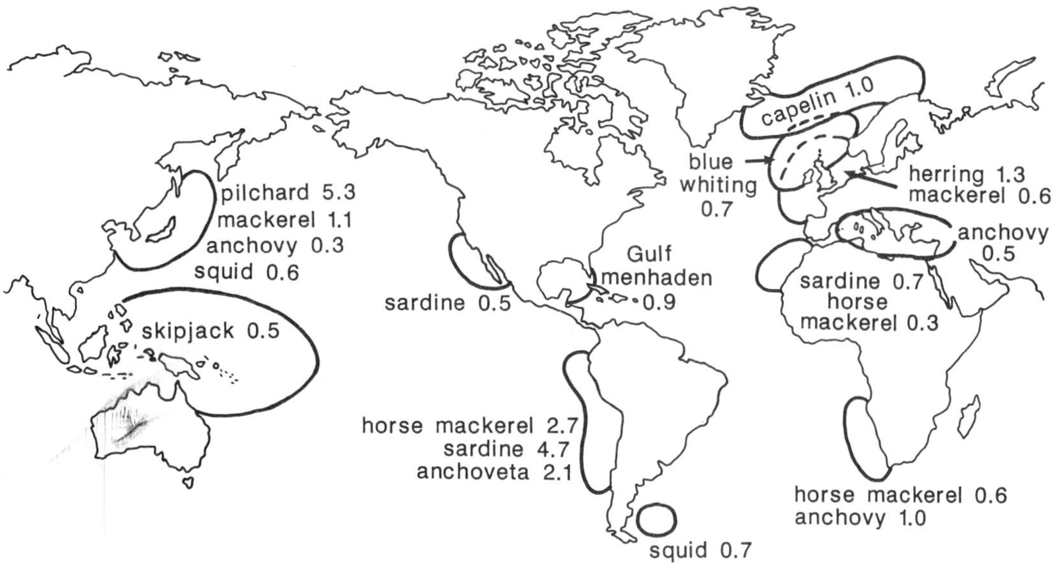

Fig. 1.1 World distribution of main pelagic resources.

Herring (Fig. 1.2)

A large proportion of the herring (*Clupea* spp.) caught are from the north-east Atlantic (*C. harengus*). After a relatively steady world catch of around 3 million tonnes up to 1963, catches rose dramatically in the mid 1960s after which there was an almost uninterrupted decline until 1980. Since than there has been a partial recovery to 1.9 million tonnes in 1987. The marked fluctuations are due largely to the collapse and recovery of two important stocks in the north-east Atlantic – the Atlanto-Scandian stock of spring spawning herring that lives in the Norwegian Sea spawning off the Norwegian coast – and the North Sea herring that spawn in the autumn. The collapse of these stocks was reflected in the collapse of several other less important stocks including those off southern Ireland and on Georges Bank in the north-west Atlantic. During the period of collapse around 1980, the reduced landings were mainly sustained by stocks in the north-west Pacific (*C. pallasi*), the Canadian Atlantic, Iceland and particularly the Baltic.

Sardines and pilchards (Fig. 1.3)

The sardines or pilchards, treated here as biologically synonymous, are represented in all three oceans. In the central tropics the various species of *Sardinella* are prominent, accounting in total for between 1 and 2 million tonnes per annum, with a slight downward trend since the peak in the late 1970s. In contrast, landings of the true sardines (*Sardinops* and *Sardina* spp.) have increased dramatically over the past decade (Fig. 1.3). Up to 1970 the southern African pilchard (*Sardinops ocellatus*) predominated, but since 1977 most of the landings have come from the north-west (*Sardinops melanostictus*) and south-east Pacific (*Sardinops sagax*), while landings of *S. ocellatus* have declined to a low level. Landings of the European pilchard (*Sardina pilchardus*) have tended to increase, mainly in the east-central Atlantic and Mediterranean. As a group, sardine landings increased by almost a million tonnes every year from 1976 to 1984, since when they have levelled off at around 12 million tonnes, almost 80% of which is caught by Japan, Chile and Peru.

Menhaden (Fig. 1.4)

The menhaden (*Brevoortia* spp.) comprise a group of species found in the tropical and subtropical west Atlantic. The most important species is the Gulf menhaden *B. patronus* found in the Gulf of Mexico. Landings of menhaden showed a slow but steady increase up to 1983 (1.3 million tonnes) and have subsequently levelled off at a slightly lower level.

Table 1.2 National landings (thousand tonnes) of the main groups of pelagic fish in 1987 (only countries landing more than 100 000 tonnes of any item in 1987 are included).

Herring	
USSR	354
Norway	345
Canada	239
Denmark	157
Sweden	120
Scotland	106
Sardinellas	
USSR	263
India	244
Indonesia	122
Senegal	110
Thailand	102
Sardines and pilchards	
Japan	4362
Peru	2469
Chile	2203
USSR	1117
Mexico	477
Morocco	275
South Korea	194
Spain	163
Menhaden	
USA	1230
Sprat	
Denmark	134
Capelin	
Iceland	810
Norway	143
Saury	
Japan	197
Blue whiting	
USSR	278
Norway	193
Anchovy	
Peru	1765
South Africa	596
Namibia	376
Chile	336
Turkey	310
South Korea	168
Mexico	161
USSR	152
Japan	141
Panama	127
Indonesia	109
Philippines	108
Horse mackerel	
Chile	1770
USSR	1551
Japan	187
Mackerel	
Japan	701
USSR	245
Scotland	195
Norway	155
Indonesia	136
South Korea	102
Skipjack	
Japan	331
USA	104
Tuna	
Japan	352
USA	179
Indonesia	157
Mexico	103

Sprat (Fig. 1.4)

Landings of sprats (*Sprattus sprattus*) from the north-east Atlantic increased markedly in the mid 1970s and have subsequently stabilised at a much lower level. The largest landings come from the North Sea, Skagerrak, Kattegat and Baltic.

Capelin (Fig. 1.4)

The capelin (*Mallotus villosus*), an arctic fish found north of the polar front, is caught off the Canadian seaboard, around Iceland and Greenland and in the northern parts of the Norwegian and Barents Seas. Landings first became prominent in the mid 1960s and increased to a peak of 4 million tonnes in 1977. They have since decreased progressively to only just over a million tonnes in 1987, partly as a result of the collapse of the Barents Sea stock.

Blue whiting (Fig. 1.4)

The northern blue whiting (*Micromesistius poutassou*) is a member of the gadidae or cod family containing a number of familiar demersal fish. It is largely pelagic in habit but tends to form extensive layers rather than isolated schools. Exploitation began in earnest in the 1970s and landings increased exponentially for about five years to reach a peak of 1.1 million tonnes in 1979 and 1980. Since then landings have decreased and have now stabilised at a lower level of around 700 000 to 800 000 tonnes. Landings of the southern blue whiting (*M. australis*) caught in the south-west Atlantic are on a much smaller scale.

Anchovies (Figs. 1.5 and 1.6)

As with sardines, anchovies are represented mainly in subtropical and tropical seas in all oceans. The predominant group of species are members of the genus *Engraulis*, but there are also a number of other genera collectively falling under the term anchovy.

For a number of years in the 1960s and early 1970s the Peruvian anchoveta (*E. ringens*) formed the largest landings of any single species of fish, reaching over 13 million tonnes in 1970 (Fig. 1.5). This was followed by a rapid decline to only 3 million tonnes by 1973. Since then landings have fluctuated around a much lower level, averaging 3.3 million tonnes in the decade to 1987. Total landings of anchovies track the landings of Peruvian anchoveta, which now account for about half the world anchovy total (Fig. 1.5).

The landings of other species of anchovy have also fluctuated (Fig. 1.6), but none have individually reached a million tonnes per year. Those of Japanese anchovy (*E. japonicus*) have remained reasonably steady at around 300 000 to 400 000 tonnes, while those of other species have tended to increase. Landings of Californian anchoveta (*E. mordax*) reached 420 000 tonnes in 1981 but have since decreased. European anchovy (*E. encrasicolus*) landings increased progressively to a peak of 840 000 tonnes in 1984, mainly from the Mediterranean and Black Seas. After fluctuating considerably in the 1970s and early 1980s, landings of southern African anchovy (*E. capensis*) increased suddenly to a record level of 970 000 tonnes in 1987.

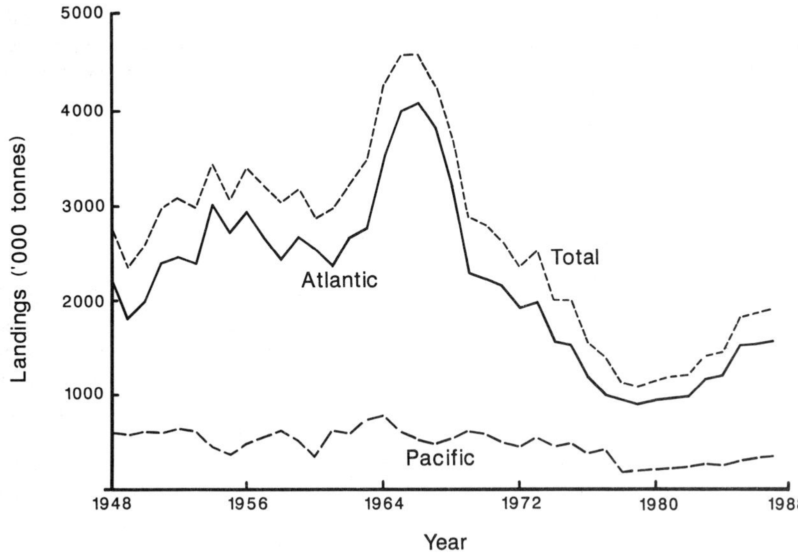

Fig. 1.2 World landings of herring (*Clupea* spp.) 1948–87.

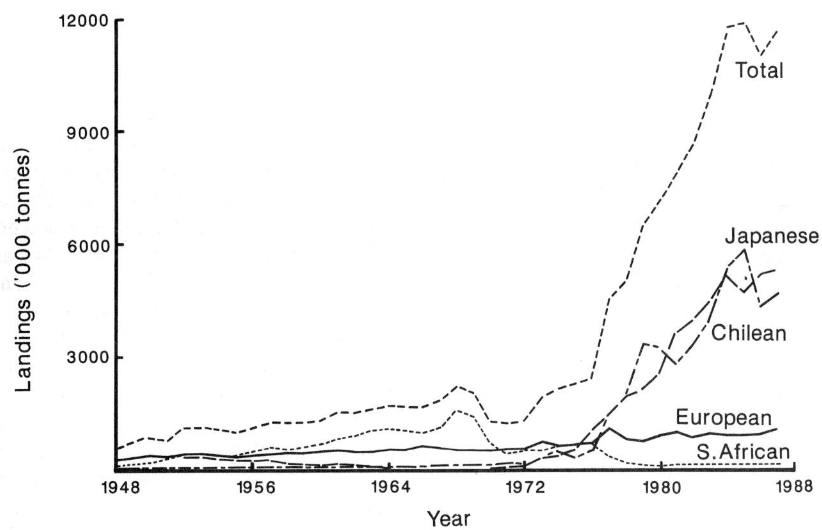

Fig. 1.3 World landings of sardines and pilchards (*Sardinella*, *Sardinops* and *Sardina* spp.), 1948–87.

Horse mackerel (Fig. 1.7)

As with sardines and anchovies, the genus *Trachurus* is represented in most subtropical areas. Total landings have increased progressively over the past 50 years to their highest recorded level of almost 4.5 million tonnes in 1987. The biggest increases in landings have taken place in the south-east Atlantic (*T. capensis*) and especially the south-east Pacific (*T. murphyi*).

In addition to species of *Trachurus* there are numerous tropical species of scad (*Decapterus* spp.), landings in total amounting to a further 400 000 tonnes.

Mackerel (Fig. 1.8)

Landings of mackerels (*Scomber* spp.) increased rapidly in the mid to late 1960s, rising to a peak of 3.6 million tonnes in 1978. Since then landings have decreased

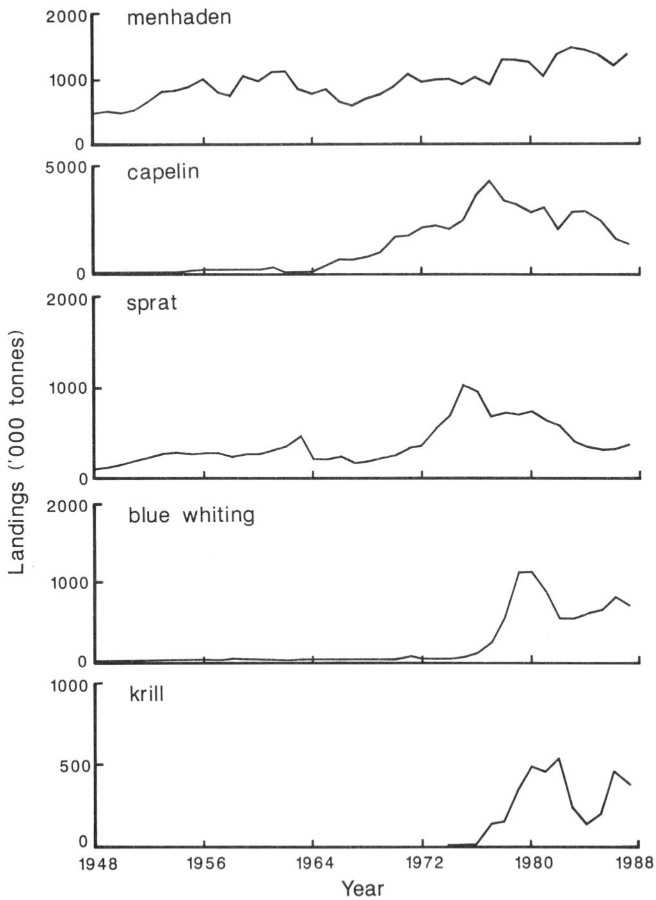

Fig. 1.4 World landings of menhaden (*Brevoortia* spp.), capelin (*M. mallotus*), sprat (*S. sprattus*), blue whiting (*Micromesistius poutassou*) and krill (*Euphausia superba*), 1948–87.

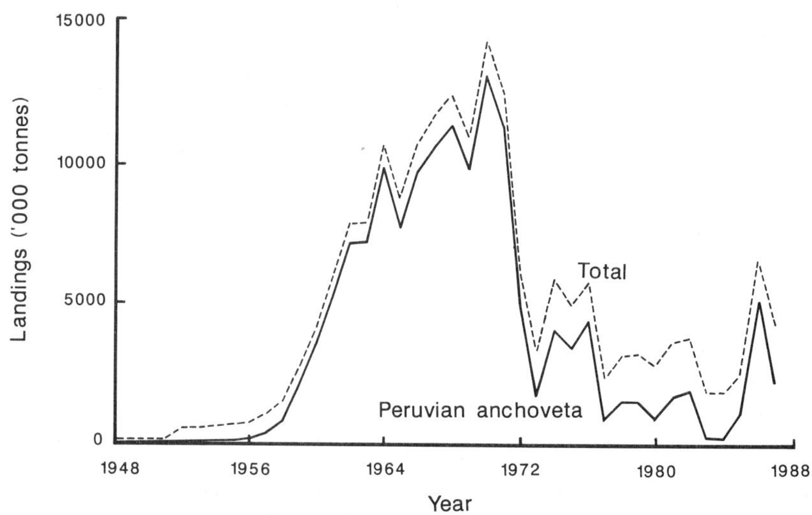

Fig. 1.5 World landings of Peruvian anchoveta (*Engraulis ringens*) and total anchovies, 1948–87.

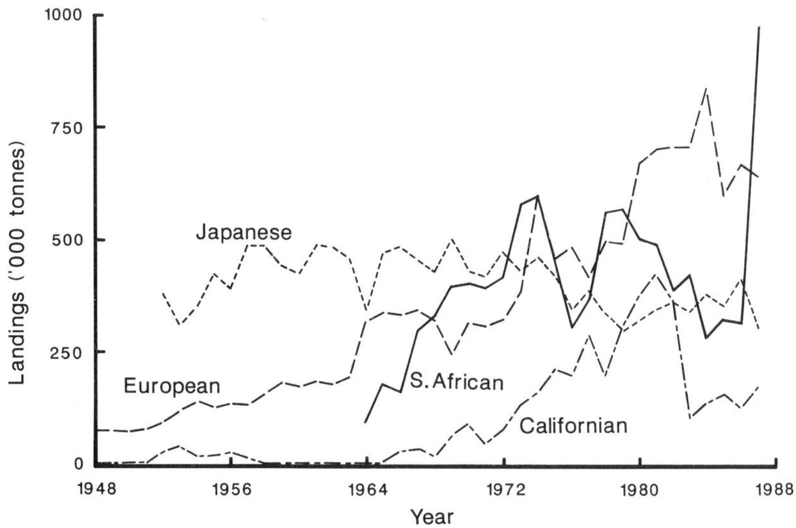

Fig. 1.6 World landings of anchovies (*Engraulis encrasicolus, E. japonicus, E. mordax* and *E. capensis*), 1948–87.

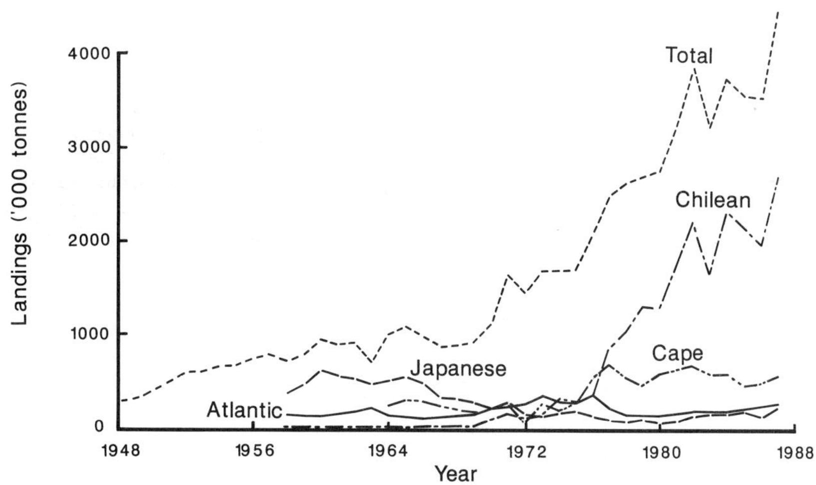

Fig. 1.7 World landings of horse mackerel (*Trachurus trachurus, T. japonicus, T. murphyi* and *T. capensis*), 1948–87.

and have recently fluctuated around 2.5 million tonnes. After increasing to over a million tonnes in the mid 1970s, landings of Atlantic mackerel (*S. scombrus*) have stabilised at around 600 000 tonnes, most of which is caught in the north-east Atlantic. Landings of chub mackerel (*S. japonicus*) have fluctuated markedly between 1.7 and 2.9 million tonnes. Most of these are caught by Japan in the north-west Pacific.

Indian mackerel (Fig. 1.9)

A distinct group of tropical mackerels (*Rastrelliger* spp.) is caught in the Indian and western Pacific Oceans. Landings are on a much smaller scale than those of *Scomber* spp., and amount in total to around 500 000 tonnes, following an increasing trend since the early 1960s.

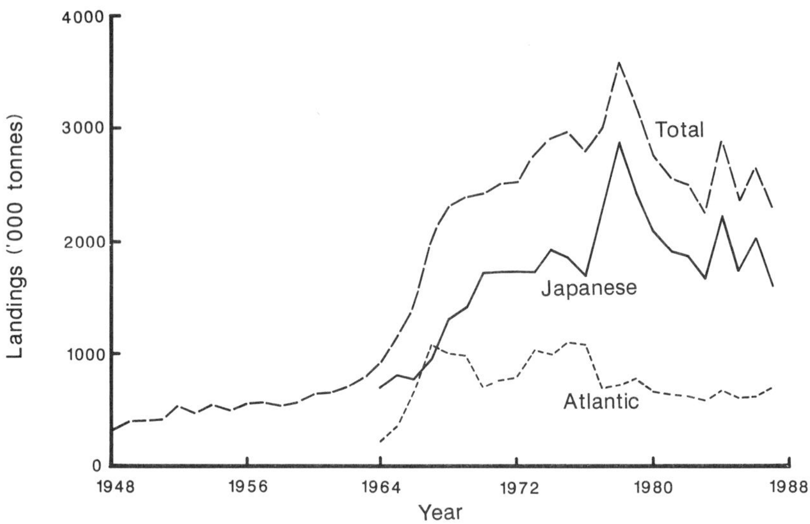

Fig. 1.8 World landings of mackerel (*Scomber japonicus*, *S. scombrus*), 1948–87.

Fig. 1.9 World landings of Indian mackerel (*Rastrelliger* spp.), 1948–87.

Skipjacks and tunas (Fig. 1.10)

Landings of both these groups have increased progressively over the past 50 years, the combined landings totalling 2.8 million tonnes by 1986, of which roughly 50% are typical tunas (*Thunnus* spp.). Species contributing most heavily to the combined total are the skipjack (*Katsuwonus pelamis*), the albacore (*T. alalunga*), the yellowfin (*T. albacares*) and the bigeye (*T. obesus*). They are caught widely in all three oceans, the largest total quantities coming from the eastern and western central Pacific.

Other species

There are many other species and groups of pelagic fish but none approach those dealt with above in terms of total landings. To complete the review of landings of the 'pelagic resource' it is relevant to mention two

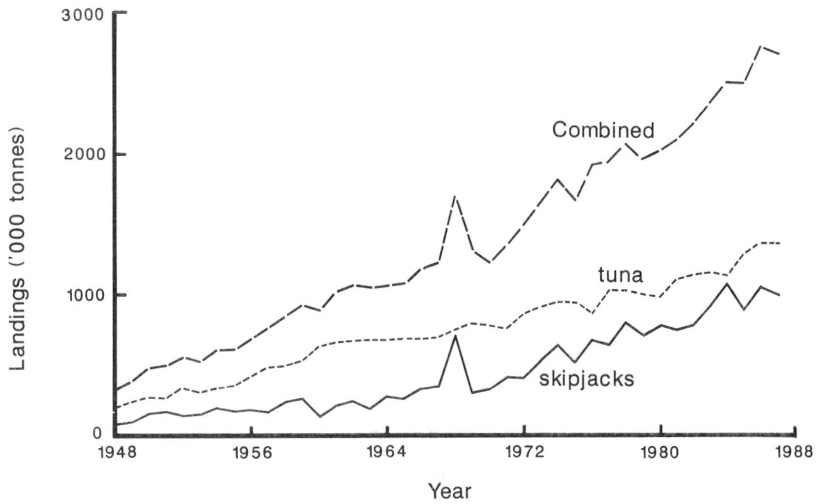

Fig. 1.10 World landings of skipjacks (*Katsuwonus pelamis*), tunas (*Thunnus* spp.) and combined 'tuna-like fishes', 1948–87.

groups of organisms other than fish – the squids and the Antarctic krill. Squids, octopus and cuttle fish contain both demersal and pelagic members and only some groups of squids can truly be claimed as pelagic. Treating the ommastrephidae as containing mostly pelagic representatives, landings have recently increased from around a million tonnes in the period 1982–86, to 1.6 million tonnes in 1987. Approximately 40% is caught in the south-west Atlantic and a further 40% in the north-west Pacific. Antarctic krill (*Euphausia superba*) have been exploited since the early 1970s but records of landings were not available until about 1974. Since then, landings have fluctuated markedly, reaching peaks of 480 000 tonnes in 1980 and 450 000 tonnes in 1986.

Total pelagic fish

By adding the main categories of pelagic fish listed in Table 1.1, it is possible to gain a picture of the potential of the total pelagic resource. Summed landings of the individual groups described are shown in Figure 1.11. They show that landings increased slowly up to 1958, rapidly up to a peak of about 25 million tonnes in 1970 and then dropped sharply to recover gradually to the even higher level of over 30 million tonnes in 1986. Much of the pattern of landings can be accounted for by the growth and decline of the Peruvian anchoveta fishery, and by the recent growth in landings of sardines and to a certain extent tunas and blue whiting. What is clear is that the major fluctuations in landings of individual species are to a large extent smoothed out by considering the combination of pelagic fish as a unit world commodity. The exception to this is the fishery for Peruvian anchoveta, the collapse of which caused a major hiccup in the development of landings.

COLLAPSE AND RECOVERY – THE STABILITY OF THE RESOURCE

The landings of pelagic fish shown in Figures 1.2 to 1.11 amply testify to the apparent instability of individual pelagic fish resources. This characteristic can be seen even more clearly by considering a number of case histories chosen from individual stocks of fish.

Norwegian spring-spawning herring

Historically, the largest stock of herring was that spawning off the Norwegian coast. Catch records available from 1925 show a rapid increase after 1945 to a peak of almost 2 million tonnes in 1966, after which there was a total collapse of the fishery (Schumacher 1980). Spawning stock sizes are plotted from 1950 onwards in Figure 1.12. After the large 1963 year class recruited to the fishery, the exploitation rate increased to a level far beyond that sustainable. Coupled with recruitment failure, the stock collapsed in two years to 2 or 3% of its size in 1950. After 1963 there was a gap of 20 years until the good 1983 year class appeared. This, however, has proved so far to be an isolated event

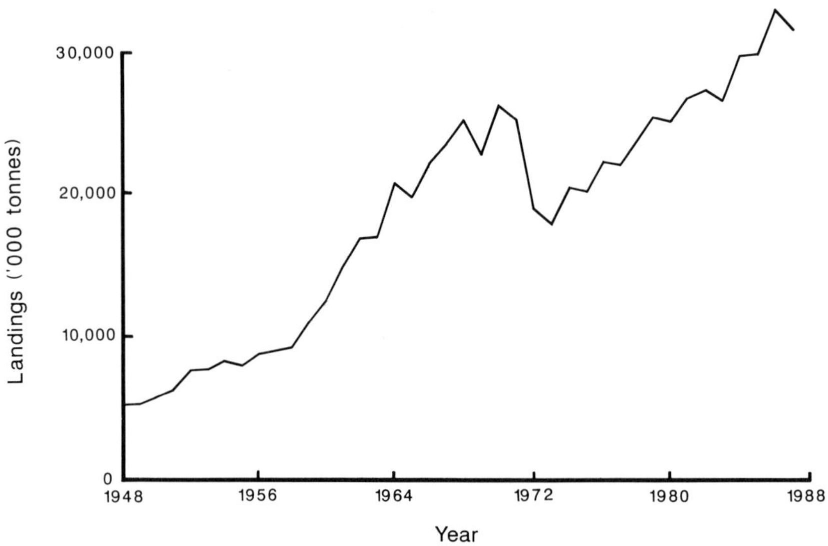

Fig. 1.11 Summed world landings of the main groups of pelagic fish (herring, sardines, anchovies, menhaden, capelin, blue whiting, sprats, tunas, mackerel, Indian mackerel, horse mackerel) and Antarctic krill, 1948–87.

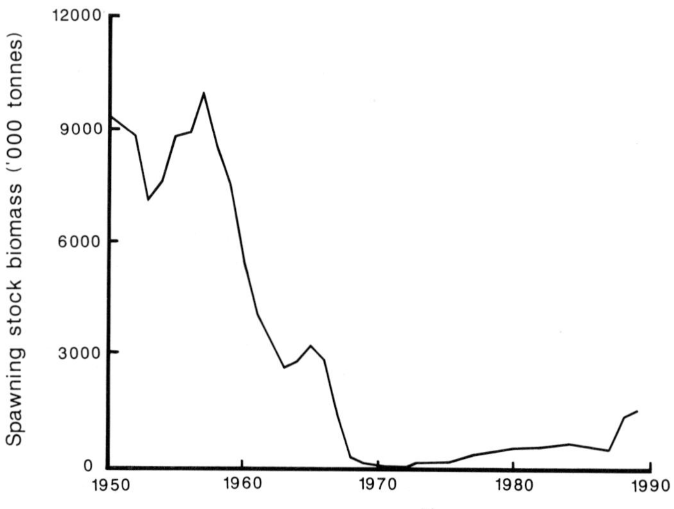

Fig. 1.12 Spawning stock biomass of Norwegian spring-spawning herring, 1950–89. Values for 1988 and 1989 are predicted values. (Dragesund *et al.* 1980, Anon 1987, Anon 1989a).

with the result that the recovery of the stock is expected to be both short-lived and incomplete.

North Sea herring

In the period 1920–60, landings of herring from the North Sea fluctuated between 280 000 and 810 000 tonnes (Schumacher 1980). In the mid 1960s landings rose steeply to a peak of almost 1.2 million tonnes as fishing effort was transferred from the depleted Atlanto-Scandian stock. This was followed by an almost continuous decline until the closure of the directed fishery in 1977 (Fig. 1.13). Since the reopening of the fishery in 1982, landings have increased to 700 000 tonnes in 1988.

The decrease in landings was associated with a major decrease in stock size (Fig. 1.14) caused by an increas-

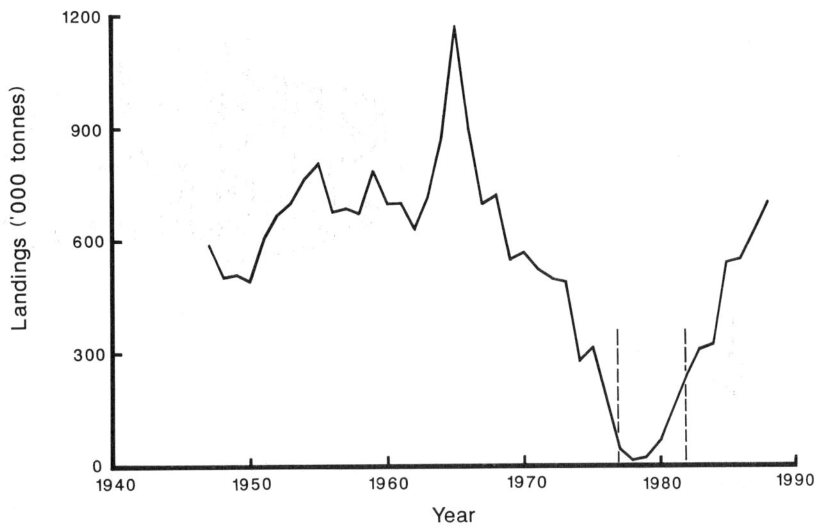

Fig. 1.13 Landings of North Sea herring, 1947–88. Fishery closed during period between dashed vertical lines. (Anon 1977, Anon 1989b).

ing exploitation rate (Fig. 1.15) and, in the last stages of the collapse, by a decrease in recruitment (Saville & Bailey 1980, Bailey, in press).

Other herring stocks

Similar stock declines have occurred in a number of other herring stocks, including those on Georges Bank in the north-west Atlantic (Brown *et al.* 1983), Icelandic spring and summer spawners (Jakobsson 1980), British Columbian herring (Hourston 1980) and the Celtic Sea stock (Molloy 1980). In some of these cases there has been at least a partial recovery in more recent years.

Sardines and anchovies

Stock collapse associated with the collapse of the fishery has also been a characteristic of other groups of shoaling pelagic fish (Saville 1980a). Some of the most notable examples are found among the sardine and anchovy stocks in the subtropics. The fishery and biomass of southern African Pilchard collapsed over the years 1960–66 (Fig. 1.16, from Butterworth 1983). Landings of Japanese sardine (Fig. 1.17) rose to a peak of over 2.5 million tonnes in the mid 1930s, thereafter declining to negligible proportions by the mid 1960s (Kondo 1980). The collapse of the Peruvian anchoveta fishery described above (Fig. 1.5) was also associated with a major change in stock size which fluctuated between 8 and 22 million tonnes in the period 1960–71 and between 1 and 6 million tonnes in the ten years up to 1981 (Csirke 1989).

THE CAUSE OF FLUCTUATIONS

The fluctuations described above have led to belief that there is something about the characteristics of pelagic fish stocks that makes them less resilient to exploitation than most stocks of demersal fish. This was the subject of a symposium held in Aberdeen in 1978 (Saville 1980a). The main conclusions of that symposium were summed up by Saville (1980b):

(1) Many exploited stocks of pelagic fish have shown a pattern of sharp increase in catches followed by an even more rapid decline, in some cases resulting in the closure of the fishery.
(2) In most cases the rate of decline has been augmented, in the final stages, by recruitment failure.
(3) In some cases environmental changes have been implicated in the declines, but as a rule fishing appears to have had the greatest effect.

The low resilience of pelagic fish stocks can be explained by a number of characteristics (Saville 1980b). Initially, a decrease in stock is not accompanied by a comparable decrease in catch rates. As the stock

Fig. 1.14 Spawning stock biomass of North Sea herring, 1947–88. Fishery closed during period between dashed vertical lines. (Anon 1987, Anon 1989b).

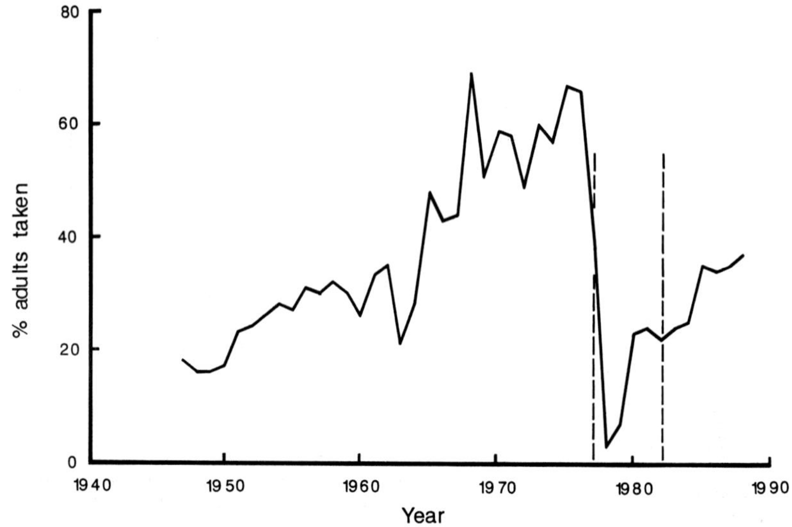

Fig. 1.15 Percentage of the North Sea herring stock taken in the fishery each year, 1947–88. Fishery closed during period between dashed vertical lines. (Anon 1977, Anon 1989b).

declines, the size of shoals probably remains much the same. The effect of this is that a unit of fishing effort takes a larger proportion of the population at low stock sizes than it does at high ones. This is illustrated in Figure 1.18 for the Peruvian anchoveta (Csirke 1989). In addition, the area of distribution of a stock may also diminish so that fishing vessels searching, for example, by sonar have less searching to do, while their catch rates are maintained (e.g. Csirke 1980 for the anchoveta, Saville & Bailey 1980 for the North Sea herring).

Fluctuations in the abundance of pelagic fish are

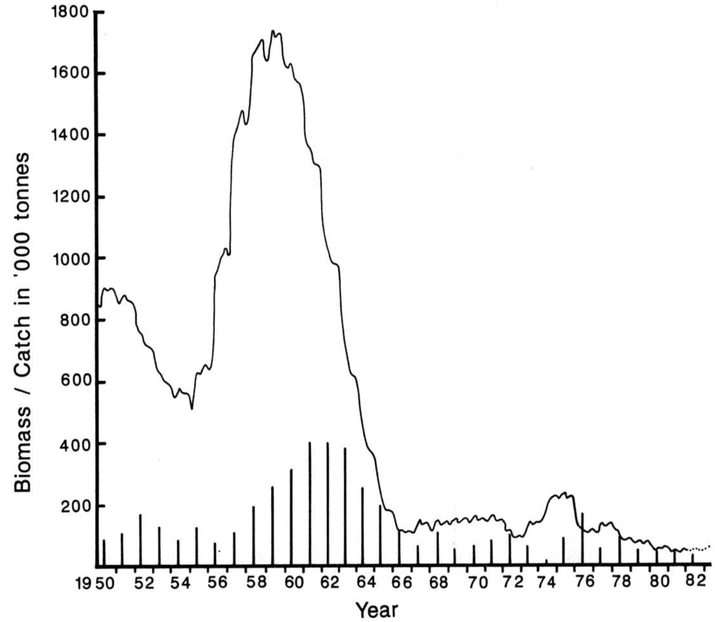

Fig. 1.16 Annual landings (vertical bars) and monthly biomass estimates (solid line) of southern African pilchard calculated by VPA. (Redrawn by permission of the Food and Agriculture Organization of the United Nations, from Butterworth 1983).

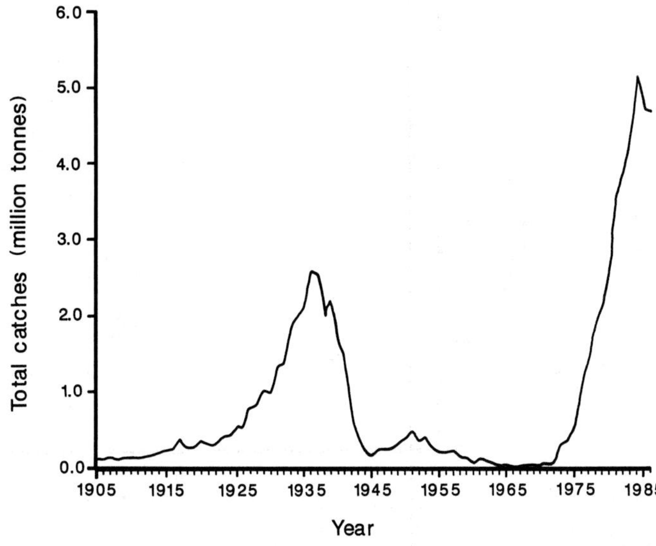

Fig. 1.17 Nominal catches of Japanese sardine in the northwest Pacific, 1905–86. (Redrawn by permission of John Wiley and Sons, from Csirke 1989)

also the result of fluctuations in recruitment, i.e. in the number of young fish entering the stock each year. The cause of reduced recruitment has been the subject of considerable debate. Since low recruitment has often been associated with low parental stock size, there has often been a presumption that they are directly related: thus, a reduced spawning stock produces fewer eggs which in turn give rise to fewer young surviving. It is clear, however, that this cannot be the only reason for poor recruitment because there is considerable variation in recruitment at any given level of stock size.

In some stocks major decreases in stock caused by decreases in recruitment are seen to be associated with extreme environmental events. In the Peruvian

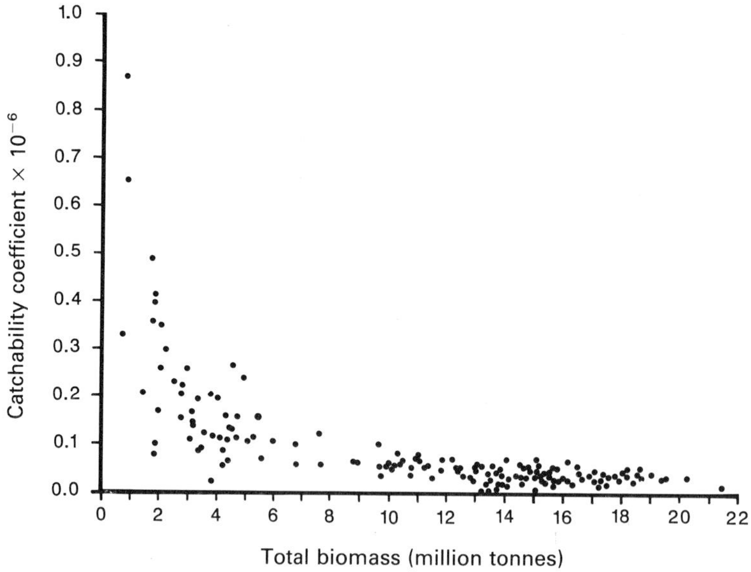

Fig. 1.18 Relationship between the catchability coefficient and the total biomass of Peruvian anchovy by month (September 1960–September 1982). (Redrawn by permission of John Wiley and Sons, from Csirke 1989).

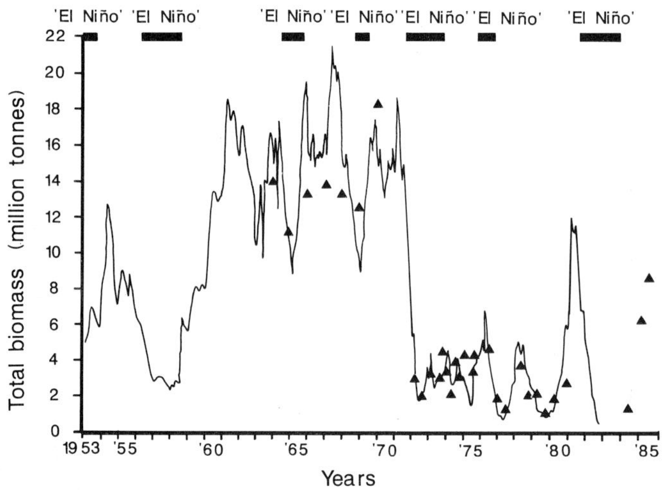

Fig. 1.19 Fluctuations in the northern and central stock of Peruvian anchovy, 1953–86. Horizontal bars signify the occurrence of the environmental anomalies termed 'El Niño'. (Redrawn by permission of John Wiley and Sons, from Csirke 1989). El Niño is a seasonal, southerly, warm current system off Peru which is interrelated with the weather systems of the region and strongly influences ocean productivity.

anchoveta, for example, the most severe decreases in stock coincide with the onset of abnormal oceanographic conditions (Fig 1.19). There are other examples, moreover, that provide at least a partial explanation for changes in the survival of the young. The low values of recruitment to the North Sea herring stock in the mid 1970s are now seen to have occurred at approximately the same time as an anomaly in the salinity in the area north-west of Scotland (Fig. 1.20), and there is some evidence of a reduction in the flow of Atlantic water into the North Sea at this time (Dooley 1987). Corten (1986) has postulated that the poor survival of herring larvae in these years may be related to the weakened current which failed to take them to their normal nursery areas in the eastern North Sea.

Some changes in the recruitment to pelagic stocks are more protracted. Recruitment to the Japanese sardine stock, for example, remained at a low level for over 30 years (Fig. 1.17). This has been put down to a long-term change in the path of the Kuroshio current to the south-east of Japan (Tanaka 1983). Kawasaki (1983) has suggested that the fluctuations off Japan may be a manifestation of a much larger

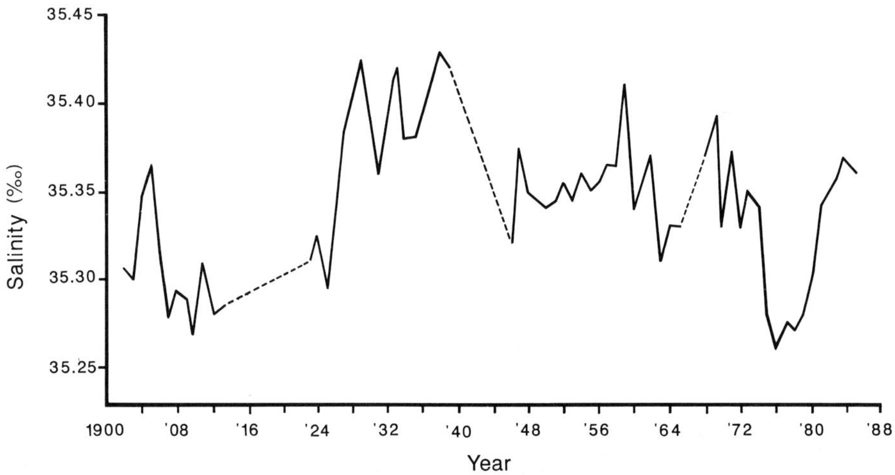

Fig. 1.20 Long-term changes in salinity of the upper 200 metres of 'North Atlantic' water in the Faroe–Shetland Channel, 1902–85. (Redrawn from Walsh and Martin 1986).

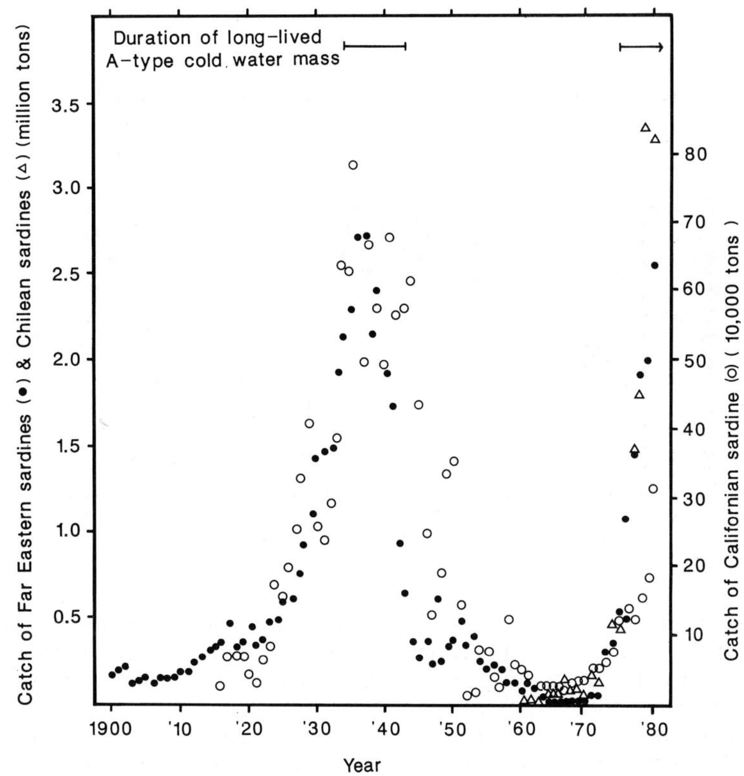

Fig. 1.21 Large-scale variations in catch of Japanese (●), Californian (○) and Chilean (△) sardines, 1900–1980. (Redrawn by permission of the Food and Agriculture Organization of the United Nations, from Kawasaki 1983).

scale pattern, and showed that the variations in catch of three species of sardines in the Pacific are in phase (Fig. 1.21). Similar fluctuations in pelagic fisheries have been recorded over even longer periods, as exemplified by the alternating periods of the Bohuslan and Norwegian herring fisheries (Devold 1963).

Changes in the abundance of pelagic fish stocks have also been explained as the effect of replacements.

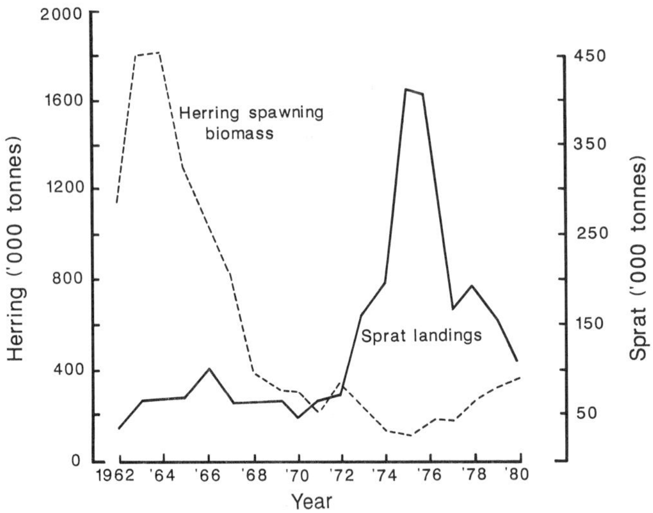

Fig. 1.22 Annual variations in herring spawning biomass and sprat landings in the North Sea (1962–80). (Redrawn by permission of the Food and Agriculture Organization of the United Nations, from Jones 1983).

Jones (1983) showed, for example, that sprat landings in the North Sea increased when the herring stock collapsed (Fig. 1.22). There are also examples of apparent replacements of one pelagic fish by another off southern Africa and California (Shelton & Armstrong 1983, MacCall 1983). In both areas the anchovy stocks increased after or as the sardine stock collapsed. However, in such cases it is not always clear whether the replacements occurred as a result of over-exploitation or as a result of environmental changes that tipped the competitive balance.

The debate about the fundamental cause of changes in pelagic fish stocks continues (Kawasaki *et al.*, in press). Nevertheless, there is little doubt that environmental factors have a major impact and that their effects can be exacerbated by overfishing. To maintain continuity in the fisheries, it is therefore essential to maintain pelagic fish stocks as far as possible above certain minimum levels to provide a buffer against short-term decreases in recruitment.

OTHER CHARACTERISTICS OF PELAGIC FISH

Migrations

The continuity of fisheries for pelagic fish depends not only on the abundance of the stocks but also on their distribution. Pelagic fish, being more dependent on the body of water they are in than on the type of sea bed, show variable patterns of migration. In some species, like blue whiting, these migrations can extend over many hundreds of miles, creating constraints on the seasonality of the fishery (Bailey 1982). There are also numerous examples in which the timing and pattern of the migration has changed.

One particularly important example in the north-west Atlantic is that of the mackerel stock that spawns to the south-west of Ireland. For a number of years in the 1970s there were three important fisheries for this stock – one off the Cornish coast in winter, one in the north-eastern North Sea in summer and one off the west coast of Scotland in the autumn (Fig. 1.23). By the mid 1980s this pattern had changed and, whereas there is still a summer fishery off the Norwegian coast, the mackerel remain in the western North Sea and the area to the north of Scotland for the entire winter. This has had major effects on the availability of mackerel to the different fleet sectors and on the seasonality of the fishery (Fig. 1.24). The causes of this change in migration pattern have not been fully elucidated, but Walsh & Martin (1986) have postulated that it may be linked to changes in the circulation pattern, because the main Cornish and west of Scotland fisheries took place during a period marked by anomalous salinity patterns (Fig. 1.25).

Biological characteristics

For processing purposes it is not only the seasonal availability of pelagic fish that is of interest but also the quality. Most species of pelagic fish undergo seasonal changes in their condition associated with seasonality of the food supply and with the maturation cycle. In

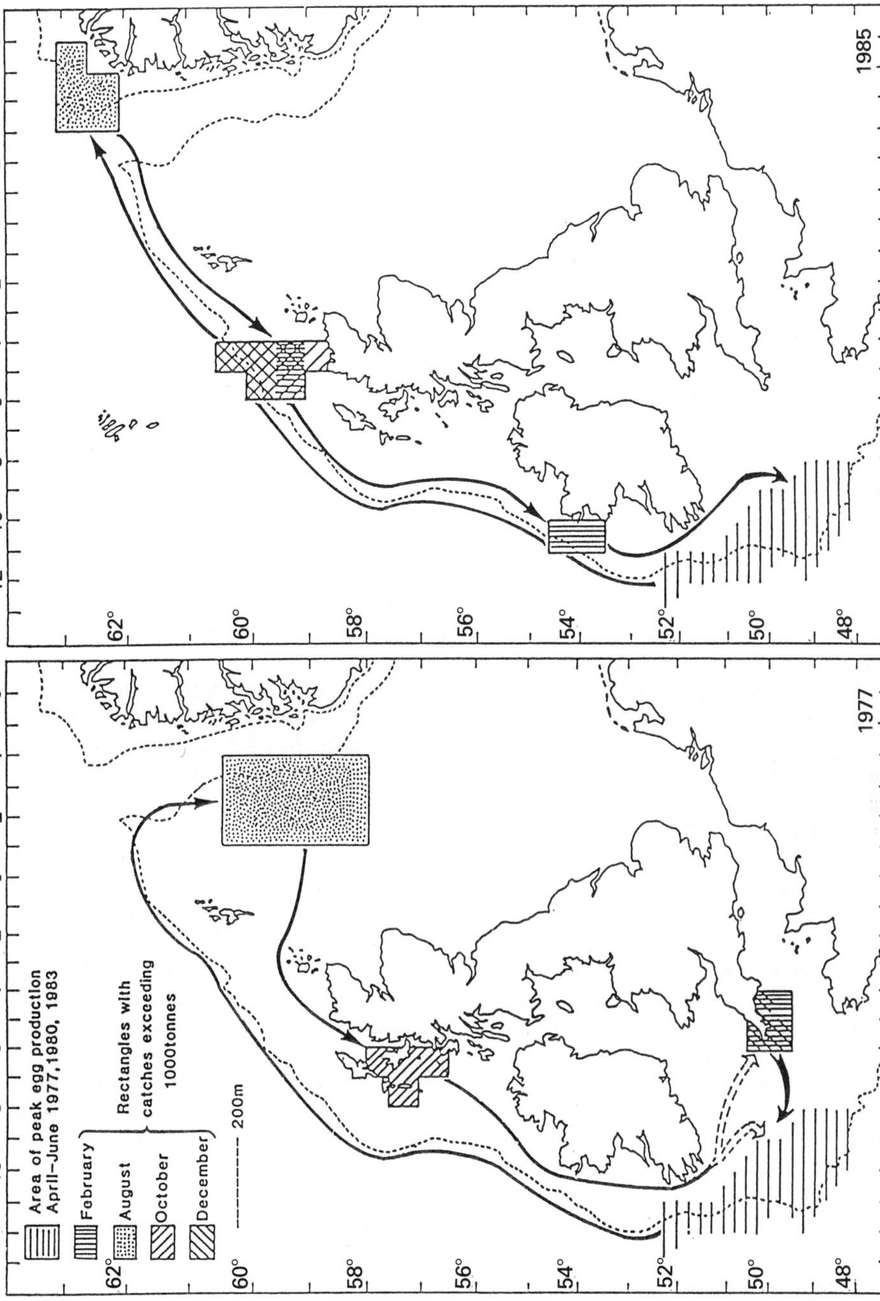

Fig. 1.23 Distribution of fisheries on western stock mackerel in selected months of 1977 and 1985 and likely migration routes. (Redrawn from Walsh and Bailey 1987).

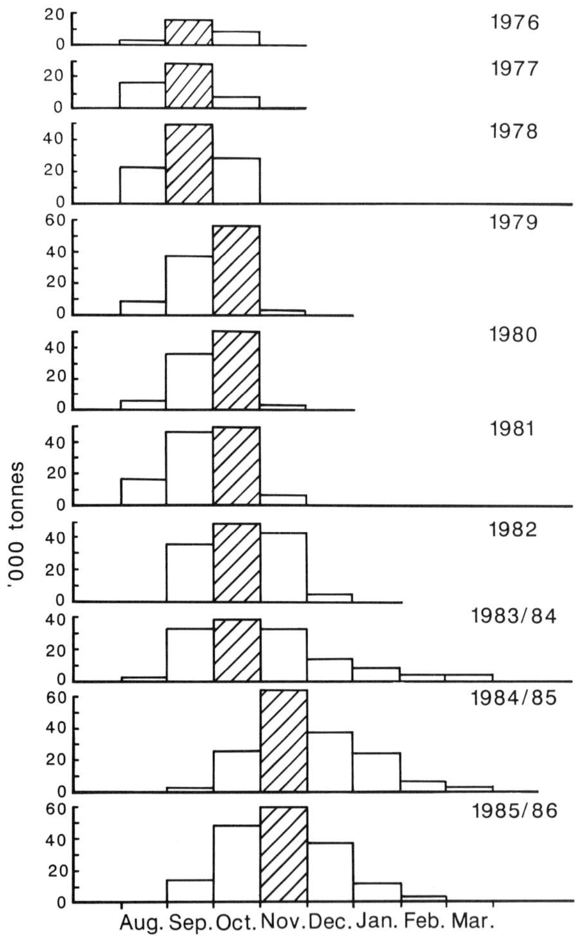

Fig. 1.24 Changes in the monthly distribution of mackerel landings in the Scottish fishery to the west of Scotland (ICES Division VIa, excluding the Clyde), 1976–86. (Redrawn from Walsh and Martin 1986).

general, these seasonal changes are predictable and well understood by the catching and processing sides of the industry, but occasionally unexpected events occur. One particularly marked example in recent years was the change in condition factor of herring in the Firth of Clyde, western Scotland. In 1986–88 the herring in the summer were between 20 and 30% lower in weight than expected (J. Morrison, personal communication). The reasons are not known but may be connected with the planktonic food supplies during those seasons.

MANAGEMENT CONSIDERATIONS

The need to control the amount of fishing on pelagic fish stocks has been discussed in the section on the causes of fluctuations above. To sum up, it is unlikely that the stability of pelagic fisheries can be maintained against longer-term environmentally-induced changes in recruitment. However, some protection of the adult stocks against high levels of fishing effort can almost certainly provide a buffer against short-term failures in recruitment. Even a short term recruitment failure accompanied by high levels of fishing effort can result in a remarkably rapid collapse of the stock because, as pointed out above, catch rates are maintained until it is too late.

There are other consequences of the high densities of pelagic fish. The fish may often be caught in very large quantities, sometimes more than the vessel is capable of handling or allowed to catch under quota restrictions. This can be a particular problem in the case of some stocks of herring which congregate in extremely dense concentrations on the spawning grounds. Similarly, on occasions the fish caught may

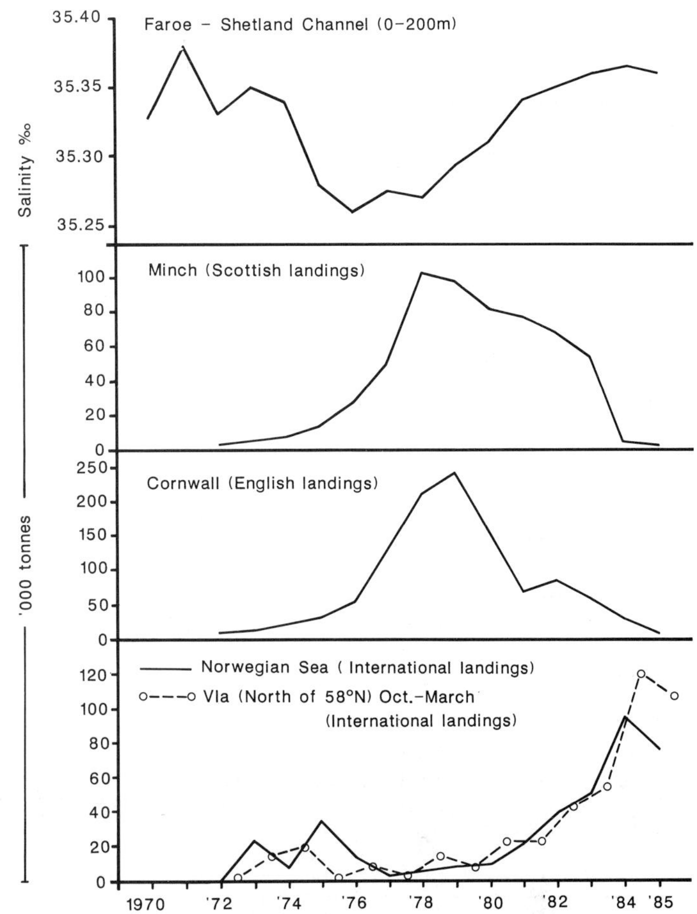

Fig. 1.25 Annual landings of mackerel in four different fisheries in relation to salinity changes in the Faroe–Shetland Channel (1970–85). (Redrawn from Walsh and Martin 1986).

not be of the size, species or quality required for the markets. All these factors can lead to large-scale discarding or wholesale slipping of catches. The extent of this varies from fishery to fishery but should never be discounted as a potential problem in heavily exploited stocks, because it tends to exacerbate the effects of overfishing.

A further potential problem occurs when grading machines are used because they provide the opportunity to catch fish and return large quantities of unwanted dead fish back into the sea.

GLOBAL PROSPECTS FOR PELAGIC FISHERIES

From all that has been summarised above it must be clear that pelagic fish stocks are largely unpredictable. Many of them, and especially those in subtropical waters, appear to be largely environmentally controlled. Elsewhere, the stocks and fisheries have shown some stability in the long term (e.g. North Sea herring) yet it is clear that even apparently stable stocks are susceptible to collapse under heavy exploitation. In all cases, recovery of the stocks concerned appears to be possible but the rate at which it occurs is unpredictable. The individual stocks are likely to continue to fluctuate, but there is no reason to suppose that the global pelagic fish resource is not likely to be stable at the current level of around 30 million tonnes per year. From what is known it appears that major increases in global supplies are unlikely. To reiterate the opinion of OECD (1989), 'it is likely that processors will put more emphasis on marketing and consumer reactions to new species and products. Alternative species and a better utilisation of the catches will be paramount in satisfying consumer demand.'

BIBLIOGRAPHY

Anon. (1977) Assessment of herring stocks south of 62°N, 1973–75. *Cooperative Research Report*, International Council for the Exploration of the Sea, Copenhagen **60**, **117**.

Anon. (1987) *Report of the Working Group on Atlanto-Scandian Herring and Capelin*, Copenhagen, 27–31 October 1986. ICES CM 1987/Assess: 8 (mimeo).

Anon. (1989a) *Report of the Atlanto-Scandian Herring and Capelin Working Group*, Copenhagen, 24–28 October 1988. ICES CM 1989/Assess: 7 (mimeo).

Anon. (1989b) *Report of the Herring Assessment Working Group for the area south of 62°N*, Copenhagen, 4–14 April 1989. ICES CM 1989/Assess: 15 (mimeo).

Bailey R.S. (1982) The population biology of blue whiting in the North Atlantic. *Adv. Mar. Biol.*, **19**, 257–355.

Bailey R.S. (In press) Changes in the North Sea herring population over a cycle of collapse and recovery. In *Long term variability of pelagic fish populations and their environment.* (Ed. by T. Kawasaki, A. Taniguchi, S. Tanaka & Y. Toba), Pergamon Press, London.

Brown B.E., Anthony V.C., Anderson E.D., Hennemuth R.C. & Sherman K. (1983) The dynamics of pelagic fishery resources off the north-eastern coast of the United States under conditions of extreme fishing perturbations. *FAO Fish. Rep.*, **291** (2), 465–505.

Butterworth D.S. (1983) Assessment and management of pelagic stocks in the southern Benguela region. *FAO Fish. Rep.*, **291** (2), 329–405.

Corten A. (1986) On the causes of the recruitment failure of herring in the central and northern North Sea in the years 1972–78. *J. Cons. Int. Explor. Mer*, **42**, 197–211.

Csirke J. (1980) Recruitment in the Peruvian anchovy and its dependence on the adult population. *Rapp. P.-V. Réun. Cons. Int. Explor. Mer*, **177**, 307–13.

Csirke J. (1989) Small shoaling pelagic fish stocks. In *Fish Population Dynamics*, 2nd edn (Ed. by J.A. Gulland), pp. 271–302. John Wiley and Sons Ltd, Chichester.

Devold F. (1963) The life history of the Atlanto-Scandian herring. *Rapp. P.-V. Réun. Cons. Int. Explor. Mer*, **154**, 98–108.

Dooley H.D. (1987) The role of hydrography in influencing biological conditions in the north-western North Sea. In *Developments in Fisheries Research in Scotland* (Ed. by R.S. Bailey & B.B. Parrish), pp. 156–64. Fishing News Books, Oxford.

Dragesund O., Hamre J. & Ulltang O. (1980) Biology and population dynamics of the Norwegian spring-spawning herring. *Rapp. P.-V. Réun. Cons. Int. Explor. Mer*, **177**, 43–71.

FAO. (1989a) *Yearbook of Fishery Statistics – Catches and Landings*, **64**, (1987).

FAO. (1989b) *Yearbook of Fishery Statistics – Commodities*, **65**, (1987).

Gulland J.A. (1971) *The Fish Resources of the Ocean*. Fishing News Books, Oxford.

Hourston A.S. (1980) The decline and recovery of Canada's Pacific herring stocks. *Rapp. P.-V. Réun. Cons. Int. Explor. Mer*, **177**, 143–53.

Jakobsson J. (1980) Exploitation of the Icelandic spring- and summer-spawning herring in relation to fisheries management, 1974–77. *Rapp. P.-V. Réun. Cons. Int. Explor. Mer*, **177**, 23–42.

Jones R. (1983) The decline in herring and mackerel and the associated increase in other species in the North Sea. *FAO Fish. Rep.*, **291** (2), 507–20.

Kawasaki T. (1983) Why do some pelagic fishes have wide fluctuations in their numbers? – Biological basis of fluctuation from the viewpoint of evolutionary ecology. FAO Fish. Rep., **291** (3), 1065–80.

Kawasaki T., Taniguchi A., Tanaka S. & Toba Y. (Eds) (In press) *Long term variability of pelagic fish populations and their environment.* Pergamon Press, London.

Kondo K. (1980) The recovery of the Japanese sardine – the biological basis of stock-size fluctuations. *Rapp. P.-V. Réun. Cons. Int. Explor. Mer*, **177**, 332–54.

MacCall A.D. (1983) Variability of pelagic fish stocks off California. *FAO Fish. Rep.*, **291** (2), 101–12.

Molloy J. (1980) The assessment and management of the Celtic Sea herring stock. *Rapp. P.-V. Réun. Cons. Int. Explor. Mer*, **177**, 159–65.

OECD. (1989) *Review of fisheries in OECD member countries, 1987.* Paris.

Saville A. (Ed) (1980a) The assessment and management of pelagic fish stocks. A symposium in Aberdeen, 3–7 July 1978. *Rapp. P.-V. Réun. Cons. Int. Explor. Mer*, **177**, 517.

Saville A. (1980b) Discussion and conclusions of the symposium. *Rapp. P.-V. Réun. Cons. Int. Explor. Mer*, **177**, 513–17.

Saville A. & Bailey R.S. (1980) The assessment and management of the herring stocks in the North Sea and to the west of Scotland. *Rapp. P.-V. Réun. Cons. Int. Explor. Mer*, **177**, 112–42.

Schumacher A. (1980) Review of North Atlantic catch statistics. *Rapp. P.-V. Réun. Cons. Int. Explor. Mer*, **177**, 8–22.

Shelton P.A. & Armstrong M.J. (1983) Variations in the parent stock and recruitment of pilchard and anchovy populations in the southern Benguela system. *FAO Fish. Rep.*, **291** (3), 1113–32.

Tanaka S. (1983) Variation of pelagic fish stocks in waters around Japan. *FAO Fish. Rep.* **291** (2), 17–36.

Walsh M. & Bailey R.S. (1987) Recent changes in the mackerel fisheries and prospects for 1987. *Fishing Prospects*, pp. 19–23. MAFF Direct. Fish. Res., Lowestoft and DAFS Mar. Lab., Aberdeen.

Walsh M. & Martin J.H.A. (1986) *Recent changes in the distribution and migrations of the western mackerel stock in relation to hydrographic changes.* ICES CM 1986/H: 17 (mimeo).

CHAPTER 2

Catching methods for pelagic fish

H.R. English and J.E. Tumilty

INTRODUCTION

The pelagic fisheries of the world are diverse and an equally diverse range of fishing methods has been evolved to make use of them. The range of methods can be claimed to extend from the simplest to the most sophisticated used by fishermen; occasionally the full range is seen in use on a single species. In spite of its title, we feel that this chapter can only be selective. Hopefully all methods of global importance are included, but we recognise that many minor methods on a global scale are of much importance to communities which have developed them. We offer our apologies to any methods so excluded.

Pelagic fish are defined in our dictionary as 'fish living in the middle depths and surface waters of the sea'. From this we gather that these fish are not reliant on food found on the sea bed, but may find their food on a variety of levels in the sea up to and including the surface. To the pelagic fisherman this immediately provides a dramatic increase in the search 'volume' in which his catch may be located, compared with his demersal fisherman counterpart. A demersal trawl operating on the sea bed would sweep an area with a maximum height of perhaps 12 m. If the vessel is fishing in 400 m of water, the demersal fisherman is interested in only 3% of the total height. The pelagic man has an interest in the other 97%!

However, pelagic fish have characteristics not applicable to demersal fish (or perhaps they are not as apparent or exploitable) and these can be used to advantage by fishermen. The most important of such characteristics are shoaling, in which large groups of the same species gather together making location possible, response to food stimulus enabling fish to be drawn to an area by use of bait, or perhaps even a basic curiosity (how else would we describe the phenomenon of fish gathering around a log in the middle of the ocean?). The fishermen have a use for all these methods in their proper place and again a mixture of simple and sophisticated methods may be seen in the same fishery.

In discussion of catching methods used for pelagic fish, we reasonably include methods of location of fish as well as the gears and vessels used in the fisheries and the methods of taking the catch aboard. In many cases, the pelagic fishery has some unique features as well as those which it shares with other fisheries.

LOCATION OF THE CATCH

Some of the species in our review have habits which allow location by the simplest of means. Species which tend to gather in shoals may also be directly visible from above the surface. Herring and sardinella species, among others, may swim close enough to the surface to affect the surface pattern by occasionally clearing the surface, particularly in shallow water. With some species an oil slick-like phenomenon may also be apparent. Experienced fishermen can readily note these changes.

Many fisheries which prosecute these species are based inshore (or even on the beach) and the method here is to site an observer on a suitable high spot, be it a hill or a building, with the intention of spotting the movement of shoals of fish out at sea. Often the disturbance observed may not be that of the target species at all, but of smaller fish on which the larger fish are feeding. A typical species here would be sardinella which in the Gulf of Oman may be taken either as a target catch in its own right or in a mixture containing larger fish. The sighting of a shoal is the signal for the fishermen to prepare their gear and set out to take their catch.

Of course, if the fishery is well offshore then the direct vision method may need mechanical assistance. The helicopter is used in various fisheries, tuna probably being the best known. Modern tuna purse seine vessels operating from the mainland USA will generally have facilities for operating a spotter helicopter which will be sent ahead of the vessel to identify the location of surfacing fish or perhaps, just as often, the location of any floating wreckage which might have acted as an attraction to fish swimming below the surface. The skipper can then plot a course to investigate likely options. If no floating wreckage exists, a small floating platform may be launched from the catcher vessel. The helicopter would be used to locate this again when the catcher vessel returned to the area after allowing a period of a few days for 'soaking'.

Aerial surveillance techniques are also used in other fisheries; one interesting application in which our own organisation was involved in the mid 1970s involved night-time operations from a small aeroplane carrying a low light level television system. This showed the bioluminescence effect created by shoals of herring and offered a significant improvement over direct vision methods alone. This drew on prior applications in the USA where the method had been used successfully in the location of menhaden and anchovy species.

Of course, there is no point in searching for fish in areas where no fish are likely to be found. Skippers of tuna vessels (of various kinds) identified relationships between position of fish, temperature of seawater and special effects such as thermoclines or upwellings, and kept continuous records of sea temperature variations when in the search area. Nowadays high technology comes to their aid with a combination of sensors mounted on satellites and onboard receiver/processor/display units which can provide maps showing sea temperature variations over wide or small areas, as required. The skipper thus has a means of identifying 'likely' areas well before he reaches one.

In the section above we have described methods which allow a fisherman to decide areas likely to be most suitable for searching, but none of them allow him to search out the actual location of fish. This is, of course, the function of the shipboard fish-finding devices, be they of the echo sounder or sonar type. Both devices use the principle of transmitting an acoustic signal from the vessel and process the echo returning from seabed or marine life forms to give a display in which distinctive features can be identified and located in relationship to the vessel. An echo sounder 'looks' (or in reality 'listens') to the sector of sea directly below the vessel and is a standard tool used in demersal fisheries; the sonar uses the same operating principle but its transponder (the combined transmit and receive function) is located in a movable housing allowing forward and sideways rotation and a degree of tilt above, at and below a horizontal plane. With a capacity to process echoes at a distance of 0.5 km or more, the device allows location of shoals of pelagic fish in good time to plan a catching strategy using whichever catching method the skipper has available. The devices are capable of indicating position (range, bearing, depth) of individual fish at closer ranges. Early equipment used paper chart recording techniques but modern systems use video display, often making use of colour as an aid to discrimination.

Another option in locating fish prior to capture is to provide a means of bringing fish to where they can be readily caught. Besides the log raft technique practised by tuna seiners, already mentioned, which seeks to provide an attraction to fish otherwise swimming randomly over large areas of featureless ocean, a further positive means is the use of light attraction, probably best known in the form of the Mediterranean lampara seine fishery. In this a small vessel carries a large gas or oil lamp which is suspended above the water and remains alight while the vessel (which does not itself carry fishing gear) is allowed to drift. Fish are attracted to the light and are then taken by use of encircling gear operated from other vessels. Other variations use a light system suspended from the catcher vessel.

A further feature of this method of fish attraction is that sometimes it is necessary to keep noise levels as low as possible (the use of gas lamps eliminates the need to run a generator on the vessel, which drifts in absolute silence), while in other uses the crew attempts to create as much noise as possible, even as far as banging on the deck or bulwarks with heavy hammers.

Another (and perhaps the most significant) method of fish attraction is to use a bait. Baited hooks, each carrying either live bait or a portion of natural or artificial bait, are used in both demersal and pelagic fisheries, but the use of live bait in a chumming activity is unique to pelagic fisheries. In this method the vessel carries stocks of bait fish in live wells (having either caught the bait itself or transferred stocks from a specialist catcher boat) and small quantities of bait are periodically thrown into the water when the target species is known to be in the vicinity. The objective is first to attract the fish, but then (if a shoal can be created) to continue feeding small amounts to convince the target species that it is in contact with a shoal of small fish and to induce a feeding frenzy. The target fish can then be caught using a variety of methods. Procurement and proper handling of the live bait is crucial to the operation and a whole pelagic industry has been established to catch and supply live bait. The target species generally associated with the live bait fishery is the skipjack tuna, but other small fisheries use the technique.

CAPTURE METHODS

As stated in the introduction, methods of catching pelagic fish are almost as numerous as the varieties of the species captured. They can range from a simple spear or hook and line, made from basic materials at minimal cost, to the largest purse seine net costing many thousands of pounds and requiring a highly sophisticated vessel from which to operate it.

The choice of capture method and level of sophistication employed is determined by a number of factors, the most significant being, of course, the financial status of the fisherman or vessel owner. Secondly, there are the behavioural characteristics of the target species, although most pelagic fish can be captured by a number of methods. These methods can be divided into four broad areas of classification:

(1) surrounding gears (purse seines, ring nets)
(2) towed gear (trawling)
(3) static gear (drift nets, long lines)
(4) other mobile gear (trolling, harpoon, pole and line).

These are only broad classifications. Some methods use a combination of techniques such as, for instance, the use of light attraction in certain surround net fisheries.

We now examine some of these methods more closely, detailing as many as possible, but it must be realised that there are many local differences throughout the world so we shall concentrate only on the principal techniques employed.

Surrounding gears

We have already mentioned behavioural characteristics of pelagic fish, in that they frequently appear in large single shoals of the same species. This particular characteristic favours the use of surrounding gears. These take different forms, but by far the most important is the purse seine. This method probably takes a higher percentage of the world's total fish catch than any other single method.

Purse seining

Purse seining is a method of particular importance for the capture of species utilised in bulk reduction processes, such as fish meal for animal feeds. It is also used extensively in the catching of species having high individual value, such as tuna, used as food fish.

The basic method of purse seining involves the setting out of a long net to form a wall of webbing

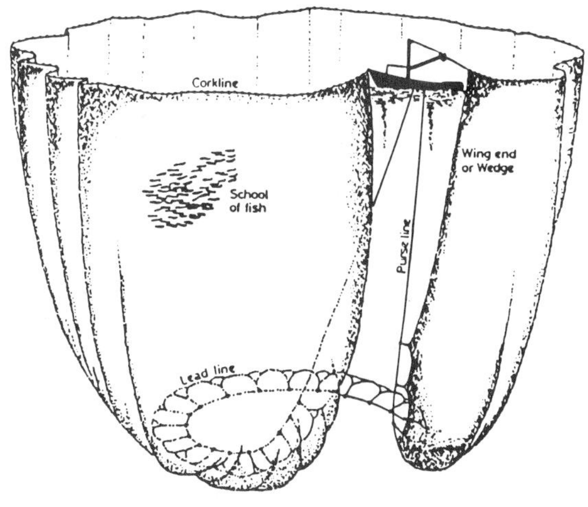

Fig. 2.1 Purse net prior to closing the purse (not to scale).

around the school of fish being taken, the top of the net usually being on the surface. When the net has encircled the fish, its bottom is pulled together so that an artificial pond of webbing holds the catch. This pond is then gradually made smaller until the fish are gathered alongside the vessel and may be taken aboard. Figure 2.1 shows the arrangement. Although not to scale it gives an indication of the very large volumes of sea that can be enclosed by the seine net.

For the most part, this method is utilised in harvesting pelagic species swimming from the surface to a depth of perhaps 130 m (70 fathoms). In some fisheries, however, the purse seine is used to catch demersal species, such as cod, swimming near the sea bed, and in such cases the wall of netting is sunk so that the bottom is on the sea bed with the top below the surface.

A number of methods of net arrangement and gear handling are in common use, the one chosen for a particular application depending largely on the characteristics and behaviour of the species being harvested.

The nets are made of large sections of netting, usually nylon, joined together and mounted to a headline, sole-rope and gables (side ropes). They may be up to 730 m (400 fathoms) long and over 180 m (100 fathoms) deep. It is important that the net sinks quickly when it is shot, and depending on the size of the net, up to five thousand lead weights each approximately 100 g (4 oz), threaded on to light lines, are attached along the sole-rope. To support this weight and the weight of the net, over a thousand floats threaded on float lines may be required on the headline. The floats are usually egg-shaped and made of cellular plastic. The shape of the net is shown in

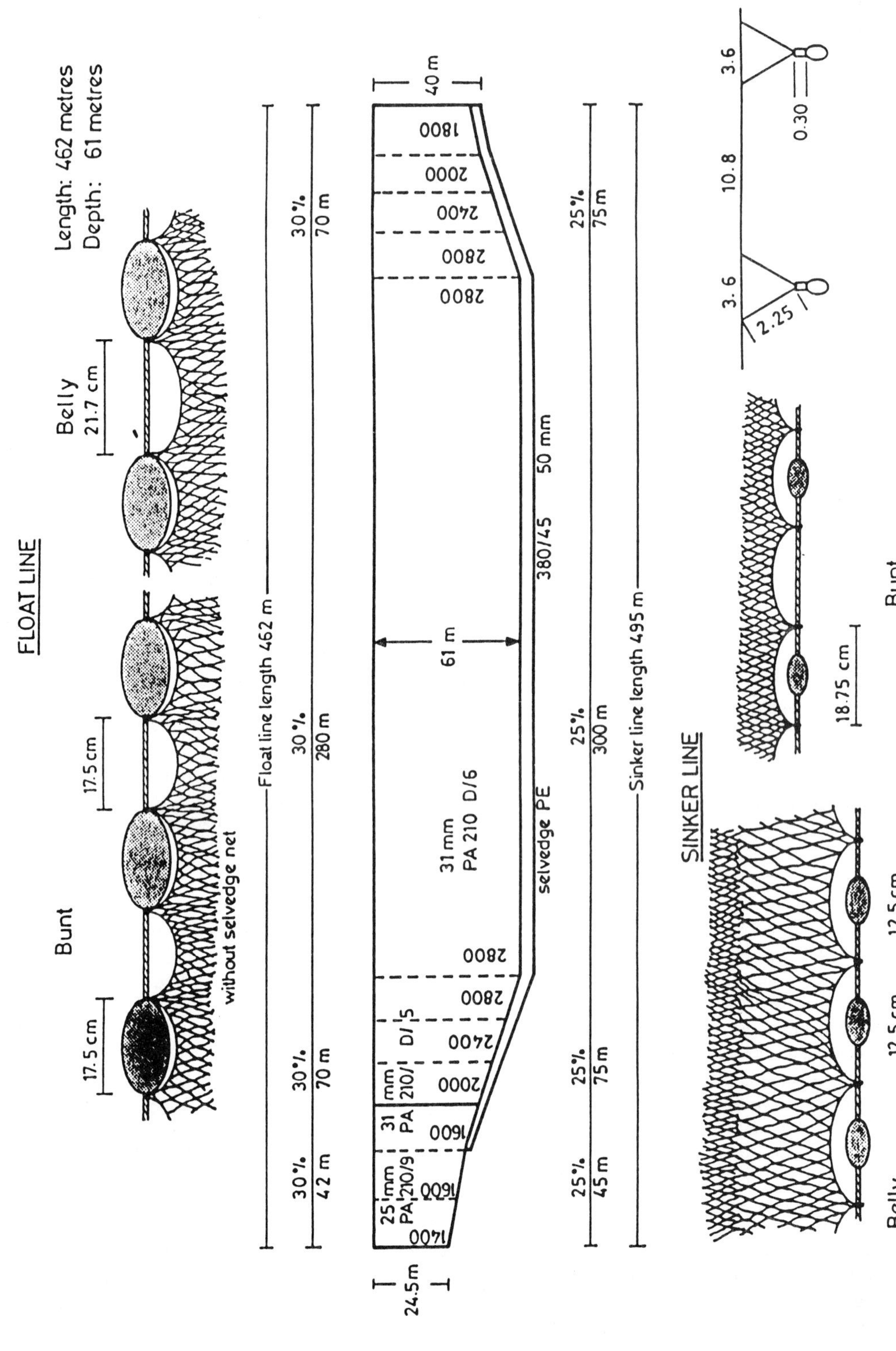

Fig. 2.2 Purse seine net.

Figure 2.2. The bunt or small mesh section is located at the left-hand end of the main sketch.

The pursing rings hang 2–4 m (1 to 2 fathoms) below the sole-rope. On typical purse seines there are between 40 and 50 pursing rings. There are various types of ring, some with a spring-loaded sliding ferrule so that the ring may be opened and unclipped from the pursing wire as it comes to hand during hauling of the net, others with a spring loaded gate for the same purpose. They are usually made of hardened stainless steel.

Top bridles up to 91 m (50 fathoms) long may be used at each end of the net. The bunt end of the pursing wire is held up by two or three buoys while the net is being shot, in order to facilitate easy retrieval of that end of the pursing wire and the marker dan as soon as the net has been set.

When the purse seine is put aboard it is stowed with the floats on the port side and the sole-rope on the starboard side (for clockwise shooting of the gear). In order to keep the pursing rings together clear of netting, and to ensure that they go away in the right order when the gear is shot, they are hung in order on a shooting bar positioned on the ship's side near to the net. Sufficient length of pursing wire is then run off the winch to allow its end to be passed through the rings on the shooting bar and then to connect it to the dan in readiness for shooting the gear.

Once the decision to fish on a shoal has been made, very careful attention must be paid to its depth, speed and direction of movement, and to the strength and direction of tide, current and wind, if the fish are not to escape before the net is pursed. The vessel moves to the starting position, the dan is dropped over the side and the bridle, with pursing wire, allowed to run out followed by netting as the vessel moves round the shoal (Fig. 2.3). The pursing wire takes pursing rings with it as the net pulls each one in turn off the shooting bar. Figure 2.4 shows the situation as the gear is paid away over the stern. The pursing rings are pulled away in turn from the shooting bar. In the background the floats can be seen on the upper edge of the net.

When all the net has been shot and the dan, buoys and the ends of the bunt bridle and pursing wire have been taken back on board and that end of the pursing wire made fast at the gallows, the pursing of the net commences without delay (Fig. 2.5). When the pursing rings are secure at the gallows and all slack wing netting is pulled in, the headline at the bunt is hooked up to a fore-boom. As the power block hauls in netting, each pursing ring in turn is unclipped from the purse wire and reclipped to a taut wire which guides the ring aft as it is pulled by the netting.

It is then unclipped from the taut wire to allow it to be hauled with netting over the power block. Hauling and stacking the net continues until the catch is concentrated close to the ship. Figure 2.6 shows this operation at an intermediate stage. Fish can then be lifted out of the net with a lifting net called a brailer, operated from a derrick, or fish and water are pumped out by a fish pump. When all the fish are on board the rest of the net is hauled and stacked and the gear made ready to shoot again.

A common feature of purse seine operations on mechanised vessels is the power block, which is used to bring the net on board again after the shooting sequence and to control the trapped volume from which the catch may be recovered on board. Originally developed in the USA for their tuna fishery, it took the form of a V-grooved powered pulley mounted from a deck boom, which could be swung through a wide arc over the ship side and at a range of heights by raising or lowering the boom. It could thus be lowered almost to sea level if necessary to allow a bight of netting to be first lifted and then lowered on to the open area of the pulley.

Lifting of the boom then applied the necessary grip between pulley and net, and the net could be hauled in by rotation of the pulley. The block would be sited to allow the slack net to be relaid in the storage/shooting bin on the vessel deck. This is the arrangement shown in Figure 2.6. It is still the popular system used in the tuna fishery.

When the method was adopted for European fisheries for herring and mackerel the combined effect of using smaller meshes in the net (and hence heavier nets) and heavier weights of catch led to design changes in which the power block was located at the ship's side on a pedestal which could be rotated or tilted to allow variations in the direction from which the netting was to be hauled. When a large catch was being taken it would not be uncommon for the 'bag' of fish to be longer than the catcher vessel itself. The new arrangement meant that a new approach had to be taken to ensure the necessary grip between powered roller and net. In some cases this has meant modifications and additions to the original V-pully concept, but a popular variant is the Triplex system in which three parallel powered rollers replace the original V-design, and the necessary grip is achieved by routing the net alternately inboard and outboard of succeeding rollers.

A variant of the power block at high level is still necessary to allow for the net being relaid in the storage bin, but this device handles only slack netting downstream of the main block and is a relatively small device.

In different parts of the world there are a number

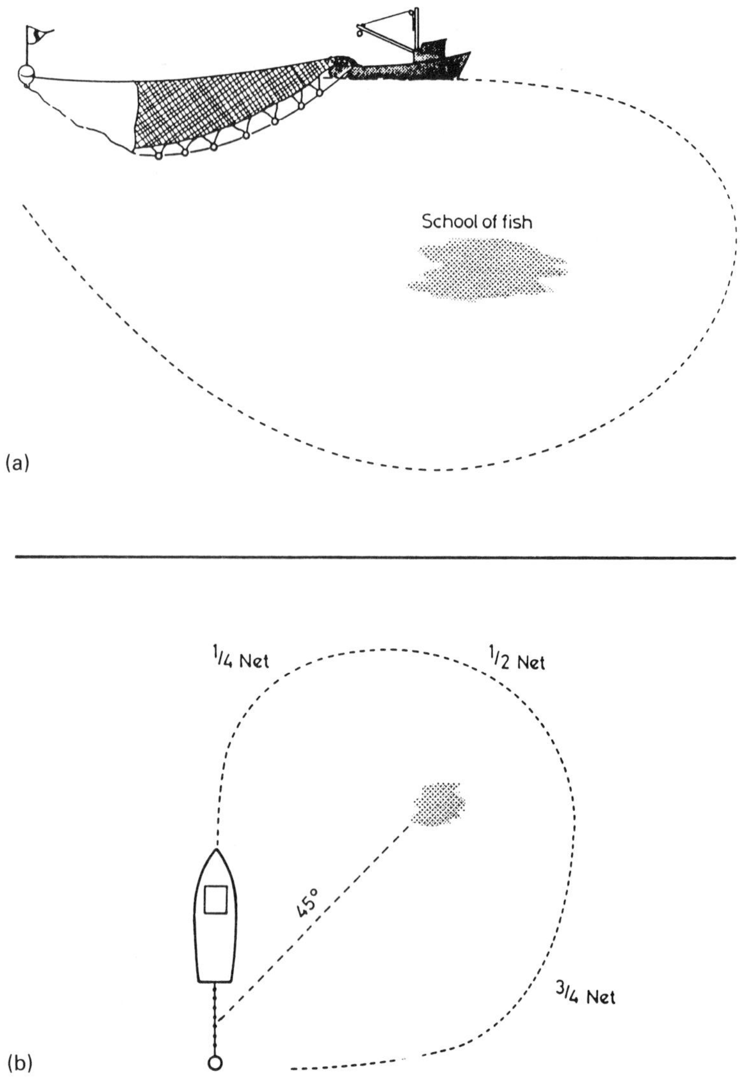

Fig. 2.3 Starting to shoot the net (not to scale). (a) Side view. (b) Plan view.

of variations on this method. The method described above utilises only one vessel; other fisheries may use two. Both the one boat and two boat systems have their advantages and disadvantages; however, purse seining with a single vessel is generally considered to be more economical, although shooting a purse seine with a two boat system may be quicker and a larger gear can be operated than with one boat systems.

In two boat purse seining two boats, each carrying about half the gear, operate together, setting each part simultaneously beginning with the middle of the seine net. In one boat purse seining the gear is set beginning with one end, around the located fish shoal. This method of operation has some influence on the construction of the gear. Generally speaking, the main part of the net of a purse seine is made of the same material with constant mesh size. Only that part of the gear where the catch will be concentrated, the so-called bunt, will be strengthened, and its position in the gear varies according to the method of operation in the two boat or one boat system.

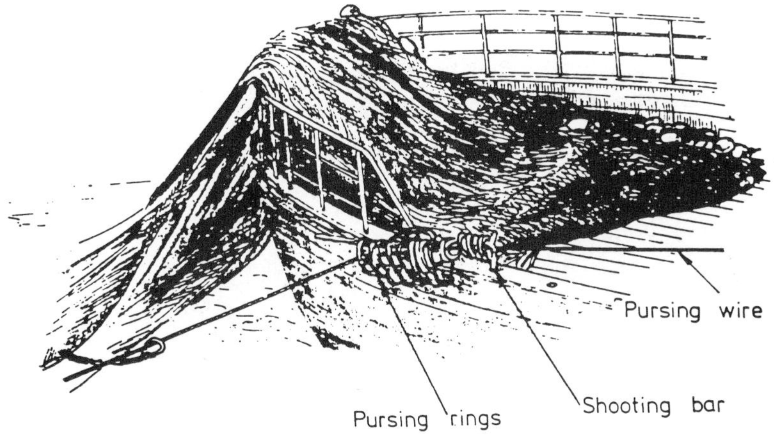

Fig. 2.4 Shooting the purse seine (Strange 1981).

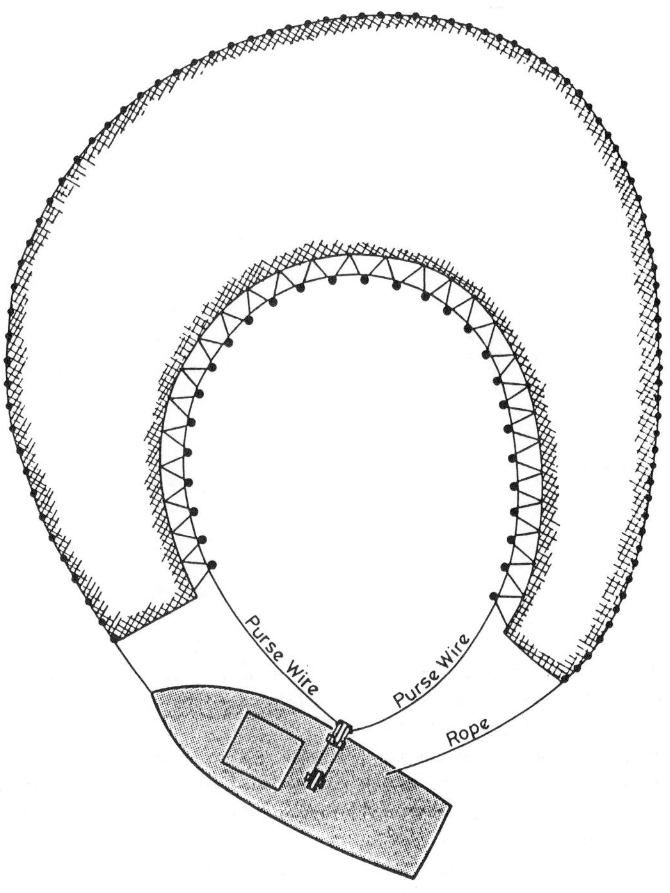

Fig. 2.5 Starting to purse. Here the boat will come astern trying to keep out of the net while the winch pulls the purse wire in, closing the bottom of the net.

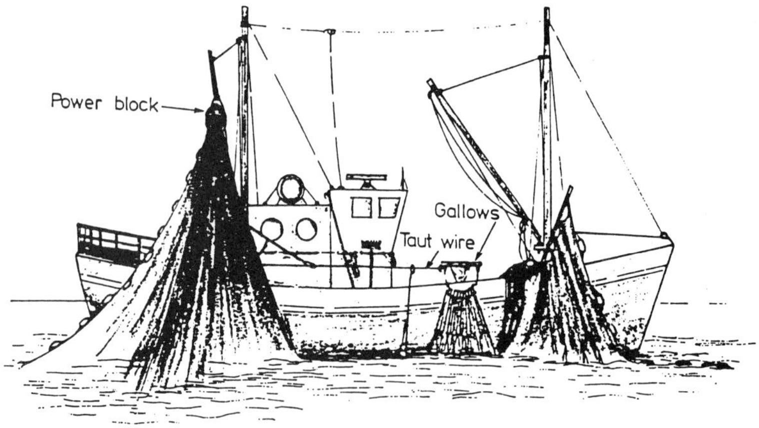

Fig. 2.6 Hauling the purse seine (Strange 1981).

The ring net
A further variation of the purse seine is the ring net. It is another encircling net used mainly to catch herring when they come off the bottom at night and aggregate in tight shoals. The ring net is particularly suited for working in relatively shallow waters near the coast and in sea lochs. It was extensively used in the Scottish west coast herring fisheries, but has declined in recent years since the introduction of the less arduous method of pair trawling. It can however be used by lower powered boats than would be necessary for pair trawling.

The net is usually made of nylon and is similar in shape to a purse seine but smaller, with a bunt in the middle to hold the catch. Mesh sizes range from about 40 mm in the top section of the bunt to 70 mm in the deepening and 100 mm in the wings.

Typical ring net boats range in size from 12.2 m (40 ft) to 18.3 m (60 ft) and work in pairs, each vessel carrying a net and gear and taking turns to shoot its net or act as the partner vessel. The boats are usually heavily fendered along one side with old car tyres for protection against impact when they come together during fishing operations.

The fishing cycle is shown in Figure 2.7. When a suitable shoal is located and both vessels are ready, the first vessel will drop its dan light over the side and proceed round one side of the shoal, paying out bridles and net. The partner vessel picks up the dan and connects the bridles to its sweep, and then moves round the other side of the shoal towing the sweep from the midship bollards.

When the two vessels come together at the end of shooting operations, the leading end of the sweep on the partner vessel is passed over to the first vessel and led through the fair lead to the ring net winch. Most of the partner crew then jump aboard the first vessel to assist with hauling the net. Usually these operations are carried out so that the vessels come together on the shore side of the shoal, taking account of wind and tide, so that when hauling the gear the net is hauled towards the shore. The bottom ropes are hauled faster than the top ropes in order to pull the bottom of the net under the shoal.

During hauling the partner vessel holds the other vessel broad side away from the net until sufficient net is hauled to bring the catch to the side. It then moves to the opposite side of the net to assist in holding up the net, which otherwise might sink below the surface owing to the weight of the catch, allowing many fish to escape. The fish are then brailed out and conveyed to the fish hold.

Lampara nets
The lampara net is yet another variation of a type of surround net. It originated in the Mediterranean and is shaped like a dustpan with wings (Fig. 2.8). While the net bag, or bunt, is made of small-meshed netting, the wings are made of coarse material with wider meshes. The dustpan shape is attained because the weighted ground-rope is much shorter than the upper line which carries the floats. This construction causes the middle part of the gear to be formed like a bag and the lampara nets are sometimes called purse seines with bag. However, in contrast to genuine purse seines, this gear has no arrangement for pursing.

The net itself is now mostly used with a single vessel. However, one or more small partner vessels carrying lights are often used to attract fish to the area around which the gear will be shot. These vessels drift silently for a time (often hours) before the master vessel shoots the gear. When shooting, the end of one

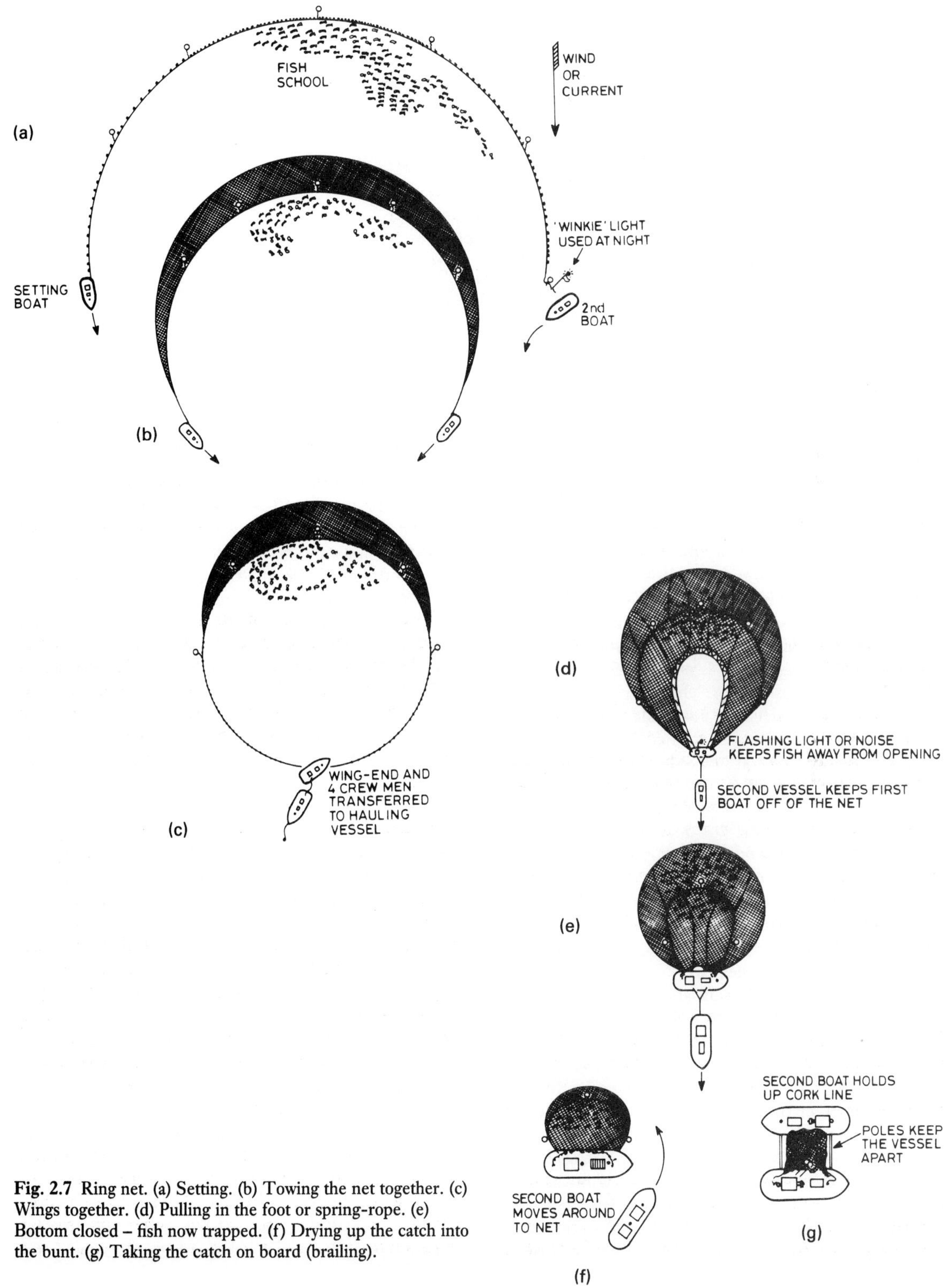

Fig. 2.7 Ring net. (a) Setting. (b) Towing the net together. (c) Wings together. (d) Pulling in the foot or spring-rope. (e) Bottom closed – fish now trapped. (f) Drying up the catch into the bunt. (g) Taking the catch on board (brailing).

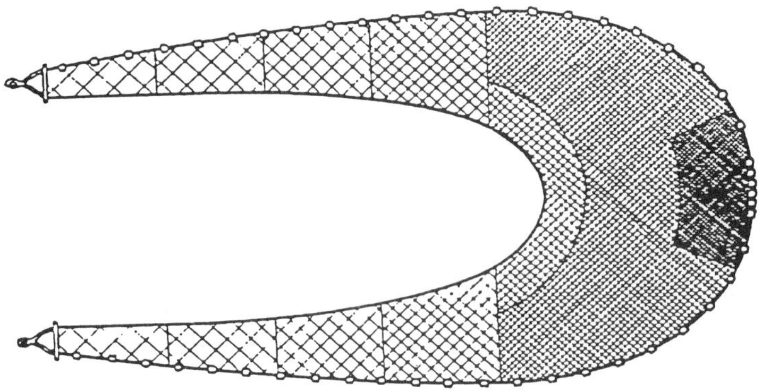

Fig. 2.8 Design of the lampara net (von Brandt 1984).

wing is fastened to a dan buoy and from that buoy the remaining net can be carried by a boat around the fish shoal. The buoy can also be replaced by a boat, or skiff, which remains at the spot or drifts slowly while the main boat encircles the fish shoal.

That boat then returns to the initial spot and begins, uniformly, to haul the two wings. The leadlines come together closing, downwards, the area surrounded by the gear before the wings are drawn together. To prevent fish or squid escaping before the gear between the two wings is closed completely, a submerged flashing light is often hung from the vessel into the water, or a line with white boards is submerged and moved up and down. Splashing or 'cherry bombs' or loud hammering on the deck or gunwhale, can serve the same purpose. When this is finished and the catch in the net is concentrated at one side of the vessel, with or without the help of an additional smaller boat, hauling can begin.

The typical lampara net was developed in the Mediterranean area, its name originating from the Italian word 'lampo', as the nets are usually employed with lights. The Italians took the Mediterranean lampara net to the Californian coast for fishing sardine and mackerel in the early 1900s and it was from these designs that purse seines were developed as the necessary methods of mechanisation became available.

As stated at the beginning of this section, these are the principal techniques involved in surround nets, but there are many variations worldwide. In particular we should, of course, mention the beach seine variant, which contains many of the elements of the lampara fishery. Typically, the net will be anchored on the beach with its bulk stored on a suitable vessel. When a shoal of fish is seen to be moving within an area that the net can encircle, the vessel is launched and powered or rowed in a semi-circle while the net is paid away. As with the purse seine fishery, sinkage is important if the net is to be effective. When the vessel returns to the shore, the net is hauled onshore from one or both ends until the catch can be taken from the net.

The arrangement does not involve pursing as such since the netting is affixed permanently to both headline and sole line. The net will generally be of parallel form (i.e. constant height) and of a constant mesh throughout, since fish can be taken in any part of the net. There would be little point in applying a tapered shape to the net since it is common practice to shoot only as much net as is needed to encircle the identified shoal. This reflects the need to close the escape route as quickly as possible, and also the high manual labour content in relaying the net correctly on the vessel in readiness for the next sortie.

Trawling

Trawling is the operation of towing a net to catch fish, and the basic requirements for operating the trawl net are power to tow the net, a means of holding the mouth of the net open while towing, a system of wires to connect the net and gear to the source of towing power and the ability to cast and haul the net.

The mid-water trawl is used to capture pelagic species that school at various levels between the sea bed and the surface. While bottom trawls are often towed for several hours at a time and fish a large area, so capturing loosely distributed fish, the mid-water trawl is often towed for only some 10–20 minutes in order to pass through and catch a particular shoal of fish. Much of the vessel's time is spent hunting for fish in schools large enough to justify shooting the net; successful mid-water trawling requires the effective use of various electronic aids, both to find the fish and to manoeuvre the vessel while catching them; the

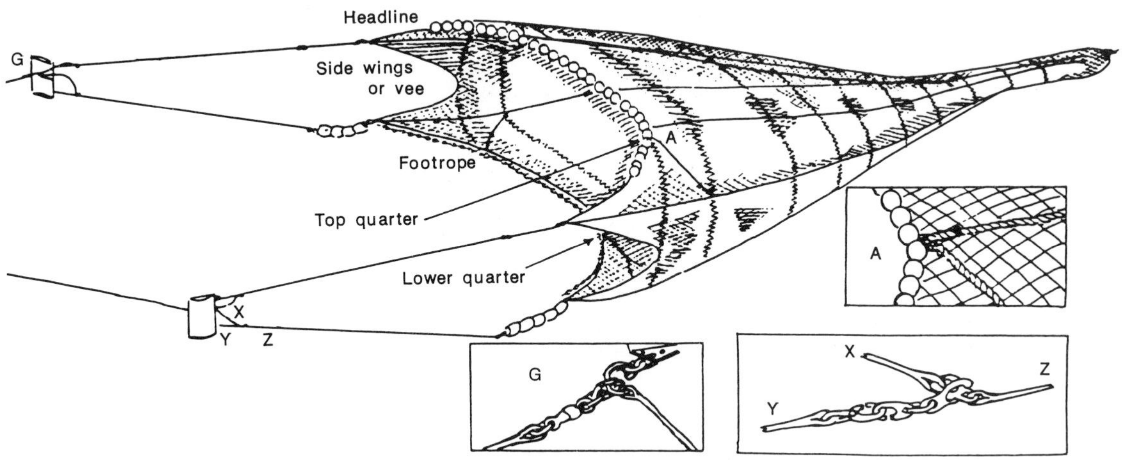

Fig. 2.9 One design of mid-water net.

net must be set at the correct depth, and the vessel proceed on a course that will ensure the net passes through the school.

Due to the density of the fish shoals through which the net passes, the catch per tow is often much greater than for demersal trawls, perhaps by a ratio of 10:1.

Both single vessel and pair trawling are of importance in mid-water; the latter is particularly useful in connection with smaller, low-powered vessels, and is especially applicable in shallower waters where a single vessel passing over the fish would tend to scare them downwards so that they may pass below the level of the net. In addition, two boat trawling appears to offer advantages in that the towing warps do not pass through and frighten the school of fish before the net reaches them. However the single boat method is still much more common than the two boat variant.

The depth at which the trawl fishes may be adjusted by varying the vessel's speed and the length of the towing warps; there may also be some changes made in the rigging of the net and gear, depending on the water depth and whether the net is fishing near the bottom or surface.

The towing speed depends on whether the species being hunted are fast or slow swimmers; for some fish 2–3 knots is sufficient, while for faster swimming species $4-4\frac{1}{2}$ knots is essential. It is important, therefore, that the size of the net be matched to the vessel's power, to suit a particular application.

The mid-water trawl net

The mid-water trawl resembles more a cone of netting, compared with the flat bag of the bottom trawl, the mouth taking up somewhat of an oval, circular, or square shape depending on the design.

During normal use, the net does not come into contact with the sea bed except in very shallow water, when the headline may lie on the surface while the footrope scrapes the bottom. The net is, therefore, of relatively light construction, enabling a vessel to tow a larger mid-water trawl than it could a bottom trawl with its heavy net and bottom gear.

There are a number of designs for mid-water nets, but generally they may be grouped into two types: those that have the same shape above and below the net centre line, ending in small wings at either side; and those that are essentially constructed from four panels of netting joined to form a box shape, with small extension gussets or wings at each corner. Figure 2.9 shows a version of the latter design.

Floats are placed at the headline, and the footrope is often weighted with chain; other weight and float arrangements depend on whether one or two boat trawling is being undertaken. As well as assisting the net in maintaining its correct geometry, the chain on the footrope assists in providing a strong echo on the recording machine from the headline transducer.

In modern mid-water trawls, very large meshes are used in the wings and front part of the trawl to decrease the water resistance of the trawl. The use of large meshes does not appear to substantially reduce their effectiveness in guiding fish into the cod end of mid-water trawling. As stated earlier, the method can be undertaken by either a single boat or two boats; in each case the trawl design is very similar. The principal difference between mid-water and demersal trawl design is that the demersal trawl is towed on the ground rope and achieves its opening by virtue of the net design and flotation. In a mid-water trawl the opposite effect can be observed.

The gear is towed on the headline with the opening achieved by the weights placed at the lower wing ends of the net. Any flotation provided is often used to keep the headline clear while shooting.

While mid-water trawling, the skipper must use his knowledge and experience to know exactly the right time to haul his gear. If he hauls early he may not have caught a viable quantity of fish, while hauling too late may result in a massive catch which at times can burst the trawl or prevent it from being handled on board. However, much of the uncertainty as to the contents of the codend is now being removed with the increased use of electronic sensors on the trawl which will give the skipper a reliable indication of his catch.

Static gear

There are a number of static methods of catching pelagic fish, the most popular involving drift nets, floating long lines and lift nets, the latter often used with light attraction techniques.

Drift netting

Drift netting is one of the oldest fishing methods for catching fish that spend part of their adult life in the upper layers of the sea. For many generations of fishermen it was the principal method of catching herring and pilchards. It is not in common use today having been almost entirely superseded by more efficient and less arduous methods, e.g. purse seining and pair trawling.

However, drift netting is a very selective form of fishing. The mesh size and rigging of the nets can select accurately the size of fish to be caught. It may thus take on a greater role in the future in any conservation-led approach to fisheries management.

The drift net consists of a curtain of lightweight netting with a border of stronger netting called guarding, its edges trimmed with soft cord, a headline carrying small floats (sometimes joined throughout its length to a second rope of opposite lay to prevent kinking), separated from the top of the net by short cords called norsels at regular intervals throughout its length and the bottom similarly attached to a sole rope lightly weighted with lead. The size of drift nets for herring is about 35 m (114 ft) long by about 15 m (49 ft) deep. The mesh size is chosen to allow the heads of marketable size fish to pass through and become enmeshed by their gills.

Traditional nets were made of cotton and were then treated with a variety of substances, including tanning solutions such as catechu or cutch (bark extracts), to preserve them. There are now no longer the materials, the facilities or the willingness of the fishermen to carry out these tasks to continue the fishery in its traditional form. The Sea Fish Industry Authority has been actively supporting moves to reintroduce drift nets into our Cornish pilchard industry and therefore undertook to find a modern synthetic equivalent to the traditional net. This project was successfully carried out and nylon nets suitably treated with a latex based solution can be substituted for the old cotton nets.

Developments in other countries, Iceland in particular, have led to the introduction of more mechanisation to reduce the labour content of the operation and the method may yet see a revival in its importance, though not to the level that it once enjoyed.

Figure 2.10 shows the seabed features of the nets and their use. The traditional method of pilchard drifting or 'driving' involved the use of considerable lengths of netting (up to 3 km), a full fleet comprising 25 separate nets. The fishery was prosecuted by vessels in the 10–15 m class and power involvements were low.

The following description is taken from a recent trials report on our pilchard fishery development but could have been written any time this century:

'Pilchard fishing is a night time operation, the fishermen aiming to have located the shoals of fish and have their nets shot just as darkness falls. It is at this time that the pilchard shoals move towards the surface and come within range of the drift nets.

Once the shoals have been located the nets are shot. This operation is usually carried out downwind with the aid of a mizzen sail. The nets are shot as the vessel sails or steams at enough speed to clear the nets away, the nets being buoyed and marked with dans. The shooting operation takes place over the starboard side of the vessel. When the last net has been shot, the warp or messenger rope, attached to the top of the nets, is made fast to the forward part of the vessel, usually at the winch. Enough warp is then paid out to allow the vessel to ride easily at the nets, the vessel's head eventually coming round into the wind under the influence of the mizzen sail.

The actual drifting operation is carried out usually downwind, but if there is no wind the vessel will move with the tide. Pilchard shoals have a tendency to move with the wind. (The pilchard is known as a 'lee' fish as opposed to the herring which is known as a 'windward' fish.) The drifting operation varies from 1 to 4 hours depending on conditions and density of shoals.

On completion of the drifting operation the nets

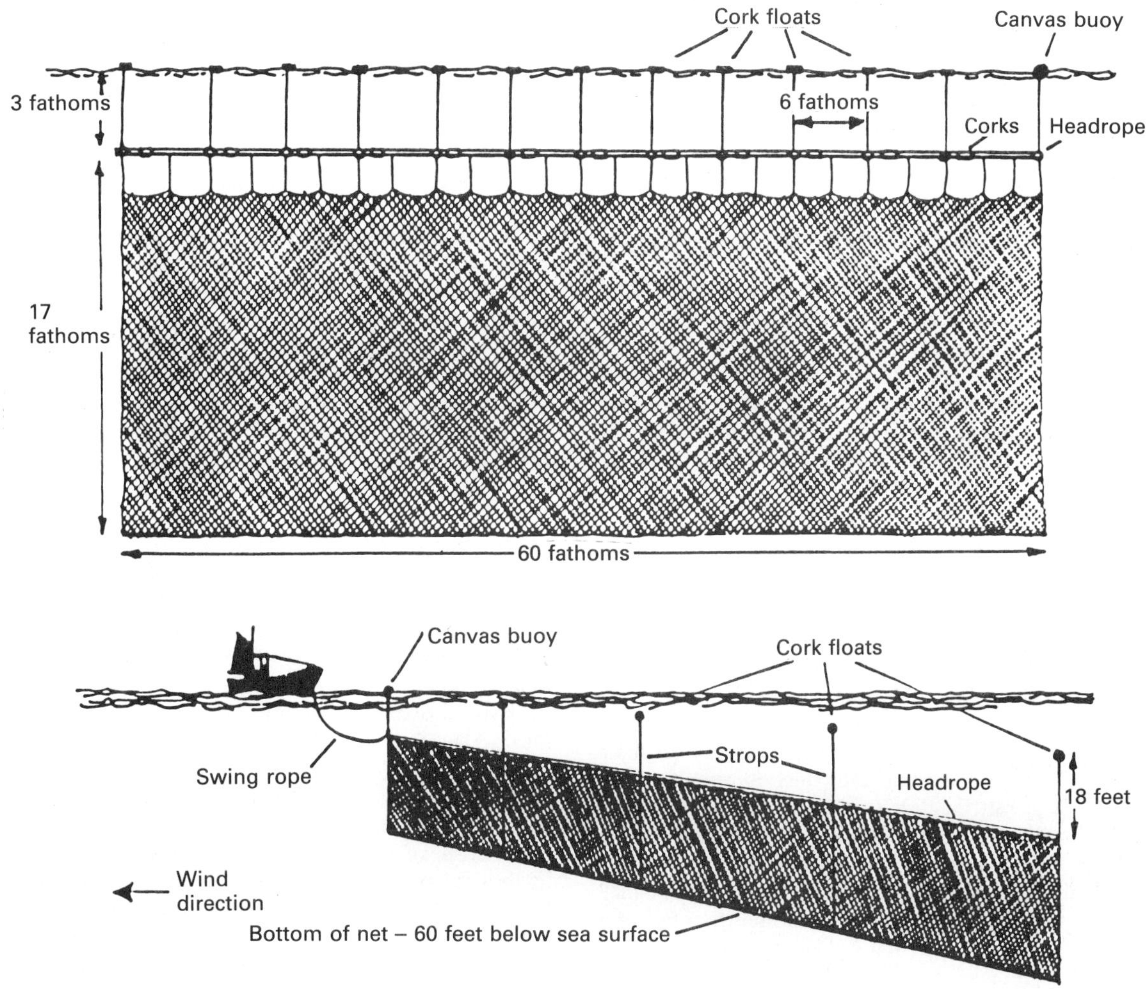

Fig. 2.10 Drift netting. (a) Drift net as used by the Looe boats during the summer season. Up to 15 nets would be joined end on together, each net 60 fathoms in length, to give a full fleet of nets, a little more than a mile of nets. In the winter an average of 10 nets at 60 fathoms each. (b) Length of nets shot with the boat laying head to wind, and pointing in the direction of the nets with the swing rope attached (Sainsbury 1971).

are hauled. The vessel, under her own power, slowly moves back along the nets enabling two men to haul the nets by hand, usually over a roller on the starboard side of the vessel. As the nets come aboard the fish are shaken out into the hold and the nets stored in a 'net room' next to the hold. Hauling time is dependent on the size of the catch, number of nets shot, and the weather/sea conditions. With a full fleet of nets in the water, the hauling operation is likely to take at least 3–4 hours. In cases of very heavy catches it is sometimes necessary to store the nets with the fish still gilled, to be cleared ashore later.'

Northern European use of the method was developed for the pilchard and herring fisheries. Drift nets are also used to catch salmon and tuna and it is believed that the Japanese use them to catch a certain species of squid.

Floating long lines and drift lines

There are wide variations in the dimensions, rigging and operation of floating long lines depending on the species, area and local tradition. The arrangement of the gear is very similar to that used for bottom set lines, with the depth of the line being maintained at the desired depth below the surface by a series of regularly spaced lines running up to surface buoys which are of assistance in the hauling operation. Anchors set at each end of the total set, and at intermediate positions as necessary, hold the gear in its original location. The vessel will meanwhile move away to attend to other gear. The species caught by this method are various, but include tuna, swordfish, mackerel and shark.

The drift line is of similar construction but of course is not anchored to the bottom. In most marine fisheries the vessel drifts with the line. There are again many different arrangements of lines and hooks to suit individual fisheries.

Perhaps the most important feature of this method of fishing is the knowledge of fish migration habits which the fisherman has to develop to use it successfully.

Hand lines

Hand lines also are used commercially in a number of pelagic fisheries. This is probably one of the simplest forms of fishing employed. The lines are rigged and operated in an endless number of local variations.

The gear principally consists of a line, a sinker, snood and hook. There may be added swivels, lures and possibly floats. In some fisheries such as the UK inshore mackerel fishery, hand gurdies are used in setting and hauling the lines. These comprise a simple reel which holds the line and which is wound to and from to attract the fish to the moving lines. These, combined with a stripper, help to take some of the labour out of the task. Powered variations of the same equipment are becoming increasingly popular on small vessels and a major Japanese squid fishery has been set up using automatic reeling or 'jigging' machinery.

Although landings of hand line caught fish are generally small they often command a good price because the fish are usually undamaged and of better appearance than fish caught by other methods.

Lift nets and stick nets

These methods again cover a vast number of local variations. They can be used to catch crabs and other crustaceans, as well as small pelagic species such as sardines. Lift nets, or blanket nets, are frequently used from the shore due to their large unwieldy size. However, medium-sized unframed blanket nets can also be operated from boats, but in this case the size of the blanket net must be smaller in comparison with the stationary ones operated from the shore.

The nets work on the principle of providing a 'box' into which the target species swims (or is lured by lights etc.). The catch is taken by lifting the base of the box to reduce the volume of the net and allow retrieval of the catch by trail net or gaff. In beach installations they can be very large, perhaps 50 m (162 ft) square.

Nevertheless, there are some boat lift nets which are not just beach lift nets placed on a vessel but are specially developed for boat fishing. There are two such highly developed fishing gears – one originated in the Philippines, the other in Japan. The Philippines have adopted single boat lift nets provided with a rim which shapes them like a box on the top. These lift nets, called 'basnigs', have undergone a special development (Fig. 2.11). They were originally nothing more than the liftable bottom of a large trap, but by about 1920 they developed into a stationary gear mounted between poles. That developed further (1924) into a gear worked by two boats and finally into a lift net spread by poles installed on a single boat. This took place about 1935. When the problem had been solved as to how such nets could be set and hauled from a single boat, the basnig became an efficient large-scale fishing gear by 1950. The fish are attracted by light and guided over the sunken net. The Philippines now take 90% of their sardines by means of this gear.

Another type of specialised boat lift net for a large-scale fishery is the Japanese 'bouke ami' or 'stick-held dipnet' shown in Figure 2.12. The blanket net is held by sticks on one side of the vessel. The fish are meanwhile attracted by strong lights on the *other* side of the vessel. Then, when a large shoal has been attracted, by switching off those lights and switching on a single lamp above the dipnet, the fish are guided beneath the keel or around the vessel into the net over its lowered side. When enough fish are gathered, the lowered side of the gear is hauled and the caught fish are brailed.

This net, too, has proved very efficient for catching pelagic fish. The fish are not only attracted by light but also by bait scattered over the netting. Then the fish are caught by suddenly lifting the net. The Pacific saury (*Cololabis saira*) is especially caught by the Japanese with this gear and it is useful for catching small species which are to be used as line bait in other fisheries. Therefore attempts have been made to mechanise this fishing method and to modify it by use of vessel power to move the net under the shoal instead of attracting the catch over the netting.

Other mobile gear

Now to come to the final category of capture methods, which is a mixed bag of a number of different methods used to catch pelagic fish.

Trolling

This method is particularly applicable to the capture of pelagic species having high individual value, and

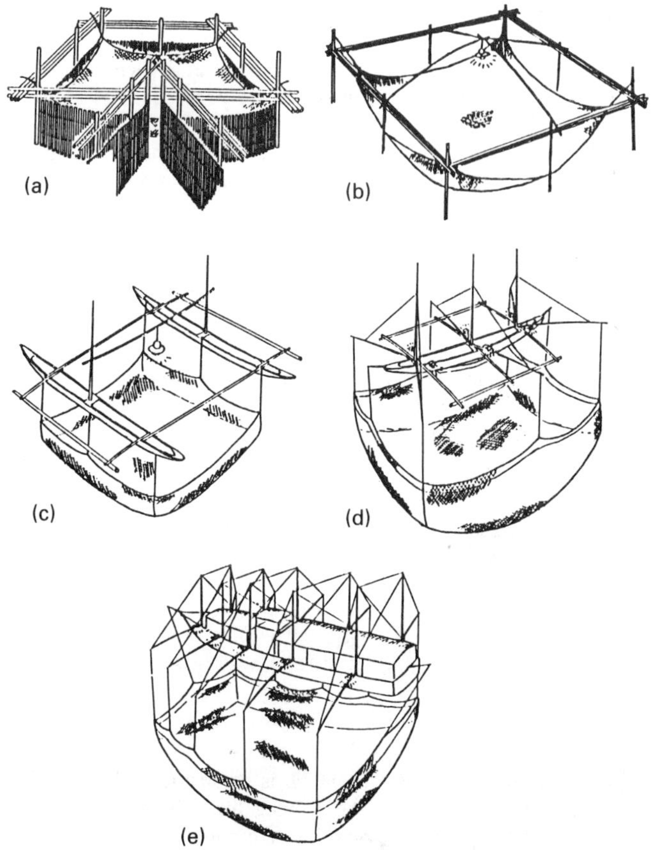

Fig. 2.11 The development of the Philippine 'basnig' for work at sea. Diagrams represent stages of development from fixed (a) and (b); twin artisanal boat working (c); single artisanal boat working (d), to motor vessel working (e) (von Brandt 1984).

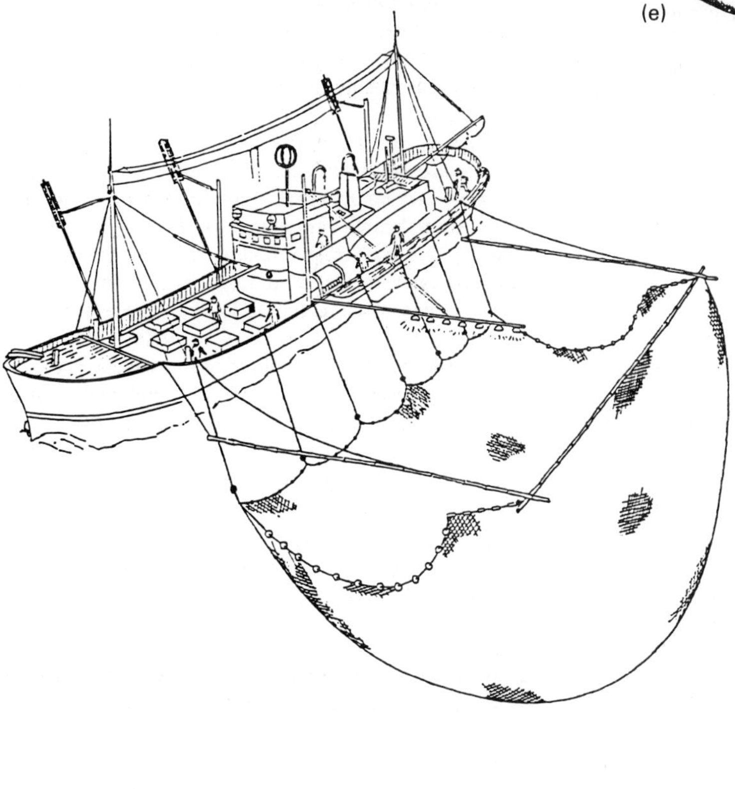

Fig. 2.12 Japanese 'saury' stick-held dip net (von Brandt 1984).

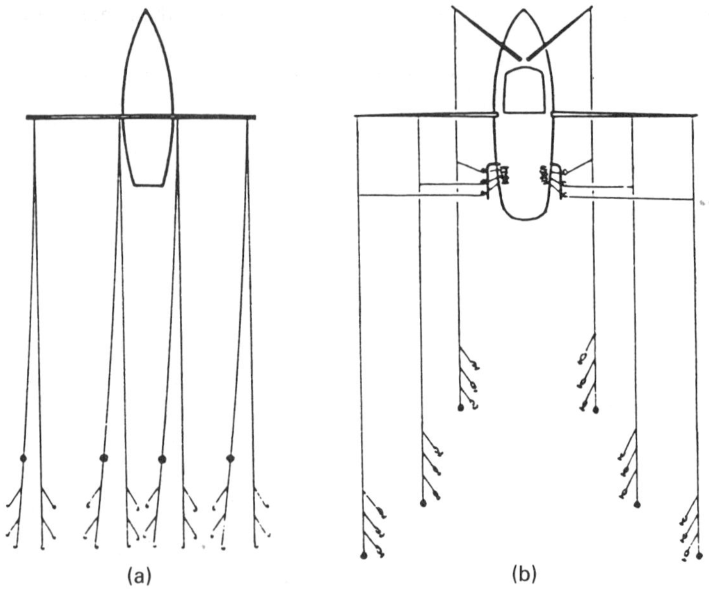

Fig. 2.13 Rigging of troll lines. (a) For mackerel in Germany. (b) For salmon in British Columbia. The addition of weights allows hooks to be trolled at different depths (von Brandt 1984).

where high quality is necessary; typical examples are the salmon and tuna. Effectiveness depends on the fish schooling.

A number of lures or baited hooks are towed astern of a slowly moving vessel, the fish being hooked after snapping at the lure, and held by the mouth until they can be brought aboard as the line is hauled in.

A typical gear arrangement for a troller is to have several trolling lines – each with a number of lures on short leaders clipped to the main line – which are towed from two or more outriggers arranged to hinge outboard from their vertically stowed positions. The storage, setting and retrieval of these lines, which are usually of piano wire, is handled by individual drums in a set of trolling gurdies. These may be placed abaft the wheelhouse or between davits mounted on each side of the vessel to support blocks for providing a satisfactory run of the trolling wires to the gurdy spools.

There are many methods in use for the towing and handling of the lines. The trolling wire usually runs from the gurdy drum or spool to a block hung from a davit at the ship's side aft and hence through the clip ring and into the water. A number of lures or baited hooks, perhaps 6–8, are snapped to the line by connectors attached to two or three fathom leaders. The trolling lines are weighted by 'cannonball' type sinkers which may weigh up to 50 pounds. Typical arrangements are shown in Figure 2.13. The lines are arranged to tow at different depths, so providing a spread of coverage, by varying the length run out and the weight.

Small vessels are typical of this method, usually being operated by a crew of one or two men; the gear arrangement is such that the helmsman can run the gurdy and handle the catch himself. Vessel lengths between 8 and 15 m are common, normally with a wheelhouse forward arrangement, allowing a clear well or working deck aft.

The vessel steams at one or two knots through the fishing area, the hooking of a fish being denoted by a jerk at the outrigger. To retrieve the fish, the appropriate gurdy spool is actuated to wind in the trolling line.

When a tag line and clip arrangement is used, the trolling line is unclipped when the tag line clip reaches the rail, and empty leaders are unclipped as hauling is continued until the gurdy is stopped, when the hooked fish can be gaffed and brought aboard. The line is then run out again from the gurdy spool, the leaders being re-attached and the tag line clipped on as it is paid out to the required scope for fishing.

If the lines are towed directly from outriggers, the gurdy is used to haul in the trolling line until the first leaders are in sight; the vessel is then turned so that the line can be reached from the stern, the fish gaffed and brought aboard, and the line run out once more with the vessel back on course.

Pole and line

Pole and line fishing is undertaken for pelagic fish which school, and is particularly applicable to such species as tuna or large mackerel-type fish having reasonably high individual value. A number of men, equipped with fishing poles, position themselves around the vessel and heave the fish aboard as they strike at the hooks.

The fishing gear consists of a bamboo pole, some $3\frac{1}{2}$–4 m (12–14 ft) in length, to which is attached a line having a feather or other lure with a barbless hook; alternatively, bait may be used. In some cases, two men using poles connected to a common line and hook may be required for large fish.

The vessel is equipped with a platform outside, and lower than, the bulwarks around the stern; in some cases the platform extends forward along each side. Live or frozen bait is mainly used for 'chumming' (attracting fish to feed by scattering bait fish) so that bait tanks or storage will be provided with individual bait positions for the fishermen. The deck or a false deck will often be arranged to slope upwards to the transom, so that fish when heaved aboard slide forward to a handling position at which they are cleaned before being placed below. The size of vessel depends principally on the area to be fished and may range from 15 m (50 ft) upwards.

When a school is reached, the pole fishermen take their positions and begin 'chumming' to bring the fish to a feeding frenzy. Each man casts his line astern as the vessel passes slowly through the school, the fish snap at the splash of the hook and upon one striking, the fisherman heaves on his pole to fling the fish behind him and on to the deck; as the hook is barbless, the fish is disengaged automatically as the line tightens, and it slides forward along the sloping deck.

We have ended this review with a description of the pole and line fishing method to illustrate our original statement on the variations in sophistication that pelagic fishermen employ. This method must surely be the simplest of all, using only a pole, a shot line and a simple hook. However, the simple method is backed up by skills and equipment that must catch the bait fish and keep it alive, locate the target species and encourage it to 'chum', and then control the vessel to maintain contact with the shoal. Plenty of sophistication there!

BIBLIOGRAPHY

Arkley K. (1987) Fishing gear trials relating to the development of a suitable synthetic replacement for the traditional cotton drift net as used in the Cornish pilchard fishery. *Internal Report 1306*. Sea Fish Industry Authority, Edinburgh.

FAO (1971) *Modern Fishing Gear of the World*, **3**. Fish Finding, Purse Seining, Aimed Trawling (Ed. by Hilmar Kristjohsson). Fishing News Books, Oxford.

Sainsbury J.C. (1971) *Commercial Fishing Methods*, An Introduction to Vessels and Gears, 2nd edn. Fishing News Books, Oxford.

Strange E.S. (1981) An Introduction to Commercial Fishing Gear and Methods used in Scotland. *Scottish Fisheries Information Pamphlet no. 1* (2nd edn revised). Department of Agriculture and Fisheries for Scotland, HMSO, Edinburgh.

Thompson D. (1978) *Pair Trawling and Pair Seining*, the technology of two-boat fishing. Fishing News Books, Oxford.

Von Brandt A. (1984) *Fish Catching Methods of the World*, 3rd edn. Fishing News Books, Oxford.

Watt Robert B. Modern purse seine fishing with winch and sonar. *Technical Report no. 288*, Sea Fish Industry Authority, Edinburgh.

CHAPTER 3

Preservation of pelagic fish quality for further processing on board and ashore

N.K. Sørensen and A. Mjelde

INTRODUCTION

Overall, the world fisheries biological potential is about 100 million tonnes and fishery resources around the world are close to maximum catch limits. Many of the recent global changes in landings appear to be more the result of climatic fluctuations of stock size than of fishery development or management processes.

There has been a steady increase in landings during the 1980s; 1987 (92.9 million tonnes) was slightly more than the record catch in 1986 (92.1 million tonnes). Much of the increase is accounted for by three pelagic shoaling species (Peruvian anchovy, south American sardine and Japanese sardine) and the semi-demersal Alaska pollock. As there are few untapped resources of conventional species left, and the present increase in landings is more dependent on the unstable, shoaling pelagic species, it is likely that the world production of fish will fluctuate around the average figure for the last 10 years (FAO 1989).

Pelagic fishing usually gives a single species catch. When caught in coastal fishing by smaller boats, mixed catches may demand species sorting as well as size grading; this also implies sex-sorting for some species, e.g. capelin. Traditionally, off-shore fishing with big vessels has been for fish meal and oil production, while the coastal fisheries have supplied the fresh fish market and the processing industry, especially canning.

Fishing vessels are increasingly specialised, and the number of processors at sea has increased very much during the last few years. Both purse seiners and big freezer trawlers operate in addition to factory vessels and boats involved in joint-venture operations and klondyking (direct landings to foreign container or factory vessels for immediate export).

The increasing demand for high quality raw material for all types of production has put pressure on fishermen to utilise all available knowledge and organise the processes on board in an attempt to optimise quality, value and efficiency to satisfy the market. Special grades of fish meal demand the highest degree of freshness, thereby requiring rapid and efficient chilling also for this type of production. The interest in using small pelagics as raw material for animal feed, either as silage or as whole fish frozen in blocks, also is a challenge for the on board handling and storage of fish. By-products are of increasing interest, e.g. roe production and collection of other raw material for biotechnological utilisation.

The choice of using ice or chilled/refrigerated seawater depends on fish species, the quantity of fish to be handled and the type and value of product to be produced on board or ashore. The use of ice when boxing pelagic fish for human consumption was long regarded as uneconomical, but the development of methods for automatic addition of ice has made this more attractive. Likewise, the improved methods of internal transport and storage using new methods for identifying and labelling each box which can be monitored and recorded by microprocessor techniques, are important developments.

Filleting of fish on board is a specialised process and increases the product value, but sets certain limits on fishing capacity and often on processing flexibility. The advantage of on board processing is that freshness can be assured. Also, the question of full utilisation of the catch is important; today dumping is too easy and fish are wasted in an effort to store only the most valuable products.

Highly specialised vessels for surimi production based on pelagic fish have been developed but, so far, this venture has not been successful due to many technical problems and limited availability of suitable fish resources.

In this chapter we will briefly discuss the small scale fisheries for food, review on board handling of fish for meal processing ashore and conclude by considering the on board processing possibilities utilising pelagic fish as raw material.

SMALL SCALE FISHERIES FOR FOOD

Chilling

Many of the pelagic resources are found close to the shores, giving small vessels the opportunity to participate in the fisheries. The need for chilling is important, particularly when temperatures are high (>15°C), while low temperatures may allow bulk transport if the distance is short. Sorting of fish is necessary ashore due to both damage and/or autolysis which increases at high temperatures. Often only a small part can be utilised for consumption.

Small vessels can improve handling and increase quality of the catch by using boxes or small insulated containers with ice or CSW. It is important to avoid anaerobic conditions in containers without circulation. Therefore a system of pumping air via perforated pipes into the bottom of the container can be used (Jensen & Hansen 1983) – the CSW-champagne method. Many such installations have been tried out and proven beneficial. To reach a practical solution to benefit the fishermen, quality must be paid for, otherwise the new investments will not be made. The FAO has been heavily involved in giving advice and in undertaking projects to increase the utilisation of small pelagics for human consumption (Coackley & Karnicki 1985, Haywood & Curr 1987, James 1988).

In some cases when the small pelagics are intended for curing/salting and sun drying, the process may start on board. Containers with brine, or chilled brine, can be used, starting the brining on board so that high quality raw material can be brought back for sun drying. The best results are achieved when small fish are chilled and brined rapidly (Jensen & Machava 1985, Sørensen & Motta 1989). Bigger fish (>25 cm) should be gutted if the brine is not chilled.

Canning

In many countries canning is the most important utilisation of small pelagics. The many different products of sprat, sardine, herring, mackerel, pilchard and tuna are well accepted and important in the market due to good storage stability and high nutritional value. Fish for canning must be of highest quality, and because most small pelagic fish are very delicate, damage due to handling often results in a high degree of post harvest losses. Also, the many old purse seiners operating without chilling make losses substantial, especially in areas with high temperatures such as India, California, Chile/Peru and South Africa.

Autolysis in combination with rough handling results in belly-bursting, which is very dependent on storage time and temperature (Gildberg 1978, Hansen 1983, Hansen *et al.* 1980, Martinez & Gildberg 1988). Reducing storage time and temperature can very much improve quality of the raw material. Chilling is of utmost importance and only a few hours delay at 15–20°C will reduce storage ability by several days (Hansen & Jensen 1982). This means that the quality of landings of small pelagic fish for canning is dependent on season and fish species, on the chilling rate on board and on the loading/unloading method. The bigger fish can often withstand more rough handling.

Disintegration of fish results in lower productivity in the processing plant. In a Mexican canning plant the quantity of sardines nobbed by machine decreased from 160 kg/person/hr to 60 kg/person/hr when the proportion of belly burst increased from zero towards 100% (Jensen 1983).

For the small fatty species, oxidation must also be avoided. For fresh consumption and canning it is not a big problem if chilling is properly done and if fish is utilised within 2–4 days as is usually the case.

In Scandinavia the catching seasons for sprat for canning and for herring for fresh consumption coincide with heavy feeding periods. Sprat for canning must be of a special size and fat content and be free from feed. The fishermen often solve the problem of feed by keeping the fish alive in the net for 1–2 days close to the vessel, but if the sea-water contains much feed, the fish eat also during storage. Then the fish may be kept in a dark bag, not allowing the sprat to see the feed, or kept at high density. The latter results in 15–20% mortality. When the fish is taken on board it is immediately chilled, e.g. in containers with ice, and transported to the factory or frozen on board. All herring intended for fresh consumption must be free from feed, eventually using the same methods as mentioned above. The fish is transported to auction or market, chilled in tanks or containers on board. Herring to be frozen on board or filleted on board may contain some feed.

LARGE SCALE FISHERIES FOR FISH MEAL

Raw material

Fish meal and fish oil are produced worldwide from different species of fish, mostly from species which are not utilised for direct human consumption. These species do not easily find their way to the human food markets, often due to small size and the appearance of the fish. The nutritional quality of these smaller pelagic species is as good as that of bigger fish and this makes them valuable resources for the manufacture of fish meal. In Scandinavia, species like capelin, blue whiting, sand eel, Norway pout, and to some extent herring and mackerel are used for fish meal and oil production. Anchoveta in Chile and Peru, menhaden and pilchard in North America and sardines in Japan are examples of other important species utilised for this purpose.

Quality of fish for fishmeal

Traditionally, good quality fish for industrial purposes (fish meal and oil) meant high fat content. Though fish oil is important (and today becoming increasingly more valuable as a nutrient) this is not the only quality criterion today. Especially during the last 10 years, we have seen a great improvement in fish quality for meal and oil production. Among the factors which have influenced this development are the environmental protection laws which forced the industry to use fresher fish; the restrictions on chemical preservatives leading to other and better ways of handling; and new consumer groups asking for better quality. In Norway the first special quality meal was developed for mink farming. Today, the fish farming industry is by far the most important customer. A typical fish feed may contain about 50% fish meal of high nutritional quality. The standard quality used in animal feeds is not good enough in fish feeds (Nygård 1988).

To produce high quality fish meals the raw material must be of good quality at the time of processing. The higher the quality of fish, the higher the quality of fish meal that can be produced. To produce a high quality meal (Norse LT-94), processing must take place before the total volatile nitrogen value (TVN) reaches the limit of 50 mg N/100 g. This means that the quality of the fish at landing should comply with standards of freshness close to those of fish for direct consumption. It is a challenge to the fishermen to meet these standards (Mjelde 1985).

Investments in fish handling on board to take care of fish and to preserve its quality have been low compared with investments in fish finding and catching equipment. This has to some extent to do with the fishermen's attitude to fish for fish meal production. Today this has very much changed and there is a growing understanding that it is necessary to use good quality fish for fish meal. The driving force in this development is profit. It must be profitable to land fish of better quality. In future the fishermen will have to be more concerned with quality than quantity. Reduced fish resources and a market demanding higher quality are important factors in this change. To maintain fish quality it is necessary to consider quality at every step from catching to unloading. This is the essential message of this chapter, but we must mention that the final result, the quality of the fish meal, is also strongly dependent on the process ashore. This calls for good coordination between the fishermen and the fish meal plants.

Catching methods influence raw material composition and quality

The catching methods may be quite different (see Chapter 2), and this may influence the final quality of the fish. For example, the content of glycogen in the fish muscle depends on the degree of struggle before death, a high glycogen level being favourable for the keeping quality of the fish. Another example is the diffusion of salt into the fish after its death. As fish may contain about 0.3% salt (as NaCl) on a wet weight basis while living, and the surrounding sea about 3.4% salt, there is a strong driving force causing salt to diffuse into the fish as soon as it is dead (Landfald 1980, Smith *et al.* 1980). Time in sea water after death has an influence on the salt content of the fish brought aboard. The two main catching methods for industrial fish, trawling and purse seining, are obviously so different that differences in fish quality are to be expected and can be observed. Trawling for hours may result in a high proportion of dead fish, while the fish is caught alive when seining. The differences, though small, may become of greater importance as fish quality has to be improved.

Effect of physical stress on the fish on board

The degrees of bacterial and autolytic activity are important when evaluating the quality of industrial fish. Bacterial growth may produce harmful and undesirable metabolites in the fish. Autolytic activity may accelerate bacterial deterioration, and may also complicate the fish meal process and increase processing costs. The effects of physical handling on board are often neglected in connection with quality, the excuse being that the fish is to be broken down mechanically anyway and 'reduced' to fish meal. This is a misunderstanding. Physical damage to the fish will accelerate the deterioration process and inevitably reduce the keeping quality of the fish. Today there is a trend towards more careful and gentle treatment of the fish throughout, from catching well into the fish meal process. Obviously in practice there have to be compromises: the huge quantities of fish to be handled in a short time and the low price of the material compared with fish for human consumption, clearly set the limitations.

Loading and draining of the fish

On smaller trawlers the catch is often taken on board in lifts of about 2 tonnes each. Pumping the catch is a common way to bring the fish on board on larger vessels. The fish-pumps are special hydraulic centrifugal pumps with capacities of up to 300 tonnes of fish per hour, depending on the ratio of fish to sea water which should be such that physical damage to the fish is avoided. This means that large quantities of sea water are pumped on board and should be drained away before the fish enters the hold; it is especially important when the fish is to be stored in bulk without water. Long experience has shown that good drainage of water, both on deck and from the holds, increases the mechanical stability of the fish mass and improves the keeping quality of the fish; the more fragile the fish, the more important is drainage. Small species caught in their feeding seasons are easily disintegrated and bacterial and autolytic breakdown may proceed very quickly if precautions are not taken.

Maintaining good fish quality on board

The measures to be taken depend on the type of fishery and on the coordination of fishing and processing activities. When the distance from the fishing ground to the processing plant is short, the fishing vessels may deliver their catches at short intervals without the need for any special precautions. An example of this is the south American anchoveta fishery. The boats deliver their catches daily, usually within five hours from the fishing grounds, and in this way good quality is maintained in spite of high temperatures (22–28°C) (Bellido 1979). If the sea temperature is low it is easier to maintain quality. During the winter season in Norway, capelin having a temperature of 3–5°C can be brought ashore 1–2 days from the time of catching and still maintain good quality.

Usually, however, some kind of preservation is necessary if good quality is to be maintained when fishing in distant waters and/or when the catching period is long. For example, Norwegian fishing for blue whiting takes place in the Atlantic; the fishing period may last for 2–5 days and the journey home takes another 2–3 days. With a fish temperature about 10°C, spoilage will take place before landing if no precautions are taken. In earlier days, effective chemical preservation would probably have been recommended (Mjelde & Urdahl 1974). Today the raw material for the manufacture of special quality fish meals cannot be chemically preserved and must be of good quality. The traditional chemical preservatives (formalin and nitrite) are thus unacceptable; the fish must be chilled.

Cooling with ice

Cooling the catch by addition of ice is in extensive use in several countries. On smaller fishing boats especially, this seems to be the best option (see Chapter 4).

On larger fishing vessels like the Norwegian purse seiners (loading capacity 1000 to 1500 tons), there are practical problems to be solved if effective icing is to be used. It has been demonstrated with capelin and blue whiting that icing these large catches may be an effective method but, since the ice has to be added to and mixed with the fish at the time of loading, problems arise due to the high pumping rates at which these boats load their catches. It is quite normal to pump fish at rates of about 300 tonnes/h. If 10–15% ice is needed, 30–45 tonnes of ice per hour must be added, requiring effective mechanical equipment. Ice conveying systems (mechanical and pneumatic) with sufficiently high capacities are available for transporting ice from ice bins to the fish holds. The very first step in this process is, however, still a problem; that is how to break up the ice and get it into the transport system quickly enough. Ice stored for some days in an ice bin is usually not free-flowing and requires mechanical power to break it up. One solution in use on a couple of Norwegian boats simply puts a digging machine into the ice bin but, if icing is to be the preferred method of preserving fish on the big purse seiners, better solutions are called for.

Cooling in refrigerated sea water (RSW)

Cooling in chilled sea water can be done in two ways – by mechanical refrigeration or by the addition of ice. The heat transfer from fish to water is effective, and more rapid cooling can be achieved than with ice alone. The system was originally designed for use with fish for human consumption, offering the advantages of low temperature (−1°C). It is labour saving and causes less damage to individual fish. These advantages

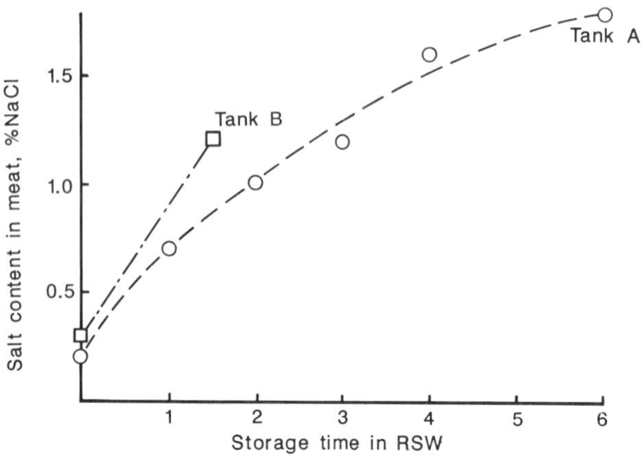

Fig. 3.1 Uptake of salt in blue whiting during cooling in RSW (from Landfald 1980).

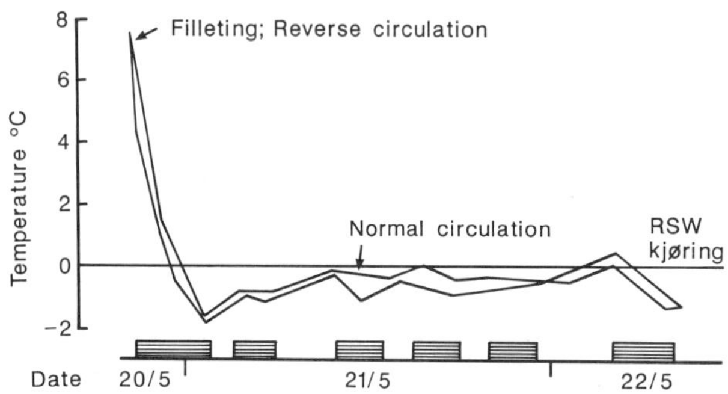

Fig. 3.2 Temperature profile in tank (volume 48 m^3) with blue whiting during cooling in RSW (from Landfald 1980).

are also of importance when handling industrial fish. Disadvantages are that the system does not work properly with all kinds of fish, especially the smaller ones, and there is some undesirable uptake of salt (Fig. 3.1). In Norway most of the larger purse seiners have RSW systems installed, which are used with species like herring, mackerel and blue whiting (Fig. 3.2).

The large purse seiners may have 6–9 tanks each with volumes from 100–300 m^3. Some or all of the tanks may be equipped with water circulating systems (pumps, strainers). Mechanical refrigeration plants are installed on most of the vessels. The refrigeration capacities are in the range 300 000 to 700 000 kcal/h. Cooling the sea water by the addition of ice (CSW) is not in common use on these large vessels. The large quantities of ice required, the size of the tanks, the difficulty of making a homogeneous ice/water mixture, and insufficient insulation, are some of the reasons why.

Ideally the RSW tanks should be fully insulated to keep heat leakage to a minimum. This is not the situation in practice since the shipbuilders, for reasons of mechanical strength, construct the tanks leaving a lot of heat bridges penetrating the insulation (Magnussen & Holt 1972). This heat leakage is, however, compensated for by greater refrigeration capacity, but the usual partial insulation reduces the heat leakage by 80–90% compared with no insulation (Kelman 1977, 1982). It has not been found economical to install fully insulated RSW-tanks on Norwegian vessels.

Under optimal conditions RSW is an efficient way of cooling fish. The effect depends on the capacity of the refrigeration plant (including heat exchanger) and the efficiency of circulation of the cold water through the bulk fish mass. The refrigeration capacity should be high enough to bring the fish temperature down to storage temperature (about 0°C) in a few hours (Fig. 3.2). The circulation rate of chilled sea water should be relatively low, but high enough to establish uniform distribution throughout the tank. The design of inlet and outlet areas influences uniform distribution but, still more importantly, sufficient water must be in the tanks; normally a ratio of water to fish of 1 : 4 is recommended. Experience has shown that the size of the tanks influences the circulation; it is much easier

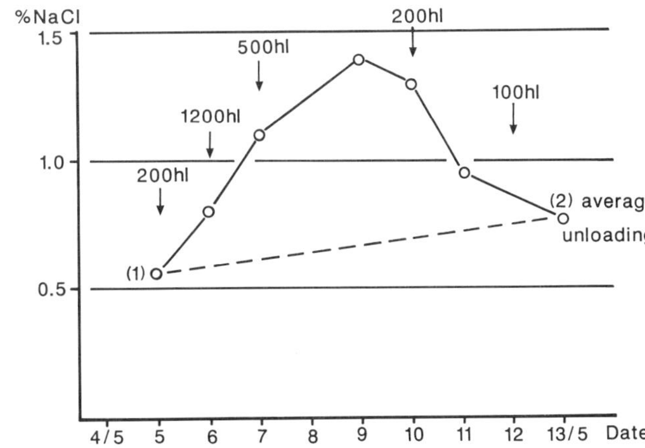

Fig. 3.3 Salt uptake in blue whiting stored in RSW on board a purse seiner. Tank filled from 5 hauls on 5 different days. (1) Average in fish from sea. (2) Average at unloading shows moderate salt uptake. Samples drawn during the fishing period show higher values of salt, indicating great inhomogeneity in the tank.

to get an even distribution of water in a small tank than in a big one. A further subdivision of the 200–300 m^3 tanks into smaller ones would probably help. Factors like the density of fish and their tendency to stick together also influence water circulation. Blue whiting cause such problems; in the course of 24 hours they settle and stick together making water circulation difficult. Forcing water through such a tank results in cooling mainly along the walls; very little cooling is obtained inside the bulk mass. Heat formed in the mass after some time cannot be removed effectively and poor quality results. A far from ideal situation (which often occurs in practice) is shown in Figure 3.3; the variation in salt concentration in the fish indicates the inhomogeneity of the fish in the tank.

Another approach to chilled sea water is in use in the menhaden fishery in the USA. Mechanically chilled sea water is sprayed over the top of the fish in an even distribution system. The chilled water then filters evenly down through the fish, is sucked off from the bottom of the holds, rechilled and pumped back to the top. The system is reported to work well in that specific fishery on boats loading 300–500 tonnes (Doody & Langseth 1979).

RSW in combination with other preservatives

Maintaining the quality of the raw material is becoming increasingly important for the manufacture of high quality fish meals. Too often we have seen RSW cooled fish being of poorer quality than desirable, with inadequate distribution and circulation of chilled water often being the main reason. This problem initiated experiments at the Norwegian Herring Oil and Meal Industry Research Institute (SSF) in Bergen, looking for possibilities to combine RSW and other preservation methods.

As mentioned earlier, the 'old' chemical preservatives are no longer acceptable for production of special quality meals. Among the 'acceptable' preservatives commonly used for preserving foods are carbon dioxide (CO_2) and some organic acids. These chemicals have too small an effect when applied alone at acceptable concentrations but, if combined with cooling, the total effect might be of practical interest.

The growth inhibitory effect of CO_2 on microorganisms is well known. The use of CO_2 in combination with RSW ought to be of interest for maintaining the quality of small pelagic fish meant for fish meal manufacture; the addition of CO_2 should, however, be straight into the circulating water. A lot of research has been done on this system (Longard & Regier 1974, Shaw & Botta 1975a and b, Reppond *et al.* 1985), and the preservative effect is beyond doubt; typically trimethylamine (TMA) formation is greatly retarded. The application of CO_2 on board fishing vessels could be a simple measure, provided the necessary safety precautions are taken.

The use of organic acids to preserve fish is well known (Windsor & Thoma 1974), especially in so-called fish silage production (Raa & Gildberg 1982). Acid preservation of fish for fish meal production is, however, quite a different matter. The acceptable acid concentration is of the order of 0.2% while fish silage requires about 2% addition. We have found the addition of this small amount of acid to the RSW system to be of practical interest. Our choice of acid is acetic acid. Propionic acid may be quite as efficient but it is less acceptable in fish feed. The resulting pH drop in the fish is quite small (about pH 6), which rules out formic acid. The addition of acid should be

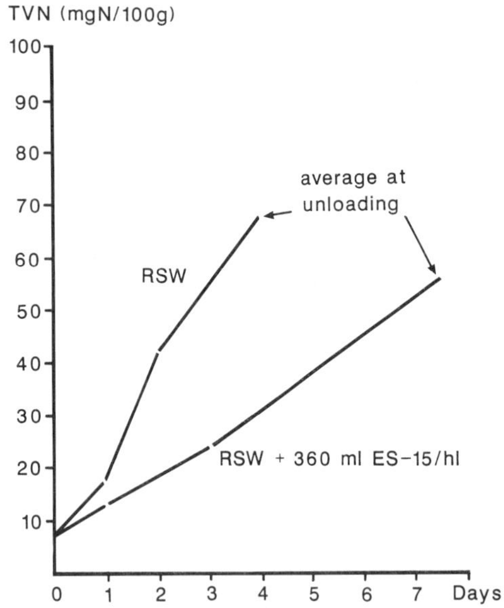

Fig. 3.4 Effect of acetic acid (ES-15) on TVN in blue whiting stowed in RSW on board (trial I).

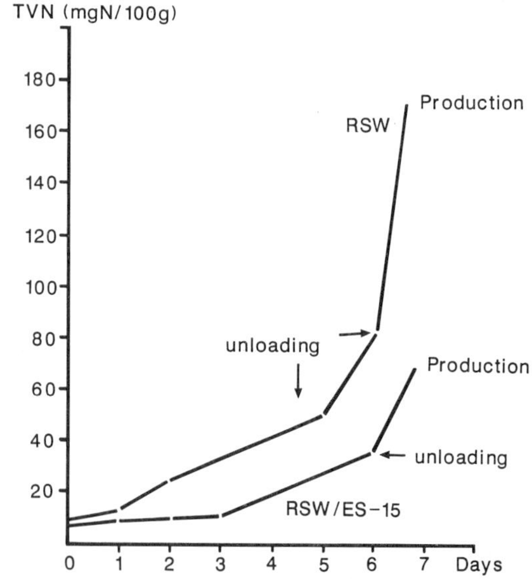

Fig. 3.5 Effect of acetic acid (ES-15) on TVN in blue whiting cooled in RSW on board. No cooling after landing (trial II).

done as soon as possible after pumping the fish on board, either by addition directly to the fish stream or to the circulating chilled water. When uniformly distributed through the whole fish mass, the acetic acid will retard bacterial growth both in the sea-water and in the fish. Our results are quite promising and the method seems to be a useful alternative, especially on fish species where RSW alone does not give the desired quality (Larsen 1988, 1989) (Figs. 3.4 and 3.5). Use of acetic acid on purse seiners requires, however, some precautions to be taken to minimise the risk of corrosion, especially on the evaporators.

Another possible way to control bacterial growth in RSW could be the use of ultra-violet radiation. It is reported to be effective on fish for human consumption (Peters *et al.* 1965).

The journey to the fish meal plant

Catches stored in RSW are usually cooled intensively during the voyage to the fish meal plant, to minimise heat leakage from the sea to the fish. Another point, probably adopted from the RSW cooling of fish for human consumption, is to replace the chilled water with fresh sea-water after some time. The reason for this is that the circulating water becomes an excellent growth medium for bacteria (Gundersen 1971). It is wise to drain this away, but the change of water should not be exaggerated since it results in an appreciable wash out and loss of dry matter from the fish.

For fish stored in bulk with or without ice there is hardly anything that can be done during the voyage to the fish meal plant. For boats with this kind of fish hold, it is worth mentioning that damage to the fish caused by the ship's movements will reduce the keeping quality of the fish. Fish stored in containers or in smaller holds will keep better than fish in large tanks or holds.

Cleaning

After the fish have been unloaded it is important that the holds are cleaned; no fish or parts of fish should be left over for the next voyage. Fish debris, blood, slime, fat etc. are excellent growth substrates for bacteria which will infect new catches. The design of the fish holds should be such as to make cleaning easy and effective. The RSW tanks pose problems with all the strainers, tubes, pumps and heat exchangers which may be difficult to clean. Fish debris can settle and bacteria grow, thus creating sources of infection. Starting up the cooling of a new catch with such an RSW system will effectively bring infection to the fresh fish and greatly accelerate spoilage processes.

Conclusions

It is increasingly important to maintain good quality for fish meal production. This calls for proper handling of the catch on board. If catches cannot be landed at the plants within certain time limits, cooling must be applied. Direct icing and use of RSW are both effective, the latter being preferred on larger fishing vessels. The main problem with RSW applied to small pelagic fish is to obtain uniform circulation of chilled water through the tanks. The preservative effect of RSW may be increased by adding an acidulant to the circulation water. Finally, the importance of cleaning the fish holds after unloading must be emphasised. In the RSW systems inadequate cleaning, leaving pockets of infection, may seriously reduce the keeping quality of the next catch.

ONBOARD PROCESSING IN LARGE SCALE FISHERIES

Grading, egg production and by-products

The low quotas for purse seiners in Norway have made it necessary for the fishermen to be flexible both regarding catching methods and utilisation of the fish raw material, when trying to increase product value.

The capelin fishery is important for Canada, Iceland, the USSR and Norway. A look at the development in Norway can serve as an example. During the affluent years (1970–80), the purse seiners mainly caught the fish during a few sets and could bring the fish to fish meal processing in a short time, a few hours to a few days depending on the area of catching and the capacity of the factory. Quality would be acceptable in winter time due to low temperatures and the low fat content of the fish. Formerly, during the summer fishery chemical preservatives had to be added, but later just chilling was used. The development of high quality fish meals for feed put increasing standards on raw material quality. Then quality was aimed for as it was paid for.

Another development in the capelin fisheries was the demand from Japan for high quality, graded female capelin and, later, capelin eggs were produced successfully (Eikje 1989). In the beginning, this fishery was prosecuted only by smaller coastal trawlers that delivered the catch within 6 to 10 hours to obtain the highest degree of freshness. Often, only part of the catch, taken from the top of the load, was used for human consumption.

The interest in onboard production of these high priced products increased rapidly as quotas decreased and, for a few years, many seiners with facilities for freezing, were sorting capelin and subsequently producing capelin roe. This development, which did not benefit the factories on shore, has been the subject of much debate. The economical aspects and the arguments for and against on board processing of capelin have been discussed by Olsen & Martinussen (1986).

Capelin stored in ice keeps well for 7–13 days (Shaw & Botta 1975a, Botta *et al.* 1978, Broch-Due *et al.* 1986), depending on the spawning condition of the fish. When stored in RSW or RSW + CO_2, the fish is rejected for consumption after 3 days (Shaw & Botta 1975b).

Capelin must be processed rapidly to avoid quality reduction in appearance. Since the season is very short, a high capacity processing line had to be developed to deal with large quantities of fish when the size of the fish and maturity of the roe is right.

Sex-sorting of capelin is easily done by hand because of the apparent difference between the mature sexes. Manual capacity was too low but could be increased by introducing size grading of the fish prior to manual sorting. This system is practised both on board and ashore, but to a higher degree of perfection ashore due to availability of manpower and space.

The sorting system is a size grading based on the fact that the width of the back of the fish is dependent on the length of the fish (Fig. 3.6). Rotating rollers or vibrating bars allow the fish to move forwards and they fall between the rods according to size. The capacity of the machines depends on their length and the type of system for moving the fish forward. A belt system works better for bigger fish. The principle is in use for most size grading of fish and, in the case of capelin, also works as sex-sorting because females are both shorter and more slender than males. Typical quality requirements for the females are: roe content (>18%); size (<40 per kg); freshness (<24 hrs); evaluation of quantity and type of stomach feed.

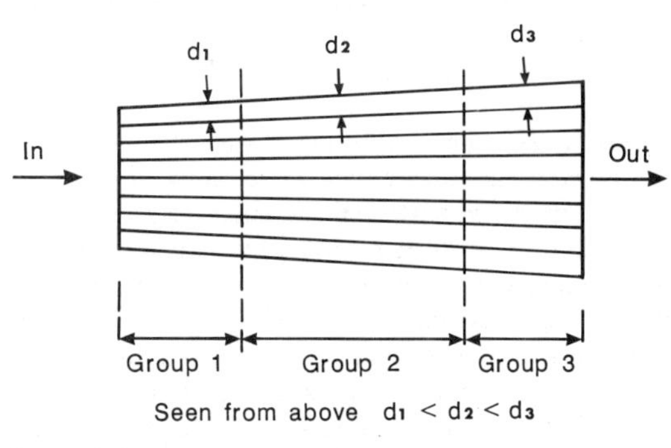

Fig. 3.6 Principle for size grading of fish.

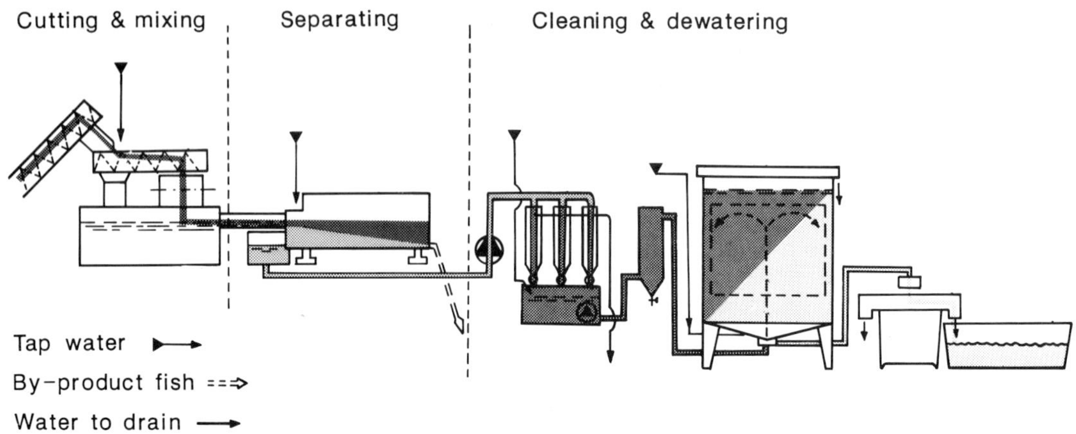

Fig. 3.7 Main steps in processing capelin roe.

With correct adjustment and close supervision in production, a sex-sorting of 95% females can be attained by the machines (Bertheussen 1981). The product is usually frozen in horizontal plate freezers as 10 kg blocks or in tunnels as 2 × 10 kg blocks, interleaved by plastic sheets before wrapping in plastic bags and cartons. It is important to secure proper freezing of the blocks, not being tempted to increase capacity by reducing freezing time. Buyers of high quality and high priced products do require certificates proving proper handling and processing of products. This will surely be part of integrated fish handling systems, stressing quality control and quality assurance, as they gain increasingly more interest.

The processing of capelin roe became a success in Norway, both for vessels and for plants ashore. It started in Iceland, where the initial idea was to collect loose eggs that appeared in the water when pumping the catches ashore. With increasing demand for quantity and especially quality eggs, the pumping became a problem. Improvements were made by squeezing the capelin to release the eggs, and then by cutting open the fish–the technique that became an industrial process in Norway.

Figure 3.7 shows the main steps in processing capelin eggs: the release and the cleaning of the eggs. Also the high capacity is characteristic for the equipment. By forcing all fish, male and female, through a set of rotating knives, the belly cavity contents, including the eggs, are set free. The slurry of eggs and particles is transported in water to a cleaning system consisting of hydrocyclones and a micro-air-flotation tank. These steps result in eggs of very high purity and quality. After dewatering, the eggs can be packed and

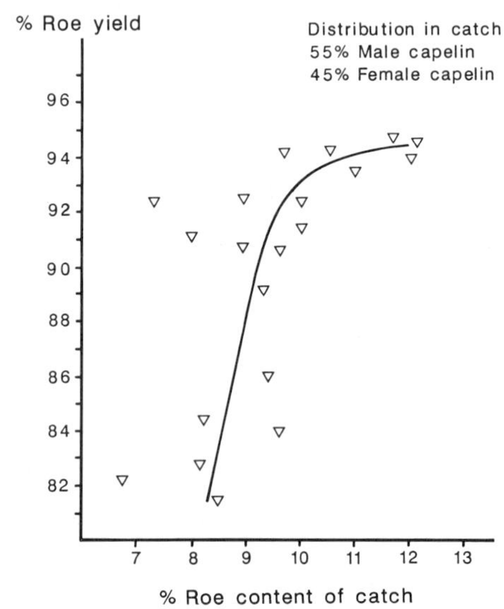

Fig. 3.8 The yield of capelin roe versus roe content in average capelin catch (sample) containing 45% female and 55% male fish (from Tangstad & Børresen 1983).

frozen in blocks or consumer packages (Tangstad & Børresen 1983).

The capacity of one line is as high as 10 tonnes capelin per hr, producing approximately 1 tonne/hr of roe, dependent on the quantity of females in the catch and the quantity and maturity of roe (Fig. 3.8).

An increasing number of vessels started on board production of sorted female capelin and capelin roe in the late 1970s. The Icelandic system of squeezing the roe free dominated, but it was regarded as second best in terms of quality. The reason is that cutting avoids most of the nematodes that are mixed with the eggs by squeezing, and micro-air-flotation almost completely removes all particles of skin, bone, blood etc. As a result the Norwegian system became predominant.

On board processing of female capelin and capelin eggs leaves approximately 50% and 90%, respectively, of the catch as by-products. Four options are available for utilisation:

- freezing – in competition with main product
- silage – dependent on storage capacity
- fish meal – dependent on space and storage capacity
- dumping – easy but an unacceptable waste of the resource.

The increasing aquaculture industry in northern Europe, and especially in Norway, has put pressure on the fish feed market (Minsås 1987). Traditionally, the by-products from the filleting industry and fatty small pelagic fish frozen on board have been utilised. Along with increasing quantities of farmed fish, the demand for nutritionally balanced and easy-to-use feed has increased. So far, the dry pelleted feeds based on fish meal, carbohydrates and additional vitamins are predominant in the market.

The by-products should be utilised as a raw material for biotechnological processes or feed, via frozen blocks, silage or fish meal. The changing fat content of capelin or capelin cuts from roe production, as well as uncertainty in supply, have made it difficult for on board processed products to keep up in the feed market. Fish feed should be an interesting niche of the market to exploit and is much preferred to throwing the by-products overboard.

The same type of by-product raw material is available on board when processing fillets from herring and mackerel. The priority for the vessel's freezing capacity is given to the fillets and the by-product ends up as waste because the market for frozen offal or silage is uncertain. Most vessels do not have a fish meal factory on board for this purpose but the biggest boats that have such plant installed can operate a profitable production as a result of the current high prices for fish meal. In spite of high freshness of raw material, the fish meal is of variable quality regarding protein and fat content, due to the variable chemical composition of the raw material and the limited opportunities to mix batches of fish meal on board. Still, a compact fish meal factory in combination with high sorting and freezing capacity seems today to be the best economical solution for new purse seiners. Governments are concerned, however, with such solutions as they do not want fish meal as the main production on board, because policing of raw material utilisation becomes difficult, just as it does for silage production.

Silage on board

On board vessels it is possible to utilise the offal and by-products as well as whole fish for silage production. The interest in silage production on board arose due to the increasing demand for fish feed in Norway from about 1980. The method is effective in preserving high quality feed as well as being a simple process giving high capacity in utilising by-products or even whole catches of small pelagics. By adding inorganic or organic acid to the fish raw material, the fish silage becomes liquid because the tissue structures are degraded by enzymes naturally present in the fish. The pH is lowered to around 4.3–4.4 in the stabilised silage, which is sufficient to prevent microbial spoilage. The theoretical basis for the silage process has been reviewed (Raa & Gildberg 1982), stressing the need for high quality raw material to produce high quality fish silage. In practice, only fresh fish suitable for human consumption, or its by-products, can be used (Stormo, personal communication).

Acid silage has a long tradition in Denmark for animal feed. Also in Poland there is a silage industry. A few years ago, fish silage gained interest in Norway in the fur industry. Defatted silage was also used in animal husbandry. The most promising possibilities are, however, within aquaculture where defatted silage can be used for white fish, and a standardised product with regard to fat, protein, and vitamins for salmonids (Raa & Gildberg 1982, Stormo 1987). Silage is a good alternative during winter particularly, when temperatures are low and the fish have a poor appetite (Landfald, *et al.* 1989). Also Canada, Ireland and other countries with developing aquaculture industries have shown interest in silage made both on board and ashore. In Denmark on board production of silage is not allowed because it is not possible to trace the origin of the raw material, and quotas can be misused, resulting in overfishing.

During processing on board, acid (1.3–2%), depending on the buffering capacity of the raw material and the strength of the acid, is automatically added before feeding to the grinder. The ground silage is pumped to the storage tank and the pH is automatically checked. Mixing when making silage and during storage is important to avoid layering of solids, fat and soluble proteins. An antioxidant may be mixed into the acid to reduce development of rancidity in the oil. Oxygen is kept out of the tank also, preventing

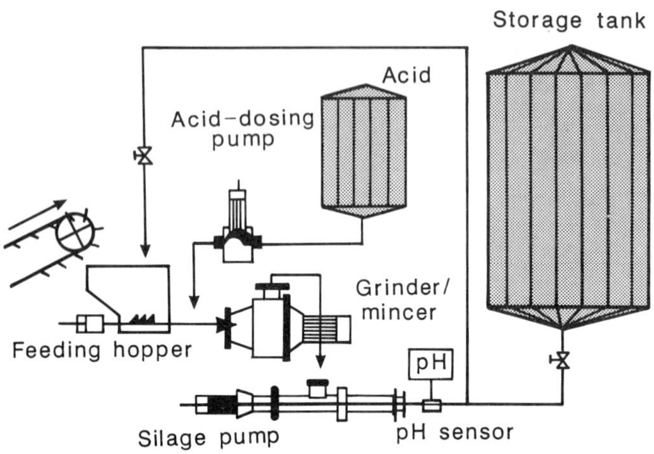

Fig. 3.9 Typical acid silage installation (courtesy Senta, Hamar).

oxidative reactions. Addition of sorbic acid is advisable to prevent growth of moulds. Capacity of the line can be from 2 to 30 tonnes of silage per hour, and this means that storage capacity is often the main limitation. The process is shown in principle in Figure 3.9.

Investment in a silage plant on board is usually 30–50% higher (£40 000 to £50 000 total) than the same line ashore. This is mainly because of special arrangements for the installation. On board it is extremely important to be aware of the corrosive *milieu* and to arrange safety precautions for the acid tank. Electrical components should always be kept dry.

The economics of producing silage on board are favourable (Stormo 1987), but it remains to be seen if more fish farmers will use fish silage. Silage cannot, in common with other preservation methods, improve the quality of a spoiled raw material. The product must be made from fresh raw material and be of consistent, high quality to make it attractive to the user (Pedersen 1987a, 1987b). Only then will there be a market for the product from the processing lines that were installed on board Scandinavian purse seiners and trawlers in the 1980s.

If aquaculture in northern Europe increases at the same pace as in recent years, feed will be a limiting factor. Silage production on board will then reduce import needs and allow utilisation of by-products otherwise wasted.

Filleting on board

Factory trawlers, especially from east European countries (USSR, Poland, GDR) have for years been producing frozen fish products, canned or salted fish. Waste has been utilised for fish meal production. These vessels have to some extent lost their market due to limited access to the resource, because of 200 mile economic zones. Limiting quotas also affects production by increasing the number of klondykers and joint ventures. This development will not be discussed. We will look at vessels that catch, handle and process on board.

Traditionally herring, mackerel and pilchard have had important markets as filleted products frozen, salted or smoked. Because the fish were caught in large quantities, bleeding and gutting were not practised on board and the quality of processed products was totally dependent on rapid handling, chilling and delivery to the shore plant.

Onboard processing is increasingly popular for new, medium sized and big vessels, both seiners and trawlers, because it allows a certain flexibility in production and the possibility of exploiting the sales value of freshness. This is most important if high value is to compensate for less quantity.

A few purse seiners do bring the catch on board by vacuum pumping, avoiding rough handling and breakage of the backbones of the fish, but capacity is less than for the hydraulic centrifugal pumps. The latter may leave the fillets with a high degree of blood spots, making them unfit for high quality consumer products. The catch is pumped to intermediate storage in RSW or CSW tanks at low temperature until processed on board. This storage may be 1–2 days dependent on fishing and weather conditions. The raw material is still regarded as high quality even without bleeding and gutting, because chilling to low temperature is rapid and the fish float without pressure in the RSW tanks. Salt uptake is not a problem during such a short period (Fig. 3.1).

The fish are transferred from the chilled tanks to standard grading machines (see capelin sex-sorting).

These graders are high capacity and robust but handle fish roughly during grading and conveying. Other electronic graders, either for size or weight, must sense single fish and do not have sufficient capacity to grade large catches of pelagic fish. The development of a high capacity feeding system to feed single fish to an electronic grader provides an interesting development project that when solved will benefit both the pelagic and bottom fish processing industries.

The fish are then fed round into the filleting machines. Several makes can be found, but VMK and Baader seem to dominate, depending to some extent on the origin of the boat in question. The wide variety of cuts and products that can be made ashore are principally also available on board, dependent on space, qualifications and organisation of the crew and mechanics.

Usually these products are headed and gutted (H&G) or H&G and tail cut, or butterfly fillets or single fillets skin on/off. Freezing in 10 or 20 kg blocks is standard. Requests for smaller packages and individual handling are reported to attract double the price in some cases but are of little interest due to reduced capacity. This type of processing is typical for the medium sized new purse seiners (total freezing capacity <100 tonnes/24 hours). They can operate with high flexibility according to the market, changing from product to product within a range of species, e.g. herring, mackerel, pilchard, blue whiting, according to quotas and season. Freezing capacity is very dependent on product; the rate for fillets (13 tonnes/24 hours) is approximately double that for freezing round herring per unit freezer. Fillets are preferably frozen in horizontal freezers.

The new generation of very large factory trawlers, from for example the Netherlands and Ireland, operate more on the basis of quantity and carry out only some processing on board, such as sorting and H&G, before freezing. The advantage is high capacity for fishing, freezing (>200 tonnes/24 hours) and storage on board. Rough sex-sorting on board can make herring roe collection more efficient ashore. Processing of herring roe products is discussed in Chapter 31.

Because of this high capacity there is little or no need for chilled intermediate storage before processing; the whole fish are often at ambient temperature before processing. This, in addition to the fact that trawling in deep waters reduces the quality of the raw material, means that the quality of the bulk production is inferior to that of products from purse seiners.

Blue whiting (*Micromesistius poutassou*) being a lean fish, is suitable for markets different from those for herring and mackerel. The same machines may be utilised, but it takes some time for adjustments to obtain acceptable yields. Fillets frozen in blocks for the fish and chips or fish sticks market are the main products. Minced fillet blocks can also be used for the same products. They are often preferred due to the smaller content of black spots (skin, belly lining etc.) and bones.

Another important factor when producing and marketing blue whiting products is the high content of nematodes in the fish flesh. Although harmless when frozen, they constitute an aesthetic problem. This was experienced when television teams focussed on the nematodes in herring as a health risk to consumers in Germany, USA and Canada. Fish exports plummeted immediately and took a long time to rebuild. To reduce the nematode problem it is necessary to cut away the belly flaps, which also reduces yield. When processing on board, there is often surplus raw material resulting in a high degree of wastage unless by-products are also processed. Freezing capacity is often the limiting factor, and a high degree of dumping occurs.

For blue whiting to be processed into fillet products freshness is an absolute requirement, the limit being 2–3 days in ice, giving time for onboard processing (Nordforsk 1981). In practice, salt uptake should not be a problem as it is not significant until 4–5 days of RSW storage (Landfald 1980). Storage in RSW is not preferred for fillet and surimi production, because the fish stick together and become difficult to handle. The use of CSW is advisable but is not much in use as fillet and surimi production is small. Another important factor regarding blue whiting is the tremendous difference in size and shape of the fish before and after spawning. This makes the post-spawned fish unsuitable for filleting at acceptable yield and capacity. In addition the fish do not shoal after spawning, making catching more difficult.

During the late 1970s interest in blue whiting, as one of the underutilised species, increased. Research programmes were run in the UK, the Nordic as well as other countries. A lot of information on the stock size and behaviour, fishing techniques, handling, processing and possible product utilisation was produced (Whittle *et al.* 1980, Nordforsk 1981). In spite of this valuable information concluding in general that blue whiting is a fish well suited for human consumption, big markets have not been identified. This fish is still mainly a resource for fish meal production. Recent investigations indicate that partly digested blue whiting protein is an excellent nitrogen source in growth media for bacteria (Gildberg *et al.* 1989).

Bulk handling and surimi production

The main resource for surimi production is Alaska pollock (*Theragra chalcogramma*) because it is readily available at low cost and possesses the basic properties for making high quality surimi. This huge resource has been utilised also for roe extraction and fish meal production. In the last few years a tremendous increase in filleting capacity on board has also been developed.

The production of surimi consists of steps for mincing and washing the fish flesh to remove water soluble proteins and impurities. In the traditional process the muscle is washed in open tanks (Suzuki 1981), but a new development uses continuous in-line washing to improve yield and reduce water consumption (Søbstad 1988). The product is bland in flavour and light in colour. In this context it is interesting to evaluate underutilised fish species as a potential for surimi production. Fish that is not readily marketable as traditional fillet products due to strong flavour and dark and/or bony flesh, might be utilised as surimi raw material. Also, small species that cannot be filleted at high capacity may have potential, provided a bulk handling process to eviscerate and debone the fish can be developed. This is dependent on technological developments where dark muscle is efficiently removed or where new products, that do not necessarily demand white coloured surimi, are developed. Yield is also important, especially when dark muscle and fat are removed, resulting in just over 10% surimi. The economics of the process must then be closely watched.

Much effort has been expended on utilising small fatty fishes as surimi raw material. In Japan, sardine (*Sardinops melanosticta*) is the most interesting species, while in the USA Atlantic menhaden (*Brevoortia tyrannus*) has been evaluated (Tokunaga & Nishioka 1988, Moore *et al.* 1988). It is not likely that these species will be processed on board as the resources are found near the coast but chilling is still essential. Menhaden must be chilled to less than 10°C within 4 hours, and then kept at 1°C until processed within 24 hours (Bimbo 1988). Herring and mackerel may be potentially useful for surimi but technical problems in processing, low yields and government interest in increasing use for human consumption cast doubt on the likelihood.

Today blue whiting is the only pelagic resource that has the potential for production of high quality surimi, based on availability and suitability (Langmyhr & Sørensen 1987, Sørensen, 1989). Another resource that may supplement blue whiting is Norway pout, but a bulk technology for gutting and cleaning this small fish must be developed.

Until now four vessels have been trying to utilise blue whiting for surimi production (1 Faroese, 2 Norwegians, 1 French) but so far with little success. They have had different processing equipment but all are based on mincing fillets. Fish should be handled as carefully but more quickly than for fillet production, because freshness is the single most important factor in obtaining high quality surimi. This evaluation is based on the fact that the functional properties of the surimi are best when processed on board.

It is important to reduce storage time and to lower temperature in order to minimise pH reduction. The functionality is closely correlated to the degree of protein denaturation, which is more rapid near pH 6 than at neutrality. This implies that only very fresh fish can be used to prepare surimi from dark fleshed pelagic fish, which tend to have a rapid drop in pH after death (Suzuki & Watabe 1986–7, Machii *et al.* 1988).

Colour is as important as the functional properties when considering marketing the product. It is not directly correlated to freshness but to flesh colour and the functioning of the equipment. An acceptable total yield is also important, demanding utilisation of filleting waste.

In several countries, projects have evaluated the gelling abilities of industrial fish species and the effects of seasonal variations and storage of the raw material (Opstvedt 1985, Verrez *et al.* 1986, Chandra 1987, Langmyhr & Sørensen 1987, Moore *et al.* 1988). The gelling properties of surimi are evaluated in several ways, e.g. shear strain at failure, compression, penetration and folding tests. The shear strain at failure is an objective test measuring protein quality and shows a sharp decrease during the first two days of chilled storage of different species (Fig. 3.10) (Langmyhr & Sørensen 1987).

Based on these results, Norway pout and blue whiting are the only pelagic species that can be utilised for production of high quality surimi. For both mackerel and herring the gelling ability is rapidly reduced during the first day of storage, as also found in Canada by Chandra (1987). Also the colour is too dark, due to the high degree of dark muscle. The fat content does not strongly influence the gelling ability of herring and mackerel, but may lead to the development of rancidity (Hultin 1988).

The 'modori' effect increases during storage, indicating increasing protease activity, due to leakage of enzymes from the stomach during storage (Lanier 1988). This effect is highly pronounced for mackerel, less for herring and blue whiting. It also suggests that storage in ice is preferable to RSW (Langmyhr & Sørensen 1987).

The use of Norway pout as raw material is interesting as its availability is complementary to blue whiting. Together with other small pelagic fishes less than 20–25 cm in length, they pose the problem of removing heads and intestines efficiently. Some filleting machines are suitable for small fish such as anchovy, but the low capacity and the need for manpower to feed

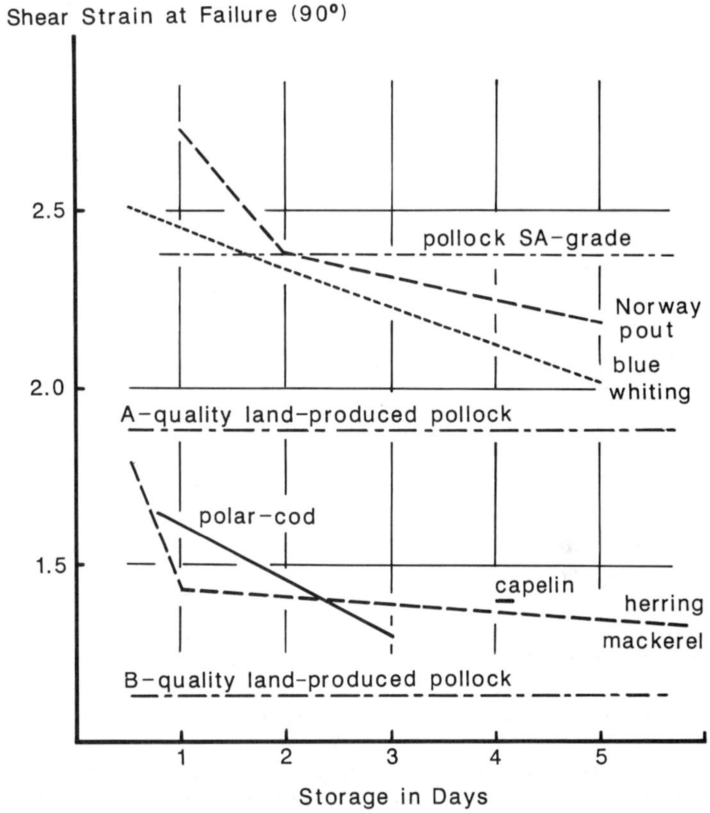

Fig. 3.10 Main results of gelling ability for surimi made from North Atlantic industrial fish.

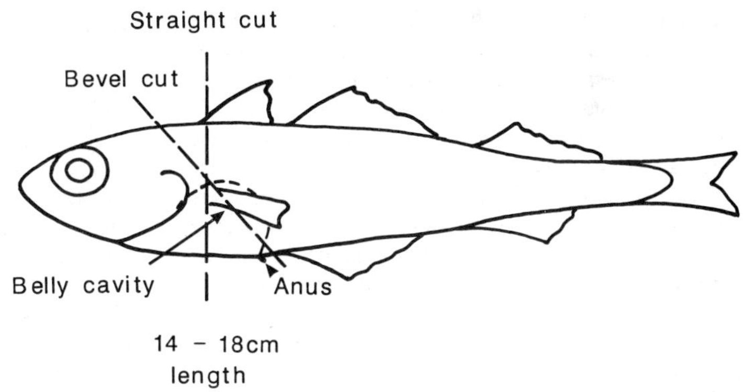

Fig. 3.11 Norway pout – possible cuts to improve gutting.

such machines make them unfit for commercial surimi production. Filleting is therefore not feasible due to the high throughput required, difficulties in orientation and sorting and general technical problems in operating the machines. Mechanical nobbing and specially angled cuts, as shown in Figure 3.11, may have increased capacity somewhat for Norway pout.

At the Norwegian Institute of Fishery Technology Research (FTFI) a long term research programme was initiated in 1975, aimed at producing minces from fatty fish, with good functional properties, low fat and light colour. The work concentrated on capelin, investigating an efficient method for gutting the fish in bulk. This also resulted in a method for effective roe extraction (Tangstad & Børresen, 1983, Broch-Due *et al.* 1986). Nevertheless, the removal of skin and fat remained a problem, and this was solved by washing the small pieces of capelin either in weak acid (pH 4) at 20°C or at neutral conditions at increased temperature, 40°C, for 45 minutes (Eide *et al.* 1982). Unfortunately the functionality of the mince was low as well as the yield.

The problem of releasing fat and skin relates to the fat deposits, which are mainly under the skin and the belly lining in capelin (Mohr *et al.* 1976). When high temperatures were used for a short period to avoid denaturation (90°C for 4–5 seconds) the removal of fat and skin was improved during subsequent washing in chilled 0.05% NaCl solution (Chandra 1987). When fat is distributed within the flesh, however, as is the case for herring and mackerel, this method may not be so effective.

Bulk gutting and removal of skin and fat therefore remained the main obstacles to obtaining minced flesh light in colour and with a high degree of original functionality, from small fatty pelagic fishes. For fish with the fat deposits underneath the skin and belly linings, however, a method may soon be available. The company FIDECO A/S has developed a method for bulk handling of small fish based on the roe process. In combination with surimi technology it is claimed to be possible to obtain capelin mince surimi with functionality as high as in the surimi made from capelin fillets (Langmyhr *et al.* 1988).

The introduction of a bulk handling system for gutting and cleaning small pelagics would be very welcome; it would make new fish resources available for human consumption. It might be installed both on board big vessels and ashore, but would still demand a high degree of freshness in the raw material. FAO projects have also tried to solve the problem, but the machine developed was not successful.

Surimi production on board is therefore dependent on traditional handling of fish, blue whiting being the only real alternative resource as it is available for filleting for 3 to 5 months a year. The product has high functionality but the colour is too dark so that the vessel can produce only low-priced, second grade surimi, which is already in surplus in the market. This is not economical for a vessel investing in highly specialised equipment unless the fillets and mince of blue whiting can be sold at high prices and/or the surimi can be utilised successfully in new products. Other quotas must also be utilised to maintain fishing opportunities over the year, e.g. herring and mackerel for filleting and/or freezing. It is difficult to achieve all these objectives on board one multi-purpose vessel.

Conclusions

In this section we have focussed on some handling and processing methods that have been introduced on board bigger pelagic fishing vessels during the last decade. Onboard handling and processing closely follow marketing possibilities regarding choice of products. There are several reasons why onboard processing has become popular:

- it offers better economy of operation
- it offers flexibility in catching as the fish can be followed
- it offers the possibility of high freshness, improved quality and higher profits.
- it reduces the uncertainty of quality control
- it offers improved working conditions on board.

The possibilities of following the fish at sea and of introducing improvements in onboard filleting technology for smaller sized fish have resulted in increased utilisation of blue whiting, capelin, Norway pout and argentine in addition to pilchard, herring and mackerel. The central message is that the basic principles are still valid regarding rapid and careful handling and chilling of the catch. New technology may facilitate and improve steps in the handling and processing line, resulting in increased utilisation of raw material, improved quality and/or more profitable production. The ability to collate, develop, communicate and apply the knowledge of good handling and processing practices must be regarded as a very important part of the technology.

BIBLIOGRAPHY

Bellido A. (1979) *Peruvian experience on preservation.* News summary 46, pp. 53–4. International Association of Fish Meal Manufacturers, Potters Bar.

Bertheussen S. (1981) Loddesortering. *Observasjoner ved produksjon av konsumlodde 1981.* Notat 6, prosjekt 663.1–10, FTFI-Foredlingsseksjonen, Tromsø. Unpublished data.

Bimbo A.P. (1988) The development of menhaden surimi in USA. *Infofish International*, **5**, 22–5.

Botta J.R., Noonan P.B. & Lauder J.T. (1978) Chemical and sensory analysis of ungutted offshore capelin stored in ice. *J. Fish. Res. Board Can.*, **35**, 976–80.

Broch-Due M., Ofstad R., Stenberg E., Sørensen N.K. & Tidemann E. (1986) *Mat av lodde.* Rapport U-33, FTFI-Foredling, Tromsø.

Chandra C.V. (1987) Laboratory and pilot scale production of surimi from selected Atlantic Canadian fish species. In *Atlantic Canada Surimi Workshop–Proceedings* (Ed. by J.F. Roache) pp. 59–83. General Education Series 4, Canada.

Coackley N. & Karnicki Z.S. (1985) Construction of onboard insulated fish containers for pirogues. *FAO Fish. Circ.*, (775).

Doody P.J. & Langseth D.C. (1979) *Preservation of raw menhaden in the United States.* International Association of

Fish Meal Manufacturers, Potters Bar. News summary 46, pp. 47–52.

Eide O., Børresen T. & Strøm T. (1982) Minced fish production from capelin. A new method for gutting, skinning and removal of fat from small fatty species. *J. Food. Sci.*, **47** (2), 347–9.

Eikje S. (1989) *Konsumlodda–Feitsildlagets og Finnmarks velsignelse*. Annual report, Feitsildfiskernes Salgslag, 1988, pp. 42–57. Trondheim.

FAO, Marine Resources Service, Fishery Resources and Environment Division (1989) Review of the state of the world fishery resources. *FAO Fish. Circ*, (710) Rev. 6.

Gildberg A. (1978) Proteolytic activity and the frequency of burst bellies in capelin. *J. Fd Technol*, **31**, 409–16.

Gildberg A., Batista I. & Strøm E. (1989) Preparation and characterization of peptones obtained by a two-step enzymatic hydrolysis of whole fish. *Biotechnology and Applied Biochemistry*, **11**, 413–23.

Gundersen J.O. (1971) Tankkjøling av sild og brisling i nedkjølt sjøvann. *Tidsskr. Hermetikkind*, **57** (6), 157–61.

Hansen P. (1983) Bulk handling and chilling of large catches of small fish. Part III: Trawl catches. *Infofish Marketing Digest*, **2**, 34–5.

Hansen P. & Jensen J. (1982). Bulk handling and chilling of large catches of small fish. Part I: Quality and storage life. *Infofish Marketing Digest*, November 1982, 26–8.

Hansen P., Jensen J. & Teutscher F. (1980) Bulk handling of purse seine catches for the sardine canning industry. *Scandinavian Refrigeration*, **9**, 143–8.

Haywood K.H. & Curr C.T.W. (1987) Project identification in small pelagic fisheries: Indonesia, India, Thailand, Morocco. *FAO Fish. Tech. Pap.*, (289).

Hultin H.O. (1988) Potential lipid oxidation problems in fatty fish processing. In *Fatty Fish Utilization: Upgrading from Feed to Food*, (Ed. by N. Davis) pp. 185–223. UNC Sea Grant College Publication UNC-SG-88-04.

James D. (1988) Increased use of small, oily pelagic species in the developing world. In *Fatty Fish Utilization: Upgrading from Feed to Food*. (Ed. N. Davis) pp. 291–303. UNC Sea Grant College Publication UNC-SG-88-04.

Jensen J. (1983) Bulk handling and chilling of large catches of small fish. Part IV: Handling on shore. *Infofish Marketing Digest*, **3**, 38–41.

Jensen J. & Hansen P. (1983) Bulk handling and chilling of large catches of small fish. Part II: Purse seine catches. *Infofish Marketing Digest*, **1**, 34–38.

Jensen J. & Machava A.M. (1985) Salting onboard of bycatch from the shrimp fishery. In *Proceedings FAO Expert Consultation on Fish Technology in Africa*, Lusaka, Zambia, pp. 81–86.

Kelman J.H. (1977) *Stowage of fish in chilled sea water*. Torry Advisory Note 73, Torry Research Station, Aberdeen.

Kelman J.H. (1982) Handling wet fish at sea. In *Fish Handling & Processing*, 2nd edn. (Ed. by A. Aitken, I.M. Mackie, J.H. Merritt & M.L. Windsor), pp. 28–41. Her Majesty's Stationery Office, Edinburgh.

Landfald B. (1980) *RSW-føring av kolmule til konsum*. FTFI-NYTT, No. 2 Mars, Tromsø.

Landfald B. & Løvland J. (1990) Ensilert fiskeavfall. *Kostnader og markedsutsikter ved bruk i fiskefor*. Rapport U-53, FTFI-Foredling, Tromsø.

Langmyhr E. & Sørensen N.K. (1987) *Surimi fra industrifisk - Norske fiskeråstoff*. Report H-79, Norwegian Herring Oil and Meal Industry Research Institute, Bergen.

Langmyhr E., Ofstad R., Eide O., Sørensen N.K. & Opstvedt J. (1988). Potential conversion of north Atlantic fatty species into surimi and surimi derived products. In *Fatty Fish Utilization: Upgrading from Feed to Food* (Ed. by N. Davis), pp. 79–117. UNC Sea Grant College Publication UNC-SG-88-04.

Lanier T. (1988) Muscle protein functional properties and protease content of surimi prepared from fatty, dark-fleshed fish species. In *Fatty Fish Utilization: Upgrading from Feed to Food* (Ed. by N. Davis), pp. 247–62. UNC Sea Grant College Publication UNC-SG-88-04.

Larsen T. (1988). *Holdbarhet på kolmule*. No. 2, 4–7. Meldinger fra SSF, Bergen.

Larsen T. (1989) *Eddiksyrekonservering–klar for praktisk bruk?* No. 1, 4–8. Meldinger fra SSF, Bergen.

Longard A.A. & Regier L.W. (1974) Colour and some composition changes in ocean perch (*Sebastes marinus*) held in refrigerated sea water with and without carbon dioxide. *J. Fish. Res. Board Can.*, **31** (4), 456–60.

Machii Y., Ohsawa N., Ando M., Horiuchi T. & Kameda W. (1988) Masking the fishy odour in surimi products. In *Fatty Fish Utilization: Upgrading from Feed to Food* (Ed. by N. Davis), pp. 225–46. UNC Sea Grant College Publication UNC-SG-88-04.

Magnussen O.M. & Holt S.E. (1972) Tankføking. Et viktig ledd i sildemelfremstillingen. *Scandinavian Refrigeration* **5**, 156–61.

Martinez A. & Gildberg A. (1988) Autolytic degradation of belly tissue in anchovy (*Engraulis encrasicholus*). *International Journal of Food Science and Technology*, **23**, 185–94.

Minsås J. (1987) *Nåværende og framtidige behov for forrressurser til havbruk*. SINTEF, Trondheim. Unpublished data.

Mjelde A. (1985) *Råstoffkvalitet*. Annual Report from Norwegian Herring Oil and Meal Industry Research Institute (SSF), pp. 12–16, Bergen.

Mjelde A. & Urdahl N. (1974) Latest results in technology of preserving and handling small pelagic fish for food and feed. In *Fishery Products* (Ed. by R. Kreuzer), pp. 74–7. Fishing News Books, Oxford.

Mohr V., Møller M., Ormberg A., Flo A., Halvorsen J. & Padget E. (1976). *Fettvev i fisk*. Report II, Norges Tekniske Høgskole, Trondheim.

Moore R.E., Anderson C.F., Bimbo A., Hurley T.W. & Meanes R.H. (1988) Harvesting and pre-process handling of food-grade menhaden. In *Fatty Fish Utilization: Upgrading from Feed to Food* (Ed. by N. Davis), pp. 163–72. UNC Sea Grant College Publication UNC-SG-88-04.

Nordforsk (1981) *Nordfisk - fra industrifisk til konsum*. Nordisk Industrifond, Oslo.

Nygård E. (1988) *Tørking. Et viktig ledd i sildemelfremstillingen*. Annual Report from Norwegian Herring Oil and Meal Industry Research Institute (SSF) pp. 11–17, Bergen.

Olsen J.A. & Martinussen T. (1986) An economic impact analysis of the onboard processing of fish in northern

Norway. In *Proceedings of the International Conference on Small Scale Fisheries Development* (Ed. by Groupe d'etude des resources maritimes). pp. 381–90, Université du Quebec à Rimouski.

Opstvedt J. (1985) A national program for studies on the value for surimi production of industrial fish species in Norway. In *Proceedings of the International Symposium on Engineered Seafood including Surimi* (Ed. by R.E. Martin & C.L. Collette). pp. 218–25, National Fisheries Institute, Washington.

Pedersen O. (1987a) Fiskeensilasje og kvalitet, varedeklarasjon og kvalitetsstandard I. *Norsk Fiskeoppdrett.* **2**, 74–6.

Pedersen O. (1987b) Fiskeensilasje og kvalitet, varedeklarasjon og kvalitetsstandard II. *Norsk Fiskeoppdrett.* **3**, 50–51.

Peters J.A., Slavin J.W., Carlson C.J. & Baker D.W. (1965) Storing groundfish in RSW: use of UV-radiation to control bacterial growth. In *Fish Handling and Preservation, Proceedings of Meeting on Fish Technology*, pp. 79–87. OECD, Paris.

Raa J. & Gildberg A. (1982) Fish silage: a review: *CRC Critical Reviews in Food Science and Nutrition*, **16** (4), 383–419.

Reppond K.D., Collins J. & Markey D. (1985) Walleye pollock (*Theragra chalcogramma*): changes in quality when held in ice, slush-ice, refrigerated seawater, and CO_2-modified refrigerated seawater then stored as blocks of fillets at −18°C. *J. Fd Science*, **50**, 985–9.

Shaw D.H. & Botta J.R. (1975a) Preservation studies of inshore male capelin (*Mallotus villosus*) stored in ice. *J. Fish. Res. Board Can.*, **32**, 2039–46.

Shaw D.H. & Botta J.R. (1975b) Preservation of inshore male capelin (*Mallotus villosus*) stored in refrigerated seawater. *J. Fish. Res. Board Can.*, **32**, 2047–53.

Smith J.G.M., Hardy R., McDonald I. & Templeton J. (1980) The storage of herring (*Clupea harengus*) in ice, refrigerated seawater and at ambient temperature. Chemical and sensory assessment. *J. Sci. Food Agric.*, **31**, 375–85.

Stormo B. (1987) *Ombordproduksjon – ensilering.* Foredrag ved Fiskeriteknisk Forum, Trondheim. Unpublished data.

Suzuki T. (1981) *Fish and Krill Protein – Processing technology.* Applied Science Publishers, London.

Suzuki T. & Watabe S. (1986–7). New processing technology of small pelagic fish protein. *Food Reviews International*, **2** (3), 271–307.

Søbstad G. (1988) The Alfa Laval surimi system. In *Fatty Fish Utilization: Upgrading from Feed to Food* (Ed. by N. Davis) pp. 331–49. UNC Sea Grant College Publication UNC-SG-88-04.

Sørensen N.K. (1989) *Surimi – Erfaringer fra Europa.* Presented at Dansk Konsumfiskmøde. FTFI-Foredling, Tromsø.

Sørensen N.K. & Motta H. (1989) Storage and processing trials with anchovy (*Stolephorus sp*) in Mozambique. In FAO Expert Consultation on *Fish Technology in Africa*, Abidjan, Côte d'Ivoire, April 1988. FAO Fish. Rep. 400, Suppl. Rome.

Tangstad W. & Børresen T. (1983) *Industriell produksjon av lodderogn.* Rapport U-23, FTFI-Foredling, Tromsø.

Tokunaga T. & Nishioka F. (1988) The improvment of technology for surimi production from fatty Japanese sardines. In *Fatty Fish Utilization: Upgrading from Feed to Food* (Ed. by N. Davis), pp. 143–59. UNC Sea Grant College Publication UNC-SG-88-04.

Verrez V., Laroche M. & Han-Ching L. (1986) *Evaluation of texture parameters for horse mackerel kamaboko* (*Trachurus trachurus*). Unpublished paper presented at meeting of West European Fish Technologists' Association.

Whittle K., Robertson I. & McDonald I. (1980) Seasonal variability in blue whiting and its influence on processing. In *Advances in Fish Science and Technology* (Ed. by J.J. Connell), pp. 378–87. Fishing News Books, Oxford.

Windsor M.L. & Thoma T. (1974) Chemical preservation of industrial fish: new preservative mixtures. *J. Sci. Food Agric.*, **25**, 993–1005.

CHAPTER 4

Shipboard handling of pelagic fish with special emphasis on fast handling, rapid chilling and the working environment

K.B. Olsen

INTRODUCTION

The most common and most modern catch handling and holding equipment and methods for chilled pelagic fish used on board vessels from 9–60 metres in length will be reviewed in this chapter, together with the short-term trend in the development of new methods. Special emphasis will be given to the developments which have taken place since the Torry Golden Jubilee Conference in 1979.

The aims of the developments and research work reviewed could be characterised by the following:

(1) *To maximise the quality of the landed fish raw material.* Quality should never be the limiting factor in the food use of chilled catches from inshore (coastal) fisheries. Different types of chilled storage in sea-water have been implemented and further refined in this period. In the offshore sector huge pelagic freezers and factory trawlers have been constructed to meet the demand for quality frozen products, and also as a consequence of exploiting still more distant fishing grounds (Anon. 1988). This subject is reviewed in Chapter 5. In the fish meal sector too, improvements in handling methods to meet market demand for quality fish meal have taken place.

(2) *To improve the working environment on board fishing vessels* by trying to eliminate those catch handling procedures which lead to physical strain and fatigue, to such a degree that no fisherman needs to leave his occupation prematurely for health reasons. Attempts have been made to meet these aims by introducing equipment and handling procedures which will eliminate heavy lifting, unsuitable working positions and rough handling of fish. By doing this the handling is speeded up, thereby starting the chilling process much earlier than was previously the case. The aim is to give the fisherman the opportunity to concentrate almost exclusively on the quality aspects of fish handling.

ENCOURAGING INVESTMENTS IN CATCH HANDLING AND HOLDING

Too often investments in catch handling and holding equipment are far below investments in equipment designed for catching large quantities of fish at the lowest possible cost. It is absurd to concentrate only on quantity, especially when fishing under a quota system with limited resources where it is important to maximise catch values as a method to survive financially. It is not enough to advise on good individual technical solutions to these complex problems. It is not even enough to subsidise investments in equipment which will improve quality. A system which correlates fish quality and first hand prices, thereby making investments profitable, will be a necessity on a short-term basis.

CATCH HANDLING AND HOLDING METHODS

Catch handling and holding methods can be divided into several so-called 'unit operations', or subdivisions of the whole process from the moment the fish are landed on board the vessel until they are unloaded. Examples of unit operations are methods for sorting, chilling, boxing and unloading.

The unit operations can be done in several ways

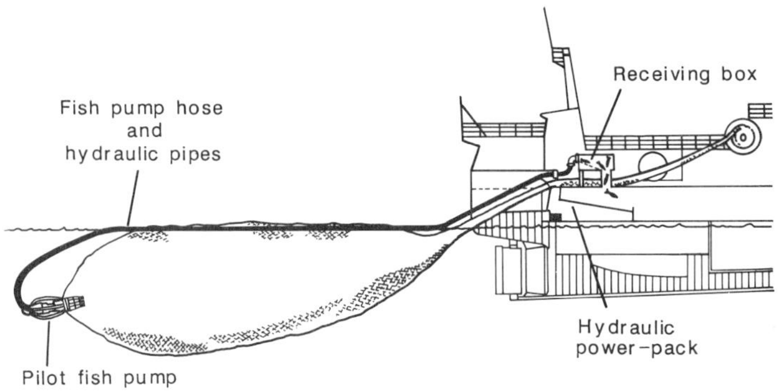

Fig. 4.1 Fish pump permanently mounted on the pelagic trawl.

which range from manual methods to fully automated operations. The most important unit operations in practical use will be described. How many operations will be used in practice and the order in which they are done depends on the species to be caught, the fishing gear used, vessel size, duration of the voyage and the markets which have to be supplied.

The most important unit operations used for handling chilled pelagic fish are: transferring catch from fishing gear to vessel; raw material holding before handling; sorting/grading; transport to hold for chilling; chilling/chilled storage; unloading.

In general, the handling of pelagic species should be fast and rational and should use careful handling operations which ensure a rapid start of chilling, a high chilling rate and a homogeneous storage temperature around −1°C to 0°C. If you cannot start the final chilling within a short time, a prechilling method could be used.

Transferring catch from gear to vessel

Trawlers and purse seiners

Midwater trawlers use tackling in lifts of up to 4 tonnes, or pumping. When using pumping, from large midwater trawls the most common practice is to attach a conventional submersible pump to the codend after bringing the trawl-bag alongside.

When bringing huge hauls (100 tons and over) on board by these methods the danger always exists of losing fish and gear if the fish start to sink after having been brought to the surface (Rasmussen 1978). How fast the catch starts to sink depends upon species, catching depth and weather conditions during hauling. A new type of submersible pump for securing large catches has therefore been invented by a Faroese skipper. Before shooting the trawl for the first time the fish pump is attached to the codend. The pump hose and hydraulic hoses are attached to the outside of the bag and led to the front of the bag. While fishing, the pump is not in operation and does not interfere with trawling (Fig. 4.1).

Large catches can be brought in by hauling the trawl partially on board with the bag remaining in the sea. The fish pump hose and the hydraulic hoses are then connected to corresponding couplings on the vessel and pumping can be started immediately. While pumping, the ship steams slowly ahead thus preventing the bag from sinking (Anon. 1989b). At the moment the pump is suitable only for relatively big trawlers.

Purse seiners use submersible pumps or brailing for bringing the catch on board. It can in practice be difficult with submersible pumps to control the correct fish-to-water ratio, so the risk of bruising the fish during pumping exists.

In recent years the so-called P/V pump (pressure/vacuum) has found increasing use when transferring catches from purse seines to the vessel. The P/V pump has also recently been used for pumping fish from mid-water trawls. But its most widespread use is for unloading fish stored in fixed tanks on board or for unloading bulk iced pelagic fish from coastal inshore trawlers (Fig. 4.2). The P/V pump principle is that an accumulation tank of 500–1500 litre size is alternately put under vacuum and pressure by a water-ring vacuum-pump. (Fig. 4.3). The fish, together with some water, are sucked through a hose and a valve into the system's tank. When the tank is full it is pressurised by changing the vacuum and pressure side connections from the tank to the pump and the fish water mix flows through a valve and a hose into a strainer. The P/V pump is claimed to handle the fish more gently than other fish pump types, but the capacity is generally lower, mostly because of the alternating way of operation.

Gill netters

Small herring gill netters (10–15 m) haul the nets with the net hauler and very often store their catch in the

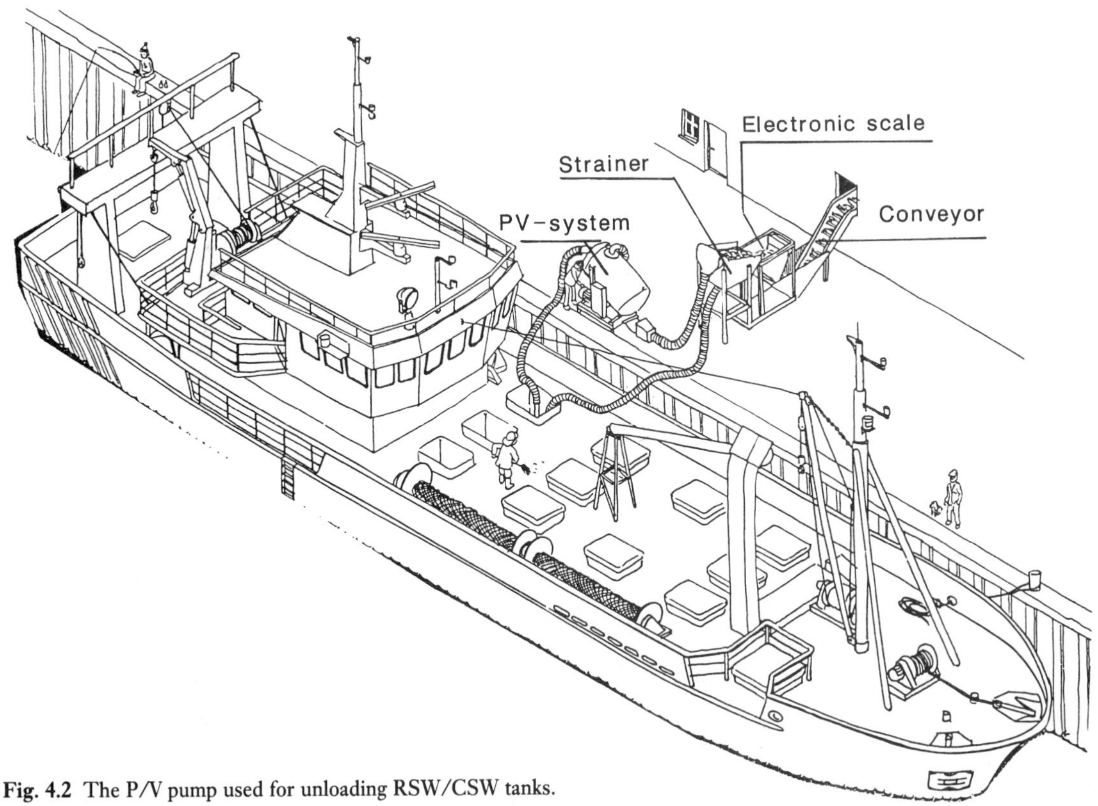

Fig. 4.2 The P/V pump used for unloading RSW/CSW tanks.

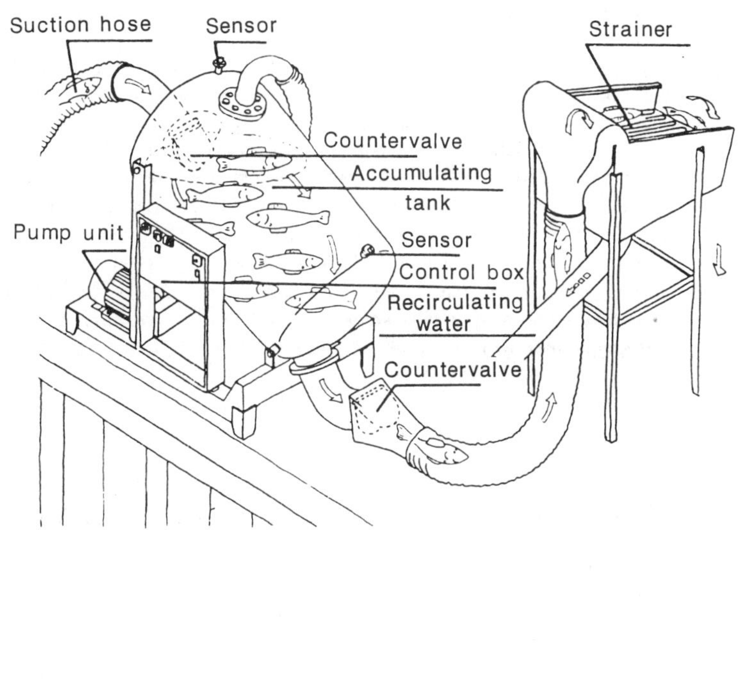

Fig. 4.3 Working principle of a P/V pump.

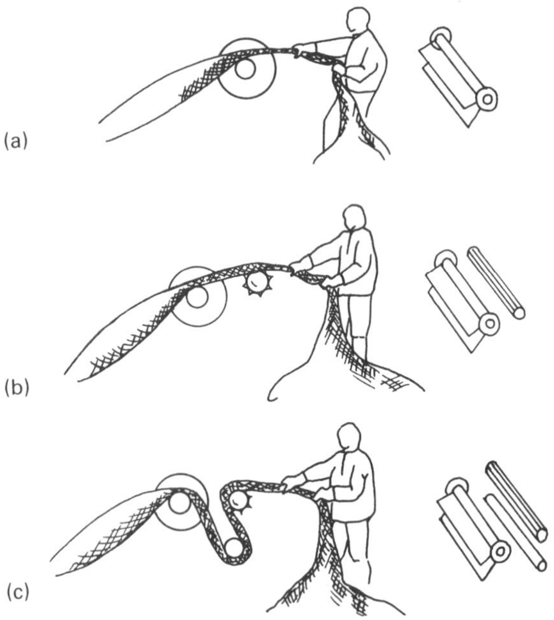

Fig. 4.4 Gill-net shaker, showing ergonomic improvements. (a) Conventional. (b) With a pulling drum. (c) With a pulling and a damping drum.

net until landing. Here the net is drawn through a net shaker by two men to free the herring from the gear. It has been shown that the violent way in which the shaker works can be harmful to the men's hands, arms and shoulders. Ergonomic precautions have therefore been suggested to overcome this problem (Blide *et al.* 1986) (Fig. 4.4).

Holding before handling

Traditional 1 m high deck pounds are used on many trawlers. Some trawlers prechill the fish in deck pounds by icing. A few use refrigerated sea water (RSW) or ice chilled sea water (CSW) prechilling tanks (Olsen 1986). Continuous addition of ice to the fish during loading of the tanks will be of advantage, for securing rapid and homogeneous chilling.

Prechilling holding systems are mostly used on pelagic trawlers which grade their catch in sizes before storing in boxes or in portable CSW containers. The prechilling tanks are unloaded by elevator or P/V pump according to handling speed (Anon. 1988). It is often essential to use prechilling when the pelagic fish are soft and feeding and therefore very liable to belly burst. If no processing is done on board, the fish are conveyed directly for chilled storage in the hold.

Sorting/grading according to size

The equipment used for grading pelagic fish can be divided according to three principles, which all grade on the basis of thickness:

(1) vibrating, inclined, diverging bars
(2) contra-rotating, inclined, diverging rollers
(3) diverging conveyors where fish are being transported along a power-driven V-belt.

Grading by thickness can meet the demand for the high capacity needed in pelagic fish handling and therefore it is generally accepted that the correlations between thickness and length or weight are not too good. (Hewitt 1980). The most important point, which is often forgotten, for making a grader work at its optimum is even feeding. This could be done with an elevator delivering to a (vibrating) water sprayed chute leading to the inlet guide chute at the machine. Sometimes it is necessary to install a manual sorting conveyor before the grading machine, for removal of larger fish and debris, for example in the fishery for argentine with by-catch of grenadier (Chapter 28) (Fig. 4.5).

Most Danish industrial fish trawlers going for sand eel, blue whiting and Norway pout use a rotating sorter for separating white fish by-catches from the industrial fish (Olsen & Hansen 1977) (Figs. 4.6 and 4.7).

Transport to the hold for chilling

CSW/RSW vessels pump their catches to a water separator. Chutes lead the catch to the hold. Very often the pumping rate is so high that no sorting of by-catches, e.g. mackerel or blue whiting, can take place in the herring fisheries. This sorting is therefore done at unloading or in the factory.

During boxing the fish are normally transferred manually from the grader buffer silos with baskets to the boxes for icing (Blide *et al.* 1984). The most advanced pelagic trawlers use a conveyor system leading the fish from the grader to the boxes. (Fig. 4.8). This system, which in the author's opinion represents 'the state of the art' for pelagic boxed fish handling, was developed in Sweden (Blide *et al.* 1984). An even more advanced Swedish system is installed in another Swedish trawler. In principle it consists of a fork lift mounted upside down on a crossbar crane. The fork lift can make a three dimensional movement of palletised fish boxes. It is possible to transport the pallets from the grading machine to the stacking

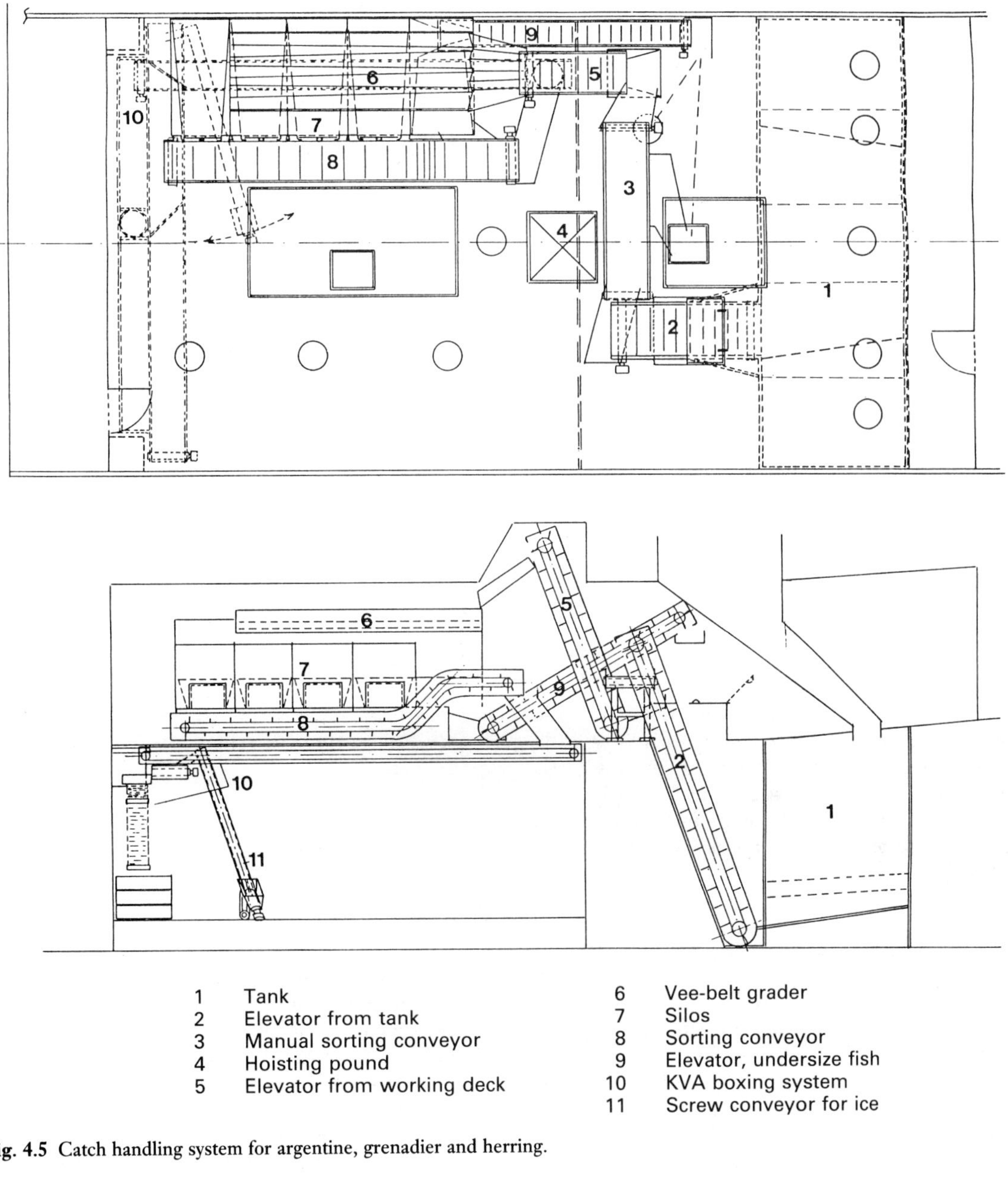

Fig. 4.5 Catch handling system for argentine, grenadier and herring.

position in the hold with only one man controlling the system by joy-sticks. (Fig. 4.9) (Blide *et al.* 1984). The newest version installed even has a one man wireless-controlled unloading system (Anon. 1989a). The boxing capacity is up to 16 tonnes/hour depending on the amount of by-catch to be sorted manually during handling.

Industrial fish trawlers make a homogeneous mixture of fish with ice according to fish temperature and storage time on board in connection with transport to the hold. The newest version still under development is controlled by a microprocessor (Olsen *et al.* 1988) (Figs. 4.10, 4.11, 4.12 and 4.13).

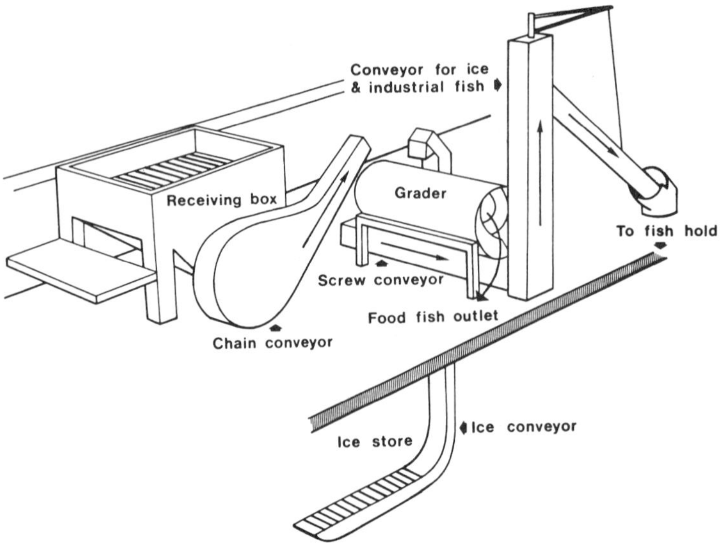

Fig. 4.6 Working deck for industrial fish trawler including rotating sorter for separating groundfish by-catch from industrial fish and a system for mixing fish and ice for chilling to 0°C.

Fig. 4.7 Rotating sorter for separating white fish by-catches from industrial fish.

Chilling/chilled storage

In Scandinavia boxing has been one of the most widespread methods used to store chilled pelagic fish, but the use of boxes is stagnating especially on trawlers, purse seiners and combined midwater trawl/purse seine vessels, which are chasing only pelagic species. The methods which gain increasing popularity are RSW or CSW in fixed tank systems or, to a much more limited extent, portable CSW containers.

Because of the ever existing need for flexibility, many trawlers are dependent on changing between catching pelagic and demersal species. Those fishermen will still prefer boxing. This is the situation for example in Denmark, where many trawlers from 10–40 m try to make a living out of fisheries by changing between pelagic and demersal species depending on season and quota regulations.

Boxing

Boxing has a big advantage over shelf storage by very much reducing static pressure on the fish and also by

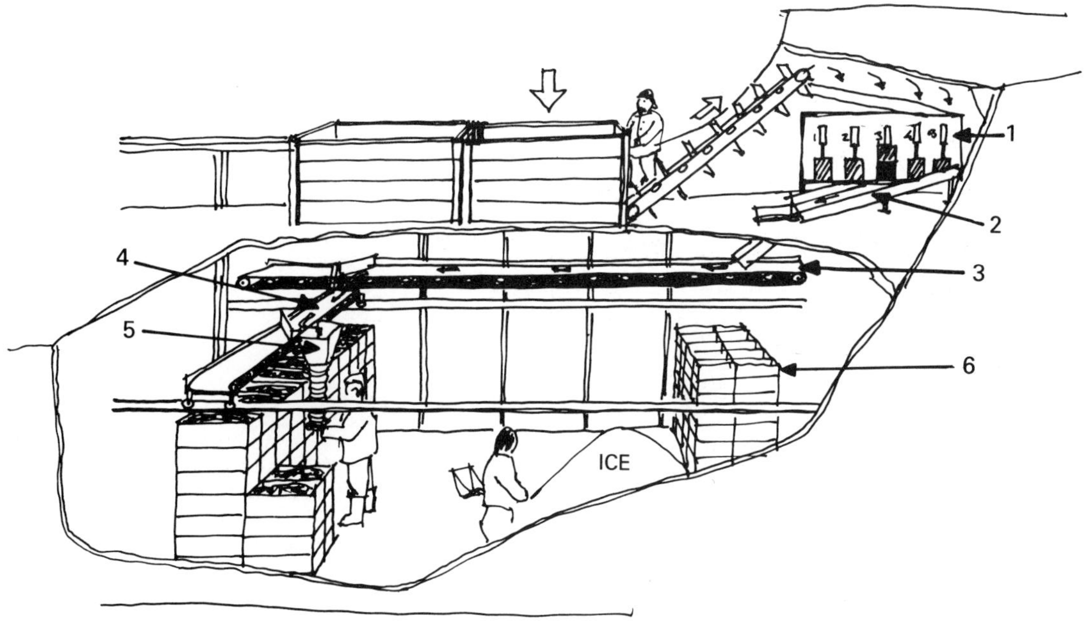

Fig. 4.8 'Polar' system. Mechanised sorting and boxing of herring. (1) Herring sorting machine. (2), (3) and (4) conveyors; (5) flexible dosing tube; (6) box store.

1 Grading machine
2 Flexible tubes for graded fish
3 Fork lift for palletized boxed fish
4 Crossbar-crane
5 Control-box with joy-sticks

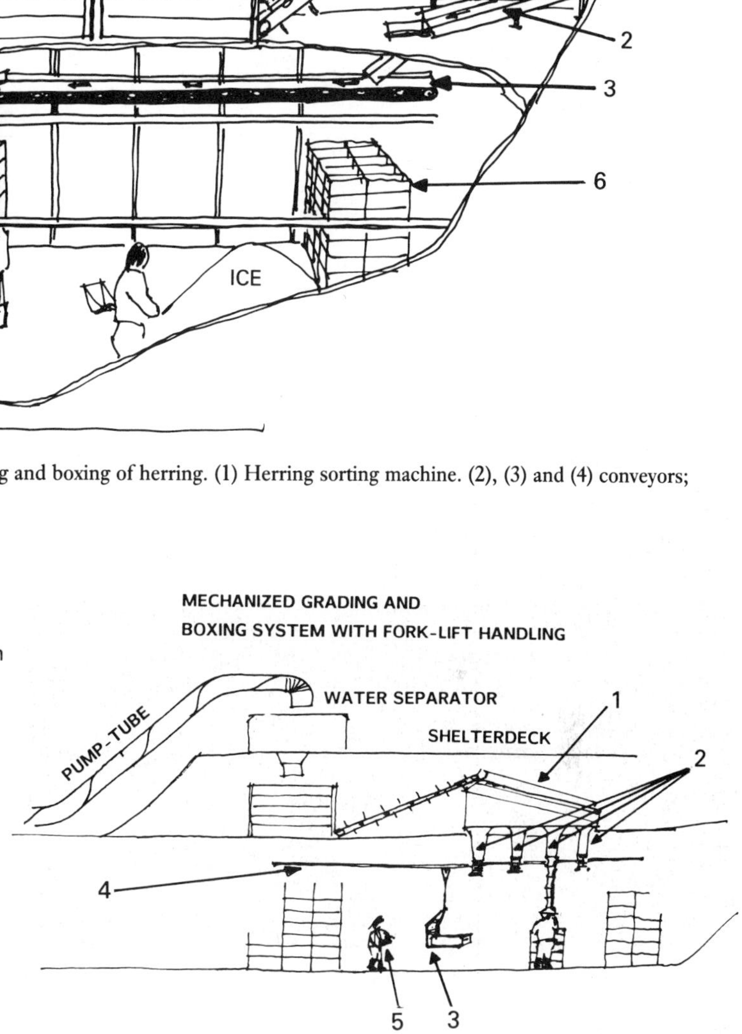

Fig. 4.9 Mechanised grading and boxing system with fork-lift handling.

facilitating unloading. In pelagic fisheries boxed fish will be untouched until they are processed, which is of great advantage from a quality point of view.

(1) Stack-only plastic boxes (Fig. 4.14) The main drawback is that the stack-only type of boxes are space consuming and can be inconvenient to work with in practice unless special methods and handling procedures have been developed, e.g. the Swedish systems mentioned earlier. As pointed out by one of the major producers of boxes, it is very important that the whole infrastructure concerning effective use of boxes both at sea and on shore is seriously considered (Strømberg 1986).

Besides all the space which is needed for the empty boxes, space for bulk storing of the ice which is re-

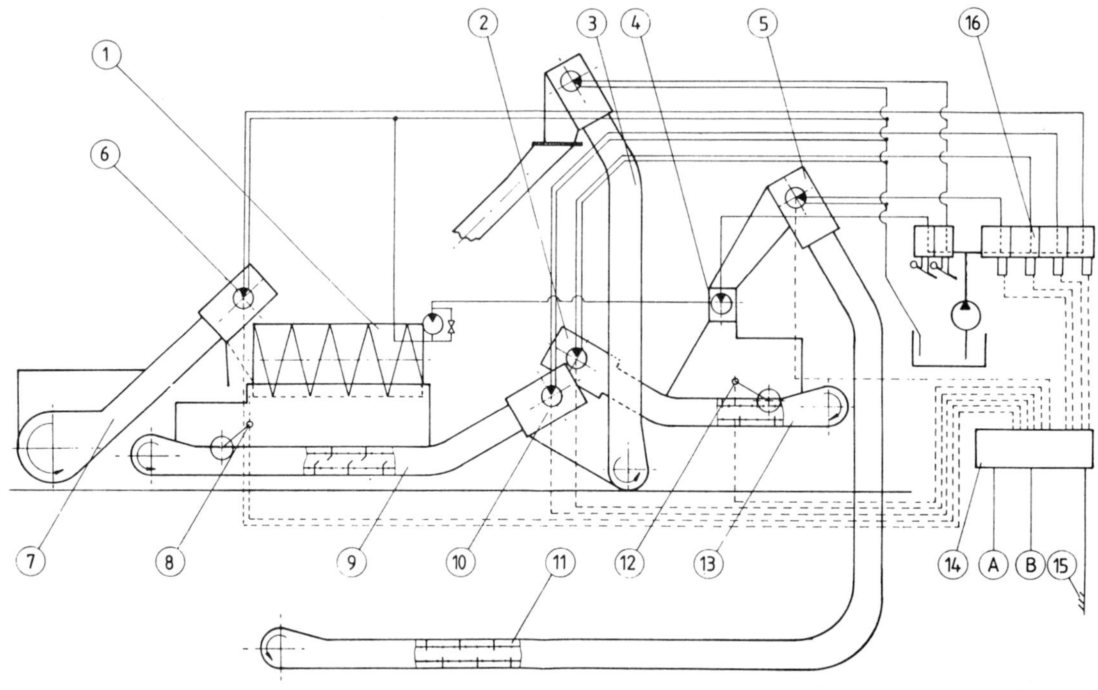

1	Sorting drum (sorts food fish from 'industrial' fish)	10	Rpm-counter, fish conveyor
2	Rpm-counter, ice conveyor	11	Ice elevator
3	Elevator for fish and ice mixture	12	Filling indicator, ice
4	Ice crusher	13	Ice dosing conveyor
5	Rpm-counter, ice elevator	14	Electronic controller
6	Rpm-counter, fish elevator	15	Thermometer, measuring sea temperature
7	Fish elevator	16	Proportional valve
8	Filling indicator, fish	A	Input: species of fish
9	Fish dosing conveyor	B	Trip – duration

Fig. 4.10 EDB-controlled icing system for industrial fish.

quired during the icing process is also needed. Altogether this makes the working conditions very cramped, especially in the beginning, until some of the ice has been shovelled into the boxes. To get enough working space in the hold it is necessary to store some empty boxes on deck equivalent to the space that the bulk stored ice occupies in the hold and the space needed when starting boxing of the first caught fish.

(2) Nest/stack plastic boxes (Fig. 4.15) To overcome some of the space problems when using stack-only boxes, the nest/stack type has been developed. When stored empty these occupy only approximately a third of the space taken up when the boxes are loaded with fish and ice. This type of box is widely used in France, Holland and Germany.

It has even been suggested that the space needed for bulk stored ice can be provided by filling the void volume in the nested boxes with ice from the harbour (Hansen 1977). However, it will be very difficult to introduce this system because it would mean changing a well established infrastructure and tradition.

(3) General remarks If a tailor made system for a certain type of plastic box is constructed the quality advantages of using boxes can be fully utilised on board. The key points to consider are:

(a) The handling rate necessary to prevent quality loss because of delayed icing. Prechilling can be of advantage.
(b) Handling methods which make it possible to

Fig. 4.11 Microprocessor controlled system for mixing of industrial fish and ice according to sea temperature and storage time.

Fig. 4.12 Metering conveyor for ice showing filling indicator.

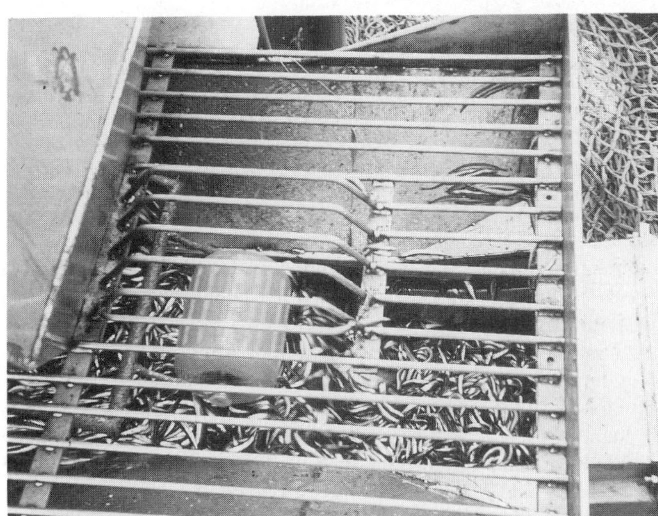

Fig. 4.13 Metering conveyor for industrial fish showing filling indicator.

Fig. 4.14 Stack-only fish boxes shown on a trawler using the handling system in Fig. 4.8.

Fig. 4.15 Nest-stack plastic boxes.

guarantee that the icing procedure is sufficient to chill the fish to 0°C and maintain this temperature until landing.

(c) The hold must be constructed in such a way that safe and easy stacking of the boxes can take place.

(d) Hold insulation of a relatively high quality should be considered. A small mechanical refrigeration plant can be of advantage, especially if it is made as a jacketed system covering the total stack of boxes and in this way removing the heat entering the hold from the surroundings.

Taking into consideration the technical development which has taken place concerning fixed RSW or CSW tanks in the handling of pelagic species, especially concerning the unloading procedures, it is in the author's view probable that the handling of pelagic fish in boxes will stagnate in the future to the advantage of RSW/CSW handling.

The possibility exists that combined inshore pelagic/demersal trawlers, which only need to store their catches for 1–3 days, could be equipped with small fixed CSW tanks to be used for the demersal fish too. The catch will be unloaded with P/V pumps.

RSW storage

This technique is a well established practice which has been refined both from a theoretical and a prac-

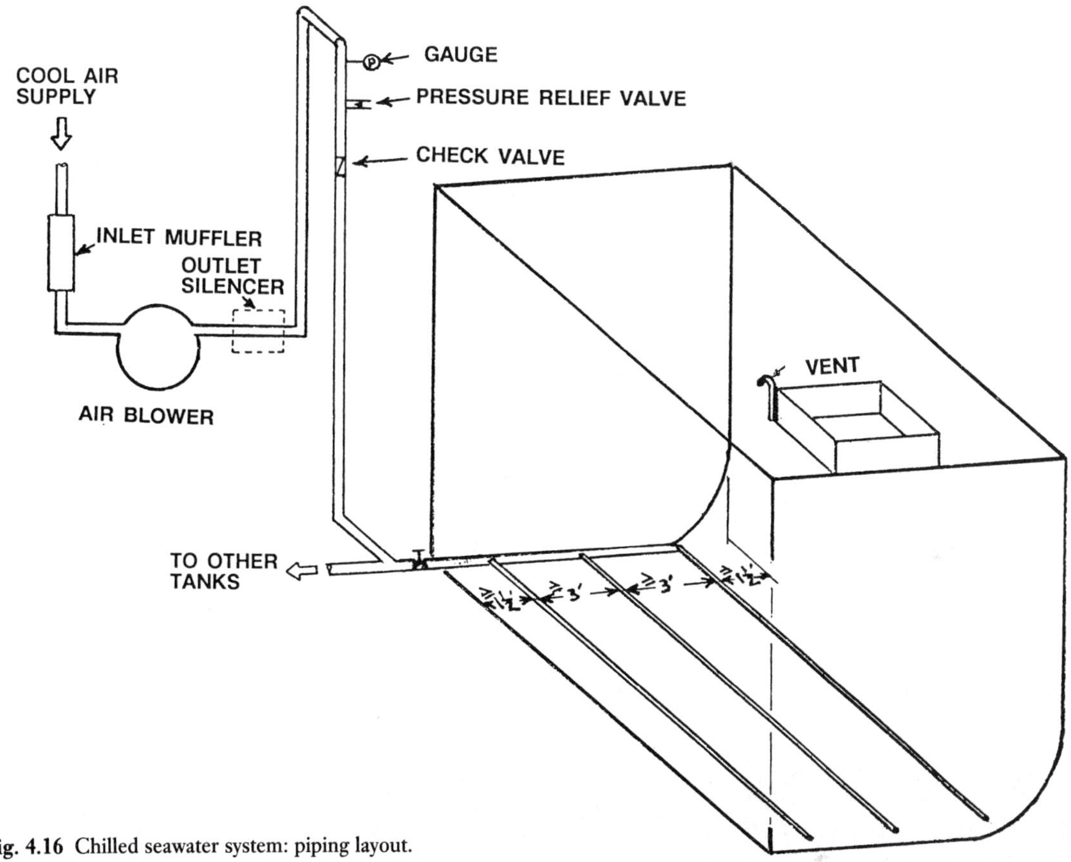

Fig. 4.16 Chilled seawater system: piping layout.

tical point of view since its introduction in the 1960s in Canada where it was developed for salmon and herring storage (Roach *et al.* 1967). In the beginning, most RSW vessels were salmon packers and because of some failures in design which were attributed to insufficient refrigeration or circulation systems, an attempt was made to establish a standard for control of RSW systems. Since vessels are different, the RSW installation has to be studied carefully in every fishery to determine its actual capability. Methods for rating each individual system and vessel, and providing general specifications and guidelines for the proper installation, have been suggested by the Canadians (Gibbard & Roach 1976).

To get maximum shelf life from RSW systems, temperature homogeneity in the region of −1°C is very important. The factors affecting temperature homogeneity were recently studied in Denmark. The most important conclusions were that the input of chilled sea water in the bottom of the tank must take place over the whole tank bottom area, and that filling capacity is dependent on fish species. It was found that maximum filling capacity for mackerel is 80% and for herring 70%. The necessary chilling rate for these species was suggested to be below 3°C within 4 hours and below 0°C after 16 hours, and the temperature should be kept between −1.5°C and 0°C until unloading. At least one temperature sensor had to be installed at the hottest spot in the tank, which in a well designed tank should be in the suction box (Chapter 6).

CSW storage

The CSW system has also been developed in Canada as a much cheaper way – from an investment point of view – to obtain rapid uniform chilling of pelagic fish. The most popular method used is the so called 'champagne' method where rapid heat transfer between fish and ice is obtained by agitation with compressed air introduced at the bottom of the tanks, instead of using circulation pumps as in RSW and some earlier CSW systems designs (Fig. 4.16) (Kelman 1977, Lee

1985). An indication of chilling rate for herring could be 2 hours from 15°C to 0°C. The idea of a CSW system is to load the well insulated tanks at the harbour with the amount of ice necessary to chill the catch to between 0 and −1°C and maintain this temperature until unloading. In theory this method is very effective but the main practical difficulty is to prevent stratification due to differences in specific gravity between sea water and ice and between water of different salinity and temperature.

On the Canadian west coast, fishermen have by experience learned to overcome this difficulty in practice by using a minimum of sea water when they start loading the tank, and by blowing air through the ice-seawater-fish mixture only during loading, stopping blowing immediately the tank is full. Thereafter they will blow only for 5–10 minutes with 3–4 hours interval. The air agitation therefore only serves as a method to overcome local temperature stratification in the tank. From the fishermen's point of view the whole idea is to obtain a uniform mixture of fish and ice and by all means to prevent stratification.

An analytical approach to estimating ice quantities in a CSW tank system has been developed. The quantity of ice required takes into consideration tank size, catch volume, time at sea, water temperature, hold insulation and hold flooding strategy (Kolbe *et al.* 1985). A proven rule of thumb, for estimation of the necessary amount of ice, is simply to observe the amount of ice left in the tank at unloading and compare it with temperature readings, which should be in the −1°C range measured in the landed fish. The starting situation should be conservative, which at a sea temperature of around 12–14°C, on a trip lasting 7 days and with 10 cm polyurethane insulation, is 25% ice by weight of the tank capacity. The amount of ice is adjusted, according to the observations, on following trips.

CSW 'champagne' systems can be used in small coastal vessels too, e.g. in fishery for small pelagic fish in India with vessels of 10–14 m length with a fish carrying capacity from 3–10 ton (Roach 1980).

Another way of loading a CSW tank, which is in practical use in Denmark, is to add the necessary amount of ice to the fish during loading by mixing a stream of fish with a controlled stream of ice. You start with a bottom layer of ice loaded in the tanks just before hauling, but by far the greatest amount of ice is added to the fish during loading. When the tank is full the voids are filled with sea water from a hose and the tank is left undisturbed with only compressed air blowing for 5–10 minutes with a 4 hour interval. The ice is bulk stored in the forward hold and shovelled into a conveyor flush with the floor. The conveyor then leads the ice to the mixing point at the deck. Stratification is no problem with this method.

Portable CSW containers

The use of this type of container for pelagic fish handling was tested at Torry in the early 1970s (Eddie & Hopper 1974). The approximately 2 m^3 heat insulated containers were loaded with the necessary amount of ice from the harbour and agitated with compressed air in a similar way as for CSW tanks. The main advantages of using this method are that the fish will be undisturbed until processed, and unloading is easy. The disadvantages can be marketing problems and reduced pay-load on existing vessels (Eddie 1980).

In recent years the Canadians have developed the system for open trap boats. Containers tailor made to the open boat look most promising. The top of the containers form part of the working platform and ice is carried in some of the containers. No air injection is used, but the containers are loaded with alternating ice–fish layers. At unloading, the special containers are unloaded into standard insulated ones for further transport and the special containers are immediately returned to the open deck boat (Slade & Way 1986).

Implementation of insulated 1 tonne capacity containers has taken place in Canada on 10–20 m decked vessels with such a low height clearance that crew members cannot stand freely and have to work in a stooped position. Eleven containers can be carried in the hold of a 17 m gillnet vessel and three on deck filled with ice. The containers are loaded through scuttles in the deck. The containers are moved to a position under the hatch by a pallet jack when unloading (Slade & Way 1986).

The two above mentioned systems are designed for cod but are suitable for pelagics too.

In Sweden portable 1.1 m^3 CSW containers (Fig. 4.17) are used to a limited extent in combination with the earlier mentioned conveyor system originally laid-out for boxing (Anon. 1986).

UNLOADING

Boxed fish

Pelagic fish iced in 20 or 40 kg boxes at sea will normally be unloaded in pallet loads of, for example, twelve 40 kg boxes per pallet. Swedish boats use hydraulic deck mounted cranes and a special pallet fork during unloading. An unloading rate of approximately 30 t/h is possible by this method (Fig. 4.18).

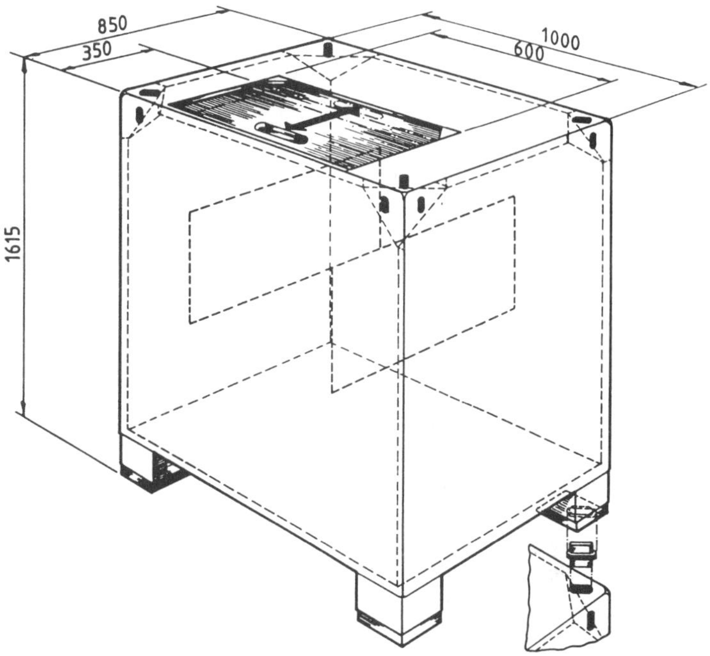

Fig. 4.17 Swedish CSW container, 1.1 m^3 volume. The containers are made as a sandwich construction with inner lining of stainless steel, outer lining of reinforced polystyrene. Insulation polyurethane foam. There are two stainless swash bulkheads in the container. There is a sliding hatch with snap lock on top of the container.

Fig. 4.18 Unloading palletised 45 kg boxes with a hydraulic crane and a special pallet fork.

Inshore (coastal) vessels

Danish coastal vessels landing their pelagic catches daily have for 1–2 years used P/V pumps for unloading. Their catches are often iced in pens in layers up to approximately 1 m height, and it is necessary only to add small quantities of water to make the pump function properly. The herring is delivered to a strainer from where a conveyor leads the fish to a size grader. The strained water is recirculated to the hold. Sorting rates of up to 30 t/h are often installed.

RSW/CSW vessels

In Scandinavia the 30–60 m vessels still use brailing when unloading their catches at a rate from 30–50 t/h. The main disadvantage of using this method is that very big hatches are needed to get reasonable unloading rates. P/V pumps have recently been introduced for unloading herring and mackerel. Thus vessels with small tanks, e.g. 30 m^3, and small hatches can also be unloaded at a rate similar to or higher than the above mentioned brailing rate. P/V pumping rates will typically be around 40–50 t/h. The fish can be transported directly in a tube system into the factory, where representative samples are taken for quality assessment.

Other pump types of the wet or dry suction principle are used in the Mexican fishery for sardines. Often it leads to a high proportion of damaged fish – Monterey sardines – which makes the discard unacceptably high for the canneries. The main purpose of this Mexican fishery is to supply a fish meal plant, where only a fraction of the catch is used for canning. It has been shown that quality catch handling with the high chilling rate obtainable in a champagne CSW system will reduce the proportion of damaged fish to an acceptable level, especially compared with the existing RSW system with a chilling rate of approximately 2°C per hour (Hansen *et al.* 1980; Jensen 1983). According to Jensen a new version of the wet suction pump has been developed in Vancouver, Canada. The syphon pump works on the principle that a liquid will flow voluntarily from a higher tank to a lower when the two are connected. As there are no moving parts in contact with the fish before they reach a cup elevator, practically no damage takes place even with very soft fish. The main drawback is that the price is substantially higher than for the later developed P/V pump.

Industrial fish

Many industrial fish vessels store their catches in bulk. In Denmark all fish are bulk iced as described earlier. These vessels are unloaded by a dry suction pump rate of up to 120 t/h depending on the free flowing characteristics of the fish. Sand eel will give the 120 t/h rate but horse mackerel will give a substantially lower rate because the bottleneck in the system will be the men shovelling the fish to the suction pipe of the pump. A representative sample is taken during unloading for price setting according to quality.

THE FUTURE

It is not easy to introduce rational handling and holding methods on existing vessels. Also today it is often necessary to be flexible and to change fishing methods at very short notice, e.g. from catching demersal to pelagic species. In the construction of new vessels, it is important during the planning stages that the skipper/owner, naval architect, mechanical engineer and fish handling technologists seriously consider catch handling and holding methods in order to reach the best possible compromise when giving the vessel its optimum design. As already mentioned (Eddie 1980) it is of prime importance that the involved technologists go to sea, observing performance and testing prototype equipment and systems under commercial conditions.

The most important point when introducing better handling and holding methods is to find the system which correlates fish quality and price. As a long-term strategy, it is vitally important to point out to the fishermen what is considered to be quality handling and holding practice.

BIBLIOGRAPHY

Anon. (1986) Swedes switch to containers. *Fish. News*, Dec. 19/26, p. 16.

Anon. (1988) Trawling with 10.000 HP. *Fish. News Int.*, August, pp. 40–1.

Anon. (1989a) Denmark delivers Sweden's biggest. *Fish. News Int.*, January, p. 37.

Anon. (1989b) More about the Pilot fish pump. *Fish. News Int.*, July, p. 53.

Blide G., Eriksson H., Kadefors R., Petersen I. & Törner M. (1984) *Trålfiske: Arbetsbelastning och Tekniske Åtgärder.* Projekt Lindholmen, Göteborg.

Blide G., Eriksson H., Kadefors R., Karlsson R., Petersen I. & Törner M. (1986) *Garn, Krok og Snörrvadsfiske, Arbetsbelastning och Tekniske Åtgärder.* Projekt Lindholmen, Göteborg.

Eddie G.C. (1980) Past, present and future in fish handling methods. In *Advances in Fish Science and Technology* (Ed. by J.J. Connell), pp. 18–28. Fishing New Books, Oxford.

Eddie, G.C. & Hopper, A.G. (1974) Containerized storage on fishing vessels using chilled sea water. In *Fishery Products* (Ed. by R. Kreuzer), pp. 69–74. Fishing News Books, Oxford.

Gibbard G.A. & Roach S.W. (1976) *Standard for an RSW*

System. Technical Report 676, Fisheries and Marine Service, Vancouver.

Hansen P. (1977) Fresh Fish Handling in Coastal Fisheries. *Scandinavian Refrigeration*, **6** (1), 31–7.

Hansen P., Jensen J. & Teutscher F. (1980) Bulk handling of purse seine catches for the sardine cannery industry. Scandinavian Refrigeration, **9** (3), 143–7.

Hewitt M.R. (1980) The application of engineering science to fish preservation. Part 1. In *Advances in Fish Science and Technology* (Ed. by J.J. Connell), pp. 175–83. Fishing New Books, Oxford.

Jensen J. (1983) Bulk handling and chilling of large catches of small fish. Part IV: Handling on shore. *Infofish*, **3**, 38–41.

Kelman J.H. (1977) *Stowage of fish in chilled sea water*. Advisory Note 73, Torry Research Station, Aberdeen.

Kolbe E., Crapo C. & Hildebrandt K. (1985) Ice requirements for chilled water systems. *Mar. Fish. Rev.*, **47** (4), 33–42.

Lee F.N. (1985) Design and operation of a chilled sea water system. *Can. Tech. Reports of Fish. and Aqua. Sci.*, No. 1363.

Olsen K.B. (1986) European fish handling and holding methods, new directions in fisheries technology. In *Proceedings of the Atlantic Fisheries Technical Conference* (Ed. by J.F. Roache). Ottawa. pp. 46–56.

Olsen K.B. & Hansen P. (1977) Handling mixed catches in a North Sea trawler. *Fish. News Int.*, **16**, 8, 49–52.

Olsen K.B., Popescu V. & Kraus L. (1988) Handling and holding of fish on fishing vessels in Denmark. In *Proceedings of World Symposium on Fishing Gear and Fishing Vessel Design*. St. John's, Newfoundland. pp. 363–7.

Rasmussen H. (1978) *Blåhvillingfiskeri. Nye erfaringer*. Sydjysk Universitetsforlag, Esbjerg. (In Danish).

Roach S.W. (1980) *A chilled sea water (CSW) system for fishing and carrier vessels engaged in the small pelagic species fishery of south-west India*. FI: DP/IND/75/038, FAO, Rome.

Roach S.W., Tarr H.L.A., Tomlinson N. & Harrison J.S.M. (1967) *Chilling and freezing salmon and tuna in refrigerated sea water*. Bulletin 160, Fish Res. Board of Can., Ottawa.

Slade F.L. & Way E. (1986) Practical applications for improving fish quality and handling procedures on inshore fishing vessels, new directions in fisheries technology. In *Proceedings of the Atlantic Fisheries Technical Conference* (Ed. by J.F. Roache). Ottawa. pp. 65–70.

Strømberg P. (1986) International trends in fish containerization, new directions in fisheries technology. In *Proceedings of the Atlantic Fisheries Technical Conference* (Ed. by J.F. Roache). Ottawa. pp. 57–60.

CHAPTER 5

Pelagic fish in the Netherlands

H. Houwing

INTRODUCTION

Pelagic fish, especially herring, have been caught by Dutch fishermen for centuries. Preservation of herring is even claimed to be a Dutch invention. At the beginning of the 14th century W. Beukelsz preserved herring by gibbing and salting. (Gibbing is removal of the perishable gills and intestines. A well gibbed herring still contains the pancreas. A former chemist of our Institute, A.F.M.G. Luypen, has proved that this cut-through pancreas is very important for the ripening process of the famous Dutch 'matjes' herring which can be produced from herring only in the 'matjes' state (state 2 according to Hjort). During this eating period the pancreas is rich in proteolytic enzymes which are able to break down part of the proteins into polypeptides, peptides and amino acids, thus resulting in the delicious ripened herring products.)

Until some decades ago herring were caught in herring fleet or drifting gill nets, a rather simple way of fishing, resulting if done properly in excellent herring. Also until recently much herring was caught in herring 'bottom' trawls. Nowadays it is either caught by pair trawling (two ships) or by pelagic trawling (one ship). In the Netherlands pelagic trawling is done only by rather big stern trawlers, which are also able to freeze the catch. In this chapter most attention will be paid to this way of catching herring, and most of the statements also apply to mackerel.

The Dutch freezer trawlers used to catch herring in the North Sea and adjacent waters until 1977. In that year fishing of herring in this area was forbidden, so they switched over to other pelagic fish such as mackerel. Since 1984 mackerel catches have been limited. Instead of mackerel, horse mackerel have been caught and, in the context of experimental fishery, other species such as squid have been brought aboard as well. Blue whiting are among the species in which the fishermen are interested. Incidentally some ships are also used for other purposes such as 'klondyking', in this case freezing the fish caught by British, American or Irish ships. From time to time ships are used only as carriers. In this chapter most emphasis is given to processing of herring and mackerel on board stern trawlers with freezing facilities. Some attention will also be paid to the technical and economic aspects of running this type of ship.

CATCH AND STORAGE

Catching begins as soon as schools of herring or mackerel are detected. Fishing is done by means of pelagic trawls. Often the skipper does not stop fishing until the net is full of fish. This may take half an hour but can last for four hours, and even then it is not certain that enough fish have been caught.

Thirteen big ships are fishing under the Dutch flag now. The overall length of the biggest one is about 110 m with a frozen storage capacity of 4400 m^3. Its 26 vertical plate freezers can freeze 160 tonnes of fish per day. When too many fish have been caught the surplus fish can be stored in refrigerated sea water (RSW) tanks with a total volume of 300 m^3. On some ships small barrels filled with herring can be frozen in freezing tunnels.

ECONOMIC ASPECTS

The Agricultural Economics Research Institute (LEI) in the Netherlands has done some research on the economic aspects of stern trawlers. Some economic data is given in Tables 5.1, 5.2 and 5.3. It is clear that the number of ships has diminished over the years. However the fishing effort has only slightly decreased. This has resulted in a somewhat larger crew per ship.

Table 5.1 Technical and financial data on freezing trawlers (all ships together) (AEI 1989).

	1983	1984	1985	1986	1987
Number	28	27	22	15	13
Power in hp	73 770	82 450	74 850	66 300	58 700
Tonnage	20 670	24 197	22 581	21 123	19 099
Crew	567	586	499	397	349
Costs in DFl. $\times 10^6$*	284	342	317	314	286

* Nautic and detection equipment *not* included.

Table 5.2 Technical and financial data averaged per ship (AEI 1989).

	1983	1984	1985	1986	1987
Power in hp	2635	3054	3402	4420	4515
Tonnage	738	896	1026	1408	1469
Crew number	20.3	21.7	22.7	26.5	26.8
Costs in DFl. $\times 10^6$	10.2	12.7	14.4	21.0	22.0

Table 5.3 Proceeds in DFl. $\times 10^6$ (all ships together) (AEI 1989).

	1983	1984	1985	1986	1987
Total	184	202	214	170	161
Fishing in EC water	179	169	177	158	137
Fishing elsewhere	3	5	16	—	7
Transport of mackerel	2	24	18	12	16
Transport of cod	—	4	4	0.5	—
Net result	5	−4	−9	−2	0.2
Salaries*	49	52	56	45	40

* Crew are paid according to proceeds.

HANDLING OF PELAGIC FISH

General

All fish caught by the big ships are frozen as whole fish, except occasionally for herring in the 'matjes' season. In most cases fish are graded on size before freezing.

Due to the fishing method hauls may be so big that nets have to be emptied in several shifts. One net may contain almost 200 tonnes of fish. While one part of the catch is transported to the next destination the next part is brought aboard. Unloading of nets may therefore last up to $1\frac{1}{2}$ hours, and further treatment more than 24 hours. Figure. 5.1 shows a flow-sheet of the processing line.

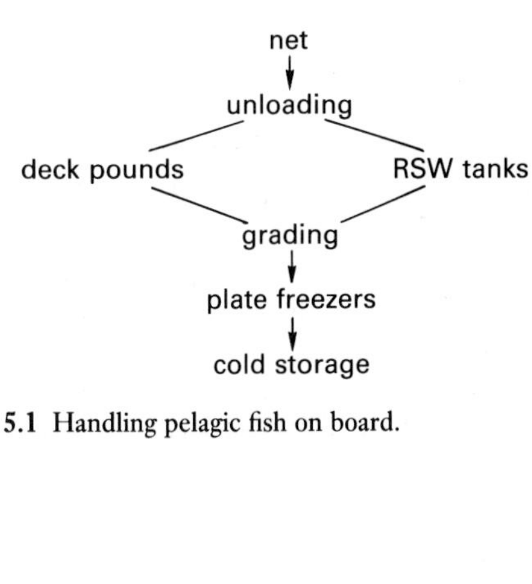

Fig. 5.1 Handling pelagic fish on board.

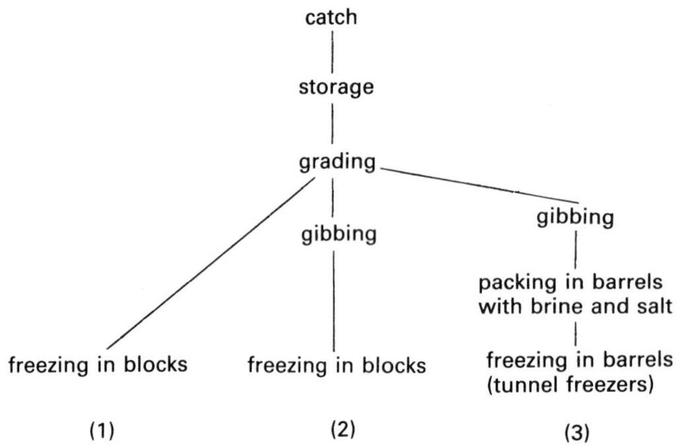

Fig. 5.2 Processing matjes herring on board.

Matjes herring

Figure 5.2 shows some possible ways of processing matjes herring. In olden times the Dutch only gibbed herring which were almost alive. This resulted in an excellent product without any visible blood on the fillets. Such an appearance is rare today. Treatment 3 in Figure 5.2 approaches this ideal appearance most closely. Moreover, this way of handling may give fillets of matjes herring with a very good taste.

GENERAL REMARKS

Pelagic fish like herring and mackerel are among the most perishable fishes, especially when they are feeding actively. This is caused by the fact that the abdominal cavity contains the pancreas, which is full of proteolytic enzymes. These enzymes may easily dissolve the belly lining, thus resulting in damaged belly flaps. The enzymes may also break down proteins at too early a stage so that the flesh softens. These fish will be hard to fillet mechanically.

If no precautions are taken in the whole processing line, from catch to freezing, the catches may easily be damaged. This is mainly due to unloading the net in shifts and storing the fish in the deck pounds before freezing. In these pounds the fish at the bottom lie under pressure. Moreover the fish can move freely. Both circumstances may cause damage.

There is a tendency towards the use of fish pumps. This procedure may cause less damage because the pressure in the nets during unloading is almost zero.

In the present situation we prefer short hauls to long ones. This preference is in conflict with the skippers' interest, especially if catches depend on the presence of a school of fish. Long hauls may have other disadvantages. When the net is unloaded the quality is not homogeneous. One may expect only E grade fish at the time of unloading. Certain fish meet some, important, criteria of A grade fish. This grade will normally be given to fish kept on ice for three or four days. This problem may make it difficult to produce batches of a homogeneous quality. This does not influence the acceptance of fillets by the consumer very much, but it can be seen as the price to be paid for industrialisation.

Freezing of lightly salted herring to be eaten uncooked, which is quite normal on the continent, has been compulsory in the Netherlands since 1969. The reason is the presence of the nematode *Anisakis simplex*, which can be a hazard to health. Initially this technique was hardly accepted by the industry, but this regulation was considered a blessing shortly after, because frozen lightly salted herring does not deteriorate like the unfrozen product. It is the exception rather than the rule that this type of herring is judged unfit for human consumption; the opposite could be observed before freezing became compulsory.

BIBLIOGRAPHY

AEI (1989) De grote Zeevisserij in de periode 1983–1987 (The large sea fishery in the period 1983–1987). *Report no. 377 of the Agricultural Economic Institute, Netherlands.*

CHAPTER 6

RSW treatment of herring and mackerel for human consumption

L. Kraus

INTRODUCTION

In order to land the best quality fresh fish, there is a need for handling systems on board that efficiently cool fish as quickly as possible to the storage temperature just above freezing point (Huss 1983). During fishing by purse seine for herring and mackerel, hauls of about 50 to 100 tons are likely. Alternative systems to the normal traditional method of icing in boxes are necessary because it takes too long and causes great loss of quality in the landed fish. RSW systems (refrigerated sea water) are very fast and effective chilling methods (Gibbard and Roach 1976).

Good boxing practice requires attention to important factors such as chilling time, storage temperature, filling of boxes and good hygiene in cleaning the boxes and fish room. Much research has shown the advantages and disadvantages of icing in boxes (Hansen 1981); the RSW cooling systems have not been tested as comprehensively and the aim of this research was to show clearly the key factors which affect the quality of fish landed from Danish pursers fitted with RSW tanks. At the request of the Ministry of Fisheries Quality Committee, the work was carried out by the Ministry of Fisheries Technological Laboratory in cooperation with the union of seine boats in Denmark during commercial fishing for herring and mackerel.

Table 6.1 Numbers of measurements on the vessels.

Vessel number	Total fishing trips with herring	mackerel
1	2	1
3	2	2
4	2	2
6	2	4
7	7	2
8	2	0
10	4	1
Total	21	12

PROCEDURE AND METHODS

During discussion (Nielsen *et al.* 1986) with crew from ten RSW purse seiners in Denmark, the construction and use of the RSW systems were examined. Based on this information, seven of the vessels were chosen to represent a wide range of different designs of RSW tanks (see Table 6.1 and Fig. 6.1).

Temperature measuring equipment

Measuring equipment consisted of seven temperature sensor elements (Electromatic-PT 100-250-3) that are fastened at 1 m intervals on a wire.

The temperature sensors were connected to two data recorders; one printed results every hour, the other monitored temperatures every tenth minute and stored these values electronically (see Fig. 6.2). On docking, the electronic data were read with the help of a PC.

Experimental measurements and sampling on board

The temperature measuring equipment was calibrated before assembly on the fishing vessel and temperature monitoring of the RSW tank water commenced as soon as the equipment was assembled on the vessel, continuing until return to the harbour. During filling of the tank with fish, a sample of about 25 kg fish was taken, mixed with 15 kg ice in a 50 litre insulated container, and topped up with sea water. This sample was marked REF and used as the reference sample for

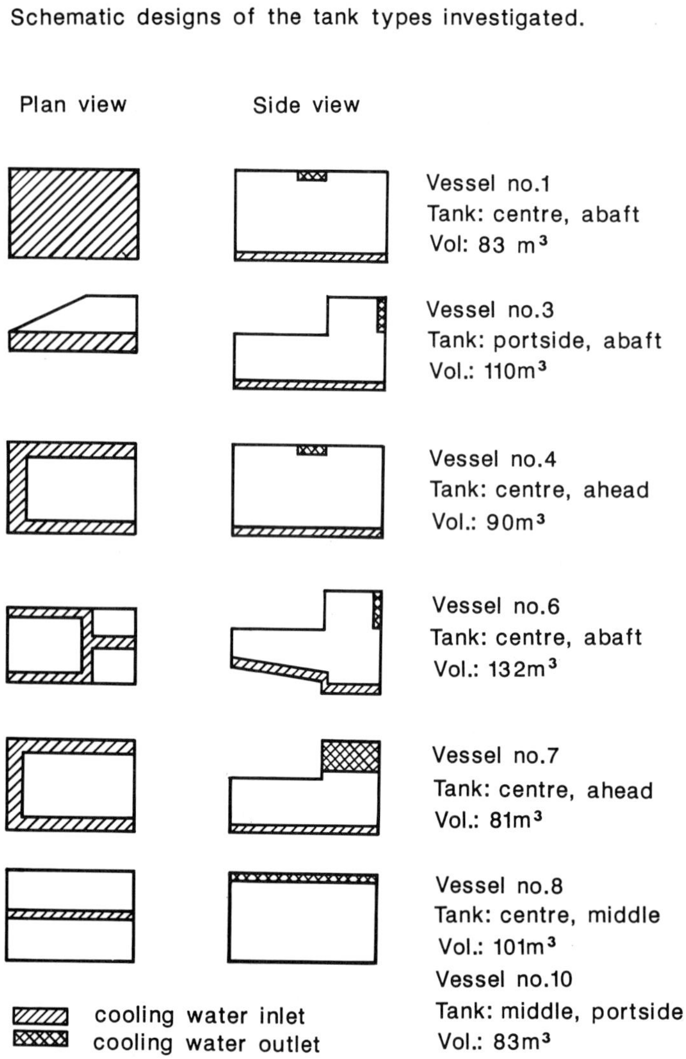

Fig. 6.1 Schematic designs of the tank types investigated.

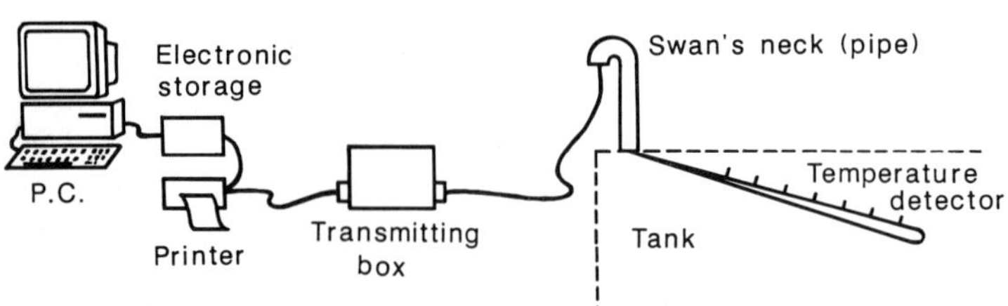

Fig. 6.2 Diagram of the measuring arrangement.

subsequent chemical and sensory analysis. At unloading, a similar sample was taken and marked RSW.

REF and RSW samples were stored in the insulated containers to simulate tank chilling and their rate of quality deterioration was assessed during subsequent storage.

The samples were analysed for peroxide value (PV), thiobarbituric acid (TBA), free fatty acid (FFA), total

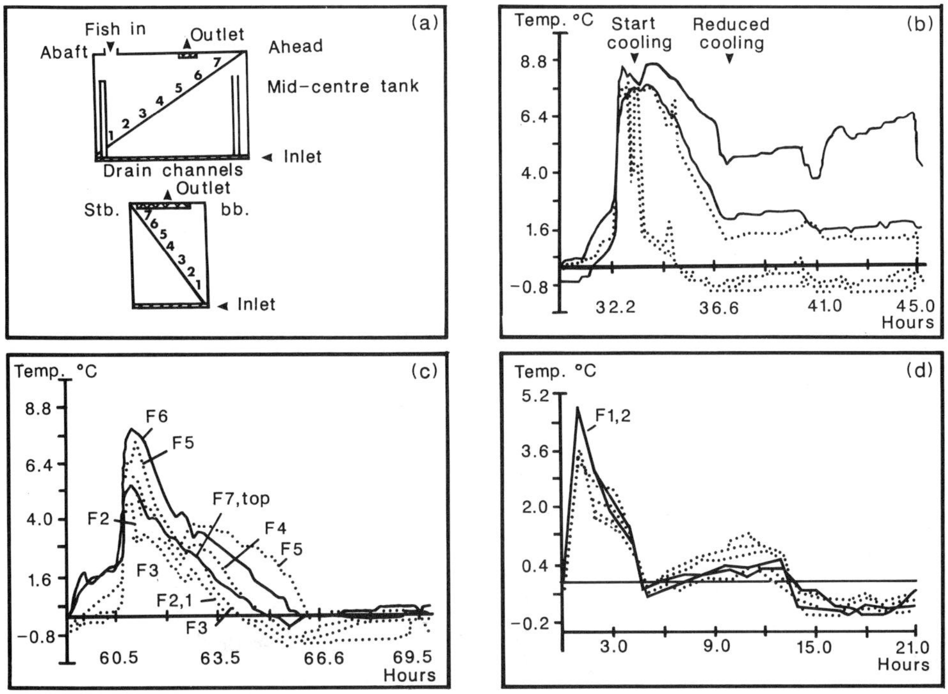

Fig. 6.3 Chilling process for vessel number 1. Tank size 83 m^3. (a) Tank-construction and the position of the temperature sensors. (b) Fishing trip number 1, 83% herring. (c) Fishing trip number 2, 72% herring. (d) Fishing trip number A, 80% mackerel.

volatile bases (TVN) and sensory scores on landing and every three days thereafter.

RESULTS

Chilling

Figures 6.3 to 6.6 illustrate the construction of the tanks on some of the vessels and also indicate the type of temperature recordings which were obtained.

Mackerel

Figure 6.3(d) represents normal uniform chilling in the case of mackerel.

The chilling process in Figure 6.4b shows the importance of the insulation in the tanks. Sensor 1, placed abaft against the machine room, shows an increase in temperature when the circulation stops. Originally the tank had been properly insulated, but at some time water had penetrated into the insulation making it ineffective.

The measurements in Figure 6.5 show clearly how important it is to avoid water loss through the drain-channels. The first measurements for this vessel when loaded with mackerel show that the second sensor from the top is not properly cooled; after 14 hours it reads 3°C. Had it been possible to block the drain-channels, more of the cooling water would have been forced up through the mackerel load resulting in quicker and more uniform cooling.

Figure 6.6 shows the importance of the design of the bottom screen, and the need for adequate pumped circulation of water to avoid the formation of uncooled areas in the tank. Alterations to the inlet and drainage channels solved some of the problems, reducing the hot areas at the top abaft in the tank. For this vessel, however, problems occurred in the distribution of the inlet water across the bottom of the tank. The inlet of cooling water should be changed to cover the whole bottom, and it should be possible to open and close the drain-channels as required and to extend the suction to the top abaft of the tank where the uncooled areas are.

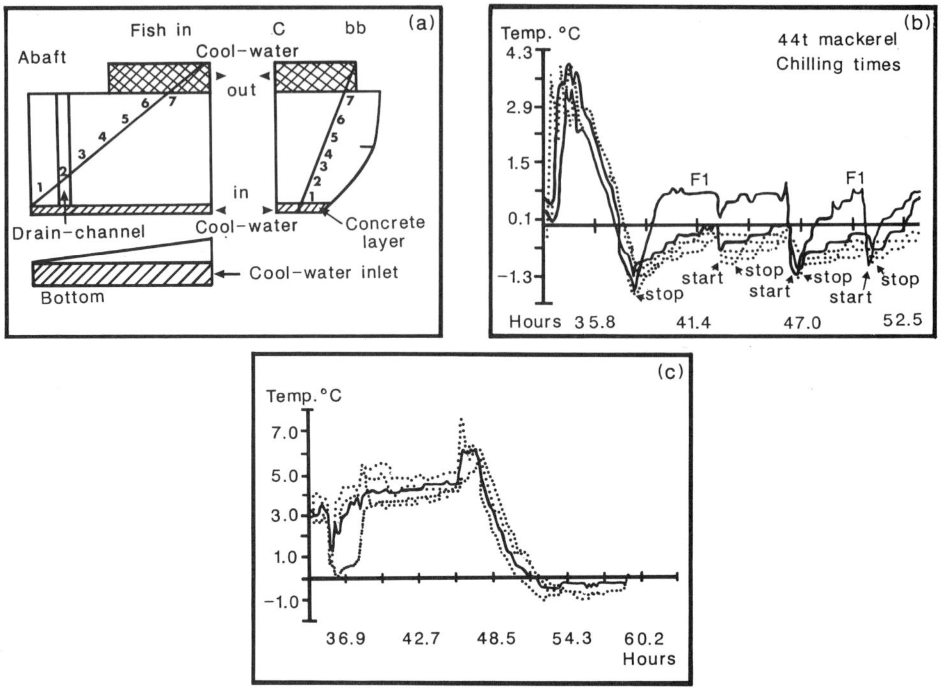

Fig. 6.4 Chilling process for vessel number 3. Tank size 110 m^3. (a) Tank-construction and the position of the temperature sensors. (b) Fishing trip number 1, 40% mackerel. (c) Fishing trip number 4, 65% herring.

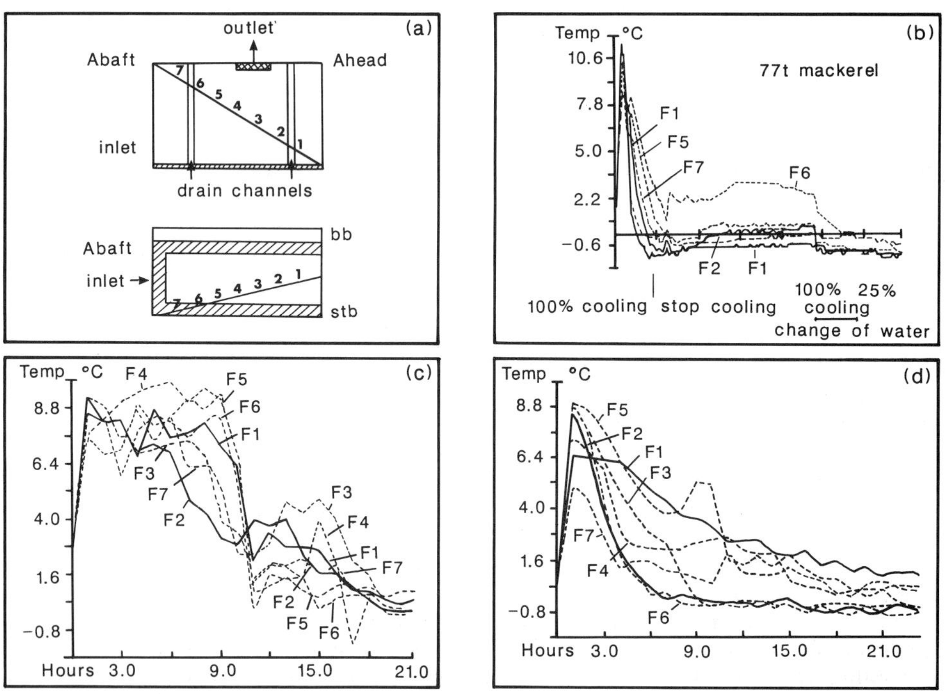

Fig. 6.5 Chilling process for vessel number 4. Tank size 90 m^3. (a) Tank-construction and the position of the temperature sensors. (b) Fishing trip number 1, 85% mackerel. (c) Fishing trip number A, 84% herring. (d) Fishing trip number B, 70% herring.

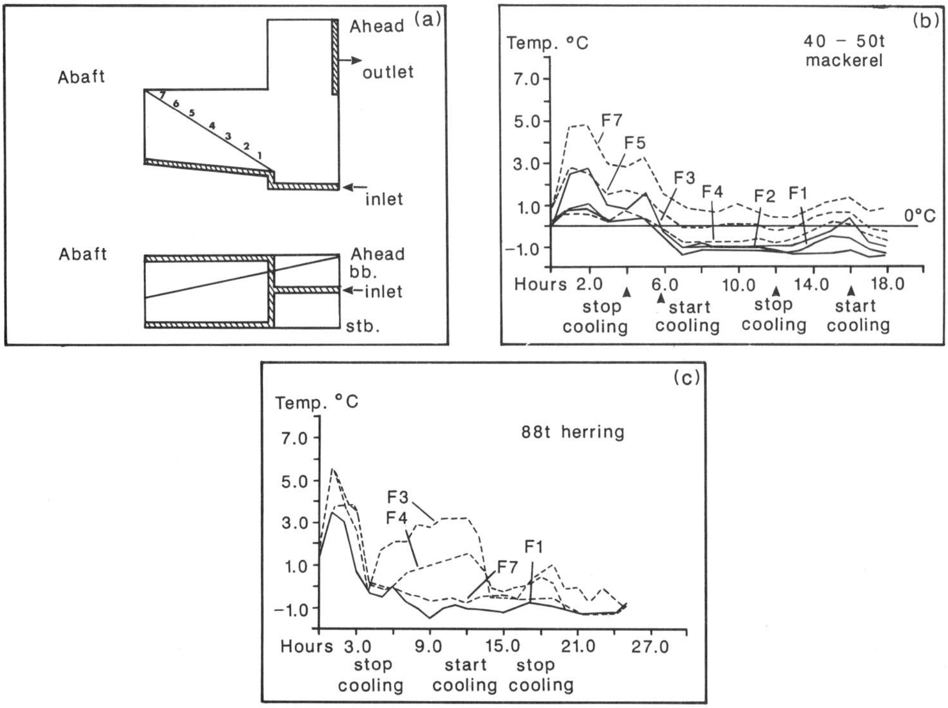

Fig. 6.6 Chilling process for vessel number 6. Tank size 132 m^3. (a) Tank-construction and the position of the temperature sensors. (b) Fishing trip number C, 75% mackerel. (c) Fishing trip number F, 73% herring.

Generally, it is not difficult to achieve uniform chilling for mackerel if the guidelines are followed and better use is made of the circulation and cooling equipment as mentioned above, resulting in better quality of the landed fish.

Herring

Figures 6.3(b) and 6.3(c) show clearly how important it is not to overfill the tanks with herring. More uniform cooling results if about 10% less herring is added to the tank.

The chilling process in Figure 6.5(c) shows the problems caused by overfilling the tank and having too small a suction area. Initially, the herring blocked the suction screen and cooling did not start until about 9 hours after loading. The results in Figure 6.6 also show how important it is to distribute the inlet cooling water uniformly over the bottom of the tank to achieve uniform cooling throughout the tank.

Figure 6.4(c) shows the importance of starting chilling immediately after the haul is pumped on board. The problem with this vessel was that its electric power capacity was too low, making it necessary to stop the refrigeration system every time the net was shot or hauled. The quality of the fish was poor. Sensory assessment immediately on landing this catch graded the herring 5.5 on a 10 point scale at which 4 was unacceptable.

Chemical analyses and sensory assessments

Figures 6.7 to 6.10 illustrate the results of the chemical and sensory analyses. The results for RSW and REF samples are shown related to the time since the fish were caught.

Mackerel

The sensory results in Figure 6.7 show generally that the mackerel was of good quality. Figure 6.8 shows that for vessel No. 3, fishing trip No. 1, higher peroxide values were observed although generally the

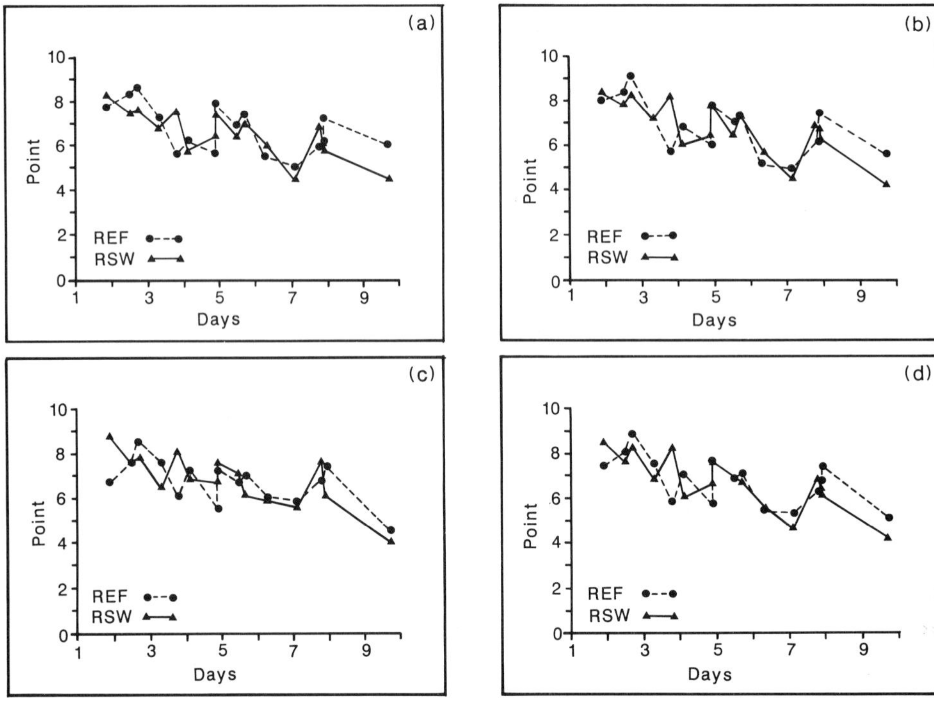

Fig. 6.7 Sensory analysis of mackerel for all purse seiners. (a) Smell of tank-handled mackerel with time. (b) Taste of tank-handled mackerel with time. (c) Consistency of tank-handled mackerel with time. (d) General impression of tank-handled mackerel with time.

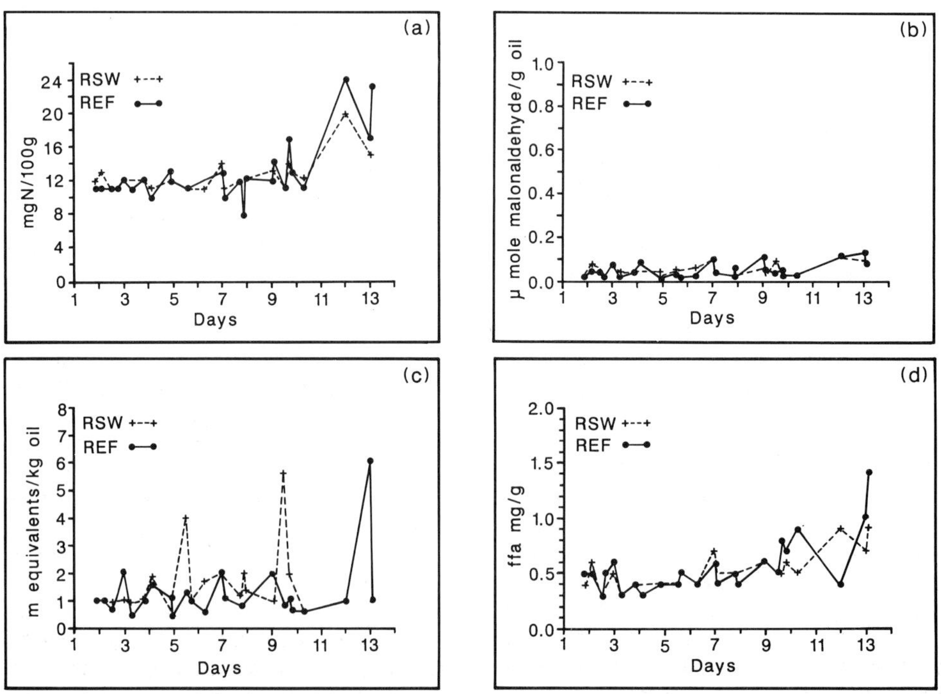

Fig. 6.8 Chemical analysis for mackerel for all purse seiners. (a) TVN in tank-handled mackerel with time. (b) TBA in tank-handled mackerel with time. (c) PV in tank-handled mackerel with time. (d) FFA in tank-handled mackerel with time.

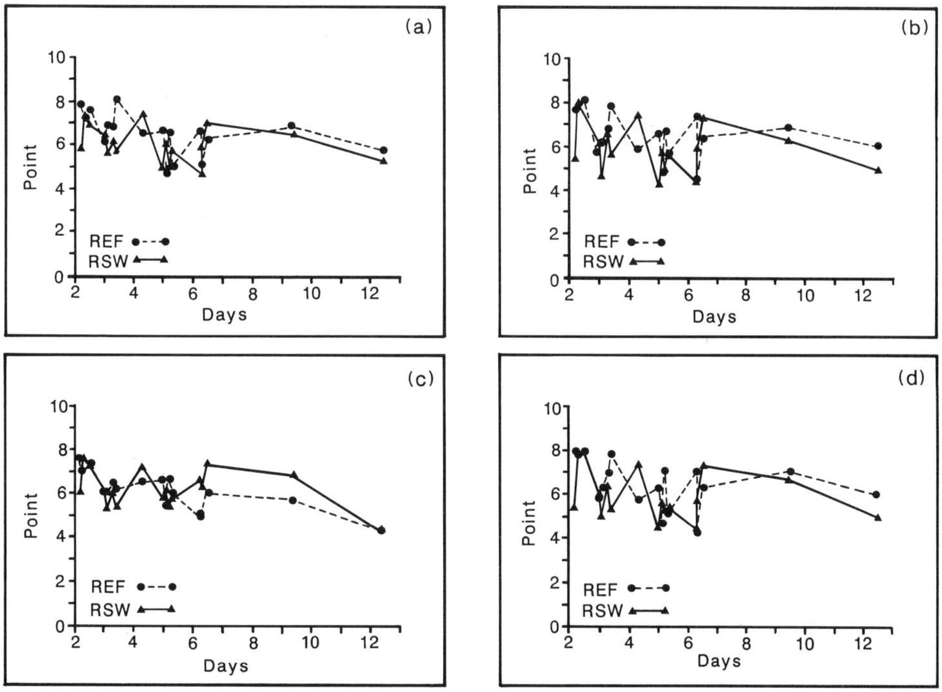

Fig. 6.9 Sensory analysis for herring for all purse seiners. (a) Smell of tank-handled herring with time. (b) Taste of tank-handled herring with time. (c) Consistency of tank-handled herring with time. (d) General impression of tank-handled herring with time.

peroxide values were less than 2. On this trip the fish filling level in the tank was low, but the tank was topped up with water. This was a small vessel which rolled heavily in bad weather making it difficult to determine when the tank was full. For this vessel the appearance of the fish suggested that they were not as fresh as those from the larger vessels. This led to the fact that the accessible quantity of dissolved oxygen in the water in the tank is higher and therefore in combination with physical damage to the fish a possible increase in formation of peroxides.

Herring

Figures 6.9 and 6.10 show a greater difference in analytical results between REF and RSW in the case of herring compared with mackerel; usually the REF samples were better. This suggests that fish species has a substantial influence on the quality of the tank-handled fish and supports the statement in FF Report No. 134, (Kraus 1987) that herring reaction proves that it has a tendency to be packed together in tanks, whereby refrigeration results are uniform.

CONCLUSIONS

General

The tank drains used today in the Danish purse seiners are of unfavourable design because some of the cooled water is forced through the tank and thereby reduces the utilisation of the cooling capacity on board the vessels.

The chemical analyses for the catch at the time of landing show that the fish generally is of good quality (Huss 1983). At present, measurement of the change in temperature with time is not suitable as an unambiguous description of the quality of RSW stored fish. The contruction and operation of the RSW systems, seasonal change in the composition of the fish and the weather, all contribute to cause variable quality from one catch to another. For each particular haul, however, the lower the temperature, without freezing the fish, achieved initially and maintained throughout storage, the better the quality of the catch. Therefore measurement of temperature with time is very useful in

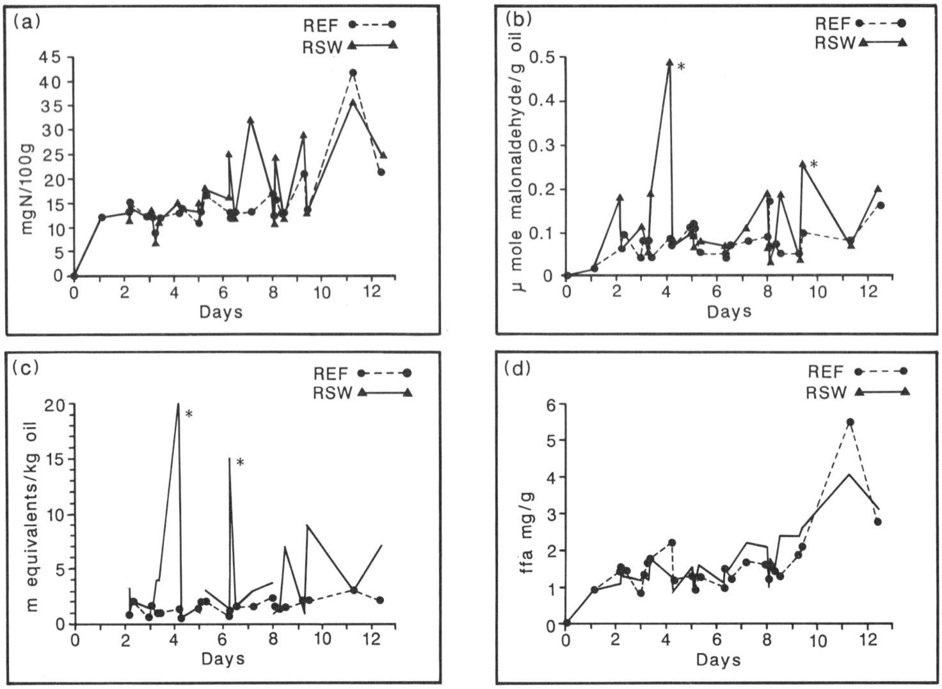

Fig. 6.10 Chemical analysis for herring for all purse seiners. (a) TVN in tank-handled herring with time. (b) TBA in tank-handled herring with time (*vessel 3, trip 4). (c) PV in herring with time (*vessel 3, trip 4). (d) FFA in tank-handled herring with time.

managing and operating RSW tank systems.

Comparison of samples of fish from the same catch, in which the samples were stored either in ice/water in the REF container, or in the RSW tank system, showed that the quick uniform cooling of the REF container invariably gave the best quality. Quick and uniform cooling is therefore of great importance in order to obtain optimum quality.

Further development of methods for quality determination is required, including sensory descriptors for tank handled fish, because the existing methods are too coarse to determine differences. The sensory methods currently used are based on box iced fish; storage in tanks affects the fish by washing out blood from the gills and descaling, which give fish a different physical appearance without changing the edible quality.

Mackerel

Measurements carried out in this project show that it is possible to land mackerel of good quality with the equipment and methods used today on board the Danish purse seiners. Modification of the bottom screen in such a way that it covers the total bottom area, together with a system for opening and closing the drains, will result in better utilisation of the cooling capacity and more uniform cooling, with further improvement of the quality of the landed mackerel.

Herring

The measurements show that the present permitted filling level of 80% fish per tank, is too high to give a uniform cooling of herring loads. Nielsen *et al.* (1986) recommended the loading of tanks with water and fish in the ratio from 1:3 to 1:4. The present work confirms that this ratio range is a more suitable guideline than the currently allowed 80% in terms of herring. Similar to mackerel, the present measurements show that the bottom of the tank has to be covered with a grating and the drains should be closed from loading the catch in the tank until the unloading.

RECOMMENDATIONS FOR THE IDEAL RSW TANK

Cooling water supply The cooled water must be distributed across the total bottom area.
Suction Suction of used cooling water has to be made from as wide a surface as possible. If the tank has a trunk, water must also be exhausted in areas outside the trunk.
Circulation system Pipes must be resistant to salt water and detergents and constructed from plastics or stainless steel. Furthermore, filters must be built-in and it ought to be possible to clean the filters by rinsing without first separating them.
Drains It is necessary that the drains should be able to be closed when the haul is loaded, and opened when the unloading of the tanks commences.
Tank surface A smooth surface is required to make the cleaning of the tank easier. Stainless steel is a suitable material, but a durable, scratch resistant, strong bonding, approved paint is suitable too. The tank lay-out has to be straight and with as few corners as possible, so that fish are not trapped and cleaning is easy.
Cleaning The tank must be self-discharging. Part or fully automatic cleaning of the tank and pipes etc. ought to be possible.
Cooling capacity The cooling capacity should be at least sufficient to satisfy the required cooling time.
Circulation rate The circulation rate must be variable, so that the rate is higher when cooling starts and reduces when the haul is cooled. It must be able to maintain a low circulation in the filled tanks and at the same time cool a new tank. The flow must be registered in each tank in order to note capacity.
Temperature measuring There must be at least one temperature sensor in each tank, which makes it possible to monitor the temperature change in the tanks and later confirm the cooling conditions. The sensor must be placed in the suction box, because this is the warmest place, for a correctly designed tank.
Recommended cooling The temperature must be below 3°C within 4 hours and 0°C after 16 hours.
Recommended temperature The temperature should be kept between −1.5° and 0°C to maintain good quality.
Recommended filling Mackerel – maximum 80%, herring – maximum 70%.

BIBLIOGRAPHY

Gibbard G.A. and Roach S.W. (1976) *Standards for an RSW System*. Technical Report 676, Fisheries and Marine Service, Vancouver.

Hansen P. (1981) *Behovet for hurtig isafkøling af fangsten.* (The need for quick ice cooling of the haul). Ministry of Fisheries, Technological Laboratory, Lyngby.

Huss, H.H. (1983) *Fersk fisk. Kvalitet og holdbarhed* (Fresh fish. Quality and quality changes). Ministry of Fisheries, Technological Laboratory, Lyngby.

Nielsen T., Holmsgaard N. & Christiansen H. (1986) *Teknisk beskrivelse af danske notfartøjers RSW kølesystemer* (Technical description of Danish purse seiners RSW systems). FF-Report 128, Ministry of Fisheries, Technological Laboratory, Hirtshals.

Kraus, L. (1987) *Fangstbehandling ved RSW-køling* (Haul treatment with RSW chilling). FF-Report 134, Ministry of Fisheries, Technological Laboratory, Hirtshals.

CHAPTER 7

Surimi from blue whiting (*Micromesistius poutassou*) produced on board m/s Uksnøy: process and quality

R. Ofstad, E. Grahl-Madsen and C. Solberg

INTRODUCTION

Blue whiting, a gadoid species caught at depths of 300–400 m, is probably the most abundant fish species in the north-east Atlantic. The estimated resource in 1988 was 4.3 million tonnes. The main catching proceeds west of the British Isles from February to April when the fish are in shoals and spawning. Generally in this period the fish are small. They have an average weight of 135 g and can be extremely soft. For the rest of the year the fish are bigger, with an average weight of 235 g, and are in a better condition, but their numbers are more sporadic. Smith *et al.* (1979) concluded from their experiments that blue whiting showed a greater degree of seasonal variability than other gadoid species, and that the overall acceptability dropped dramatically once the fish had spawned. The texture of the muscle was poor then, and the water content in the muscle could be as high as 92–93%. However, products formed from this fish, such as fish fingers, were still acceptable.

In 1987 the reported catches of blue whiting were 664 000 tonnes. Only a small portion was used for human consumption, either as frozen fillets or as mince. Blue whiting is an interesting resource for surimi production due to its low price and the abundant resources.

The Norwegian trawler m/s Uksnøy installed on board an Alfa-Laval surimi plant like the one described by Søbstad & Zetterling (1987) and started to produce blue whiting surimi in November 1987. The surimi quality has been fluctuating and the processing line has been modified several times. The purpose of this investigation was to test the quality of the surimi produced on board m/s Uksnøy as a function of the raw material, storage time, equipment and washing procedures, and to compare the quality from the full scale operation with the one attained by a laboratory scale process.

MATERIALS AND METHODS

Raw material

The raw material used for surimi production in these experiments was caught west of the British Isles, Porcupine-banks, during March 1989. The fish were mixed, both in-spawning and post-spawned. The water content of the fish flesh was 81–82%. Whole fish were stored in refrigerated sea water (RSW) from 5 to 35 hours before processing. The temperature in the fish-water mixture measured in the buffer tank before sorting was 6–7°C. Resolution of *rigor mortis* was observed after about 10 hours.

At each sampling three fillets were removed just before mincing and stored at −30°C for 3 months before performing analysis of: moisture content; pH (in a 1:1 mixture with 0.15 M KCl); protein by the Kjeldahl method (AOAC 1980) and total volatile nitrogen (TVN), trimethylamine (TMA) and trimethylamineoxide (TMA-O) by the microdiffusion method of Conway and Byrne (1933).

Surimi process methods

Blue whiting surimi was produced in a modified Alfa-Laval process as shown in Figure 7.1. Only fish of size 20–42 cm were used. After sorting, filleting (VMK 15/66) and skinning in a Baader 51, the minced flesh was prepared by passing through a Bibun mechanical meat separator with 3 mm drum orifices. The fish mince was mixed with water (ratio 1:4) in a Japanese type leaching tank. After about 10 minutes gentle agitation, more water was added to the mince-water mixture. It was then pumped into the first retention cell, where the fish/water ratio was 1:6. The temperature of the water was 8–10°C and the temperature in the fish mince-water mixture was 9–12°C. The fish

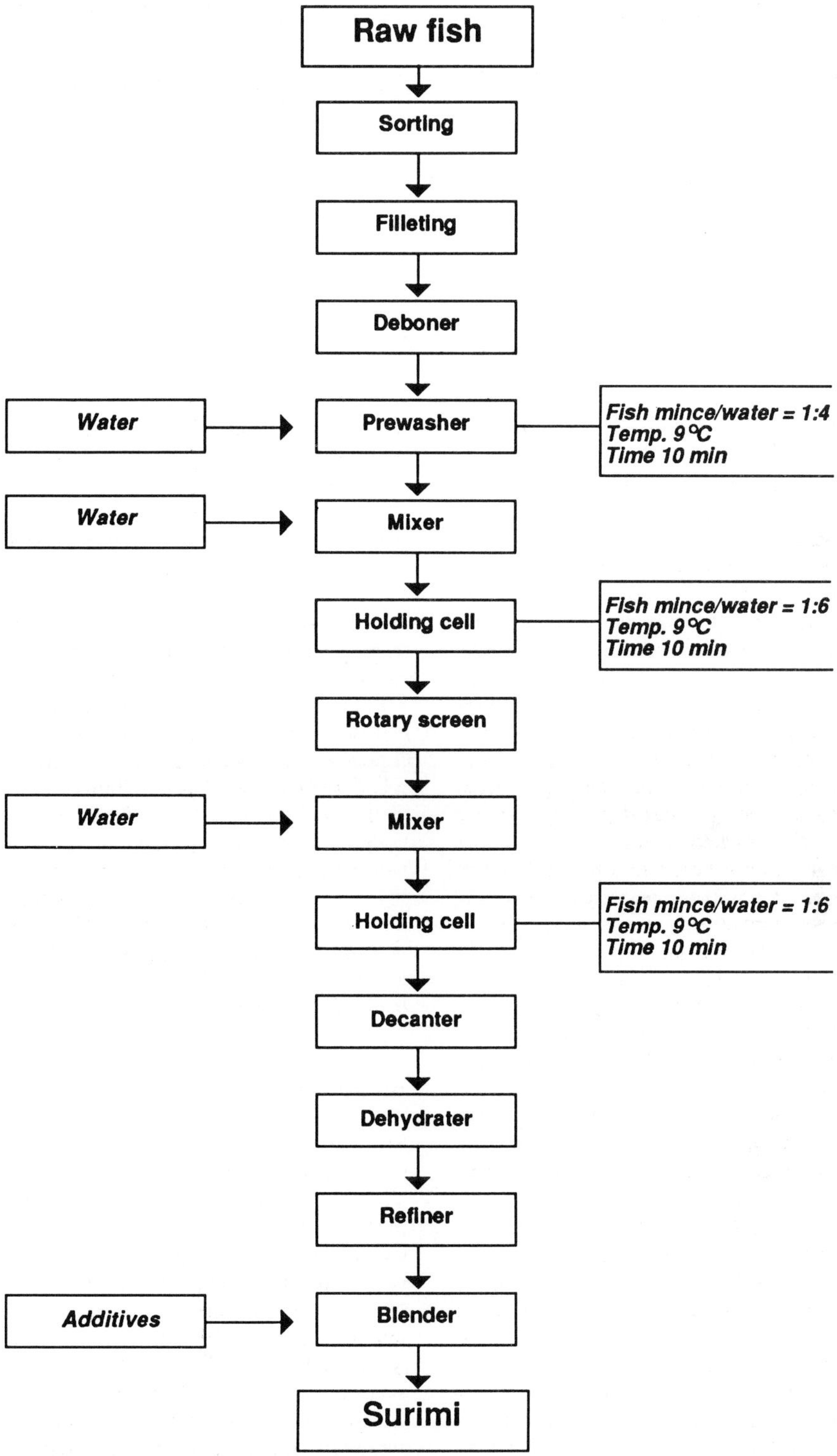

Fig. 7.1 Flow scheme for the manufacture of surimi on board m/s Uksnøy.

mince was washed twice and screened in a rotary sieve between the washing steps, before dewatering first by a decanter centrifuge and then by a twin screw press. After dewatering, the mince was passed through a Beehive refiner. The temperature in the surimi after addition of cryoprotectants was 14–16°C.

Surimis produced after about 5, 14, 19, 21, 31 and 34 hours' storage after catch were analysed. Samples of surimi were also collected after each step in the process, 14 and 20 hours after catch. In addition 7 samples were collected before and after the refiner. Fish mince collected before the screw press was dewatered in a similar way to that used in the laboratory method. These surimi samples were not strained.

Experiments W1, W2 and W3 were carried out in a full scale operation to test the effect of different washing procedures on the surimi quality. The raw material was stored for 8, 12 and 16 hours respectively. Three samples from each experiment were analysed. The leaching procedures were:

W1 The Alfa-Laval process with one retention cell, but modified with a prewasher. The washing time was about 20 minutes.
W2 The modified Alfa-Laval process, with prewasher and two retention cells as already described. The washing time was about 30 minutes.
W3 A Japanese-like process. The fish mince was mixed with 4 parts (w/w) of water and gently agitated for 10 minutes in the prewashing tank. The fish mince was screened in the rotary sieve after each of the three washing steps, dewatered in the twin screw press and strained in the Beehive refiner. The washing time was about 30 minutes.

Laboratory scale surimi was produced by mixing 3 kg of separated fish flesh with 12 litres of water. After gentle agitation for 5 minutes the water was decanted through a screen. The third and final washing was carried out in 0.15% NaCl. The fish mince was dewatered under light pressure for about 45 minutes. The temperature in the surimi after addition of cryoprotectants was 14–15°C.

Surimi quality

Laboratory analyses were performed on the surimis after 3 months' frozen storage at −30°C. The moisture and protein content and the pH of the surimi were determined as described above. The gelling properties were determined by either the punch test or both the punch test and the torsion test. Deformation (cm), breaking force (gf) and gel strength (gf × cm) were measured in the punch test using a 5 mm bulb (Alaska Fisheries Development Foundation 1987) in a KGS Systems Universal Testing Machine. Stress at failure (kPa), strain at failure and modulus of elasticity (E-module, kPa) were measured in the torsion failure tests as described by Lanier *et al.* (1985). Lightness (L*), blue/yellow colour (b*) and red/green colour (a*) were measured with the Minolta Chromameter CR100 on preheated surimi specimens.

Test specimens were prepared by mixing salt with 3 kg of sliced, tempered (−1°C) surimi, final NaCl concentration 3%, in a silent cutter (Dito Sama K55) until the temperature reached 5°C. The surimi paste was degassed twice before stuffing into sausages or steel casings. Tests were performed on both surimi with native moisture content and surimi with the moisture content adjusted to 78%. The native moisture content of the laboratory scale surimi was 79%. Surimi gels were cooked at three different temperatures: 30/90°C, 60/90°C and 90°C.

Scanning electron microscopy

Electron microscopy studies were performed on two of the surimi gel specimens. The samples were frozen in liquid freon, freeze-etched at −80°C and examined in a CRYO-SEM (Oxford Hexland CT1500) instrument.

Statistical analysis

One-way analysis of variance, ANOVA, was used to test the significance of the differences between different treatments. The statistical program Statpro was used.

RESULTS AND DISCUSSION

Storage

The chemical composition of the minced fillets and the surimis are given in Table 7.1. TVN and TMA increased after about 20 hours' storage at 6–7°C and the pH of the surimi decreased from 6.9 to 6.7. A corresponding increase in TMA and a drop in the overall quality of blue whiting caught in the spawning season (March 1979) and stored at 10–14°C is reported. Blue whiting from the same period stored either on ice or in chilled sea water (1°C) did not show any increase in TMA until 6–7 days after catch, and the sensory quality was evaluated as good until 5–9

Table 7.1 Total volatile nitrogen (TVN) and trimethylamine (TMA), in minced blue whiting fillets, and the chemical composition of surimis.

	Minced fillets		Surimi		
Storage time hours	TVN mg/100 g	TMA mg/100 g	pH %	Water %	Protein %
5	13.2	1.7	6.91	73.9	17.6
10	14.0	2.7	—	—	—
14	14.9	3.1	6.91	72.3	18.6
19	14.5	2.5	6.63	75.1	16.5
21	—	—	6.94	73.3	—
31	17.5	4.2	6.73	75.5	16.4
34	—	—	6.70	72.8	18.6

days after catch. Fish caught in February and April showed the same tendency, but the storability was better in February and poorer in April (Smith *et al.* 1979).

The surimi quality due to gelling properties (Fig. 7.2) and colour (Fig. 7.3) were highly dependent on the storage time. After resolution of *rigor mortis*, 10 hours after catch, the gel strength, breaking force and deformation increased to optimal values and decreased considerably after 20 hours' storage. The lightness (Fig. 7.3) was dependent on the moisture content and increased by adding water to the surimi. Maximum lightness ($L^* = 77.5$) was achieved about 20 hours after catch when the moisture content was adjusted to 78% in the surimi specimens. The yellowness (b^* value) increased from 6.0 to 9.5 with the storage time, while the a^* value seemed to be constant (approx. 3.5). The low L^* value was probably partly due to very small spots of skin and black belly lining.

The fillets were not sorted after deskinning and remainders of skin and black belly lining were homogenised within the fish flesh both in the mincer and in the decanter.

Even though some increase in the TVN and TMA contents and a decrease of the pH values of the surimis were observed, it does not seem likely that these relatively small changes could explain the rather big reduction of the gelling properties observed. Salt absorption during storage in RSW might also have a negative influence on surimi quality. Blue whiting has been reported to attain a salt concentration of 0.5–1.0% in the fish flesh after 24 hours' storage in RSW (Landfald 1980). Other experiments have shown (Keay 1985) that SA grade, i.e. the quality of pollack-surimi produced at sea, was achieved only when the fish was either stored on ice (5–6 days) or in chilled fresh water (4 days). When the fish was stored in RSW the surimi quality was less acceptable and comparable only to surimi produced in shore plants in Japan. However, Langmyhr and Sørensen (1987) have reported that even if the fish were stored in refrigerated sea water until two days after catch, the values of strain at failure (torsion test) were comparable to those of SA grade pollack-surimi. The Alaska Fisheries Development Foundation (1987) reports that the quality of surimi processed from RSW-stored Alaska pollack was the same as that from ordinary stored fish.

The maximum deformation of pregelled surimi specimens was 1.4 cm while maximum gel strength was 1950 gf × cm. The breaking force and also the gel strength were highly dependent on the moisture content; the gel strength was reduced to 1000–1200 gf × cm by adjusting the moisture content to 78%. Pregelling at low temperature, 30°C, increased both gel strength and deformation considerably. The average increase in deformation was 0.2 cm while the gel strength was more than doubled in the optimum period. Differences between the sample preheated at 60°C and the one heated directly at 90°C were not found.

CRYO-SEM micrographs of the surimi specimens confirmed the differences in surimi quality due to storage time. Surimi produced after 14 hours, which had high breaking force and deformation, showed a networklike structure, whereas that prepared from raw material which had been stored for 34 hours, and which had low breaking force and deformation, showed a less homogeneous, amorphous structure (Fig. 7.4).

Processing

The penetration values of gels prepared from surimi samples collected after each step in the process 14 and 20 hours after catch are given in Table 7.2. The deformability of the surimi prepared from 14 and 20 hours post-catch fish decreased from 1.59 to 1.38 cm and from 1.50 to 1.22 cm respectively on passing through the twin screw press. Even though only a small increase in the temperature (<1°C) was observed, these results may indicate a reduction of the deformability property caused by too strong a mechanical shear in the twin screw press. However, these results are based on only one sample and must therefore so far be considered as preliminary. The L^*, b^* and a^* values were all improved during each step of the surimi process (Table 7.2).

Although the temperature in the fish mince increased on average by 4°C, from 10–12°C to 14–16°C, while passing through the Beehive refiner,

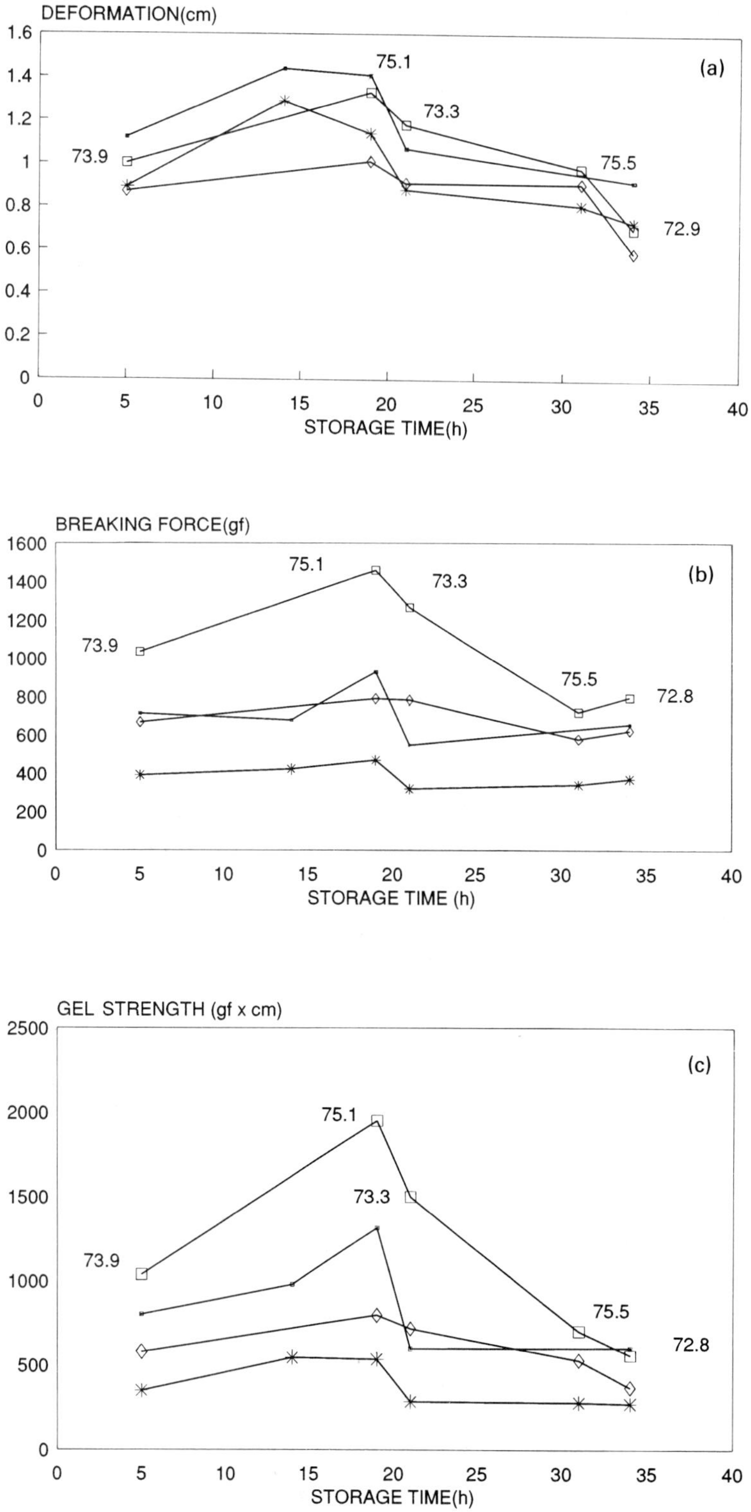

Fig. 7.2 (a) deformation, (b) breaking force and (c) gel strength of surimi made from raw material stored for 5–35 hours in RSW.

- ■ adjusted to 78% moisture content, heated at 30/90°C
- * adjusted to 78% moisture content, heated at 90°C
- □ at natural moisture contents given in figure, heated at 30/90°C
- ◊ at natural moisture contents given in figure, heated at 90°C

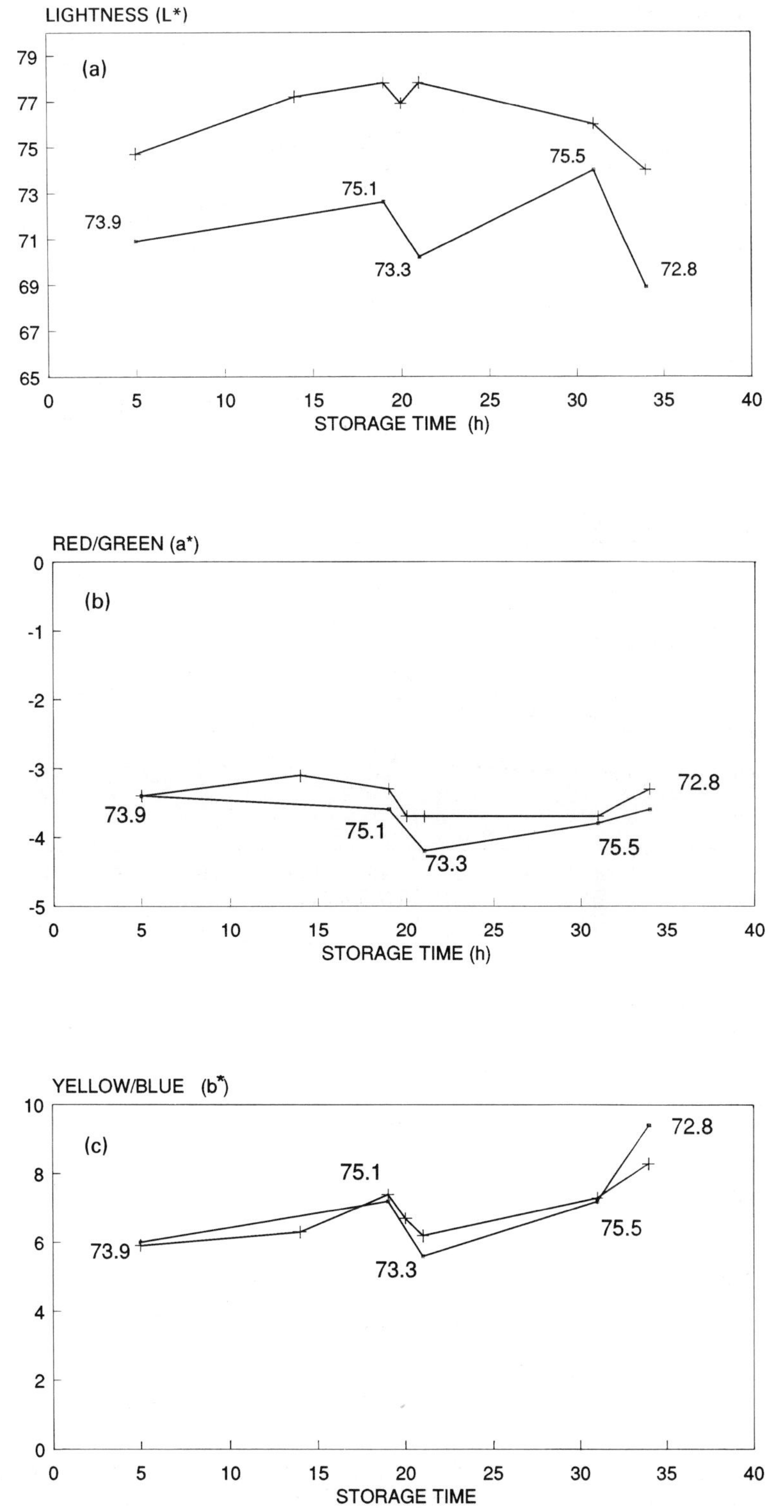

Fig. 7.3 (a) L*, (b) a* and (c) b* values of surimi made from raw material stored for 5–35 hours in RSW.

- ■ adjusted to 78% moisture content
- † at natural moisture contents given in figure.

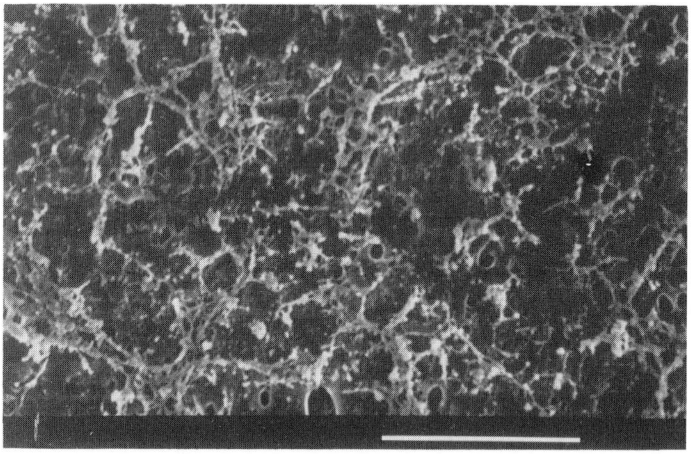

(a)

(b)

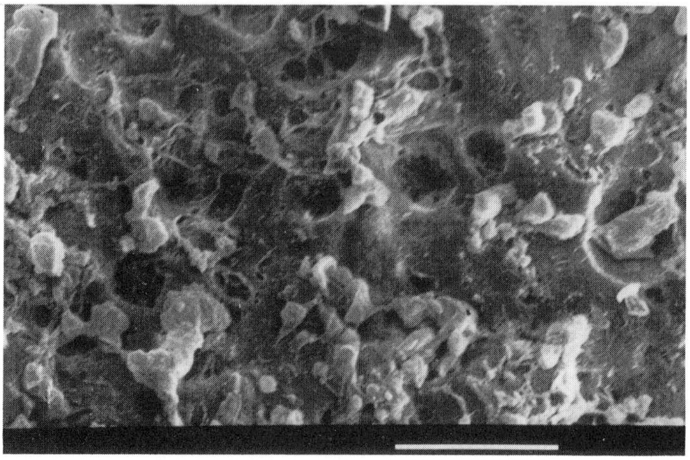

Fig. 7.4 CRYO-SEM micrographs of the surimi specimens manufactured from blue whiting stored for (a) 14 hours and (b) 35 hours. The bars indicate 10 μm.

Table 7.2 Punch test and colour measurements of surimi samples after each step in the process 14 and 20 hours after catch.

	Deformation (cm)			Breaking force (gf)			Gel strength (gf × cm)					
Component	30/90	60/90	90°C	30/90	60/90	90°C	30/90	60/90	90°C	L*	a*	b*
14 hours, moisture content 79%												
Rotary sieve	1.53	1.33	1.39	668	420	421	1025	562	587	75.3	−3.7	7.0
2. Holdcell	1.50	1.34	1.39	428	312	284	642	424	395	76.1	−3.5	6.5
Decanter	1.59	1.31	1.34	581	357	320	929	471	428	77.1	−3.0	6.1
Screw press	1.38	1.13	1.22	578	348	353	802	394	433	76.2	−3.4	6.5
Refiner	1.44	1.14	1.29	677	369	423	979	425	545	77.9	−3.3	5.6
20 hours, moisture content 81%												
Rotary sieve	1.31	1.04	1.11	591	317	311	789	330	346	75.3	−3.6	7.4
2. Holdcell	1.52	1.15	1.28	612	272	294	933	318	379	75.8	−3.4	6.5
Decanter	1.50	1.32	1.45	377	226	217	577	301	316	76.5	−3.2	6.4
Screw press	1.22	1.01	0.94	518	315	253	638	318	240	76.6	−3.2	6.4
Refiner	1.45	1.34	1.33	416	328	282	611	441	378	77.2	−3.1	6.3

Table 7.3 Punch test and colour measurements of surimi samples collected before (BR) and after refining (AR). The moisture content of the surimi specimens was adjusted to 78%.

	30/90°C			60/90°C		90°C		
	BR		AR	BR	AR	BR		AR
Replicates	7		6	7	7	7		7
Deformation (cm)	1.19 ± 0.04	*	1.26 ± 0.06	0.96	0.94	0.90 ± 0.05	***	1.07 ± 0.06
Breaking force (gf)	616 ± 34	n.s.	619 ± 49	393	396	366 ± 32	n.s.	379 ± 18
Gel strength (gf × cm)	743		834	386	381	354		394
Stiffness (gf/cm)	536		566	419	432	388		402
L*						74.5 ± 0.5	***	75.9 ± 0.4
a*						−3.5 ± 0.1	*	−3.4 ± 0.1
b*						6.6 ± 0.3	n.s.	6.7 ± 0.3

Mean ±95% confidence limits.
n.s. = not significant at 95% level.
* The difference before and after refiner was significant at 95% level.
*** The difference before and after refiner was significant at 99.9% level.

there was a moderate improvement (Table 7.3) in gelling properties following refiner separation of the surimi.

The deformation of the 90°C samples also increased significantly as a result of refining. The average increase of seven samples was from 0.90 to 1.07 cm. The improvement was less when preheated to 30°C, but still significant. The breaking force, the gel strength and the stiffness seemed all to be slightly improved. However, the difference between the breaking forces was not significant, and the other indices have not been tested statistically. The colour was clearly improved. The L* value increased significantly, from 74.5 to 75.9, and the a* value showed a small significant decrease. The b* value did not change. Unfortunately, differences in the gelling properties caused by different storage times of the raw material had to be disregarded in the statistical analyses.

The results, given in Table 7.4, indicate that the leaching of soluble materials was essentially complete within two cycles of the Alfa-Laval retention cell system. Both the gelling properties and the colour were improved by washing twice (W2) compared to once (W1). Washing three times (W3) in the Japanese leaching tank, however, did not appear to produce any increases. The experiments W1, W2 and W3 were carried out with raw material stored respectively 11, 16 and 8 hours, which should be within the optimum storage time. The differences between W1 and W2 can therefore not be explained by different storage times of the raw material.

The deformation values between W1 and W2 increased significantly, both when preheated (30/90°C) and when cooked directly at 90°C - on average from about 1.10 to 1.30 cm. The difference between W2 and W3 was not significant. However, the breaking force decreased considerably by washing twice compared to once. The difference between W1 and W2 was significant both for the 30/90°C and the 90°C samples. Whereas the difference was not significant between W2 and W3 for the 90°C sample, the breaking force of the W3 sample was significantly higher than that of the W2 for the preheated samples. This reduction of the breaking force due to increased number of washing cycles is difficult to explain. One explanation may be the enhanced washing time and then swelling of W2 and W3 compared to W1. The original moisture content of W1, W2 and W3 was 75.6%, 78.0% and 78.9% respectively.

The gel strength showed no significant differences, except between W2 and W3 and this was probably caused by a small reduction of both the deformation and the breaking force values. The stiffness decreased significantly between W1 and W2, and showed a small significant increase between W2 and W3 when the samples were preheated. Corroborating results were found by the torsion test. Stress at failure and E module were significantly different between W1 and W2, but not between W2 and W3. Whereas the breaking force was only moderately reduced, the stress at failure was almost halved by washing twice instead of once. The values of strain at failure were, unlike the deformation values, significantly different between W1 and W2.

Table 7.4 Comparisons of penetration, torsion and colour values given as the average of 2 or 3 replicates of surimi washed by three different methods, W1, W2 and W3. The moisture content of the surimi specimens was adjusted to 78%.

	30/90°C					60/90°C			90°C				
	W1		W2		W3	W1	W2	W3	W1		W2		W3
Replicates	2		3		3	2	2	2	2		3		2
Punch test													
Deformation (cm)	1.11 ± 0.06	***	1.34 ± 0.07	n.s.	1.39 ± 0.04	1.09	1.19	1.19	1.06 ± 0.03	***	1.29 ± 0.06	n.s.	1.22 ± 0.04
Breaking force (gf)	685 ± 55	*	570 ± 87	*	666 ± 58	569	458	473	532 ± 25	***	470 ± 27	n.s.	430 ± 51
Gel strength (gf × cm)	770 ± 97	n.s.	782 ± 161	n.s.	940 ± 105	624	544	567	621 ± 69	n.s.	592 ± 52	*	511 ± 45
Stiffness (gf/cm)	614 ± 21	***	420 ± 44	*	477 ± 30	521	387	395	479 ± 43	***	361 ± 18	n.s.	398 ± 49
Torsion test													
Strain at failure	2.43 ± 0.14	n.s.	2.56 ± 0.10	n.s.	2.56 ± 0.17	2.19	2.08	2.17	2.14 ± 0.12	n.s.	2.17 ± 0.09	n.s.	2.17 ± 0.09
Stress at failure (kPa)	121.4 ± 6.2	***	76.4 ± 11.5	n.s.	79.2 ± 9.4	90.9	47.9	50.4	90.2 ± 8.9	***	53.4 ± 10.3	n.s.	55.5 ± 2.5
E-module (kPa)	50.0 ± 1.6	***	29.8 ± 4.1	n.s.	30.6 ± 2.4	41.6	23.0	22.9	42.3 ± 5.9	***	24.7 ± 5.1	n.s.	25.6 ± 0.1
Colour test													
L*									74.9 ± 0.4	***	77.3 ± 0.6	n.s.	77.2 ± 0.4
a*									−3.9 ± 0.3	n.s.	−3.8 ± 0.3	n.s.	−3.7 ± 0.3
b*									7.9 ± 0.2	n.s.	7.7 ± 0.4	n.s.	8.0 ± 0.6

Mean ±95% confidence limits.
n.s. = not significant at 95% level.
* The difference was significant at 95% level.
*** The difference was significant at 99.9% level.

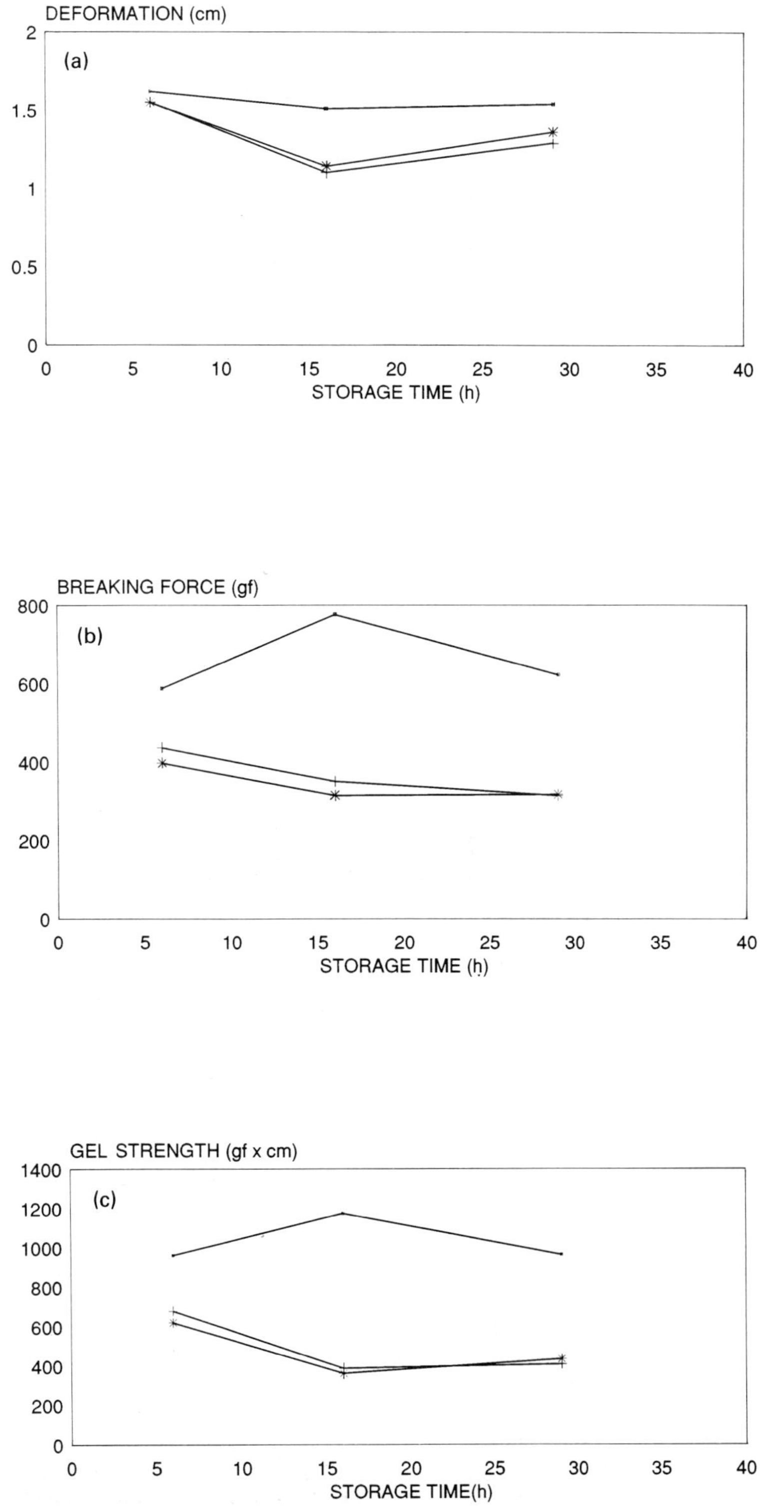

Fig. 7.5 Punch test values of surimi produced in laboratory scale from raw material stored from 5 to 30 hours in RSW (79% moisture content).

■ heated at 30/90°C
† heated at 60/90°C
* heated at 90°C

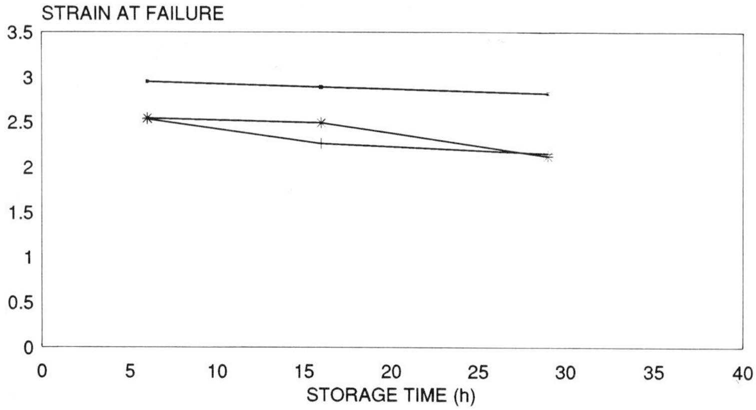

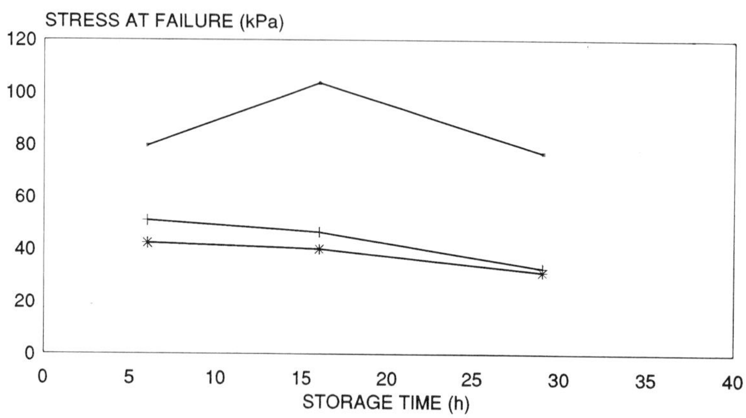

Fig. 7.6 Torsion values of surimi produced in laboratory scale from raw material stored from 5 to 30 hours in RSW (79% moisture content).

■ heated at 30/90°C
† heated at 60/90°C
* heated at 90°C

Minced blue whiting washed on a laboratory scale after 6, 16 and 30 hours' storage showed about the same tendency as material produced in the full scale process. The gelling properties were clearly reduced between 16 and 30 hours' storage of the raw material and the gelling properties increased dramatically by pregelling at 30°C. The results from the punch and torsion tests are given in Figures 7.5 and 7.6. The moisture content in the surimis was adjusted to 79%. Maximum deformation (1.6 cm) and strain at failure (2.95) were attained 6 hours after catch, while maximum breaking force (790 gf) and stress at failure (104 kPa) were attained after 16 hours' storage. In general the gelling properties of the surimi washed in laboratory scale were better than the properties of the ordinarily produced surimi.

These results corroborate experimental findings from previous works indicating that it is possible to produce surimi with quite good gelling properties even from spawning fishes, but the colour is too dark to attain first grade classification. Comparing the different surimi qualities measured on board m/s Uksnøy during the season November to May indicates that the surimi quality may be even better in January, when the fish are bigger and in good condition. The lightness could probably be improved by sorting out fillets contaminated with portions of skin and black belly lining.

A main problem seems to be the very limited storability, at least in RSW. The temperature in the RSW tanks on board m/s Uksnøy was higher than expected. In accordance with previous results it is possible that optimum gel strength could be achieved over a longer period at lower storage temperatures.

Washing of the blue whiting seemed to be complete

within two cycles, but the washing time should probably be reduced. The results may also indicate that the gelling properties of blue whiting are easily reduced by a too strong mechanical stress.

ACKNOWLEDGEMENT

The experiments of different washing procedures were performed by Peter Høiby of Alfa Laval, who has kindly permitted us to use his results.

BIBLIOGRAPHY

Alaska Fisheries Development Foundation (1987) *Surimi it's American now*. Project summary 1982–7, **III**, pp. 1–35.

AOAC (1980) *Official Methods of Analysis*, methods 2055–61. Association of Official Analytical Chemists, Washington.

Conway E.J. and Byrne A. (1933) An absorption apparatus for the microdetermination of certain volatile substances. I. The microdetermination of ammonia. *Biochem. J.* 27, 419–29.

Keay J.N. (1985) Surimi: The European perspective. In *Proceedings of the International Symposium on Engineered Seafood Including Surimi* (Ed. by R.E. Martin & R.L. Collette), pp. 619–38. National Fisheries Institute, Washington.

Landfald B. (1980) *RSW-føring av kolmule til konsum*. FTFI-nytt, no. 2, Institute of Fishery Technology Research, Tromsø.

Langmyhr E. & Sørensen N.K. (1987) *Surimi fra industrifisk, norske råstoff.* SSF-rapport H-79, Norwegian Herring Oil and Meal Industry Research Institute, Bergen.

Lanier T.C., Hamann D.D. & Wu M.C. (1985) *Developments of methods for quality and functionality assessment of surimi and minced fish to be used in gel-type food products*. N.C. State University, Raleigh.

Smith J.G.M., Hardy R., Thomson A.B., Young K.W. & Parsons E. (1979) Some observations on the ambient and chill storage of blue whiting (*Micromesistius poutassou*). In *Advances in Fish Science and Technology* (Ed. by J.J. Connell). Fishing News Books, Oxford.

Søbstad G. & Zetterling T. (1987) Recent technological advances in the surimi industry. In *Proceedings of Atlantic Canada Surimi Workshop* (Ed. by J.F. Roache), pp. 14–31. General Education Series 4, Canada.

CHAPTER 8

A techno-economic model for designing large factory vessels and its application to the Antarctic krill-finfish fishery

J.K. McElroy

INTRODUCTION

The potential development of Antarctic krill has attracted world-wide attention throughout the last two decades. Several hundred articles, popular and scientific, have been written. Few of these deal with the development of a krill fishery in economic terms, from harvesting through processing to marketing. Most notable among these are three FAO reports prepared by Eddie (1977), Grantham (1977), and Budzinski *et al.* (1985), and the various reports of McElroy (1980a,b, and c, 1982, 1983).

With the exception of McElroy (1983), all the systems considered by these authors have been relatively simple single vessel, single or two product systems, using estimates of average catch rates, throughput rates, yields, costs, etc.

Krill is unusual, however, in undergoing rapid autolysis after capture, allowing only about 4 hours post-harvest for processing into food products. Hence, frequent trawls must be made and the vessel's processing capacity must be large. Furthermore this fishery is generally considered to be seasonal, so there is the question of compatibility of vessel, catching, processing, freezing and storage systems with potential 'alternate season' fisheries.

This chapter has two aims:

(1) to present a computer model that selects preferred combinations of processes and capacities under fluctuating conditions of supply;
(2) to present the results of this model as applied to a large factory trawler whether fishing for krill year-round or in a two-season, krill-finfish fishery.

The original cost and price data applied to 1979. These results are presented and a brief re-examination with 1989 data is made.

A non-linear optimisation model, which incorporates a simple queuing simulation routine, is used to maximise the net revenue produced from the range of processing options tested. Sensitivity analysis is used to monitor the robustness of the optimum solution, given a range of technical and economic conditions.

The model is given six krill products and, theoretically at least, four fish products to select from. The products are:

(1) *For krill* Whole krill cooked and uncooked, krill tails, krill surimi, krill mince and krill meal,
(2) *For finfish* Whole gutted head-off fish, fish fillets, fish mince and fish meal.

Flow charts of the main krill processes are given in Appendix 1 at the end of this chapter. Details of the yields, capacities, space and quality requirements, together with the cost and revenue functions for each process, are also given, together with a selection of 'typical' catch distributions for the krill fishery (Appendix 2 of this chapter). This work is based on the author's PhD thesis (McElroy 1983).

Problem formulation

The basic problem is to maximise net revenue from processing for a given krill catch distribution (taken here as deterministic) over a fixed number of periods of one hour each. Different products can be produced from this raw material. In addition, each product can be produced from a number of different processes, each of which has a separate cost and revenue function.

The objective function of this multiple period problem is of the following general form:

Maximise z

$$= \sum_{i=1}^{i=q}\left[\sum_{j=1}^{j=p}\left[\sum_{t=1}^{t=n}[a_{ij}P_{ijt} - c_{ij}P_{ijt}] - FC_{ij}\right]\right] - K \quad (1)$$

where

- a_{ij} = yield factor = price per unit for product i produced by process j
- c_{ij} = yield factor = cost per unit for product i produced by process j
- P_{ijt} = amount of material used in the manufacture of product i produced by process j in period t
- FC_{ij} = fixed cost of producing product i by process j over the total period (a function of capacity)
- K = a function of the cost of fishing over the whole time period
- n = the number of time periods
- i = refers to the different products produced which may be represented by the following: whole, uncooked krill (WU), whole cooked krill (WC), mince (including surimi) (M), tailmeats (L) and meal (MC).
- j = refers to the different processes which may be used to produce any of the above products.

The specification of this problem may be simplified somewhat. We may assume that only one process is used to produce a given product at any point in time. Also, the 'cost of fishing' function (K) does not affect the determination of the optimum processing configuration. Consequently it may be left out of that narrower specification of the objective function. Needless to say, it will still figure in determining the overall profitability of the system. Also, we may reduce the specification of the objective function to a single period problem by handling the allocation of material to the different processes in each period within a separate simulation routine (see below). Continuing with this multiple period specification of the objective function, we now have:

$$\text{Maximise } z = \sum_{i=1}^{i=q}\left[\sum_{t=1}^{t=n}[a_iP_{it} - c_iP_{it}] - FC_i\right] \quad (2)$$

subject to the following constraints on capacity, space and quality of material.

Capacity constraints

Capacity constraints on each process

$$P_{WUt} \leq WU \quad (3)$$

$$P_{WCt} \leq WC \quad (4)$$

$$P_{Mt} \leq M \quad (5)$$

$$P_{Lt} \leq L \quad (6)$$

$$P_{MCt} \leq MC \quad (7)$$

where WU, WC, M, L and MC refer to the maximum capacity of each process. (In practice we would now need to introduce a separate term for surimi. For simplicity, it is left out here.)

Overall capacity constraint

$$H_t \leq C_t \quad (8)$$

where

$$H_t = P_{WUt} + P_{WCt} + P_{Mt} + P_{Lt} + P_{MCt} \quad (9)$$

C_t = amount of material available to be processed in period t

which is given by

$$C_t = B_{t-1} + X_t \quad (10)$$

where

- X_t = amount of material delivered on board in period t
- B_{t-1} = amount of material held in buffer store from the end of the last period.

Space constraint

$$S - (S_{WU} . WU) - (S_{WC} . WC) - (S_M . M) - (S_L . L) - (S_{MC} . MC) \geq 0 \quad (11)$$

where

- S = total space available
- S_{WU} = functional relationship of the unit space requirement for different capacities of a whole, uncooked krill processing line. In the base case this is assumed to be a constant.

Similarly for S_{WC}, S_M, S_L and S_{MC}.

Quality constraint

This constraint refers to the length of time after capture during which krill is still suitable for processing. As unprocessed material can be stored from one period to the next, constraints are needed to link one period to the next. However, the amount of material that can be held at any time is limited also by the storage capacity available. This situation can be handled with the following constraints. By way of example, if we assume that material no more than four hours old can be processed, it is necessary to incorporate the following storage constraints to link two consecutive periods.

$$B_t = E_{t-3} + E_{t-2} + E_{t-1} + E_t \quad (12)$$

and

$$B_{t+1} = E_{t-2} + E_{t-1} + E_t + E_{t+1} \quad (13)$$

where

B_t = the amount of material held in buffer storage in period t

E_t = the amount of material of age t held in buffer storage.

Thus when moving from period t to period (t + 1) an amount of material equal to E_{t-3} will be discarded. One such constraint, therefore, is required in each period.

Buffer storage constraint

The amount of material held in buffer storage in period t is given by:

$$B_t = B_{t-1} + X_t - H_t \quad (14)$$

where the overall storage constraint is given by

$$B_t = [O, B_{max}] \quad (15)$$

B_{max} = capacity of buffer store.

If $\quad B_{t-1} + X_t - H_t > B_{max} \quad (16)$

the excess material must be discarded.

As this formulation stands we have an LP problem which will ensure that the maximum economic use is made of the material available in all periods. However, the age of material processed (and hence its yield or value) was seen as being of importance to decisions on capacity. In order to be able to analyse this relationship, a separate simulation routine was written to handle the allocation of material to the different processes on the basis of certain decision rules. This routine requires information on the space requirement per unit input per period for the five different processes, on the capacity of the buffer store and on the distribution of catch over time.

Feeding this information into the objective function, the resulting problem is then of much lower dimension than the original LP (because the original multiple period problem has in effect been reduced to a single period problem). In this reduced form, the objective function is subject to only one main constraint, namely space (equation 11).

Figure 8.1 represents a flow diagram illustrating different processing routes that material may flow through on board a factory trawler fishing for krill. In the model, the food processing capacities (Y_1 to Y_4) and the freezing capacities (F_1 and F_2) are constrained

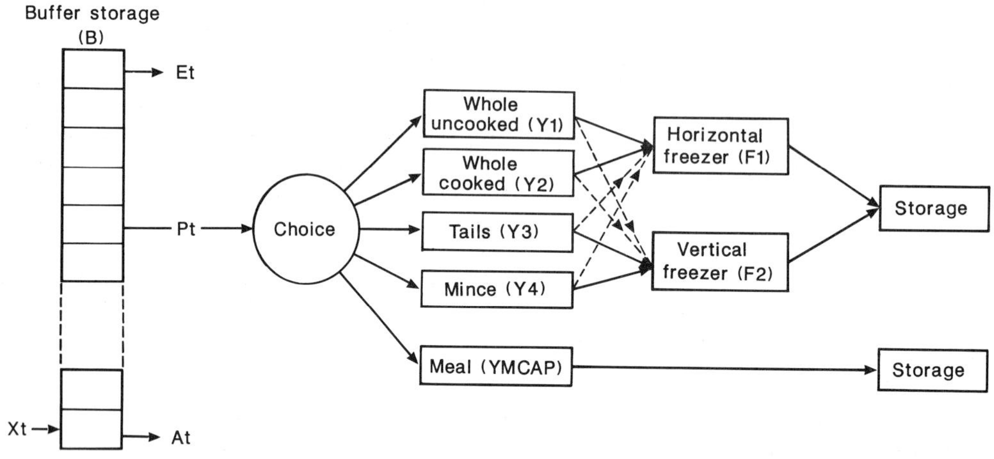

Key: Xt – catch in period t allocated to buffer storage.
Pt – amount removed from buffer storage for processing in period t.
Yn, Fm – capacities of food and freezing processes respectively (model variables).
B, YMCAP – capacities of buffer storage and meal processing respectively (model parameters).
At – amount discarded as being too old for processing in period t.
Et – amount discarded as being in excess of buffer storage capacity in period t.

Fig. 8.1 A schematic diagram of the source, sinks and possible routes for the flow of krill on board a factory trawler.

by the area of the factory deck, and so represent variables whose values are determined (optimised) internally by the model (endogenous variables). Freezing capacity is determined simply by matching it with the combined capacities of the food processes feeding material to it. Thus no attempt is made here to optimise freezing capacity to cover both the krill and fish fishing seasons.

The exogenous variables are the buffer store and meal processing capacities. Technically, the former is part of the factory deck while the latter is normally situated beside the engine room *below* the factory deck. In this treatment the buffer store was fixed by inspection at 40 t (see Results page 100), while the capacity of the meal plant could be varied independently for each run of the model. In this way the direct effect of changes in processing capacity on the age distribution of material processed could be monitored. In addition, the following decision rules were adopted to monitor the flow of material through the system.

Krill of up to 4 hours and 18 hours old may be processed for food and meal respectively. Krill is offered to processes Y_1 to Y_4 in turn. Only krill which is between 4 and 18 hours old, or which is excess to the food processing capacity of the factory deck in the next three periods, is available to be processed to meal. Only whole krill is processed to meal. Krill which is either excess to buffer storage capacity or too old to process, is discarded. Finally a decision rule is required to determine the order in which material is used up. In this application, emphasis is put on processing the maximum amount of material, subject to constraints on age and capacity, rather than processing the freshest material first. Consequently the principle adopted is first in, first out (FIFO), i.e. the oldest material is processed (or discarded) first.

Thus far we have considered the basic formulation of the model in linear terms. However, such restriction is not necessary as this problem was solved by using a NAG library routine (E04UAF) for constrained non-linear optimisation problems. Thus it is possible (and would be more realistic) to study the effect of non-linear functions relating capital cost, operating cost, revenue and/or space to processing capacity. But as the data from which such relationships are built was, and still is, either not available to the author or does not exist at present, no use was made of this facility in this study. The approach adopted was to design and cost individual processing lines to handle reasonably large (about 1.0 t/h) throughputs for each process and to use these figures as fixed per unit of capacity. Such an approach at least is reasonable and follows generally established West European practice where factory trawlers install several processing lines of modest capacity to handle the catch, rather than rely on a single, large-capacity line.

RESULTS

For a given process, the amount of revenue earned depends principally on how much of the catch is processed. There are five main factors which affect this, two of which concern the catch distribution itself. The five are:

- the average amount of the catch
- the degree of variance in the catch rate over time
- processing capacity
- quality constraints which determine the yield with age (here assumed fixed) and the upper age limit of material suitable for processing
- buffer storage capacity

General results that flow from the catch distributions used

Mean catch rate and the effect of variance

In principle we expect that the quantity of material processed will increase in relation to the catch rate. The relationship is linear for uniform rates of catch.

Generally, however, the catch rate is not uniform but highly variable from one period to the next. Thus for a given capacity, the greater the variance in the catch rate, the greater tends to be the shortfall between the actual quantity processed and that predicted on the basis of a uniform catch rate. Furthermore, this shortfall rises steeply as the average catch rate approaches the processing capacity (Fig. 8.2). For example, taking a value of 0.65 for the coefficient of variation in the catch rate, simple calculations could overestimate the amount processed by an average of about 28% (equal to 1 t/h with a processing capacity of 3.6 t/h).

Catch rate and processing capacity

Figure 8.3 and Table 8.1 illustrate that, for a given catch rate, an increase in capacity tends to result in a less than proportional increase in the quantity of material processed. Indeed three distinct phases in this relationship can be identified:

(1) Between 0 and 75% of the mean catch rate, increases in processing capacity result in an approximately proportional (i.e. near linear) increase in the amount of material processed.
(2) Between 75 and 150% of the mean catch rate, the

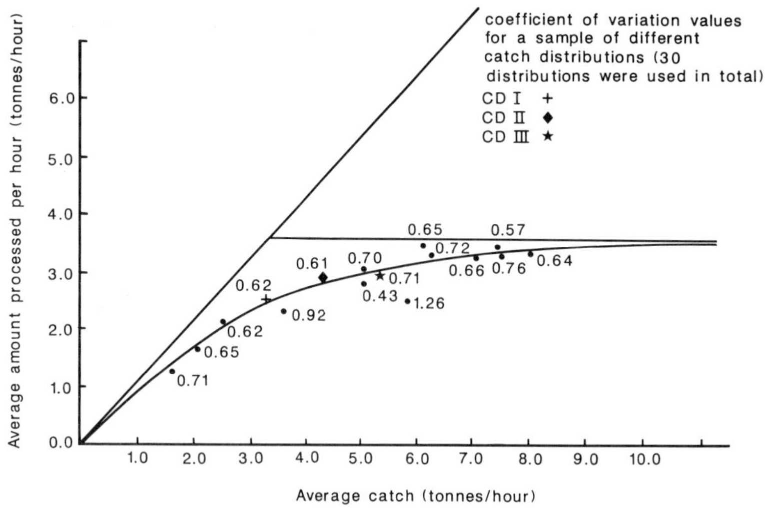

Fig. 8.2 Amount processed versus catch rate for a 3.6 t/h capacity food processing factory deck.

Table 8.1 Proportion of the catch processed at different capacity levels relative to the mean catch rate for catch distributions I, II and III.

No.	Catch distribution description	Mean catch rate t/h ≡ (t/24 h)	Relative capacity level 75%	100%	150%
I	average	3.16 (75)	57	73	87
II	above average	4.17 (100)	61	76	92
III	high	5.20 (125)	63	76	94

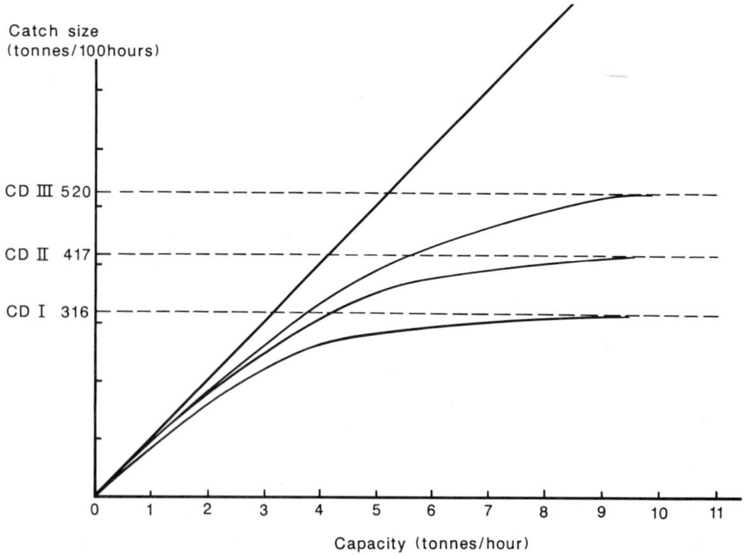

Fig. 8.3 The effect of processing capacity on the proportion of the catch that is processed within four hours of capture for three different levels of catch (model result).

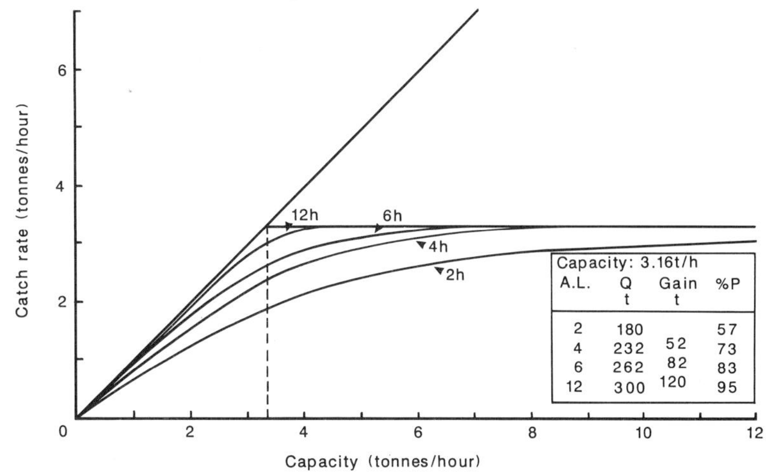

Fig. 8.4 The effect of variation in the age limit of material for processing on the proportion of the catch that is processed at different capacity levels for catch distribution I (316 t/100 h).

rate of increase in the material processed falls relatively quickly to a comparatively low level (i.e. from >70% to <30%).

(3) For increases in capacity above this point, the rate of increase in material processed continues to fall towards zero, which is reached here at capacity levels some 2 to 3 times the mean catch rate (i.e. 9 to 10 t/h).

Assuming scale economies apply to the process, we might reasonably expect that the optimum capacity for a factory vessel will occur somewhere within this second phase. The main point is that while the optimum capacity level will depend on specific cost and revenue functions, it may equally be *below* as above the level of the mean catch rate! (In fact anywhere between 0 and 10 t/h for the average catch distribution.) This contrasts with the extreme case when the catch rate is uniform. Then there are only two possible values for the optimum capacity, 0 and a value equal to the mean catch rate.

Processing capacity and quality

Figure 8.4 illustrates the dramatic effect that a quality limitation (measured as a maximum time limit) has upon the proportion of the catch which can be processed at different levels of capacity. Clearly the shorter the time limit, the smaller the proportion of the catch processed for a given level of capacity. For CDI (3.16 t/h), with an age limit of 2 h, only 57% of the catch is processed at the catch-equivalent processing capacity while at the age limit accepted here for food products (4 h) this rises to nearly 75%. Raising the limit to 12 h (e.g. for meal) increases the proportion processed to an average of 95% of the catch (Table 8.2).

Table 8.2 The effect of variations in quality standards (measured in terms of age) upon the amount of the catch processed at the catch-equivalent capacity level (3.16 t/h).

Age limit (h)	Quantity processed (t/100 h)	Percentage processed %
2	180	57
4	232	73
6	262	83
12	300	95

It is of particular importance to know how the capacity required to process a given proportion of the catch changes with variations in the quality constraint. For instance, quality limits on food products might conceivably range, depending upon the method of buffer storage, from 2 to 6 hours. Over this range the capacity required to process, say, 75% of the catch would double – from 2.5 to 5 t/h. This example serves to illustrate how significant quality limitations are to the economic viability of a foodfish factory trawler fishing krill.

Catch rate and quality

The general effects have already been noted. Of particular note here though is the observation that as the age limit is lowered (say towards 2 hours) there will be considerable pressure to take shorter, faster hauls and/or to use two nets in an attempt to keep the processing lines continuously supplied with raw material.

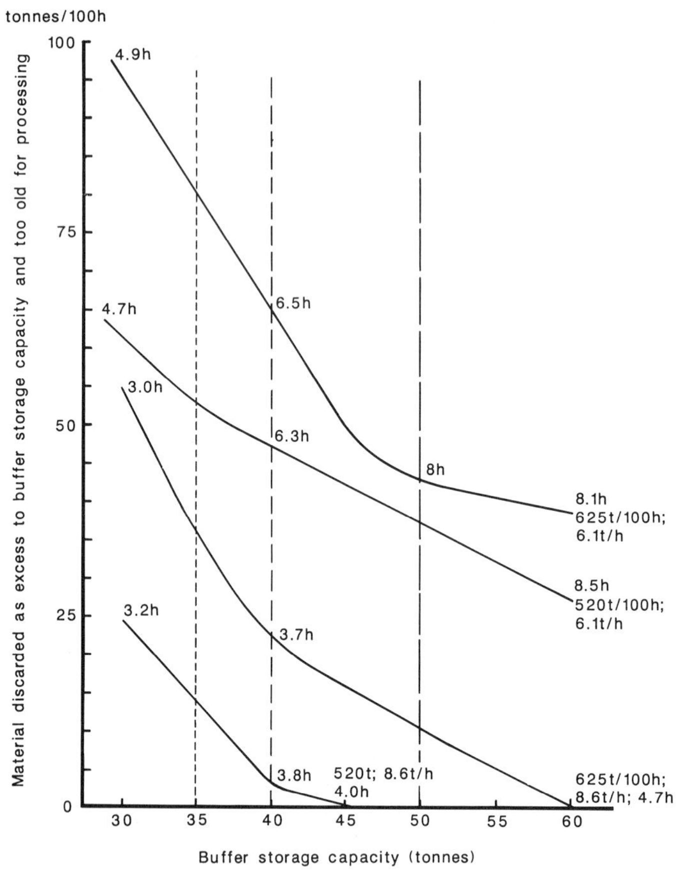

Fig. 8.5 The influence of buffer storage capacity on the amount of material discarded for two processing capacities (6.1 and 8.6 t/h) at high and very high catch rates (5.2 and 6.25 t/h). (Values for the average age of material processed at different storage capacities are indicated.)

Catch rate and its variance, processing capacity, quality and buffer storage capacity

The buffer store acts as a reservoir for whole catches or parts of catches that cannot be processed immediately. It is possible to set an upper limit on buffer storage capacity which depends upon the specific quality constraint and processing capacity that apply, such that

$$\underset{(t)}{\text{Maximum buffer storage capacity}} = \underset{(t/h)}{\text{processing capacity}} \times \underset{(h)}{\text{age limit}}$$

However, this approach is too simplistic, for it fails to take account of the fact that space is limited. And while it is possible to optimise for buffer storage, food and meal processing capacities for a vessel with a fixed factory area, in this application buffer storage capacity was fixed at an appropriate level which from trial and error seemed to satisfy reasonably well the different considerations contingent upon it: namely the desire to keep to a minimum the amount of material discarded, particularly for the highest catch distribution, while attempting to keep the average age of material processed to meal at a reasonably low level. As Figure 8.5 demonstrates, an appropriate buffer storage capacity depends on the catch distribution and processing capacity. In most circumstances a capacity of 40 t was found to be appropriate. This value was adopted.

So far we have treated processing capacity as one entity. However, it is split between food and meal processing. Significantly, the quality constraints that apply to these two product categories differ substantially. Thus material which is too old for food processing (limit 4 hours) may still be used for the production of meal (limit 18 hours). There is an important consequence which follows from this difference, which principally concerns the level of variation in the catch rate.

For any given catch rate, the greater the variance the more the distribution of material is skewed in favour of meal production and away from food production. This reduces the amount of revenue produced –

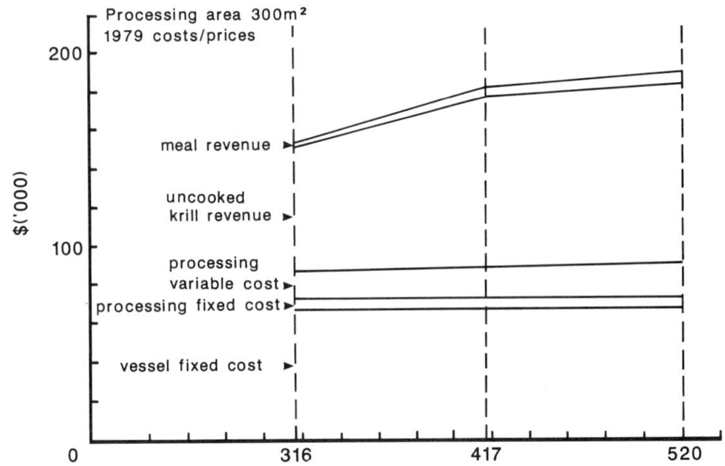

Fig. 8.6 The economics of producing raw frozen whole krill and meal.

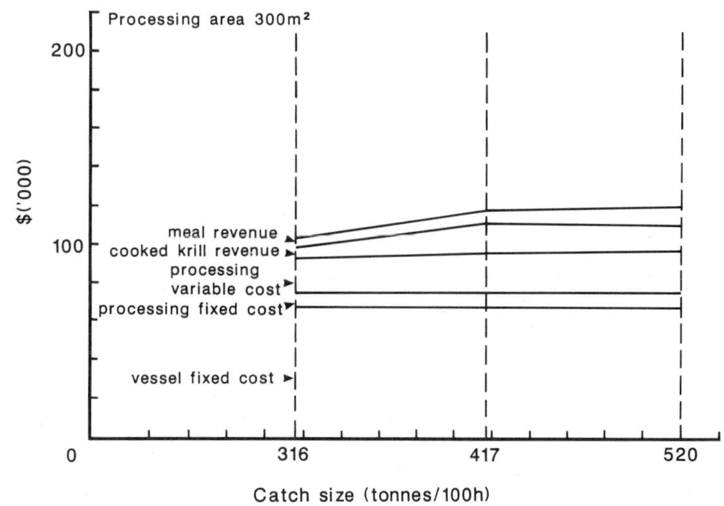

Fig. 8.7 The economics of producing cooked frozen whole krill and meal.

even if the total amount of material processed in each case remains the same.

ECONOMICS OF FACTORY TRAWLER OPERATIONS PRODUCING FOOD PRODUCTS AND MEAL FROM KRILL (1979 COSTS AND PRICES IN US$)

In considering the results that follow, no limitation was imposed on the proportion of the catch that would be suitable for processing into a particular food product. Such considerations, together with those of market size, might suggest the adoption of more than one food processing technology on board the factory trawler. However, only one food process is generally selected by the model.

Figures 8.6 to 8.10 detail the baseline results obtained for raw and cooked whole krill, krill surimi, krill mince and tail meats. With respect to the krill-finfish fishery, i.e. assuming a total processing and freezing area for krill alone of 300 m^2, all products, except tail meats and krill mince, can be produced economically at or above the 'average' krill catch rate of 75 t/d. Within the range considered (75–125 t/d), while krill mince becomes economic to produce slightly above the level of 75 t/d, tail meats never do. With the whole 500 m^2 of factory deck given over to krill products,

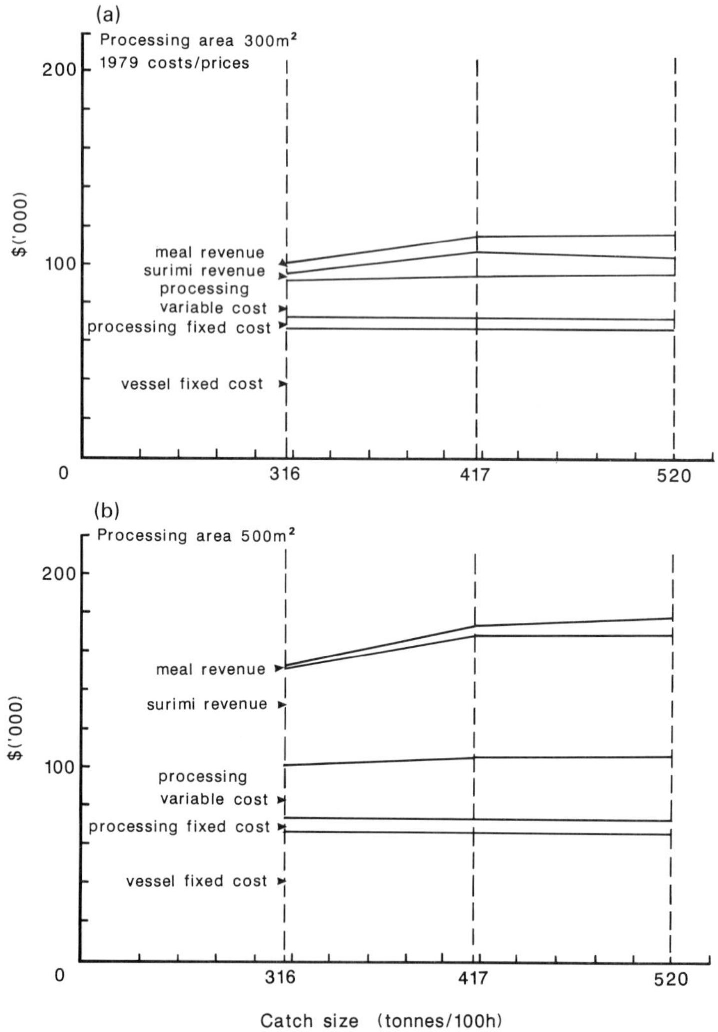

Fig. 8.8 The economics of producing frozen surimi and meal. (a) Processing area 300 m^2. (b) Processing area 500 m^2.

each process is more economically attractive. However, krill tail meats still fail to produce a profit.

The 500 m^2 'model' is equivalent to either:

(1) a 'year-round' krill fishery,
(2) a compatible krill-finfish processing technology (i.e. effectively only surimi and mince), or
(3) a processing deck equipment 'refit' at the end of each season.

Raw krill

300 m^2 Raw krill provides the largest net revenue of between \$16 000 and \$24 000 per day (representing a 75–108 per cent profit rate). With a capacity of 3.75 t/h for raw krill, the revenue from krill meal is small even at high rates of catch.

500 m^2 Processing (centrifuge) and freezing capacity (vertical plate freezer) rises by two thirds (ratio 5:3) to 6.25 t/h and 5.0 t/h respectively. Net revenue rises proportionately less steeply at the 'average' catch rate (to \$21 400/d, +34%) but equally steeply at the 'high' catch rate (to \$40 410/d, +68%). This represents an overall profit rate of between 98–171%. Only 5–15% of the catch is processed to meal.

Cooked krill

300 m^2 Next in order of profitability is cooked krill (capacity 2.4 t/h) which produces a modest net revenue, including the income from meal, of \$2000 to \$5500 per day (equivalent to an 8–18% profit rate). Despite nearly three fifths of the catch being pro-

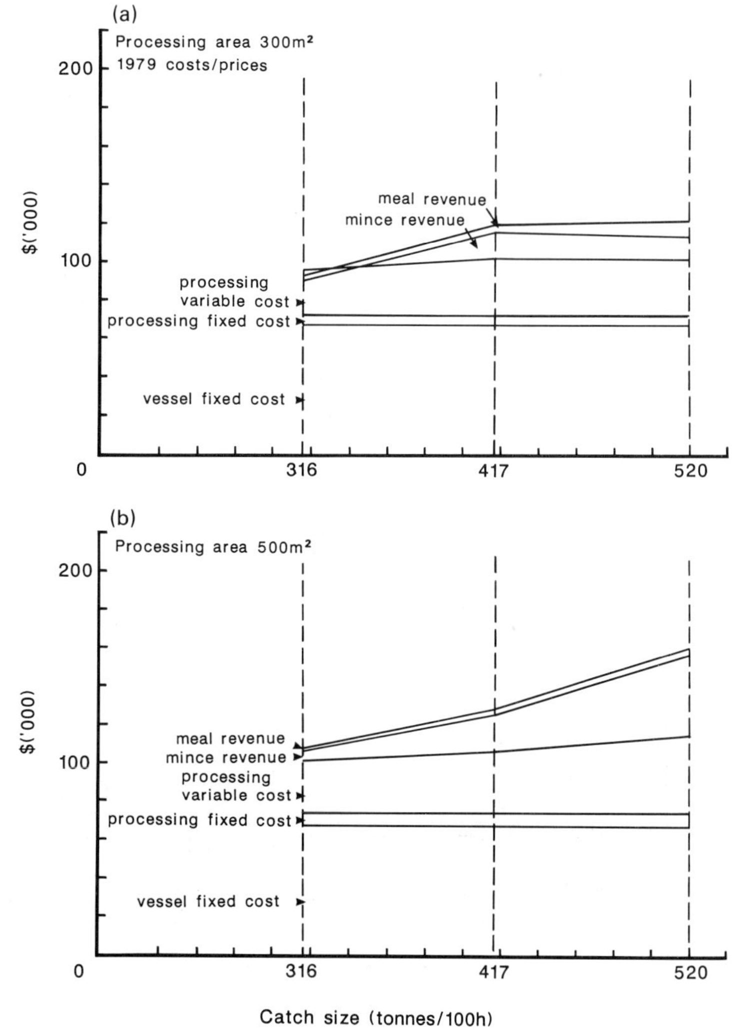

Fig. 8.9 The economics of producing frozen krill mince and meal. (a) Processing area 300 m². (b) Processing area 500 m².

cessed to meal at the 'high' catch rate, the revenue from meal remains comparatively small.

500 m² If all 500 m² is used, cooked krill processing capacity rises to 4.0 t/h (and freezing capacity (VPF) to 3.6 t/h), absorbing three quarters of the 'average' catch, and net revenue jumps to between $9200 to $17 900 per day (equivalent to a 36–66% profit rate).

Both the above processes (incorporating centrifuges or cooking baths) are unique to krill processing. With fixed costs rising by between 0 (200 day fishing) and roughly 22% (150 day fishing), both of these product systems would still be viable in a year-round krill fishery (i.e. utilising all 500 m² processing capacity).

Devoting the full 500 m² factory deck area to a single process in a two-season, krill-finfish fishery would be possible if the vessel undertook a refit of its processing deck (say 10–14 days) between fishing seasons. An increase in the profit rate in moving up from 300 m² to 500 m² of between 5–8% in each fishery (equivalent to say $220 000 to $350 000/year) should be sufficient to cover the net costs and lost revenues of the bi-annual refit.

Krill surimi

300 and 500 m² Surimi production appears to be a marginal activity at and even above the assumed 'average' catch rate when only part (1.85 t/h out of 3.09 t/h) of the available processing area is given over to it. However, as this process is applicable to many species of fish, it is reasonable to consider that the whole food processing area may be devoted to its

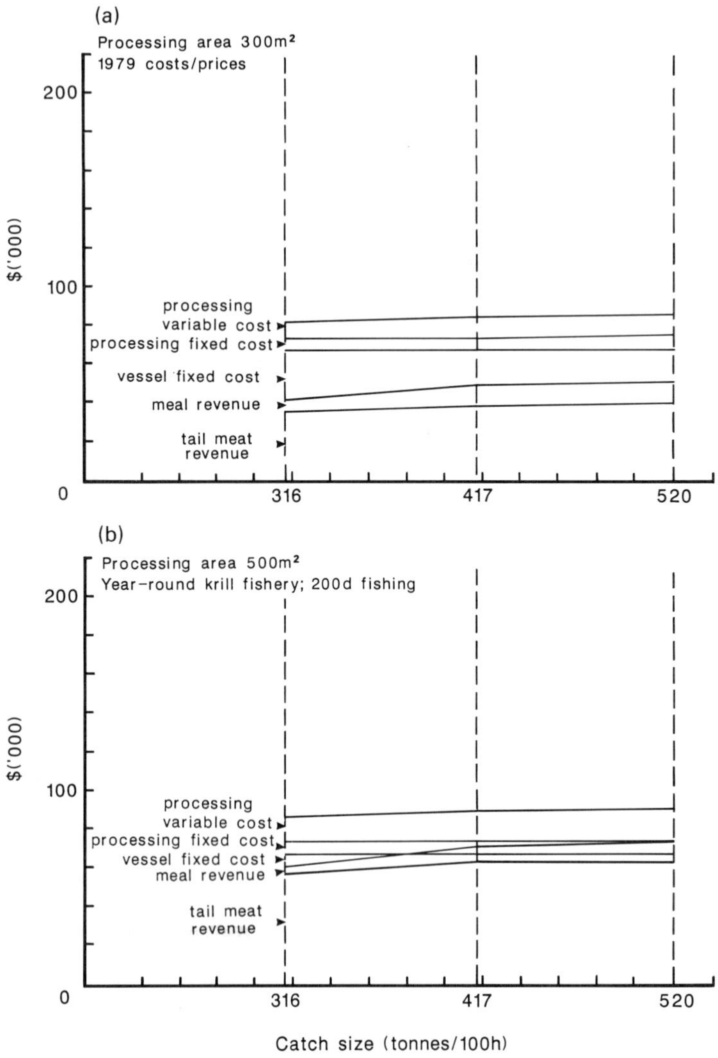

Fig. 8.10 The economics of producing tail meats and meal. (a) Processing area 300 m^2. (b) Processing area 500 m^2.

production (thus avoiding the need of a bi-annual refit). In this case, the krill fishery becomes a considerably more attractive proposition, yielding net revenue values of \$12 500 to \$17 000 per day (a 52–67% profit rate). This clearly demonstrates that, even if only part of the processing deck is permanently assigned to this process, it should be profitable in both the krill and finfish fisheries.

Krill mince

300 and 500 m^2 Raw krill mince is simpler to process than surimi, and has a correspondingly lower value. It is a marginally economic process at the 'average' catch rate, even when the whole processing area is given over to it (\$1400/d, or 5.7%). It becomes more profitable at the 'high' catch rate largely because, with its greater processing capacity (6.25 t/h), it can handle a greater share of the catch coming on board. At the 'high' catch rate, net profit rises to \$10 750/d (a 39% profit rate).

Tail meats

300 and 500 m^2 The production of tail meats at the rates of throughput and yield currently pertaining does not appear to be justified, even when the vessel is operated in the krill fishery for the whole year at the highest possible utilisation (200 day fishing). The picture is even less encouraging with only part of the

Table 8.3 Key input costs and average prices in the krill model.

Unit prices (ex-vessel)	1979 US$/tonne	1989 US$/tonne	% change
Whole uncooked krill	750	850	13
Whole cooked krill	600	700	17
Surimi – krill	1100	1100	0
– Alaskan pollock	1200–1600	1350	
Mince – krill	450	450	0
– fish	250	400	
Tail meats – krill	1850	1850	0
– shrimp (count 300–500)	3000–4000	2200	
Meal – krill (52%)	240	270	12.5
– fish (65%)	400	400	
– soya (48%)	205	210	
Transport costs (cif)			
Frozen W. Europe	150	175	17
Meal W. Europe	100	70	−30
Fuel oil	106	100	−6
Vessel costs			
Capital cost	$10 m	$12 m	20
Vessel fixed cost ($/100 h)	67 000	80 400	20
Processing cost (index)	100	120	20

Sources 1979 data from McElroy (1983). 1989 data based on Infofish International and Globefish Highlights (various issues).
Note The meal % figures refer to protein content which is the main factor in determining price. The 52% protein content of krill reflects the average length of time before processing it. The 1979 transportation costs for meal were overestimated.

processing area given over to the production of krill tail meats. [The latest findings (1988/9 season) of the Poles appear to support these results (FNI 1989).]

In summary, while the introduction of the four hour quality constraint on the use of krill for food production has a substantial effect on the earnings achieved by the factory vessel, the production of krill products for human consumption still appears to be worthwhile for whole krill products and the minces. However, krill meal production, even with the highest catch rate considered here, though of significance, does not appear to add substantially to vessel earnings.

Though not considered explicitly here, it would appear that unless the additional processing capacity possible with a larger vessel can increase net earnings significantly, it is unlikely to be warranted or justified, particularly when the out-of-season fishery (i.e. other than a year-round krill fishery) is considered.

Update to 1989

No index of fishing vessel costs and revenues exists for the krill or similar fishery. We can construct a 'proxy' cost index based on the major cost items (vessel, processing equipment, fuel, transportation and crew wages). An index of product prices is easier though no more reliable to construct. It can be derived using changes in market prices of 'reference' products as a guide.

Table 8.3 presents the main cost and price items for 1979 and 1989. The striking feature is the decline in the US dollar price for shrimp, largely as a result of the rapid expansion in world trade of farmed shrimp. Meanwhile the meal price has moved little. Vessel, equipment and labour costs are assumed to have risen by 20% overall, while energy (fuel oil) costs have actually fallen slightly.

Taking these figures as being of the right order, we can make the following points. Whole krill cooked and uncooked remain the most profitable products from krill. Tail meat production becomes still more uneconomic, while for surimi and krill mince the former remains profitable ($7400 to $12 100 per day, a profit rate of 26–40%) while the latter is now uneconomic except at the 'high' rate of catch (−$5100 to $5300 per day, a profit rate of −18 to +16%) (Figs. 8.11 and 8.12).

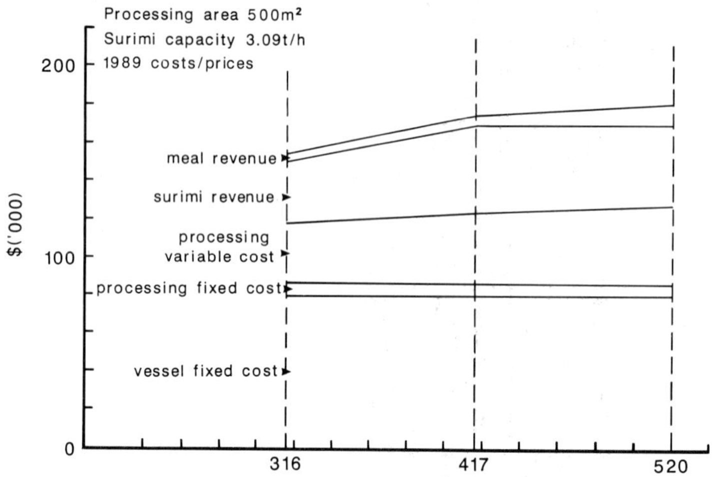

Fig. 8.11 The economics of producing frozen surimi and meal.

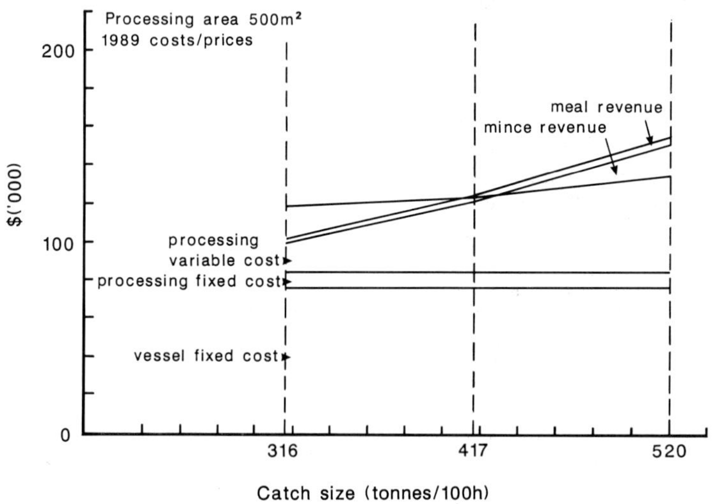

Fig. 8.12 The economics of producing frozen mince and meal.

Krill meal

Krill meal represents a low revenue generating 'sink' for the 'excess-to-food-processing-capacity' part of the krill catch. Figure 8.13 illustrates the effect of meal capacity on the age and quantity of material processed to krill for a 'high' catch rate and 'low' food processing capacity. In general, a meal processing capacity of about 2.5 t/h (60 t/d) is favoured.

CONCLUSIONS

(1) A linear programming approach (either multi-period or with a simulation model attached) offers a quick, simple yet effective way to optimise processing capacity on board a factory vessel. In this instance a non-linear optimisation model from the NAG (Numerical Algorithms Group) library was chosen to provide insight into the effect of adopting different quality criteria (time after capture) upon the amount of material processed. Given that the relationship between processing capacity and space is treated as linear for each process, the model chooses only one food process as optimal, and meal.

(2) Of the six processing technologies evaluated, four were economic in both 1979 and 1989 with a processing area of 500 m^2. They were in order: whole krill uncooked and cooked, krill surimi and meal. Krill mince was marginal, being economic roughly at the 'average' catch rate (75 t/d) in 1979 but only above the

Total catch 500t/in 100 hours
Amount available for meal : 325t

Capacity (t/hr)	1.25	2.5	5.0
Amount used (t)	120	240.5	325
Average age of material (hrs)	9.1	6.1	3.1
S D	3.9	3.0	1.4

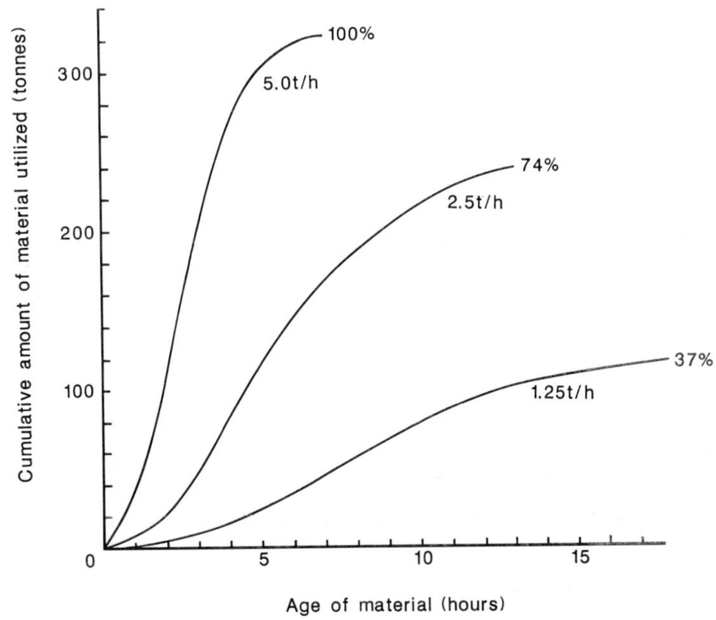

Fig. 8.13 The effect of meal processing capacity on the age and quantity of material processed.

'above average' catch rate (100 t/d) in 1989, while tail meat was never economic.

(3) The whole krill products represent end-products only in the East (Japan). In the West, they would be considered as an intermediary product for further processing, and hence their price would be lower. In the absence of an economic tail meat process, surimi-based products would perhaps have the greatest appeal in the West.

(4) The supply of shrimp from aquaculture has tended to depress shrimp market prices during the past five years, which has made the Antarctic krill fishery of less economic value. The outlook is for shrimp prices to remain near current levels in the short to medium term (say to the mid-1990s). This will tend to make the krill fishery of lower economic value over the next few years unless there is a significant breakthrough in krill processing technology (in terms of product yield per unit processing space).

(5) Notwithstanding, the three krill products which have been marketed in Japan for most if not all of the last ten years (whole krill uncooked and cooked, and krill meal), on the basis of costs and prices employed here, represent the commercial exploitation of Antarctic krill!

APPENDIX 1

PROCESS FLOW CHARTS FOR THE MAIN KRILL PRODUCTS

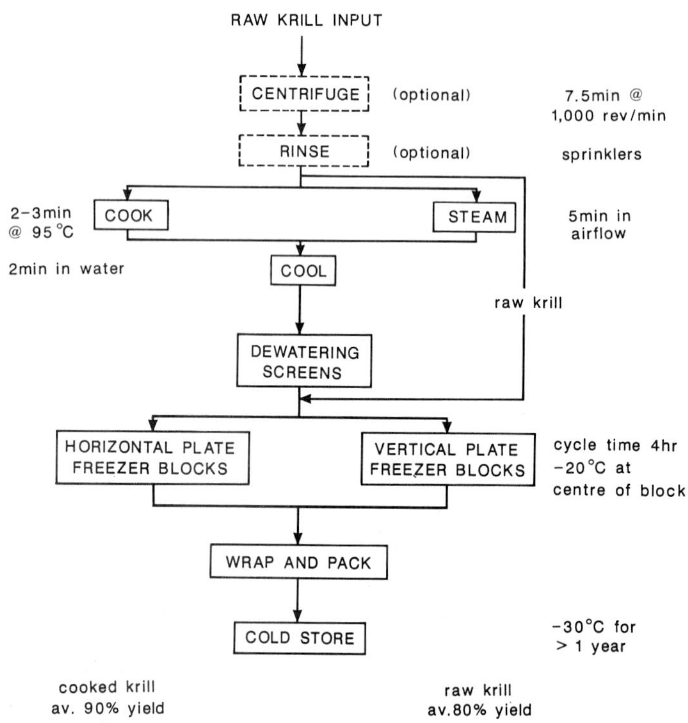

Fig. A1.1 Process flow diagram for raw and cooked frozen whole krill.

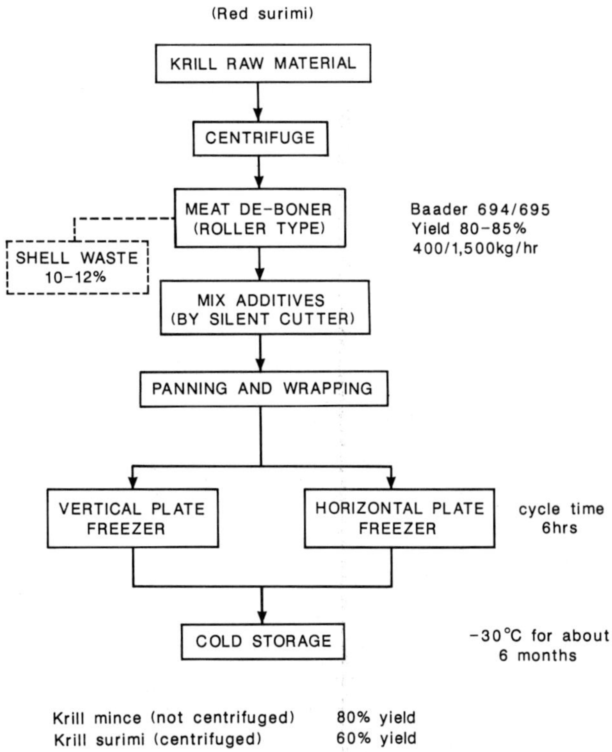

Fig. A1.2 Process flow diagram for krill mince (surimi) production.

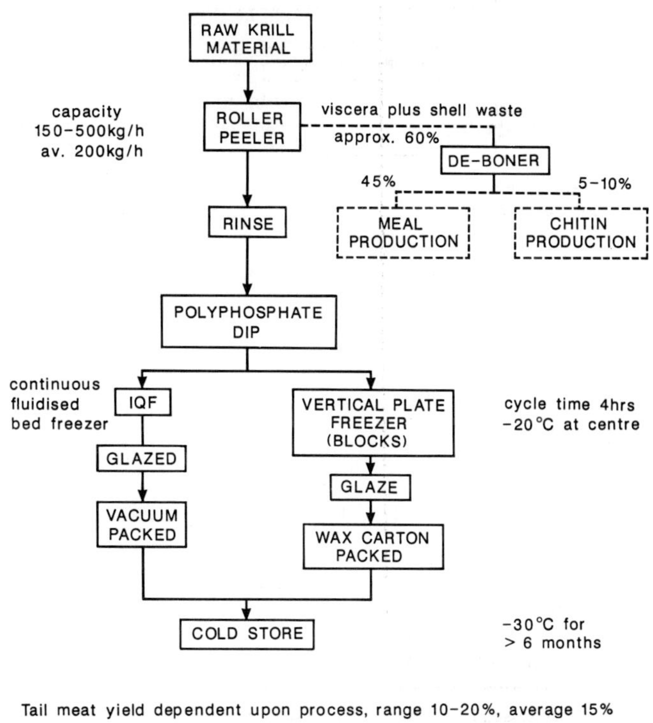

Fig. A1.3 Process flow diagram for peeled tail meat production.

APPENDIX 2

MODEL INPUT DATA

CATCH DISTRIBUTIONS

The catch distributions used in this study were developed by the author, paying attention to the two features of particular importance to a process designer: the distribution of catches by size and the distribution of catches over time. A review of the limited information on krill catch distributions (McElroy 1983) together with the important features of catch distributions obtained for other pelagic species, such as herring (Coverdale 1972) and blue whiting (Curr 1977, 1981), enabled guidelines to be adopted in constructing the krill catch distributions used in the study. The main points are:

(1) Haul size ranges between 0 and 70 t, but hauls of less than 10 t are preferred (more than 50%).

(2) A direct relationship between haul size and variance seems to apply generally in pelagic fisheries (Curr 1981, Shepherd 1981). Here too average haul size ($\overline{x}$) and its variance (σ) were taken to be directly proportional to the coefficient of variation (v). Based on information provided by the Polish gear technologist R. Stefan (personal communication 1979), the shape of the krill catch distribution was similar to that obtained by the Poles for blue whiting (cf. Curr 1977, 1981). On this basis a reasonable value for the coefficient of variation would be about 0.7.

$$V = \frac{\sqrt{\sigma}}{\overline{x}} = 0.7$$

(3) Average daily catch rate for a full-scale operation is taken to be about 75 t/d (maximum feasible >200 t/d), from an average of about 9 (maximum 16) hauls/day (average towing time 1 hour).

(4) Higher catches or catch rates tend to occur within distinct periods of the day, e.g. early morning (0400–1000 h) and, though less commonly, early evening (1600–2000 h).

Two methods were used to generate catch distributions. One involved selecting feasible catch patterns for three specific krill catch rates (designated 'average', 'above average', and 'high', with values of 75, 100 and 125 t/d). The other involved adopting several feasible probability distributions for catch size and inter-haul times, and using random numbers to generate a number of krill catch distributions. The catch distributions were then fed into the computer model and the results compared for certain fixed food and meal processing capacities. From this study, the three catch distributions representing 'average', 'above average' and 'high' were chosen as being reasonably representative, given the above guidelines.

Details of these catch distributions are summarised in Table A2.1, while the main characteristics of the distributions are presented in Figures A2.1 and A2.2.

OTHER INPUT DATA

(1) Buffer storage capacity

While the appropriate buffer storage capacity was shown to depend upon catch rate and total processing capacity, a capacity of 40 tonnes was found to be satisfactory under most circumstances. This value was adopted in the base case.

Table A2.1 'Average' 'above average' and 'high' krill catch distributions.

	Total catch (t)	Catch mean (t/h)	Catch S.D. (t/h)	Co-var.	Haul mean (t)	Haul max. (t)	Haul S.D. (t)	Co-var.	Mean catch/ 24 h (t/24 h)	Max* catch/ 24 h (t/24 h)	Min* catch/ 24 h (t/24 h)	No. of hours with zero catch
CDI	316	3.16	5.34	1.69	8.778	26	5.469	0.62	75.84	87	65	64
CDII	417	4.17	6.494	1.56	10.42	28	6.342	0.61	100.1	135	59	60
CDIII	520	5.2	8.337	1.60	12.38	39.5	8.751	0.71	124.8	152.5	75.5	58

* Assessed from consecutive 24 hour periods

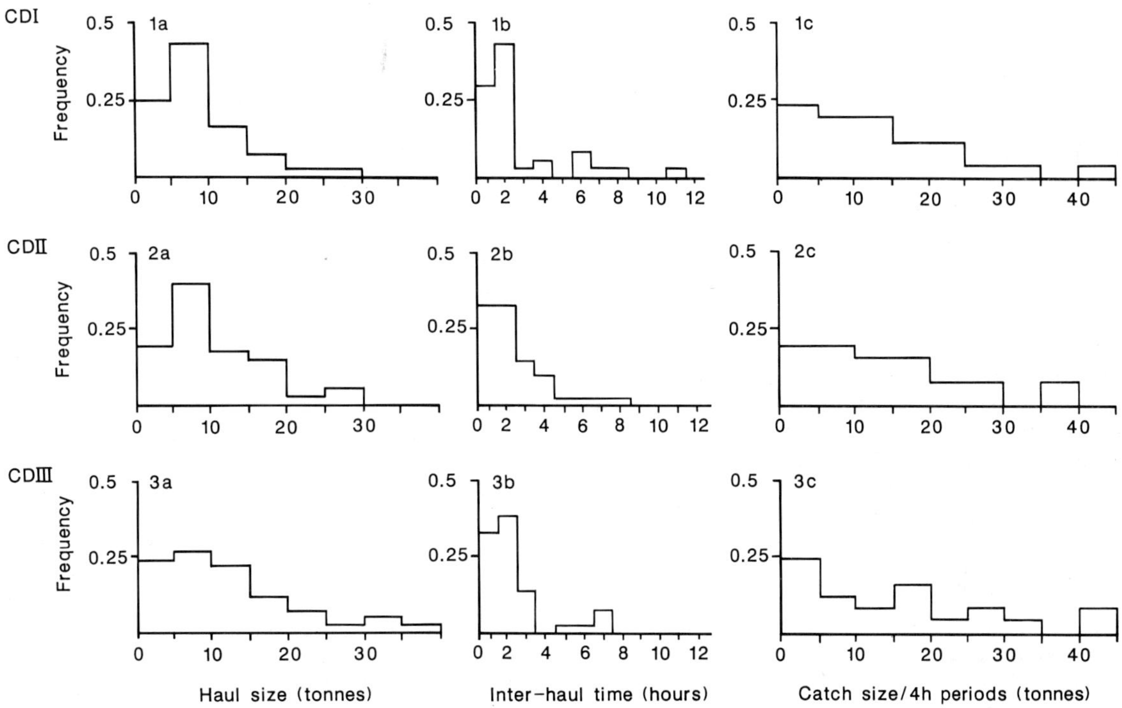

Fig. A2.1 Breakdown of the model catch distributions.

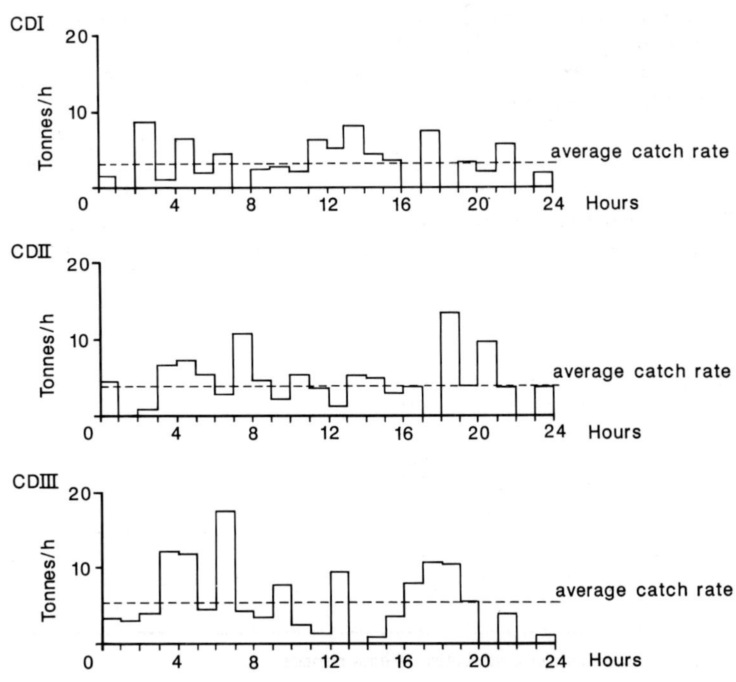

Fig. A2.2 Variations in catch rate over 24 hours.

(2) Floor area

The floor area of the processing deck, which included the areas occupied by the processing and freezing capacities, was taken to be 500 m^2. This area was reduced to 300 m^2 where the vessel was used in the two separate krill-finfish fisheries and the process was specific to krill.

(3) Product yields

The following product yield values were adopted:

Whole uncooked krill (centrifuged)	80
Whole cooked krill (not centrifuged)	90
Krill mince (surimi) (centrifuged)	60
Krill mince (not centrifuged)	80
Tail meats	15
Meal without stickwater recovery	15
Meal with stickwater recovery	21

(4) Product prices/tonne (ex-vessel)

(In US$ at prevailing 1979 costs/prices)

	Low $/t	Average (base case)	High $/t
Whole uncooked	500	750	1000
Whole cooked	350	600	850
Surimi	850	1100	1350
Mince	250	450	650
Tail meats	1600	1850	2100
Meal	210	240	270

Transport costs to a W. European market assumed in the base case.

Transport costs

Frozen	W. Europe	$150/t
	(S. America	$50/t)
Meal	W. Europe	$100/t
	(S. America	$30/t)

(5) Model fixed costs and variable costs – definitions

All costs unrelated to the processing of the raw material are taken as model fixed costs. These include a capital charges element which varies according to the economic life of the vessel (15 years) or processing machinery (generally 8 years); and a maintenance and insurance charge which is taken as 10% of the capital cost over the full year.

Fixed costs are split up into vessel fixed costs and processing fixed costs. Direct vessel costs are taken as part of vessel fixed costs. Only processing variable costs therefore vary with output.

A distinction is made between common processes which are shared with fin fish and those which are specific to krill. With the former, fixed costs (capital charge, insurance, maintenance) are spread over the full year. In the latter, the full fixed cost element must be met by the krill fishery.

(6) Uncooked whole krill

(In US$ at 1979 prevailing costs)

Capital cost

Capital cost of centrifuges etc. per tonne input per hour	$50 000
Capital charge per year	$7 735
Fixed cost per 100 h/tonne input	$530/100 h
Variable costs per tonne input	$20
Space requirement per tonne input	40 m^2

(7) Cooked whole krill

Capital cost

Capital cost of equipment (cookers/dewatering screens, etc) per tonne input per hour	$160 000
Capital charge per year	$24 750
Fixed costs per 100 h/tonne input	$1 700/100 h
Variable costs per tonne input	$60
Space requirement per tonne input	80 m^2

(8) Krill surimi

Capital cost

Capital cost of equipment per tonne input per hour	$150 000
Capital charge per year	$23 200
Fixed cost per 100 h/tonne input	$800/100 h
Variable cost per tonne input	$100
Space required per tonne input	120 m^2

(9) Krill mince

Capital cost

Capital cost of equipment per tonne input per hour	$75 000

Capital charge per year	11 600
Fixed costs per 100 h/tonne input	\$400/100 h
Variable costs per tonne input	\$60
Space requirement per tonne input	40 m^2

(10) Tail meats

Capital cost

Capital cost of peeling machinery per tonne input per hour	\$250 000
Capital charge per year	\$38 675
Fixed cost per 100 h/tonne input	\$2650
Variable cost per tonne input	\$50
Space requirement The space required for 5 peeling machines and associated equipment, per tonne input	175 m^2

(11) Meal (see McElroy 1982 for details)

Capital cost

Essentially three sizes of meal plant are considered in the base case: 0, 2.5 and 5.0 t/h. It is assumed that stickwater is not recovered, giving a meal yield of 15%. The capital costs of the two plants are:

2.5 t/h	\$350 000
5.0 t/h	\$600 000
Fixed costs per 100 h (full year use):	
Capacity 2.5 t/h – cost per tonne capacity	\$670/100 h
Capacity 5.0 t/h – cost per tonne capacity	\$530/100 h
Variable costs per tonne input:	
2.5 t/h	\$9.2
5.0 t/h	\$8.6
Average assumed	\$9.0

(12) Freezing plant

Space requirement

The space required per tonne of input per hour (4 hour freezing cycle) for a vertical plate freezer including working area was given as 50 m^2 (I. Hutchison pers. comm.). The floor area occupied by the freezer itself would be 20 m^2. The corresponding area for a horizontal plate freezer was 70 m^2.

Capital cost

The capital cost (including installation) of the vertical plate freezer per tonne input per hour	\$140 000
Capital charge per half year	\$6 740
Fixed costs per 100 h/tonne input VPF	\$580/100 h
Fixed costs per 100 h/tonne input HPF	\$600/100 h
Variable costs	
Variable costs per tonne input VPF	\$40/t input
Variable costs per tonne input HPF	\$40/t input

(based on the labour and power requirements for an operation with a freezing capacity of 100 t/24 h).

(13) Factory trawler costs and technical details

Vessel details based on Polish vessel, type B 418.

Vessel costs

Capital cost	\$10.0 m
Fixed cost per 100 h	\$40 000/100 h
Direct costs	\$/day at sea
Crew (\$50/shift/man × 30)	1500
Fuel (9.0 t/24 h) \$106/t (excludes processing energy costs)	950
Fishing gear, stores, etc.	800
Management expenses, etc.	1500
Total	4750
Direct costs per 100 h fishing (\$4 750 × 135/100 × 100/24)	\$27 000/100 h
Total vessel cost/100 h fishing (Operating costs based on a similar sized vessel operated by Nordsee GmbH)	\$67 000/100 h

BIBLIOGRAPHY

Budzinski E., Bykowski P. & Dutkiewicz D. (1985) Possibilities of processing and marketing of products made from Antarctic krill. *FAO Fish. Tech. Pap.* (268).

Coverdale I.L. (1972) *The application of a chilled sea water container system for the Scottish inshore fishing industry.* M.Sc. thesis, University of Hull.

Curr C.T.W. (1977) *The effect of catch rate on the economics of selected blue whiting fishing operations.* Field Report 483, Industrial Development Unit, Sea Fish Industry Authority, Hull.

Curr C.T.W. (1981) A mathematical model used for pre-feasibility studies of fishing operations. In *Applied Operations Research in Fishing* (Ed. by K.B. Haley), pp. 161–72. NATO Conference Series, Plenum Publ. Corp., New York.

Eddie G.C. (1977) The harvesting of krill. *Southern Ocean Fisheries Survey Programme*, GLO/SO/77/2, FAO, Rome.

FNI (1989) Polish trawlers processing krill. *Fish. News Int.* March.

Grantham G.J. (1977) The utilization of krill. *Southern Ocean Fisheries Survey Programme*, GLO/SO/77/3, FAO, Rome.

McElroy J.K. (1980a) The potential of krill as a commercial catch. In *The Management of the Southern Ocean* (Ed. by B. Mitchell & R. Sandbrook), pp. 60–80. International Institute for Environment and Development, London.

McElroy J.K. (1980b) The economics of harvesting krill. In *The Management of the Southern Ocean* (Ed. by B. Mitchell

& R. Sandbrook), pp. 81–103. International Institute for Environment and Development, London.

McElroy J.K. (1980c) Economic evaluations of krill fishing systems. In *Advances in Fish Science and Technology* (Ed. by J.J. Connell), pp. 314–21. Fishing News Books, Oxford.

McElroy J.K. (1982) *The production and use of krill meal: a techno-economic appraisal.* Paper presented at scientific committee meeting of International Association of Fish Meal Manufacturers, Windsor; issued in extended form as Marine Resources Research Unit, Research Paper 30, School of Economics, Portsmouth Polytechnic.

McElroy J.K. (1983) *The Antarctic krill fishery: a techno-economic investigation.* PhD. thesis, University of Stirling, Scotland.

Shepherd J.G. (1981) Matching fishing capacity to the catches available: a problem in resource allocation. *J. Agric. Econ.*, **32** (3), 331–40.

CHAPTER 9

Composition and deterioration of pelagic fish

J.R. Burt and R. Hardy

INTRODUCTION

The term pelagic, when applied to fish, relates to mode of life and feeding habit but these characteristics are linked to the anatomical, physiological and biochemical properties of those species which have become adapted to a pelagic existence for much, if not all, of their life cycles. In turn, the biochemistry and physiology of such fish are based on, or alternatively give rise to, distinctive chemical compositional features in various tissues but particularly in the musculature. Finally, it is the chemical constituents which are acted upon post-mortem by autolytic and bacterial enzymes, to produce the various decomposition compounds responsible for the characteristic spoilage and putrefactive odours and flavours.

Pelagic fish, in terms of composition and structure, can be divided into two major categories: the fatty pelagics and the non-fatty pelagics. The latter resemble the so-called white, non-fatty, demersal species of fish in almost every respect, apart from the fact that for appreciable periods of time they shoal in large numbers and can be caught at or near the surface of the sea. They also tend to have a more migratory habit than other non-fatty species.

The fatty pelagic species of fish, on the other hand, make up the group which is more commonly described as pelagic. They have high, and largely seasonally variable, lipid contents; they tend to be migratory in habit; and they have high proportions of dark muscle with its associated aerobic type of energy metabolism.

For the purposes of this chapter, the non-fatty blue whiting and the crustacean krill are to be considered as pelagic species along with the more usually regarded fatty fish.

STRUCTURE

The differences between the two main types of muscle (dark and white) have been reviewed (Love 1970, 1980) and the relationship between type of swimming activity and the proportions and distribution of dark muscle within the musculature described in detail.

Dark muscle is more vascular than white and contains much higher concentrations of myoglobin and cytochrome C, these three factors contributing to the difference in appearance between the two types of muscle. Dark muscle contains many more mitochondria than white muscle does and thus is better supplied with the enzymes and metabolic systems associated with these sub-cellular particles. As a consequence dark muscle can function aerobically as well as anaerobically, while the energy metabolism of white muscle is almost entirely dependent on anaerobic glycolysis. Generally, dark muscle can provide propulsion on a continuous basis, but the white is liable to early exhaustion and can therefore be used primarily only intermittently when chasing prey or attempting to avoid predators.

Structurally the two types of muscle are similar in that both are arranged into fibrils, but the dark muscle fibres are thinner than the white ones (Braekkan 1959) and they also have different patterns of innervation (Bone 1966).

In terms of chemical composition the differences between the two types of muscle are just as marked. Within a species, dark muscle is characterised by having higher concentrations of lipid than white muscle has (see Table 9.1) and another major difference is that, despite its dependence on anaerobic glycolysis for providing energy, white muscle usually contains much less glycogen than does dark muscle, even in the resting state.

Pelagic fish are more active than demersal ones and the former have, on the whole, higher proportions of dark muscle; the more active the fish, the higher the ratio of dark to white muscle. Furthermore, it would appear that the relative amounts of dark and white muscle vary along the length of the fish, there being more dark muscle in regions associated with higher activity such as towards the tail and in proximity to

Table 9.1 The lipid contents of white and dark muscle of typical pelagic and demersal fish (After Love 1988).

	Atlantic mackerel		Atlantic cod	
	White muscle	Dark muscle	White muscle	Dark muscle
Total lipids				
% wet tissue	2.3	13.1	0.7	1.8
% total lipids				
Phospholipids	24.0	19.0	87.7	87.1
Triglycerides	70.9	76.8	1.6	1.5
% wet tissue				
Phospholipids	0.6	2.5	0.6	1.6
Triglycerides	1.6	10.1	<0.1	<0.1

certain very active fins (Love 1970, 1980, 1988). Because of this variability, comparisons of the amounts of dark muscle between pelagic and non-pelagic fish have to be made cautiously but in one study (Greer-Walker & Pull 1975) the proportion of dark muscle in comparable regions of each fish (one third of the fish length from the tail) ranged from 30% for the pelagic Japanese mackerel (*Scomber colias*) to 11–17% for several non-pelagic species (cod, *Gadus morhua*; haddock, *Melanogrammus aeglefinus*; saithe, *Pollachius virens*; whiting, *Merlangius merlangus*). In one extreme example, Japanese sardine (*Sardinia* (*Sardinops*) *melanosticta*), dark muscle has been estimated to constitute 48% of the fish's body weight (Fujikawa & Naganuma 1936).

Figure 9.1 illustrates the distribution of dark muscle in certain pelagic and non-pelagic species. In most species, the dark muscle lies as a subcutaneous layer in the region of the lateral line; it is thickest just under the lateral line and tapers off dorsally and ventrally. In the pelagic species shown the layer is noticeably thicker in all comparable regions than it is in the non-pelagic ones. In addition to the subcutaneous layer, some of the most active fish, like the tunas and species related to them, have blocks of deep-seated dark muscle situated near the vertebrae. These are also illustrated in Figure 9.1.

The lipid stores which are used by pelagic fish for providing energy are situated not only in the dark muscle but also in depot cells elsewhere. A layer of these is usually found under the skin but they can also occur within the white muscle. Table 9.1 gives comparative figures for the dark and white muscles of typical pelagic (Atlantic mackerel) and non-pelagic (Atlantic cod) fish.

The preponderance of the structural phospholipids in the albeit low levels of lipid in both types of cod muscle, can readily be seen, as can the relatively high concentrations of triglycerides in the pelagic fish. Bearing in mind the fact that there can usually be several times as much white as dark muscle, even in a fish as active as mackerel, the importance of the triglyceride lipid depots in mackerel white muscle to the total muscular lipid reserves of that fish is very evident.

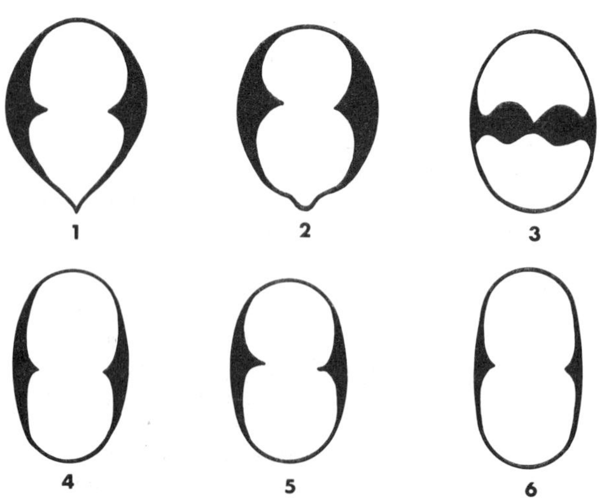

Fig. 9.1 Sections through the bodies of several species of fish showing the depth of dark muscle. (1) Herring; (2) mackerel; (3) tuna; (4) haddock; (5) cod; (6) whiting. (From Love 1970, 1988.)

While on the subject of lipid distribution, another major difference between most pelagic (particularly the fatty ones) and non-pelagic fish lies in the relative importance of the liver as a fat depot in the latter. The non-fatty pelagic fish, such as blue whiting, resemble non-pelagic fish in this respect.

An important aspect of structure relates to the yields of edible flesh which can be obtained from different species. On these figures are made the calculations which determine whether or not particular fisheries can be exploited, or whether or not certain processing operations can be done economically. Table 9.2 gives

Table 9.2 Yields from whole fish (%).

Name	Trunk	Fillet	Edible flesh	Source
Amberjacks				
Seriola grandis	65	29–55		Bykov 1985, Vlieg 1985.
Anchovies				
Anchoa nasus	64			Golovin 1977.
Engraulis anchoita			64	Bykov 1985.
Engraulis encrasicolus	66		61–64	Carteni & Aloj 1934, de las Heras & Mendez 1952, Gimenez & de las Heras 1943, Bykov 1974, 1985.
Engraulis japonicus	57–64		60	Suzuki 1981, Bykov 1985.
Engraulis mordax	59–63	51	62	Miyauchi & Steinberg 1970, Bykov 1985.
Engraulis ringens	64			Golovin 1977.
Capelin				
Mallotus villosus	60–82	40–69	65	Konstantinova & Minder 1975, Bykov 1985.
Herrings				
Clupea harengus	66–76	43–58	61–80	Clark & Clough 1926, Reay *et al.* 1943, Popa *et al.* 1968, Keay & Hardy 1974, Ravichander & Keay 1976, Bykov 1985.
Clupea pallasi	64–77	50–60	42–67	Miyauchi & Steinberg 1970, Kizevetter 1971, Bykov 1974, 1985.
Jack and horse mackerels				
Trachurus capensis	52		42–63	Bykov 1985.
Trachurus declivis		46		Pickston *et al.* 1982.
Trachurus japonicus	56–70	52–58	54–61	Kizevetter 1971, Bykov 1985.
Trachurus novaezelandiae			73–74	Popa *et al.* 1968.
Trachurus symmetricus			53–62	Kizevetter 1971, Maciejczyk 1974, Golovin 1977.
Trachurus trachurus	57–73		46–81	Arevalo 1948, Goncalves Ferreira 1951, Bykov 1985, Torry Research Station unpublished data.
Mackerels				
Rastrelliger brachysoma			54–66	Sulit *et al.* 1953, Intengan *et al.* 1956.
Rastrelliger chrysozonus			45–64	Sulit *et al.* 1953, Intengan *et al.* 1956.
Rastrelliger kanagurta	64–76	46–57	62	Chari 1948, Kizevetter 1971, Bykov 1985.
Scomber japonicus	61–85	46–47	50–82	Miwa *et al.* 1958, Tulsner 1965, Kizevetter 1971, Maciejczyk 1974, Golovin 1977, Bykov 1985.
Scomber scombrus	60–75	33–68	48–75	Plimmer 1921, Clark & Clough 1926, de las Heras & Mendez 1952, Coppini 1967, Popa *et al.* 1968, Leu *et al.* 1981, Bykov 1985, Bimbo.
Menhaden				
Brevoortia tyrannus	46–78	17–56	8–38	Bykov 1985, Regier *et al.* 1985, Bimbo.
Pilchards				
Sardina pilchardus	66		60–64	Gimenez & de las Heras 1943, Goncalves Ferreira 1951, IEO 1961, Viviani *et al.* 1968, Bykov 1985.
Sardinops caerulea	65		51	Bykov 1985.
Sardinops melanosticta	56–71	43–56	51–58	Kizevetter 1971, Vakhrusheva & Nikiforova 1979, Suzuki 1981.

Table 9.2 *Continued*

Name	Trunk	Fillet	Edible flesh	Source
Sardinops neopilchardus	65–73			Bykov 1985.
Sardinops ocellata	61–66			Bykov 1985.
Sardinops sagax			53	Maciejczyk 1974.
Sardinellas				
Sardinella allexia	64			Bykov 1985.
Sardinella aurita	62–79		43–71	Bykov & Rusin 1971, Bykov 1985.
Sardinella eba	70			Bykov 1985.
Sardinella fimbriata	68–69		57–76	Setna *et al.* 1944, Chari 1948, Sulit *et al.* 1953, Intengan *et al.* 1956, Bykov 1985.
Sardinella gibbosa	64–67	51–55		Bykov 1985.
Sardinella longiceps	59–68		47–70	Santos & Ascalan 1931, Chari 1948, Sulit *et al.* 1953, Revankar *et al.* 1981, Bykov 1985.
Sardinella perforata	–		47	Intengan *et al.* 1956.
Sauries				
Cololabis saira	74–75	54–61	60–72	Tsuchiya *et al.* 1953, Miwa *et al.* 1958, Kizevetter 1971, Bykov 1985.
Scomberesox saurus	77–79			Bykov 1985.
Scads				
Decapterus kiliche	63–75		55	Bykov 1985.
Decapterus macrosoma			40–71	Sulit *et al.* 1953, Intengan *et al.* 1956.
Decapterus maruadsi	60–66	51–57		Bykov 1985.
Decapterus polyaspis	67	56		Bykov 1985.
Decapterus rhonchus	71	53	55	Bykov 1985.
Decapterus russelli	63–66			Bykov 1985.
Sprat				
Sprattus sprattus	66–80		53–81	Plimmer 1921, de las Heras & Mendez 1952, Skachkov & Yudina 1975, Bykov 1985.
Tunas				
Auxis thazard	65–68		45–71	Sulit *et al.* 1953, Yaroslavtseva 1966, Perova *et al.* 1980, Bykov 1985.
Euthynnus alleteratus			53–74	Busson *et al.* 1953, Yaroslavtseva 1966, Perova *et al.* 1980, Bykov 1985.
Katsuwonus pelamis	76		46–66	Kizevetter 1971, Perova *et al.* 1980, Vlieg *et al.* 1983, Balogun & Talabi 1985, Bykov 1985, Balogun *et al.* 1986.
Sarda chiliensis			57–58	de Haan 1974, Golovin 1977.
Sarda orientalis		70		Bykov 1985.
Sarda sarda	68			Bykov 1985.
Thunnus alalunga		32	54–78	Kizevetter 1971, Bykov 1985, Diez Ezkerra & Whittle 1986.
Thunnus albacares	56–72		54–79	Chari 1948, Intengan *et al.* 1956, Tulsner 1965, Kovalchuk 1970, Kizevetter 1971, Bykov 1985.
Thunnus maccoyii			49	Yaroslavtseva 1966.
Thunnus obesus	67–75		62–77	Tulsner 1965, Kizevetter 1971, Bykov 1985.
Thunnus thynnus	77		62–64	Kizevetter 1971, Bykov 1985.
Blue whitings				
Micromesistius australis	61–76	42–71	51–59	Kizevetter 1971, Minder & Gritskevich 1975, Kakuda & Kitagawa 1979, Bykov 1985.
Micromesistius poutassou	60–66	20–44	35–58	Minder & Gritskevich 1975, Whittle *et al.* 1980, Bykov 1985, Christians & Ritter, Torry Research Station unpublished data.
Krill				
Euphausia superba			28	Budzinski *et al.* 1985, Bykowski 1986.

a selection of figures which have been obtained for many of the species under consideration; the figures are categorised under three main headings as percentage yields of trunks (headed and gutted fish), fillets, or 'total' edible flesh (obtained by dissection or by the use of deboning equipment). Some of the data relate to skin free/bone free flesh while others relate to muscle which may include some skin and, occasionally, bones. Reference will need to be made to the original papers cited in order to determine the size of the samples involved and the exact conditions of preparing them.

The great variability exhibited by these figures reflects, in some instances, the relative efficiency with which the different cutting or separating operations have been done, but in others they are a true reflection of the natural changes in body proportions which are associated with age, season, migration, sexual maturation and feeding activity. These latter factors have long been recognised as having a greater effect in pelagic than in non-pelagic fish.

CHEMICAL COMPOSITION

The chemical differences between dark and light muscle have been summarised by Love (1970, 1980) in two extensive tables. Many of the substances mentioned are enzymes and the differences in these will be reflected first in metabolic differences between the two types of tissue, and then in the modes of life of those fish which have small or large proportions of dark muscle. Many of the other substances found at different levels in dark and white muscle, although probably of great significance and importance to the various fish themselves, are not known to be of any concern to the fish catcher, processor or consumer and will therefore not be considered here. Furthermore, many of the reports cited conflict and this may be the result of true inter-species differences or of natural variability within a species.

Attention here will be focussed mainly on those compounds and groups of compounds which are, or are likely to be, of technological or commercial importance and which are present at different levels of concentration in pelagic and non-pelagic fish.

As has been mentioned in the section on structure, the differences in appearance between dark and white muscle can be explained on chemical grounds, myoglobin, haemoglobin and cytochrome C being the compounds mainly responsible. It is thus understandable why the flesh of pelagic fish, particularly the fatty species, are reportedly so dark and, to some assessors/consumers, unattractive in the minced state.

Pelagic fish are used in a variety of ways, and interest in the proximate composition of a species will depend on the context in which it is being regarded. For instance, many species are caught for industrial purposes and the whole of the fish is consigned for reduction to fish meal and fish oil. Chemical composition on a whole body basis is also of interest in those species of small fatty pelagic fish which are eaten whole as whitebait. Alternatively, it is the composition of the edible portion which is usually of more concern to the consumer and nutritionist, but the term 'edible portion' requires further definition. In some instances it can include skin and small bones; in some it will be fillet with or without skin; in some it will be machine-separated flesh free of skin and bones; and in some it will be headed and gutted fish trunks such as are used in canning.

The causes of the variability in bodily proportions mentioned earlier are also responsible for variability in chemical composition and in particular in the lipid content of fatty species. The respective uses of muscle and liver as sites for deposition of reserve lipids by fatty and non-fatty species are additional reasons for variability.

Table 9.3 summarises some of the data that are available giving the proximate composition of pelagic species. The figures in the table have been culled from a large number of studies going back over several decades. Many of the values given in the literature relate to single analyses while others have been derived from studies which, in some instances, have continued through several seasons over two or more years and represent many thousand analyses. It would be impossible to give an appropriate weighting to each value reported and to try and give a set of single figures which would best describe each fish. Rather, we have chosen to give a range of values for each parameter in the belief that this would be most useful.

Those requiring more detailed information concerning the proximate composition of a particular fish from one area at any given season would need to consult the appropriate literature. For that reason we have erred on citing as many papers as possible. Naturally, where only one figure is available, it has been cited in the belief that any figure is better than none. At all events, the figures cited must be regarded as indicative only and should be used purely for guidance. In the interests of clarity, ash contents have not been quoted since there is a large measure of similarity between species in this context. As a crude approximation, whole fish contain 2–4% ash and muscle contains 1–2%.

Water is the major chemical constituent of all fish flesh, and fatty pelagic fish have significantly lower

Table 9.3 Proportions (%) of lipid, water and protein in whole pelagic fish and the edible portions thereof (see text). (Figures rounded to nearest 1%.)

Species	Whole fish			Edible portion			Sources
	Lipid	Water	Crude protein (N × 6.25)	Lipid	Water	Crude protein (N × 6.25)	
Amberjacks							
Seriola dumerili				2	75	23	Bykov 1985.
Seriola grandis	1–4	72–75	20–21	1–2	73–77	20–24	Bykov 1985, Vlieg 1985.
Seriola lalandi				1–3	77	20–23	van Wyk 1944, Tulsner 1965, Evans & Wessels 1984, Simmonds & Heydenrych 1984.
Seriola quinqueradiata							
wild				1–13	66–76	19–25	Shimizu *et al.* 1973, Saeki & Kumagi 1979.
farmed				1–29	51–74	16–25	Shimizu *et al.* 1973, Sakaguchi 1976, Saeki & Kumagi 1979, Shimeno *et al.* 1985.
Anchovies							
Anchoa nasus				3	77		Golovin 1977.
Anchoa sp.	2		17				Thompson 1966.
Engraulis anchoita				15	64	19	Bykov 1985.
Engraulis capensis	2–9	68–76	18–23				Simmonds & Tanner 1980.
Engraulis encrasicolus	3–13	74	20	1–8	71–77	18–23	Cartini & Aloj 1934, Gimenez & de las Heras 1943, de las Heras & Mendez 1952, Fraga 1955, Acara 1957, Viviani *et al.* 1968, Cuisa & Giaccio 1969, Bykov 1985.
Engraulis japonicus				1–18	62–75	13–21	Hayashi & Takagi 1978, Bykov 1985.
Engraulis mordax				1–9	71–79	17–19	Bykov 1985.
Engraulis ringens	2–18	65–78	18–24	4	75		Barrera & Aguilera 1968, Golovin 1977.
Stolephorus commersonii	0–2	77	15–18	0		21	Sulit *et al.* 1953, Intengan *et al.* 1956, Ayyappan *et al.* 1976.
Stolephorus indicus	2		19				Sulit *et al.* 1953.
Stolephorus malabaricus	4		19				Ayyappan *et al.* 1976.
Stolephorus tri	1		15				Chari 1948.
Capelin							
Mallotus villosus	1–24	64–83	12–15	1–23	61–84	12–17	Ackman *et al.* 1963, Andrews 1969, MacCallum *et al.* 1969, Kizevetter 1971, Konstantinova & Minder 1975, Eaton *et al.* 1975, Jangaard 1974, Bykov 1985.

Herrings							
Clupea harengus	1–24	60–81	17–21	1–35	42–81	13–23	Atwater 1892, Milroy 1906, Milroy 1909, Johnstone 1914, Clark & Almy 1918, Lovern & Wood 1937, Kordyl 1951, Braekken & Probst 1953, Brandes & Dietrich 1953, 1954, Bramsnaes *et al.* 1954, Taarland *et al.* 1958, Jason 1965, Popa *et al.* 1968, Stoddard 1968, Dambergs 1969, Exler *et al.* 1975, Lang & Christen 1983, Bykov 1985, Herring Industry Board unpublished data.
Clupea pallasi	2–24	58–80	10–20	1–35	56–82	15–21	Dill 1925, Hart *et al.* 1940, Levanidov 1950, Schmidt 1952, Sasa 1958, McBride *et al.* 1959, Kizevetter 1971, Bykov 1974, Kondo 1974, 1975, Stansby 1976, Bykov 1985.
Jack and horse mackerels							
Trachurus capensis	1–9	70–80	17–22	0–9	74–81	17–21	van Wyk 1944, Black & Schwarz 1950b, Simmonds & Tanner 1980, Simmonds & Seaman 1981, Evans & Wessels 1984, Bykov 1985.
Trachurus declivis & novaezelandiae				1–22	60–78	17–25	Popa *et al.* 1968, Pickston *et al.* 1982, Vlieg 1982.
Trachurus japonicus				1–13	68–78	19–24	Kizevetter 1971, Toyomizu *et al.* 1976, Tashiro *et al.* 1981, Saeki & Kumagi 1984, Bykov 1985.
Trachurus symmetricus				3–6	74	20–22	Dill 1925, Kizevetter 1971, Maciejczyk 1974, Golovin 1977.
Trachurus trachurus	3–25			0–23	66–80	15–22	Arevalo 1948, Goncalves Ferreira 1951, Acara 1957, IEO 1961, Paetow *et al.* 1966, Podsevalov & Perova 1975, Smith *et al.* 1980c, Bykov 1985.
Mackerels							
Rastrelliger brachysoma				1–7	72–75	21	Sulit *et al.* 1953, Intengan *et al.* 1956, Keay *et al.* 1972.
Rastrelliger chrysozonus	0–2	79	17				Sulit *et al.* 1953.
Rastrelliger kanagurta	1–8			1–9	67–78	18–25	Chari 1948, Venkataraman & Chari 1951, 1953, Nazir & Magar 1963, Madhavan *et al.* 1970, Gopakumar & Nair 1971, Kizevetter 1971, Balachandran & Vijayan 1976, Mukundan *et al.* 1981, Abdullah & Yu 1985, Bykov 1985, King & Poulter 1985.

Table 9.3 *Continued*

Species	Whole fish			Edible portion			Sources
	Lipid	Water	Crude protein (N × 6.25)	Lipid	Water	Crude protein (N × 6.25)	
Scomber japonicus				1–33	54–77	14–25	Dill 1921a, Ono *et al.* 1959, Jaffe 1960, Paetow *et al.* 1966, Ueda 1976, Golovin 1977, Simmonds & Tanner 1980, Pickston *et al.* 1982, Hale & Brown 1983, Evans & Wessels 1984, Simmonds & Heydenrych 1984, Bykov 1985.
Scomber scombrus	0–25	74	18–19	1–29	54–81	14–24	Plimmer 1921, Cartini & Aloj 1934, Quagliariello 1934, Stansby & Lemon 1941, de las Heras & Mendez 1952, Braekkan & Probst 1953, Bramsnaes *et al.* 1954, Acara 1957, Butler 1958, Taarland *et al.* 1958, Sohn *et al.* 1961, Jason 1965, Coppini 1967, Popa *et al.* 1968, Cuisa & Giaccio 1969, Hardy & Keay 1972, Maciejczyk 1974, Podsevalov & Perova 1975, Lyubavina *et al.* 1975, Exler *et al.* 1975, Ke *et al.* 1976, Maslennikova 1976, Smith *et al.* 1980b, Leu *et al.* 1981, Bykov 1985, Bhuiyan *et al.* 1986.
Menhaden							
Brevoortia tyrannus	2–28	56–78	14–18	4–21	58–80	15–19	Dubrow *et al.* 1976, Durbin *et al.* 1983, Bykov 1985, Bimbo.
Pilchards							
Sardina pilchardus				1–23	57–78	17–21	Carteni & Aloj 1934, Hickling 1945, Goncalves Ferreira 1951, Herrera & Munoz 1957, IEO 1961, Lisac 1963, Abdelmoulek *et al.* 1980, Bykov 1985.
Sardinops caerulea				1–26	56–81	16–22	Dill 1921b, 1926, Bykov 1985.
Sardinops melanosticta	1–23	58–82	15–18	3–32	51–75	16–23	Hayashi & Takagi 1977, Vakhrusheva & Nikiforova 1979, Yamada 1981, Kurokawa 1983, Bykov 1985.
Sardinops neopilchardus				1	55–81	17–22	Bykov 1985.
Sardinops ocellata	1–17	64–75	15–22	4–10	68–73	19–21	Black & Schwarz 1950a,b, Simmonds & Tanner 1980, Bykov 1985.
Sardinops sagax				1–9			Maciejczyk 1974, Pazos *et al.* 1984.
Sardinellas							
Sardinella allexia	1–8	66–76	18–22	1–8	66–76	18–22	Watanabe 1963, Bykov 1985.
Sardinella aurita	2–18	60–77	17–20	0–20	58–77	17–22	Busson *et al.* 1953, Herzberg 1965, Herzberg & Pasteur 1968, 1969, Bykov & Rusin 1971, Diouf *et al.* 1982, Hale & Brown 1983, Hale 1984, Bykov 1985.

Sardinella eba				1–4	76–79	18–23	Busson *et al.* 1953, Bykov 1985.
Sardinella fimbriata				1–2	76–78	19–22	Khorana *et al.* 1943, Setna *et al.* 1944, Chari 1948, Sulit *et al.* 1953, Intengan *et al.* 1956, Bykov 1985.
Sardinella gibbosa				1	74–77	21–23	Bykov 1985.
Sardinella longiceps	2–18	60–75	3–18	1–17	56–80	18–21	Santos & Ascalan 1931, Chari 1948, Sulit *et al.* 1953, Intengan *et al.* 1956, Vasavan *et al.* 1960, Sen & Chaluvaiah 1968, Madhavan *et al.* 1970, Gopakumar & Nair 1971, Shenoy & Pillai 1971, Sen & Revankar 1972, Madhavan *et al.* 1974, Pillai 1974, Joshi & Saralaya 1981, Bykov 1985, Krishnakumar *et al.* 1985.
Sardinella perforata				1		18	Intengan *et al.* 1956.
Sardinella sirm	1–2	75	22–23	1–3	74–76	21–23	Chinivasagam & Vidanapathirana 1985, Nandanie *et al.* 1985.
Sauries							
Cololabis saira				1–28	54–72	18–25	Hata & Tashiro 1953, Tsuchiya *et al.* 1953, Hirao *et al.* 1955, Miwa *et al.* 1958, Kizevetter 1971, Hara *et al.* 1984, Bykov 1985.
Scomberesox saurus	0	75	19	1–11	65–73	22–25	Podsevalov & Perova 1973, Simmonds & Tanner 1980, Bykov 1985.
Scads							
Decapterus kiliche				1–11	66–75	21–22	Bykov 1985.
Decapterus macrosoma				1–2	74	19–22	Sulit *et al.* 1953, Intengan *et al.* 1956.
Decapterus maruadsi				0–4	72–75	21–23	Harao *et al.* 1955, Tsao *et al.* 1980, Bykov 1985.
Decapterus polyaspis				5	71	22	Bykov 1985.
Decapterus punctatus				2–3	74	22–23	Hale 1984.
Decapterus rhonchus				1–5	71–77	19–22	Busson *et al.* 1953, Tulsner 1965, Bykov 1985.
Decapterus russelli				0–1	74–77	20–24	Bykov 1985.
Sprat							
Sprattus sprattus	1–25	61–80	13–17	2–20	65–76	13–17	Plimmer 1921, de las Heras & Mendez 1952, Bramsnaes *et al.* 1954, Taarland *et al.* 1958, Viviani *et al.* 1968, Hardy & Mackie 1969, Krassowska *et al.* 1969, Skachkov & Yudina 1975, Podeszewski & Stodolnik 1976, Bykov 1985.
Tunas							
Auxix thazard				1–16	61–73	18–26	Sulit *et al.* 1953, Yaroslavtseva 1966, Paetow *et al.* 1966, Perova *et al.* 1980, Bykov 1985, Plete *et al.* 1985.
Euthynnus affinis				3–5	69–71	18–19	Mukundan *et al.* 1979, Plete *et al.* 1985.
Euthynnus alleteratus				0–13	63–74	22–27	Busson *et al.* 1953, Perova *et al.* 1980, Bykov 1985.
Katsuwonus pelamis	9	64	23	1–13	57–75	18–29	Karrick & Thurston 1968, Kizevetter 1971, Gopakumar & Nair 1972, George 1975, Perova *et al.* 1980, Vlieg *et al.* 1983, Balogun & Talabi 1985, Bykov 1985, Balogun & Talabi 1986, Balogun *et al.* 1986.

Table 9.3 *Continued*

Species	Whole fish			Edible portion			Sources
	Lipid	Water	Crude protein (N × 6.25)	Lipid	Water	Crude protein (N × 6.25)	
Sarda chiliensis				2–11	73	21–23	Clark & Clough 1926, de Haan 1974.
Sarda orientalis				16	63	19	Bykov 1985.
Sarda sarda				2–9	67–73	23	Van Wyk 1944, Bykov 1985.
Thunnus alalunga				0–21	56–75	19–29	Dill 1925, Clark & Clough 1926, Butler 1958, Dewberry 1959, Karrick & Thurston 1968, Exler *et al.* 1975, Simmonds & Seaman 1981, Bykov 1985, Gallardo *et al.* 1986, Diez Ezkerra & Whittle.
Thunnus albacares				0–18	62–76	21–27	Clark & Clough 1926, Chari 1948, Intengan *et al.* 1956, Butler 1958, Dewberry 1959, Karrick & Thurston 1968, Kovalchuk 1970, Pirazzoli *et al.* 1980, Simmonds & Seaman 1981, Bykov 1985.
Thunnus maccoyii				1–3		24–25	Jowett & Davies 1938, Yaroslavtseva 1966.
Thunnus obesus				0–4	73–76	21–24	Yaroslavtseva 1966, Simmonds & Tanner 1980, Bykov 1985.
Thunnus thynnus				1–21	58–74	18–28	Clark & Clough 1926, Braekkan *et al.* 1955, Karrick & Thurston 1968, Simmonds & Seaman 1981, Bykov 1985.
Blue whitings							
Micromesistius australis				0–3	80–82	14–19	Kizevetter 1971, Podsevalov & Perova 1973, Kakuda & Kitagawa 1979, Bykov 1985.
Micromesistius poutassou	2–12	73–79	14–16	0–1	78–82	14–20	Podsevalov *et al.* 1966, Dagbjartsson 1975, Minder & Gritskevich 1975, Bussmann 1977, 1978, Kuzmicheva & Rekhina 1977, Svensson 1980, Karl 1984, Afolabi *et al.* 1984, Bykov 1985, Torry Research Station unpublished data.
Krill							
Euphausia superba	1–12	73–87	10–17	1	77	19	Budzinski *et al.* 1985, Bykowski 1986.

Table 9.4 Fatty acid composition of commercial samples of oils from pelagic fish. (After Young 1986.)

	Capelin	Herring	Sprat	Mackerel	Menhaden	Sardine/ Pilchard	Horse mackerel	Anchovy
C14:0	7	7	—	8	9	8	8	9
C16:0	10	16	16	14	20	18	18	19
C16:1	10	6	7	7	12	10	8	9
C18:1	14	13	16	13	11	13	11	13
C20:1	17	13	10	12	1	4	5	5
C22:1	14	20	14	15	0.2	3	8	2
C20:5	8	5	6	7	14	18	13	17
C22:6	6	6	9	8	8	9	10	9
Iodine value	94/164	95/160	125/147	136/167	150/200	141/212	158/193	178/200

water contents than non-pelagic species. This has one important economic consequence in industry: fuel costs related to the evaporation of water constitute a major part of the overall costs of fish meal manufacture. Hence the lower the water content in the raw material for fish meal, the greater will be the margin of profit, quite apart from the intrinsic commercial value of the higher yields of fish oil.

It has long been known that the sum of the lipid and water concentrations in fatty fish is approximately 80% (Love 1970) and that for herring at least this relationship holds better for whole fish than for fillets (Brandes & Dietrich 1954). A similar divergence from this simple rule of thumb was found by Iles & Wood (1965) for herring minus guts and gonads. The data of both these pairs of workers show that as lipid content increases, the sum of lipid plus water rises slightly as well. The relationship between water and lipid contents has been used in the development of a fat meter which actually measures water content (Chapter 12). Many other species exhibit a similar co-relationship between fat and water content and the figures in Table 9.3 illustrate this rough complementarity.

Data of the sort contained in Tables 9.2 and 9.3 have been used to calculate the gross nutritional value of various species of fish so that the contribution of fish to nutrition in various countries can be estimated (FAO 1989).

It is beyond the scope of this chapter to consider the detailed composition of all the lipids of pelagic fish, but current concern and interest in the health and nutritional properties of polyunsaturated fatty acids, particularly of the n−3 series, has prompted the inclusion of some data relating to them. Phospholipids are particularly rich sources of n−3 polyunsaturated fatty acids, and the consumption of any fish, even non-pelagic ones, can make a useful contribution in nutritional terms. Where pelagic fish are of importance in this regard is in the context of the preparation of isolates for dietary supplements or pharmaceutical purposes. Table 9.4 lists the fatty acid compositions of commercial samples of fish oils.

The figures here show up a quirk of nature: most of those with greater proportions of the more unsaturated fatty acids (those with 5 and 6 double bonds) come from fish that live in the southern hemisphere – for example off the coasts of southern Africa, Chile and Peru. Fats from fish from the northern hemisphere are not so well endowed in this respect and it is believed that this difference reflects differences in the lipids of the organisms which constitute the food of these two groups of fish (Ackman 1980). There is, however, a disadvantage associated with this phenomenon and that is the greater reactivity of southern fish meals compared with that of the meals which originate in the north: the instability of the residual lipids in the former render the associated meals more liable (in the absence of stabilisation with antioxidants or by other means) to damage by spontaneous over-heating. The question of autoxidation will be covered more fully later on.

Proteins form the next major group of compounds to be considered and, bearing in mind the recent explosive interest around the world in the possible use of pelagic fish mince in surimi based products, the paucity of information on the basic properties of pelagic fish proteins is remarkable. It is beyond the scope of this chapter to review the many studies which have been conducted into the gelling ability of muscle from different species. Suffice it to say that the muscle of pelagic fish would appear in general to have poorer capacities for producing high grade surimi-derived products than that from demersal fish (see other chapters on this). Whether this arises from intrinsic differences in the proteins from the two groups, or from the presence of higher levels of lipid in the pelagic fish, is a moot point (Shimizu *et al.* 1981, Torry Research Station unpublished data). The proportions of actin, myosin and tropomyosin in the muscles of pelagic and non-pelagic fish would appear to be similar, as are the patterns of the protein amino acids.

Differential scanning calorimetry (DSC) is a powerful investigative tool which has much value in studying some of the fundamental properties of materials. In the case of fish muscle proteins, DSC has been used to good effect in studies of the thermal stability of the different protein constituents (Hastings *et al.* 1985) of fish muscle. The results of such studies would appear to demonstrate that thermal stability is more related to the temperature of the environment in which the fish live than to the fact that they live a pelagic or demersal existence. These two factors are not necessarily unrelated since in most parts of the ocean the surface layers of the sea are warmer than the deeper ones.

Still on the subject of proteins, the sarcoplasmic proteins exhibit a greater consistent degree of difference between species than do any other compounds or groups of compounds. When separated by electrophoresis and allied techniques, such as isoelectric focusing, the sarcoplasmic proteins of each species can be used to give a unique set of 'fingerprints'; the pattern obtained for each species is quite different from that of all others (Laird *et al.* 1982, Mackie 1980). There are, however, similarities between related species, i.e. those which have been grouped together by taxonomists in the same genus or the same family, and such electrophoretic techniques have been used as aids in classifying as well as in the elucidation of life histories, by associating particular larval organisms with adults.

All fish, be they pelagic or demersal, rely on adenosine triphosphate (ATP) for energy and the levels of the various purine nucleotides and their breakdown products are similar throughout all muscle tissues. Other minor constituents again show a similarity of occurrences. The only major exception to this worthy of mention is the amino acid histidine. The muscle of pelagic fish, particularly those which are dark-fleshed, fatty and active, contains much more free histidine than that of white fleshed sluggish species although the histidine bond in muscle protein will be similar in all species (Love 1980). This is significant in the availability to decarboxylating enzymes of the free amino acid during post mortem storage. The histamine thus produced is often, but not invariably, associated with the so-called 'scombrotoxin' (see later).

DETERIORATION

In comparison with most other animal products fish deteriorate or spoil rapidly and pelagic fish tend to be more susceptible than other species (Whittle *et al.* 1990). Thus the storage life of sprats or herring is about 5 days in ice, whereas the equivalent storage life of cod is 12 to 15 days. The reasons for this are not entirely clear but the following factors are believed to be important.

(1) Size

Many of the major commercial pelagic species such as anchoveta, mackerel and herring are small and so the surface area to weight ratio is greater than in larger fish. As the major deteriorative processes occur predominantly at the surfaces (Shewan 1971), if all other factors are the same, small fish will tend to spoil more rapidly than large fish.

(2) Metabolism

Pelagic fish tend to be very active in their life-style and have very active enzymes in their tissues (Gildberg 1978). Many of these remain active post mortem causing hydrolysis of proteins and lipids as well as other chemical changes that influence the flavour and eating quality, usually in an adverse manner. (Sikorski *et al.* 1990).

(3) Catching and handling practices

Modern fishing operations can catch pelagic fish in very large quantities and unless extreme care is taken, physical damage will occur from the sheer weight and compression of fish in the net (Smith *et al.* 1980a). This damage is exacerbated at certain times of the year when the fish are soft because of spawning or feeding. This physical damage is not only undesirable from the viewpoints of acceptability and machine processing, but it can disrupt the viscera and cells of other tissues thus releasing bacteria and enzymes which will accelerate deterioration.

In addition, pelagic fish, particularly the smaller species, are rarely eviscerated immediately on capture and are often inadequately chilled. Under these conditions autolysis can occur rapidly in the viscera releasing bacteria and enzymes which then invade the flesh (Gildberg 1978, Whittle *et al.* 1990).

In common with all species of fish there are four major deteriorative processes: physical, enzymic, microbiological and chemical. In many instances the processes are interactive and synergistic.

Physical processes

Because of the shoaling nature of pelagic fish and as a result of modern catching methods and gear, catch rates are high and considerable physical damage can occur. Further physical damage may ensue from bulk storage, leading to quality loss and spoilage as described earlier (Kelman 1982).

In addition, dehydration may occur through storage on deck and excessive liquor loss through bulk storage. The release of these liquors in bulk storage not only causes a loss in yield, but because they will irrigate the fish and have high concentrations of bacteria and enzymes, they will induce faster spoilage in the bulk stored fish.

Enzymic changes

There are three main types that are of importance, namely those that induce rigor mortis, hydrolysis and flavour changes.

Rigor mortis

The rigor reaction is believed to be caused by the slow decline of oxidative enzymic reactions and is initiated by the release of calcium ions (Buege & Marsh 1975) and adenosine triphosphate (ATP) (Bate-Smith & Bendall 1956). Usually small pelagic fish go into rigor rapidly, 2 to 4 hours after death, although this will be moderated by the method of catching, the nutritional status and the temperature (Jones *et al.* 1965, Poulter *et al.* 1981, Yamada *et al.* 1983). The larger pelagics take longer to go into rigor, sometimes up to one day, and stay in rigor for a longer period.

Processing pre-rigor and in-rigor fish causes problems, particularly in the production of fillets. Deformation and shrinkage occur and poorer yields are obtained. However, rigor is often taken to be an indicator of freshness by the consumer and retailer and so a number of methods are being assessed that will extend rigor and possibly help to replace the sale of live fish (Iwamoto *et al.* 1987).

The simplest method of extending rigor mortis is to cool the fish rapidly and maintain it at temperatures just above the freezing point, −1°C to −2°C. Under such conditions the pre-rigor period as well as the time taken for the fish to remain in rigor is increased. Some tropical fish cooling, instead of prolonging the pre-rigor state shortens it, so that the fish immediately goes into a rigor like state (Curran *et al.* 1986). This is called cold shock. Fish so treated give lower fillet yields – 1–2% – and have a higher drip loss, 1–2%, when compared with equivalent fish that have not been so treated. Cold shock can be prevented by storage of the fish at 20°C for 6 hours immediately after catching and then chilling the fish.

Hydrolysis

Proteolysis and lipolysis are the two major hydrolytic reactions.

Proteolysis Proteolytic reactions in pelagic fish are marked and this is particularly true when the enzymes associated with the viscera are present. In krill, for instance, autolysis caused by the proteolytic enzymes is so rapid that the fish become unsuitable for processing after about 8 hours post mortem and chilling has only a modest effect in extending the time during which the fish are suitable for processing. In fatty pelagics, such as herring, proteolysis occurs most markedly in the belly causing them to burst and release enzymes and bacteria to the surrounding fish (Gildberg & Raa 1979, Fik & Magadzia 1983, Whittle *et al.* 1990). These and the endogenous flesh enzymes will autolyse the flesh tissue, which causes softening and the release of lower molecular weight components such as peptides and amino acids (Sakaguchi *et al.* 1984). In turn these latter components will cause discoloration and flavorous changes and may also provide better media for bacterial growth (Mackie 1983, Hashiguchi *et al.* 1984, Hobbs 1987).

Lipolysis The lipids in fish are most commonly present in the form of phospholipids in non-fatty species, and triglycerides, together with phospholipids, in the fatty species (see above). Both lipid classes are hydrolysed by the endogenous enzymes at rates that are dependent on species, nutritional and maturation status, tissue and temperature (Ackman 1980, Hardy 1980).

The major products are fatty acids, glycerol and various amine and phosphoric acid derivatives (Lovern & Olley 1962, Wu & Toyomizu 1974). The liberated fatty acids are suspected of causing protein denaturation but the evidence is not strong (Sikorski *et al.* 1976). In the taste panel assessment of spoiling fatty pelagic fish, soapy and non-oxidative rancidity flavours are often identified and it is suspected that such flavours are caused by the fatty acids liberated by lipolysis. Suggestions have also been made that the fatty acids produced are more susceptible to lipid oxidation (Toyomizu *et al.* 1985).

There is some evidence that fish lipases are relatively heat stable and when fatty fish are converted into fish meal and oil the lipases will continue to hydrolyse the oil even after processing, unless they are removed rapidly and efficiently by centrifugation (Banks 1967). Acids so produced are not desirable and have to be removed before converting the oil to margarine and other edible fats (Young 1986).

Flavour changes

As indicated, both proteolysis and lipolysis can induce the formation of flavorous substances (Jones 1967, Burt 1976). Just how important these are in the spoilage of whole fish is not fully known. In certain products made by causing fish to hydrolyse, however, very strong flavours are developed and are considered to be desirable. Examples of such products are the fish pastes and sauces that are made in south-east Asia. Complete identification of the components responsible for such flavour changes has not been made, but a number of flavorous lipid and protein hydrolytic products have been isolated (Mackie *et al.* 1971). In some of this work it has been noted that peptides formed from protein hydrolysates are flavorous when their molecular weights are less than 2000 daltons, and the nature of their flavour depends on their amino acid composition. Bitter flavours result when there is a high hydrophobic acid content and sweet flavours when the hydrophilic acid content is increased (Mackie 1982).

In addition to such flavorous changes it is often noticed in the storage of fish that fresh flavours are lost, possibly through enzymic action before the development of off-flavours derived from bacterial action (Cuero *et al.* 1985). Other flavorous substances are produced by the action of enzymes. Thus, in the enzymic degradation of ATP, flavour potentiating substances are produced as well as the bitter tasting hypoxanthine. Similarly in the enzymic degradation of trimethylamine oxide, a reaction which is most observable in frozen storage, dimethylamine and formaldehyde are produced which may contribute to both flavour and textural changes (Connell & Shewan 1980).

Microbiological processes

In healthy live fish the flesh is sterile and the bacteria that are present reside predominantly in the viscera and to a much lesser extent on the skin and gills. The types of organisms present on the skin and gills resemble closely those of the environment, so that in colder water psychrophiles predominate and in warmer waters mesophiles will be the major types. Within such classes the actual species present will vary quite markedly (Hobbs 1987).

The visceral flora is conditioned not so much by the environment or indeed by the food as by the conditions that exist in the stomach (Shewan 1977). As such, the organisms in the viscera tend to be facultative anaerobes such as *Vibrio* and *Aeromonas* spp. with, albeit in small numbers, strict anaerobes such as some species of *Clostridium.*

Microbial action and spoilage are conditioned by the types of organisms present, the temperature, the environment and the substrate.

Organisms present

The organisms present are derived from two sources: those that are present in or on the fish and those that are introduced by subsequent handling. Whether they will flourish depends on the factors identified above, although there is some evidence that the presence of large numbers of one species may swamp and limit the growth of another. Major species that induce observable spoilage changes belong to the *Pseudomonas* and *Alteromonas* genera (Hobbs 1987).

Temperature

The organisms found on or in cold water or temperate water fish are adapted to low temperatures (0–15°C) and thus will grow readily at these or even higher temperatures. In practice this means that even when well-handled, small pelagic fish will deteriorate rapidly so that the storage life or acceptable shelf life at 0°C is approximately 5 days and at 20°C less than 1 day.

Tropical fish species on the other hand have present organisms that are not only adapted to higher temperatures but also are not normally associated with the rapid development of malodours. Such fish would not, therefore, be expected to deteriorate as rapidly at temperatures between 0°C and 30°C as those obtained from cooler climates. This is found to be true in many instances and chilling such fish gives extended shelf lives of up to 2 weeks. Where it is not observed, other and so far unidentified factors must be involved (Poulter *et al.* 1981).

Environment

Undoubtedly the environment in which the fish have been living and to which they have been exposed post harvest, is important not only in influencing the types of organisms that are present or are introduced but also in conditioning the growth of the organisms. As most organisms grow on the surface, environmental conditions that affect surfaces or increase surface areas, such as surface drying, washing by spraying or melting ice, removal of oxygen and various forms of butchery, will all have an effect.

In this content, it is interesting to note that the *Alteromonas* and *Pseudomonas* strains that are most associated with undesirable organoleptic changes are aerobes, and thus removal of oxygen should have a marked inhibitory effect on their growth. This is not observed because they can use other components of the substrate such as trimethylamine oxide to replace oxygen (Hobbs & Hodgkiss 1982, Hobbs 1987).

Substrate

There are major differences in the growth of bacteria and in the production of undesirable substances within and between species, that cannot be attributed to the types of organisms present or to temperature or environment. It is postulated that this is due to physical and chemical compositional variations in the substrate, i.e. the fish. It is believed that in the living fish the skin is coated with a slime that is inhibitory to bacteria and reduces bacterial invasion. Lysozyme has been found in some slimes and may be a common inhibitory component (Shewan 1977). In addition some species, such as tuna, have thick skins which also moderate bacterial invasion (Liston 1980).

Pelagic fish, particularly the smaller fatty pelagic species, vary quite markedly in their chemical composition not only with regard to the lipids but also in their minor chemical constituents (Ikeda 1980). Furthermore, post mortem changes and processing or lack of it can further influence the composition through physical changes such as dehydration, enhanced enzymic reactions caused by visceral contamination or cell disruption, and chemical reactions which will compete for oxygen or/and introduce growth or inhibitory factors (Liston 1980, Hobbs 1987).

It is probable that a combination of a number of these factors cause certain small pelagic fatty fish to spoil more quickly than small non-fatty fish.

Products of bacterial action

Little detailed quantitative work has been done in this area (Connell & Shewan 1980, Hobbs 1987) but it does appear that with one major exception, scombrotoxin, the products of bacterial action are similar in composition, but possibly not in content, to those produced in non-pelagic fish. It is beyond the scope of this review to discuss the general subject of the detailed contributions of individual compounds to overall flavour, apart from indicating that without a more detailed knowledge of the chemical nature and sensory properties of these substances, little progress will be achieved in the quantitative assessment of bacterial spoilage. Scombrotoxin will be discussed briefly because of the impact it has on the utilisation of scombroid fishes.

Scombrotoxin Scombrotoxin is occasionally produced in scombroid fish that have been stored for relatively long periods at high ambient temperatures. Nearly all who eat the fish that contain the toxin are affected and the symptoms are mouth tingling, flushing, headaches, palpitations, nausea and diarrhoea. These effects are reminiscent of histamine intoxication and, because histamine was found present in samples of toxic fish, it was assumed that the toxin was histamine (Turnbull & Gilbert 1982). The assumption is not well founded for the following reasons:

(1) Histamine is not always present in toxic samples.
(2) Spoilage in other species produces histamine but these are rarely associated with toxic incidents.
(3) Ingestion of histamine on its own or in the presence of fish, including spoiled but non-toxic scombroid fish, does not produce the full toxic symptoms.
(4) Histamine can be developed readily in laboratory-induced spoilage of mackerel but, to date, no toxin development has been attained.

It must be concluded that the mechanism of production and the chemical nature of the toxin are not known (Chapter 17 and Clifford *et al.* 1989).

Chemical changes

Autoxidation

The most important chemical change is that caused by lipid oxidation and it is the major deteriorative factor in frozen fish (Hardy 1980). The major lipids, triglycerides and phospholipids, contain highly unsaturated fatty acids (PUFA) and it is these that undergo oxidative attack. A large number of products are formed, some of which have characteristic flavours, called oxidative rancidity; others are coloured and give rise to discoloration and some react with other components such as proteins, causing denaturation (Fujimoto & Kaneda 1973, Ikeda 1980).

Lipid autoxidation occurs through a free radical chain mechanism. If free radicals are present in the lipid, together with oxygen and reactive unsaturated fats, then autoxidation will ensue (Bateman 1954). There are a number of components in fish that can produce free radicals either directly, such as trimethylamine oxide, or indirectly, such as lipoxidases, so initiation of the reaction occurs readily. Because the primary products of autoxidation are hydroperoxides which readily decompose to produce more free radicals, then, after initiation, oxidation will continue without additional free radical input (Hardy 1980).

In pelagic fatty fish, triglycerides predominate and it is these lipids that are most susceptible to autoxidation. Structural phospholipids are present in all species but even though they contain more PUFA per molecule, they oxidise less readily than the associated triglycerides because they are constrained within membranes and are inhibited in mobility and ability to participate in the oxidation reaction (Hardy *et al.* 1979).

Autoxidation is influenced by temperature, composition and environment.

Temperature In general, increasing the temperature increases the rate by a factor of 2 to 3 for every 10°C rise in temperature. At temperatures above 0°C, however, other processes such as microbial metabolism may compete for the available oxygen and also swamp the fish with metabolic products so that oxidation is, or appears, less important at temperatures above 0°C (Hardy 1980, Ames & Poulter 1987).

Composition Autoxidation is affected markedly not only by the content and composition of the lipids but also by the presence of other components. Some, as indicated earlier, act as initiators; others, such as the haem proteins, decompose hydroperoxides to produce more free radicals and increase the oxidation rate (Castell & Bishop 1969).

Other components, such as tocopherols, ubiquinones and amines, act as inhibitors by reacting with free radicals and decreasing the oxidation rate. In addition, during processing, substances may be introduced by design, for example salt, acids or antioxidants, or inadvertently, for example heavy metals from the processing machinery. These substances may further inhibit or accelerate autoxidation.

The overall effect of these compositional variations makes it difficult to predict accurately how susceptible to autoxidation any fish or product will be (Hardy 1980).

Environment Autoxidation in the flesh is influenced by the environment of the lipid. In whole fish oxygen has to diffuse through the skin and, particularly where the skin is thick, the oxygen concentration at the lipid sites is reduced. Conversely, when the lipid is freely exposed to air such as on cut surfaces (fillets) or coated surfaces (fish meal) oxidation is rapid. Dehydration in cold stores also permits oxygen easier access to the lipid sites and enhances autoxidation.

The physical state of the lipids is also important. As mentioned previously, some lipids such as the phospholipids are not mobile and cannot take part so readily in the free radical chain mechanism. Much of the depot lipid is contained within fat cells. If these cells are small, approximately 1 μ, oxidation once initiated in the cell should proceed more rapidly than in larger cells (Smith & Ewart 1948). This may be one of the explanations for the observations that depleted fatty fish often oxidise relatively more rapidly than lipid rich fish.

Overall, small fatty pelagics are more susceptible to oxidation than larger fish; for instance, sardines stored at −10°C for 1 month can develop marked rancidity, whereas tuna stored in a similar manner would be unaffected (Ames & Poulter 1987).

Other chemical changes

As expected in such a complex reactive matrix, a number of chemical changes occur in addition to oxidation. The importance of these changes in pelagic fish has not received detailed study but it is known that endogenous components, sometimes made more mobile by enzymic and microbial actions, together with exogenous substances produced by these or other chemical reactions, such as oxidation, will interact (Ikeda 1980). Thus amino groups react with sugars or polyhydroxy compounds to give brown products (non-enzymic browning). Free fatty acids, formaldehyde from trimethylamine oxide degradation and other carbonyls, will react with proteins causing denaturation (Mackie 1984). Other forms of oxidation or co-oxidation with lipids such as the oxidation of sulphur compounds, haem pigments, carotenoids and vitamins, cause loss of nutrients and flavour and colour changes (Hardy 1980, Burt 1988, Chow *et al.* 1988). The haem pigments in many of the pelagics are particularly subject to change not just by oxidation but also by lysis, often causing unacceptable diffuse colour changes in the flesh.

Maintenance of quality

If quality is to be maintained, spoilage or deteriorative processes must be reduced or prevented from occurring.

Control of physical process

The physical deteriorative processes are induced by poor practices in catching and handling on ship and ashore. Large trawl catches must be avoided and, if bulk storage is to be used rather than boxing and icing, chilled sea water or refrigerated sea water should be used. All these methods when used with careful handling will minimise damage due to compression and dehydration, as well as cooling the fish rapidly (Eddie 1980). In frozen storage, dehydration can be reduced by glazing the fish and can be prevented by freezing the fish in water (Hardy *et al.* 1973).

Control of enzymic processes

Enzymic processes are less amenable to control in fish flesh. The rate of enzymic reactions is dependent on temperature and thus lowering the temperature of storage does on the whole reduce enzymic reactions. There are, however, some indications that the rate of some enzymic reactions actually increases around the freezing point (Lovern & Olley 1962). This is not uncommon in a number of other processes in food and the phenomenon is explained by the observation that as the water freezes, the remaining liquid substrate becomes more concentrated and reactions will occur more rapidly as a result of mass action effects. This is believed to be one of the reasons why superchilling, i.e. storage between 0°C and −1.5°C, is less effective than might have been expected.

Even when the storage temperature is −30°C, enzymic reactions still occur, albeit very slowly. Most enzymes are denatured by heat and so another method of control is to cook the food prior to chilling or freezing. With one exception this works well but considerable care has to be taken to prevent subsequent bacterial contamination and oxidation. The exception is with lipases. These enzymes are said to have remarkable thermal stability and can function even after treatment at temperatures in excess of 100°C (Dominova & Pronichkina 1977).

Few chemicals are used to control enzymic processes. Sulphur dioxide or sulphites are used to control polyphenol oxidase reactions which give rise to melanin pigment formation, black spot, in crustacea (Anon 1986). There is some evidence also that the high concentrations of salt used in many processes inhibit enzymic reactions.

Control of bacterial changes

The growth of bacteria can be influenced, retarded or stopped by microbial, biochemical, chemical and physical methods (Hobbs 1987).

The most commonly used microbial method uses lactic acid producing bacteria together with sugar or starch (Blood 1975, Moon *et al.* 1982). Under appropriate conditions the lactic acid bacteria flourish and inhibit other bacterial growth, both because of the competitive effect and because of the effect of the lactic acid and other antibiotic substances which seem to retard growth directly, and also by reducing the micro environmental pH.

Biochemical methods utilise enzymes. With an appropriate substrate, oxidases generate a micro environment, usually on the surface, that is inimical to bacterial growth. Perhaps the one most experimented with is glucose oxidase. It is postulated that the inhibitory effect is caused by the production of glucuronic acid and hydrogen peroxide (Ingram 1980, Wesley 1982).

A very wide range of chemicals are used to inhibit bacterial growth, including salt, smoke constituents, acids, alkalis, sulphites, carbon dioxide (CO_2), ozone, and antibiotics (Hobbs 1987). Of these the most commonly used are salt, smoke and CO_2. When sufficient salt is used most bacteria, with the exception of halophiles, are inhibited, probably because of the osmotic effect of the salt (Dusseault 1958). Wood smoke is chemically complex and many of its constituents such as the phenols, terpenes, acids and carbonyls have bacteriostatic properties and together they can exert a major inhibitory effect on bacteria (Liston 1980). However, the modern smoke cures tend to incorporate into the fish only very low concentrations of smoke with a consequent reduction in protection from bacterial attack (Cann 1988).

It has been known for a long time that CO_2 inhibits bacterial growth (Kolbe 1882) and recently, using modern packaging techniques, CO_2 has been used to extend the shelf life of a variety of fish including some pelagic species. The extension of shelf life tends to vary from species to species and is relatively modest, perhaps 20% at any given temperature between 0°C and 20°C (Barnett *et al.* 1971, Cann *et al.* 1983). The detailed mechanism of inhibition is not known.

Of the remaining chemicals perhaps the most effective and widely used are acid marinades, particularly those containing benzoic and sorbic acids (Doesburg *et al.* 1969, Magnusson & Moller 1985). The latter is particularly effective in inhibiting mould growth and appears to have an additional effect over and above that induced by lowering the pH (Chung & Lee 1982).

Physical methods of control are also numerous and varied and include chilling, freezing, packaging, irradiation, pasteurisation and sterilisation (Hobbs 1987).

Chilling in its various forms, such as icing, chilled sea water or refrigerated sea water, slows down bacterial growth by a factor of approximately 4 for every 10°C reduction in temperature. Alternatively, for many species the rate of deterioration between 0°C and ambient temperature can be described by the equation:

$$R_T = R_0(1 + 0.1\ T)^2$$

where R_T = rate at temperature T°C
R_0 = rate at temperature 0°C
T = temperature in °C

The actual deterioration rate is species dependent (Olley & Ratkowsky 1973, Storey & Owen, unpublished data); thus the shelf life for smaller pelagics caught in cold or temperate waters is approximately 4–5 days at 0°C, whereas tuna has a shelf life of several weeks at 0°C.

Freezing and storage at temperatures below −5°C stop bacterial action and cause some reduction in the viability of the bacteria present. Modern packaging in hermetically sealed bags prevents adventitious contamination and, when used in association with chill storage, will curtail bacterial action (Sumner &

Gorczyca 1981). The effect is enhanced when vacuum packaging or CO_2 gas packaging is used. The overall influence of packaging in extending the shelf life of small fatty pelagic fish is modest.

Irradiation using doses of 3 kGy, when used in conjunction with vacuum packaging will increase the acceptable shelf life by as much as 200–300%. The bacterial load is reduced by a factor of 10^3 but proportionally more of the organisms responsible for organoleptic changes are killed. An additional advantage of irradiation with doses of 3 kGy is the elimination of *Salmonella* (Anon 1981, Hobbs 1987, Whittle *et al.* 1990).

Pasteurisation and sterilisation are effective in extending shelf life because heating above 65°C kills spoilage organisms. The most common process used is sterilisation, which reduces the load of heat resistant bacterial spores by a factor of 10^{12} (McLay 1982). Fatty pelagic fish which are so treated are not only highly acceptable but in hermetically sealed impermeable containers have long shelf lives that are not conditioned by bacterial spoilage and can be stored at ambient temperatures.

Pasteurisation is less frequently used because many organisms, particularly pathogenic spore forming bacteria such as some *Clostridium* species, are relatively unaffected by the process. Consequently, storage at temperatures above +3°C and in the absence of other modifiers of bacterial growth such as salt and acids, will permit these organisms to flourish. As the organisms responsible for spoilage are killed, those products that do not require further cooking may become toxic even though the products may appear to be acceptable to the consumer (Eklund 1982). Notwithstanding this, there has been a resurgence of interest in pasteurised products, particularly in commercial catering (Falkenburg 1989).

Control of chemical changes

The prime method of control of chemical deterioration is to reduce the temperature of storage. In conjunction with other aids such as packaging and chemical inhibitors, most if not all chemical deteriorative processes can be slowed down so that spoilage due to these is of little concern in commercial practice.

In frozen storage, lipid oxidation in fish is reduced by a factor of 2 to 3 for every 10°C reduction in temperature. Small fatty fish such as herring, sprats or mackerel will become rancid in 1 to 2 months at −10°C but can be kept for up to 9 months at −30°C (Hardy 1980).

The rate of oxidation can be inhibited or prevented by controlling the availability of oxygen. The simplest way of doing this is to freeze the fish in water or to glaze them with water. Diffusion of oxygen to the lipids is inhibited and substantial extension of shelf lives, approximately 100%, can be achieved (Smith *et al.* 1980c). This method of control is most suitable for bulk storage. Consumer products are more acceptably and better stored under vacuum in flexible packs constructed from materials of low oxygen permeability. Under such conditions the oxidation rate is so small that oxidation is almost undetectable, even after several years storage at −30°C (Hardy 1980, Ames & Poulter 1987).

This method of control is, however, expensive and not entirely satisfactory for those products which are not consumed immediately after the pack is opened, so other methods are being examined and used (Hiremath & Sreenivasan 1974, Bilinski *et al.* 1981). On the whole, such methods rely on the use of antioxidants which fall into three general categories: those that remove free radical precursors, those that inhibit free radical production and those that react with free radicals.

The major free radical precursors are lipid hydroperoxides. The concentration of these can be reduced, without producing reactive free radicals, by the addition of sulphur compounds such as sulphur dioxide and possibly some amino acids, and by catalases.

A number of substances, such as haem compounds, are present in fish or are adventitiously added to fish during processing, for example heavy metals, and will induce the formation of free radicals from hydroperoxide precursors. Under certain conditions, particularly with heavy metal contaminants, their action can be curtailed by the addition of chelating agents such as phosphates, citrates and ethylene diamine tetracetic acid (Farragut 1974; Hiremath & Sreenivasan 1974).

A number of substances will react with free radicals. Some, such as the tocopherols and ubiquinones, are present in the fish, but there are also a wide range of synthetic materials available which have been approved for food use, such as the gallates, butylated hydroxytoluene and butylated hydroxyanisole, or for use in animal feeds, such as ethoxyquin. All are capable of reducing the rate of lipid oxidation.

Unfortunately there are problems in introducing these substances to the lipid oxidation sites in whole fish and so their efficacy is poor and use is minimal (Sweet 1973, Hardy 1980). In minced fish or fillets there is some evidence that cocktails or mixtures of these various types of inhibitors can be effective and this is most noted in fish meal, where marked reductions in lipid oxidation can be achieved (Thorisson 1987).

ACKNOWLEDGEMENTS

The authors acknowledge the assistance given by Dr A Aitken and several other colleagues.

BIBLIOGRAPHY

Abdelmouleh A., Ktari M.H. & Hadj Ali Salem M. (1980) Seasonal variations in the overall chemical composition of the sardine, *Sardina pilchardus* (Walbaum 1972) from the Bizerta (Tunisia) region. *Bulletin of Institut National Scientifique et Technique d'Oceanographie et de Peche de Salammbo*, **7**, 91–101.

Abdullah M.I. & Yu S.Y. (1985) The effect of freezing and frozen storage on the quality of chub mackerel (*R. kanagurta*). *FAO Fish. Rep.* 317, supplement, Spoilage of tropical fish and product development.

Acara A. (1957) The oil content variation of anchovy, horse mackerel and mackerel in the Black Sea. In *Proceedings and Technical Papers, General Fisheries Council for the Mediterranean*, **4**, 167–74.

Ackman R.G. (1980) Fish lipids part 1. In *Advances in Fish Science and Technology* (Ed. by J.J. Connell), pp. 86–103. Fishing News Books, Oxford.

Ackman R.G., Jangaard P.M., Burgher R.D., Hughes M.L. & MacCallum W.A. (1963) Observations on the oil and component fatty acids of the oil from Newfoundland capelin. *Journal of the Fisheries Research Board of Canada*, **20** (3), 591–6.

Afolabi O.A., Oke O.L. & Whittle K.J. (1984) Quality assessment of laminated fillet blocks from blue whiting (*Micromesistius poutassou*). *Food Chemistry*, **13**, 277–84.

Ames G.R. & Poulter R.G. (1987) Frozen storage life of tropical fish species. *Infofish Int.* **6**, 39–41.

Andrews G.L. (1969) *Capelin analyses*. Progress Report, Atlantic Coast Station, **59**, 6–7. Fisheries Research Board of Canada.

Anon. (1981) Wholesomeness of irradiated food. *Report of Joint FAO/IAEA/WHO Expert Committee*. WHO Technical Report Series 659, HMSO, London.

Anon. (1986) *Studies on the use of sulfites to control shrimp melanosis*. Technical Paper 46, Florida Sea Grant College.

Arevalo A. (1948) Study of the variation in the chemical composition of horse mackerel, *Trachurus trachurus* (L). *Boletin del Instituto Espanol de Oceanografia*, **8**, 1–13.

Atwater W.O. (1892) The chemical composition and nutritive value of food fishes and aquatic invertebrates. In *Commissioner's Report, US Commission of Fish and Fisheries, 1888*, pp. 679–868. Government Printing Office, Washington.

Ayyappan M.P.K., Shenoy A.V. & Gopakumar K. (1976) Proximate composition of 17 species of Indian fish. *Fishery Technology*, **13**, 153–5.

Balachandran K.K. & Vijayan P.K. (1976) Processing aspects of Indian mackerel. A review. *Fishery Technology*, **13** (2), 81–7.

Balogun A.M. & Talabi, S.O. (1985) Proximate analysis of the flesh and anatomical weight composition of skipjack tuna (*Katsuwonus pelamis*). *Food Chemistry*, **17**, 117–23.

Balogun A.M. & Talabi S.O. (1986) Studies on size distribution of skipjack tuna (*Katsuwonus pelamis*): effect on chemical composition and implications for its utilization. *Journal of Food Technology*, **21**, 443–9.

Balogun A.M., Talabi S.O., Sorinmade S.O. & Olokun O. (1986) Product technology of skipjack tuna caught in Nigeria's EEZ. FAO Fisheries Report 329, supplement, pp. 291–9. In *Proceedings of the FAO Expert Consultation on Fish Technology in Africa*.

Banks A. (1967) Deteriorative changes in fish oils. *Fish Oils* (Ed. by M.E. Stansby), pp. 148–63. Avi Publishing Co., Westport.

Barnett H.J., Nelson R.W., Hunter P.J., Baker S. & Groninger H. (1971) Studies on the use of carbon dioxide in refrigerated brine for the preservation of whole fish. *Fishery Bulletin*, US Department of Commerce, **69**, 433–42.

Barrera G.C. & Aguilera M.E.M. (1968) Monthly variation in the contents of water, protein, extractable lipids, their reactivity and tocopherol contents, in Chilean anchovies (*Engraulis ringens*). *Anales de la Facultad de Quimica y Farmacia*, **20**, 24–9.

Bateman L. (1954) Olefin oxidation. *Quarterly Reviews of the Chemical Society*, **8**, 147–67.

Bate-Smith E.C. & Bendall J.R. (1956) Changes in muscle after death. *British Medical Bulletin*, **12**, 230–5.

Bhuiyan A.K.M.A., Ratnayake W.M.N. & Ackman R.G. (1986) Effect of smoking on the proximate composition of Atlantic mackerel (*Scomber scombrus*). *Journal of Food Science*, **51**, 327–9.

Bilinski E., Jonas R.E.E. & Peters M.D. (1981) Treatments affecting the degradation of lipids in frozen Pacific herring. *Canadian Institute of Food Science and Technology Journal*, **14**, 123–7.

Bimbo A.P. Zapata Haynie Corporation, Reedville, Va, USA. Personal communication.

Black M.M. & Schwartz H.M. (1950a) South African fish products. XXX. Seasonal changes in the composition of the pilchard (*Sardina ocellata* Jenyns). *Journal of the Science of Food and Agriculture*, **1**, 182–5.

Black M.M. & Schwartz H.M. (1950b) South African fish products. XXXI. The composition of pilchard oil and maasbanker oil. *Journal of the Science of Food and Agriculture*, **1**, 248–51.

Blood R.U. (1975) *Lactic acid bacteria in marinated herring*, pp. 195–208. The British Food Manufacturing Industries Research Association, Leatherhead.

Bone Q. (1966) On the function of the two types of myotomal muscle fibre in elasmobranch fish. *Journal of the Marine Biological Association of the UK*, **46**, 321–49.

Braekkan O.R. (1959) A comparative study of vitamins in the trunk muscles of fishes. *Fiskeridirektoratets skrifter: Serie Teknologiske undersøkelser*, **3** (8).

Braekkan O.R. & Probst A. (1953) Vitamins in Norwegian fish, I. Contents of nicotinic acid, riboflavin, pantothenic acid, vitamin B12 and vitamin A in the whole fish and different organs of herring (*Clupea harengus*) and mackerel

(*Scomber scombrus*). *Fiskeridirektoratets skrifter: Serie Teknologiske undersøkelser II*, 13.

Braekkan O.R., Hansen K. & Skogland T. (1955) Vitamins in Norwegian fish, II. *Fiskeridirektoratets skrifter: Serie Teknologiske undersøkelser III*. 3, 1–18.

Bramsnaes F., Mogensen E. & Birno K.E. (1954) Fat content of herring, sprat and mackerel 1949–52. *Meddelelser fra Fiskeriministeriets Forsøgslaboratorium*, February, 1–16.

Brandes C.H. & Dietrich R. (1953) Uber die Fettverteilung im Körper des Herings. *Veröffentlichungen des Instituts fur Meeresforschung in Bremerhaven*, **2** (1) 109–21. Die Bestimmung des Fettgehaltes im essbaren Anteil des Herings auf Grund der Korrelation Fett und Wasser. *Veröffentlichungen des Instituts fur Meeresforschung in Bremerhaven*, 2 (1) 122–34. Zusammenfassende Betrachtungen uber den Fett und Wasser-Gehalt im essbaren Anteil des Herings. *Fette Seifen Anstrichmittel*, **55**, 533–41.

Brandes C.H. & Dietrich R. (1954) A new nitrogen factor for the calculation of the protein content in the edible part of the herring. *Fette Seifen Anstrichmittel*, **56**, 499–503.

Budzinski E., Bykowski P. & Dutkiewicz D. (1985) Possibilities of processing and marketing of products made from Antarctic krill. *FAO Fisheries Technical Paper* 268, FAO, Rome.

Buege D.R. & Marsh B.B. (1975) Mitochondrial calcium and post-mortem muscle shortening. *Biochemical and Biophysical Research Communications*, **65**, 478–85.

Burt J.R. (1976) Hypoxanthine: a biochemical index of fish quality. *Process Biochemistry*, **11**, 17–19.

Burt J.R. (1988) The effect of drying and smoking on the vitamin content of fish. In *Fish Smoking and Drying* (Ed. by J.R. Burt). Elsevier Applied Science. London.

Bussmann B. (1977) Tests on new processing possibilities for blue whiting. *Allgemeine Fischwirtschaftszeitung*, **29** (18) 4–5.

Bussmann B. (1978) Evaluation of the processing characteristics of blue whiting. *Lebensmittelchemie und Gerichtliche Chemie*, **32**, 86.

Busson F., Postel E. & Giraud P. (1953) Food value of fish caught off the coasts of the Cape Verde peninsula. *Médecin tropical*, **13**, 534–7.

Butler C. (1958) Nutritional value of fish in reference to atherosclerosis and current dietary research. *Commercial Fisheries Review*, **20** (7) 7–16.

Bykov V.P. (1974) Opportunities for upgrading fish with lower market value. In *Fishery Products* (Ed. by R. Kreuzer), pp. 153–6. Fishing News Books, Oxford.

Bykov V.P. (1985) *Marine Fishes*. Russian Translation Series 7, A.A. Balkema, Rotterdam.

Bykov V.P. & Rusin A.A. (1971) Technochemical composition of some species of fish of the Campeche Bank. *Rybnoe Khozyaistvo* **12**, 72–3.

Bykov V.P., Makarov O.E., Tishin V.E. & Kvan E.A. (1971) Technological studies on some fish species of the Indian Ocean. *Trudy VNIRO*, **72**, 123–42.

Bykowski P.J. (1986) Antarctic krill – possibilities of fishing and utilization. *Studia i materialy*. Seria D (15), Sea Fisheries Institute, Gdynia.

Cann D.C. (1988) Modified atmosphere packaging of fishery products. *Infofish Int.* **1**, 37–9.

Cann D.C., Smith G.L. & Houston N.C. (1983) *Further studies on marine fish stored under modified atmosphere packaging*. Torry Research Station, Aberdeen.

Carteni A. & Aloj G. (1934) Chemical composition of marine animals of the Gulf of Naples. *Quaderni della Nutrizione*, **1**, 49–63.

Castell C.H. & Bishop D.M. (1969) Effect of hematin compounds on the development of rancidity in muscle of cod, flounder, scallops and lobster. *Journal of the Fisheries Research Board of Canada*, **26**, 2299–309.

Chari S.T. (1948) Nutritive values of some of the west coast marine food fishes of the Madras Province. *Indian Journal of Medical Research*, **36**, 253–9.

Chinivasagam H.N. & Vidanapathirana G.S. (1985) *Quality changes and bacterial flora associated with trench sardines (Amblygaster sirm) under delayed icing conditions*. FAO Fisheries Report 317, supplement, pp. 1–7. FAO, Rome.

Chow C.J., Ochiai Y., Watabe S. & Hashimoto K. (1988) Autoxidation of bluefin tuna myoglobin around freezing point. *Bulletin of the Japanese Society of Scientific Fisheries*, **54**, 473–8.

Chung Y.M. & Lee J.S. (1982) Potassium sorbate inhibition of microorganisms isolated from seafood. *Journal of Food Protection*, **45**, 1310–13.

Clark E.D. & Almy L.H. (1918) A chemical study of food fishes. *Journal of Biological Chemistry*, **33**, 483–98.

Clark E.D. & Clough R.W. (1926) *Nutritive value of fish and shellfish*. Bureau of Fisheries Document 1000, pp. 502–26. Department of Commerce, Washington.

Clifford M.N., Walker R., Wright J., Hardy R. & Murray C.K. (1989) Studies with volunteers on the role of histamine in suspected scombrotoxicosis. *Journal of the Science of Food and Agriculture*, **47**, 365–75.

Connell J.J. & Shewan J.M. (1980) Sensory and non-sensory assessment of fish. In *Advances in Fish Science and Technology*, (Ed. J.J. Connell), pp. 56–65. Fishing News Books, Oxford.

Coppini R. (1967) Study of variations in the chemical composition of the flesh of mackerel from the middle western Adriatic, particularly with regard to lipids. *Proceedings and Technical Papers, General Fisheries Council for the Mediterranean*, **8**, 395–9.

Cuero R.C., Valverde P.J., Gamboa J.H. & Restrepo J. (1985) Storage life and characteristics of *Scomberomorus sierra* and the sardine, *Opistonema libertate*, in RSW, ice and freezing storage. *FAO Informes Pesca*, **340**, 140–54.

Cuisa W. & Giaccio M. (1969) Report on nutritive value and market prices of the more important fish species of the Adriatic Coast. *Quaderni di Merceologia*, **8**, 1–40.

Curran C.A., Poulter R.G., Bruten A., Jones N.R. & Jones N.S.D. (1986) Cold shock reactions in iced tropical fish. *Journal of Food Technology*, **21**, 301–10.

Dagbjartsson B. (1975) Utilization of blue whiting (*Micromesistius poutassou*) for human consumption. *Journal of the Fisheries Research Board of Canada*, **32**, 747–51.

Dambergs N. (1969) Isopropanol-water mixtures for the production of fish protein concentrate from Atlantic herring (*Clupea harengus*). *Journal of the Fisheries Research Board of Canada*, **26**, 1919–23.

de Haan P.W. (1974) Alternative uses of bonito in Peru. *Fishery Products* (Ed. by R. Kreuzer), pp. 227–33. Fishing News Books, Oxford.

de las Heras A.R. & Mendez Isla M.C. (1952) Contribution to the chemical study of Spanish fish. *Anales de Bromatologia*, **4**, 403–10.

Dewberry E.B. (1959) Tuna canning. *Food*, **28**, Mar., 87–9.

Diez Ezkerra F. & Whittle K.J. Torry Research Station, Aberdeen, Scotland. Unpublished results.

Dill D.B. (1921a) A chemical study of certain Pacific Coast fishes. *Journal of Biological Chemistry*, **48**, 73–82.

Dill D.B. (1921b). A chemical study of the California sardine (*Sardinea caerulea*). *Journal of Biological Chemistry*, **48**, 93–103.

Dill D.B. (1925) The proximate composition of certain Pacific Coast fishes. *Industrial and Engineering Chemistry*, **17**, 629–30.

Dill D.B. (1926) A comparative study of the chemical composition of the sardine (*Sardinea caerulea*) from California and British Columbia. *Ecology*, **1**, 221–8.

Diouf N., Gning D., Faye A.A., Samba A., Kandji P., Karnicki Z.S., Lima dos Santos C.A.M. & Barhoumi M. (1982) *Study of the preservation of sardinella and sea bream by ice and refrigerated sea water*. FAO Fisheries Report 268, supplement, pp. 15–26. FAO Expert Consultation on Fish Technology in Africa.

Doesburg J.J., Lamprecht E.C., Elliot M.C. & Reid D.A. (1969) The use of sorbic acid in salted fish. *Journal of Food Technology*, **4**, 339–52.

Dominova S.R. & Pronichkina A.V. (1977) Variations of the lipids in canned Caspian sardines in oil during storage. *Rybnoe Khozyaistvo*, **3**, 58–60.

Dubrow D., Hale M. & Bimbo A. (1976) Seasonal variations in chemical composition and protein quality of menhaden. *Marine Fisheries Review*, Sept. 12–16.

Durbin A.G., Durbin E.G., Smayda T.J. & Verity P.G. (1983) Age, size, growth and chemical composition of Atlantic menhaden, *Brevoortia tyrannus*, from Narragansett Bay, Rhode Island. *Fishery Bulletin*, **81** (1), 133–41.

Dusseault H.P. (1958) The salt tolerance of bacteria from lightly-salted fish. In *Microbiology of Fish and Meat Curing Brines* (Ed. by B.P. Eddy), pp. 61–6. HMSO, London.

Eaton C.A., Ackman R.G., Tocher C.S. & Spencer K.D. (1975) Canadian capelin 1972–3. Fat and moisture composition, and fatty acids of some oils and lipid extract triglycerides. *Journal of the Fisheries Research Board of Canada*, **32**, 507–13.

Eddie G.C. (1980) Past, present and future of fish handling methods. In *Advances in Fish Science and Technology* (ed. by J.J. Connell), pp. 18–28. Fishing News Books, Oxford.

Eklund M.W. (1982) Significance of *Clostridium botulinum* in fishery products preserved short of sterilization. *Food Technology*, **36**, 107–12.

Evans A.A. & Wessels J.P.H. (1984) *Fatty acid composition of various fresh and canned products and one smoked fish product*. Annual Report, **38**, 115–17. Fishing Industry Research Institute. Cape Town.

Exler J., Kinsella J.E. & Watt B.K. (1975) Lipids and fatty acids of important finfish: new data for nutrient tables. *Journal of the American Oil Chemists' Society*, **52**, 154–9.

Falkenburg J. (1989) The big chill. How safe are 'sous vide' refrigerated foods? *FDA Consumer*, **23** (7), 32, 35–6.

FAO (1989) Yield and nutritional value of the commercially more important fish species. *FAO Fisheries Technical Paper no. 309*. Food and Agriculture Organization of the United Nations, Rome.

Farragut R.N. (1974) *Effect of some antioxidants and EDTA on the development of rancidity in Spanish mackerel during frozen storage*. NOAA (US) Technical Report NMFS SSRF 6501V.

Fik M. & Magadzia A. (1983) Effect of storage in ice on activity of proteolytic enzymes and enzymatic hydrolysis of proteins in Baltic sprat and herring. *Proceedings of 16th International Congress of Refrigeration*, Paris. **3** (2), 639–43.

Fraga F. (1955) Seasonal variation of the chemical composition of the anchovy (*Engraulis encrasicolus*). *Investigación pesquera*, **2**, 21–31.

Fujikawa K. & Naganuma H. (1936) Chemical composition of sardine, *Sardinia melanosticta* (C&V) from Tyôsen. I, Comparative study on dark muscle and white muscle. *Bulletin of the Japanese Society of Scientific Fisheries*, **5**, 95–102.

Fujimoto K. & Kaneda T. (1973) Studies on the brown discoloration of fish products. 5., Reaction mechanism in the early stage. *Bulletin of the Japanese Society of Scientific Fisheries*, **39**, 185–90.

Gallardo J.M., Martin R.P. & Aubourg S. (1986) *Thermal treatments and their effects on the chemical constituents during canning of albacore* (*Thunnus alalunga*). COST 91 bis subgroup 1, first workshop on data collection and analysis, Berlin, (EEC 1986).

George C. (1975) Biochemical differences between the red and white meat of tuna and changes in quality during freezing and storage. *Fishery Technology*, **12**, 70–4.

Gildberg A. (1978) Proteolytic activity and frequency of burst bellies in capelin. *Journal of Food Technology*, **13**, 409–16.

Gildberg A. & Raa J. (1980) Tissue degradation and belly bursting in capelin. In *Advances in Fish Science and Technology* (Ed. by J.J. Connell), pp. 255–8. Fishing News Books, Oxford.

Gimenez J.C. & de las Heras A.R. (1943) Composition and energy value of some fish eaten in Madrid. *Trabajos del Instituto Espanol de Oceanografia*, **17**.

Golovin A.N. (1977) Technological properties of some species of fish from the south-east part of the Pacific Ocean. *Rybnoe Khozyaistvo*, **8**, 69–71.

Goncalves Ferreira F.A. (1951) Composition and food value of some species of fish. *Boletim da Pesca*, **8** (33), 89–102.

Gopakumar K. & Nair M.R. (1971) Phospholipids of five Indian food fishes. *Fishery Technology*, **VIII** (2) 171–3.

Gopakumar K. & Nair M.R. (1972) Fatty acid composition of eight species of Indian marine fish. *Journal of the Science of Food and Agriculture*, **23**, 493–6.

Greer-Walker M. & Pull G.A. (1975) A survey of red and white muscle in marine fish. *Journal of Fishery Biology*, **7**, 295–300.

Hale M.B. (1984) Proximate chemical composition and fatty acids of three small coastal pelagic species. *Marine Fisheries Review*, **46** (1) 19–21.

Hale M.B. & Brown T. (1983) Fatty acids and lipid classes of three underutilized species and changes due to canning. *Marine Fisheries Review*, **45** (4–6) 45–8.

Hara M., Ito K. & Hata M. (1984) Difference of lipid contents in migrations of Pacific saury. *Tohoku Journal of Agricultural Research*, **34** (3–4) 65–72.

Hardy R. (1980) Fish lipids, Part II. In *Advances in Fish Science and Technology* (Ed. by J.J. Connell), pp. 103–11. Fishing News Books, Oxford.

Hardy R. & Keay J.N. (1972) Seasonal variations in the chemical composition of Cornish mackerel, *Scomber scombrus* (L), with detailed reference to the lipids. *Journal of Food Technology*, **7**, 125–37.

Hardy R. & Mackie P. (1969) Seasonal variation in some of the lipid components of sprats (*Sprattus sprattus*). *Journal of the Science of Food and Agriculture*, **20**, 193–8.

Hardy R., McGill A.S. & Gunstone F.D. (1979) Lipid and autoxidative changes in cold stored cod (*Gadus morhua*). *Journal of the Science of Food and Agriculture*, **30**, 999–1006.

Hardy R., Smith J.G.M. & Young K.W. (1973) Influence of packaging in fish processing and technology. *Proceedings of conferences in conjunction with the 8th International Exhibition for the Food and Allied Industries*. BPS Exhibition Ltd, 4 Seaford Court, 220–222 Great Portland Street, London SW1.

Hart J.L., Tester A.L., Beall D. & Tully J.P. (1940) Proximate analysis of British Columbia herring in relation to season and condition factor. *Journal of the Fisheries Research Board of Canada*, **4** (5) 478–83.

Hashiguchi M., Suzuki K. & Matsumoto F. (1984) Studies on freshness and rottenness of fresh fish. I, Changes in k value and deterioration of total lipids of fresh fish during chilled storage. *Journal of the Japanese Society of Food Science and Technology*, **3**, 1–9.

Hastings R.J., Rodger G.W., Park R., Matthews A.D. & Anderson E.M. (1985) Differential scanning calorimetry of fish muscle: the effect of processing and species variation. *Journal of Food Science*, **50**, 503–6.

Hata M. & Tashiro M. (1953) Biochemical studies on skipper (*Cololabis saira*). II, Moisture and crude fat content. *Bulletin of the Japanese Society of Scientific Fisheries*, **19**, 518–24.

Hayashi K. & Takagi T. (1977) Seasonal variation in lipids and fatty acids of sardine, *Sardinops melanosticta*. *Bulletin of the Faculty of Fisheries, Hokkaido University*, **28**, 83–94.

Hayashi K. & Takagi T. (1978) Seasonal variations in lipids and fatty acids of Japanese anchovy (*Engraulis japonica*). *Bulletin of the Faculty of Fisheries, Hokkaido University*, **29** (1) 38–47.

Herrera J. & Munoz F. (1957) Biological aspects of the chemical composition of the sardine (*Sardina pilchardus* Walb.) from Castellon. *Investigaciones Pesqueras*. **7**, 33–48.

HIB. Herring Industry Board (UK), unpublished.

Herzberg A. (1965) Preliminary data on the fat content of *Sardinella aurita, Mullus barbatus* and *Scomber japonicus*. In *Proceedings of General Fisheries Council of the Mediterranean*, **8**, 417–20.

Herzberg A. & Pasteur R. (1968) Seasonal changes in fat content of *Sardinella aurita, Euthynnus alleteratus* and *Scomber japonicus* in the eastern Mediterranean. *Rapport Committée International Mer Méditerranée*. **19** (2) 257–9.

Herzberg A. & Pasteur R. (1969) Proximate composition of commercial fishes from the Mediterranean Sea and the Red Sea. *Fishery Industrial Research*, **5**, 39–65.

Hickling C.F. (1945) The seasonal cycle in the Cornish pilchard, *Sardina pilchardus* Walbaum. *Journal of Marine Biological Association UK*. **26**, 115–38.

Hirao S., Yamada J. & Kikuchi R. (1955) Vitamin A in fish meat. III, Individual fluctuation in the vitamin A content in fish meat. *Bulletin of the Japanese Society of Scientific Fisheries*, **20**, 853–9.

Hiremath G.G. & Sreenivasan N. (1974) Studies on the prevention of quality loss in frozen seer fillets during storage by use of additives. *Journal of Agricultural Science*, **13**, 88–92.

Hobbs G. (1987) Microbiology of fish. In *Essays in Food and Agricultural Microbiology* (Ed. by J.R. Norris & G.L. Pettipher), pp. 199–226. John Wiley, Chichester.

Hobbs G. & Hodgkiss W. (1982) The bacteriology of fish handling and processing. In *Developments in Food Microbiology* (Ed. by R. Davies), pp. 71–117. Applied Science Publishers, Barking, Essex.

IEO (1961) Instituto Espanol de Oceanografia. Seasonal variations in the chemical composition of the Mediterranean and Atlantic sardines. *Proceedings and Technical Papers, General Fisheries Council for the Mediterranean*, **6**, 175–82.

Ikeda S. (1980) Other organic components and inorganic components. In *Advances in Fish Science and Technology* (Ed. by J.J. Connell), pp. 111–24. Fishing News Books, Oxford.

Iles T.D. & Wood R.J. (1965) The fat/water relationship in North Sea herring (*Clupea harengus*) and its possible significance. *Journal of the Marine Biological Association of the UK*. **45**, 353–66.

Ingram G.A. (1980) Substances involved in the natural resistance of fish to infection – a review. *Journal of Fish Biology*, **16**, 23–60.

Intengan C.L., Alejo L.G., Concepcion I., Corpus V.A., Salud R.D., Rosario I.d., Gomez R. & Henson J. (1956) Composition of Philippine Foods, V. *Philippine Journal of Science*, **85**, 203–13.

Iwamoto M., Yamanaka H., Watabe S. & Hashimoto K. (1987) Effect of storage temperature on rigor mortis and ATP degradation in plaice *Paralichthys olivaceus* muscle. *Journal of Food Science*, **52**, 1514–17.

Jaffe W.G. (1960) The food value of fresh fish, fish preserves and fishmeal. *Zeitschrift fur Lebensmitteluntersuchung und -forschung*, **113**, 472–79.

Jangaard P.M. (1974) The capelin (*Mallotus villosus*). Biology, distribution, exploitation, utilization and composition. *Bulletin of the Fisheries Research Board of Canada*, 186, Fisheries and Marine Service, Ottawa.

Jason A.C. (1965) Effects of fat content on diffusion of water in fish muscle. *Journal of the Science of Food and Agriculture*, **16**, 281–8.

Johnstone J. (1914) The fat content of Irish Sea herring. *Lancashire Sea Fisheries Laboratory Report*, **23**, 154–61.

Jones N.R. (1967) Fish flavours. In *Symposium on Food: The Chemistry and Physiology of Flavours* (Ed. by H.W. Schultz, E.A. Day & L.M. Libbey), pp. 267–95. The Avi Publishing Co., Westport.

Jones N.R., Burt J.R., Murray J. & Stroud G.D. (1965) Nucleotides and the analytical approach to the rigor mortis problem. In *The Technology of Fish Utilization* (Ed. by R. Kreuzer), pp. 14–20. Fishing News Books, Oxford.

Joshi V.R. & Saralaya K.V. (1981) Studies on the effect of precooking in fish (*Sardinella longiceps*). Canning, II, On major chemical components. *Mysore Journal of Agricultural Science*, **15**, 142–50.

Jowett W.G. & Davies W. (1938) *A chemical study of some Australian fish*. Pamphlet 85, pp. 1–40. Council for Scientific and Industrial Research, Melbourne.

Kakuda K. & Kitagawa S. (1979) Objective and sensory tests for food quality and freezing stability of fish caught by the 'Kaiyo-Maru', a stern trawler. II. *Bulletin of the Tokai Regional Fisheries Research Laboratory*, **100**, 53–65.

Karl H. (1984) Canning of blue whiting. II. *Fleischwirtschaft*. **64**, 487–9.

Karrick N.L. & Thurston C.E. (1968) Proximate composition and sodium and potassium contents of four species of tuna. *Fishery Industrial Research*, **4**, 73–81.

Ke P.J., Nash D.M. & Ackman R.G. (1976) Quality preservation in frozen mackerel. *Canadian Institute of Food Science and Technology Journal*, **9**, 135–8.

Keay J.N. & Hardy R. (1974) *Application of bone separation techniques to pelagic and underutilized species and the preparation of derived products*. Second technical seminar on mechanical recovery and utilization of fish flesh (Ed. by R.E. Martin), pp. 88–100. National Fisheries Institute, Washington.

Keay J.N., Rattagool P. & Hardy R. (1972) Chub mackerel of Thailand (*Rastrelliger neglectus*, Van Kempen): a short study of its chemical composition, cold storage and canning properties. *Journal of the Science of Food and Agriculture*, **23**, 1359–68.

Kelman J.H. (1982) Handling wet fish at sea. In *Fish Handling and Processing* (Ed. by A. Aitken, I.M. Mackie, J.H. Merritt & M.L. Windsor), pp. 28–41. HMSO Edinburgh.

Khorana M.L., Sarma M.L., Rao P.S. & Giri K.V. (1943) Investigations on the food value of fish and other marine products. II, The protein and mineral contents. *Indian Journal of Medical Research*, **31**, 25–7.

King D.R. & Poulter R.G. (1985) Frozen storage of Indian mackerel (*Rastrelliger kanagurta*) and big eye (*Priacanthus hamrus*). *Tropical Science*, **25**, 79–90.

Kizevetter I.V. (1971) *Technological and chemical characteristics of commercial fish of the Pacific Ocean basin*. TNIRO, Vladivostok.

Kolbe H. (1882) Antiseptische eigenschaften der kohlensaure. *Journal fur praktische Chemie*, **26**, 249–55.

Kondo H. (1974) Studies on the lipids of herring. I, The lipids of the north-eastern Kamchatka herring. *Bulletin of the Faculty of Fisheries, Hokkaido University*, **25**, 68–77.

Kondo H. (1975) Studies on the lipids of herring. II, The lipids of the northern Okhotsk herring. *Bulletin of the Faculty of Fisheries, Hokkaido University*, **26**, 289–301.

Konstantinova L.L. & Minder L.P. (1975) On the question of the chemical composition and biochemical properties of capelin. *Trudy PINRO* **36**, 125–39.

Kordyl E. (1951) *Chemical composition of the Baltic cod and herring in relation to the degree of sexual maturity*. Report 6, pp. 145–57, Sea Fisheries Institute, Gdynia.

Kovalchuk G.K. (1970) The technochemical properties of the yellowfin tuna of the Indian Ocean. *Rybnoe Khozyaistvo* **2**, 65–9.

Krassowska T., Nagiello K. & Lewczuk Z. (1969) Fat content of the Baltic sprat. *Przemysl Spozywczy*, **23**, 340–1.

Krishnakumar S., Hiremath G.G., Menon N.R. & Shetty H.P.C. (1985) Preservation of *Sardinella longiceps* in iced and chilled seawater. Part 1, Changes during storage with particular reference to bacterial load and nitrogeneous compounds. *Fishery Technology*, **22**, 126–31.

Kurokawa T. (1983) Chemical component of sardine caught in the coast of Nagasaki Prefecture. *Bulletin of the Japanese Society of Scientific Fisheries*, **49**, 1741–46.

Kuzmicheva G.M. & Rekhina N.I. (1977) The effect of stabilizers on the amount of soluble protein in fish mince. *Rybnoe Khozyaistvo* **10**, 71–3.

Laird W.M., Mackie I.M. & Ritchie A.H. (1982) Differentiation of species of fish by isoelectric focusing on agarose and polyacrylamide gels – a comparison. *Journal of the Association of Public Analysts*, **20**, 125–35.

Lang K. & Christen U. (1983) The composition of the muscle meat of a number of sea fish of nutritional importance. *Fleischwirtschaft*, **63**, 967–70.

Leu S-S., Jhaveri S.N., Karakoltsidis P.A. & Constantinides S.M. (1981) Atlantic mackerel (*Scomber scombrus* L.): seasonal variation in proximate composition and distribution of chemical nutrients. *Journal of Food Science*, **46**, 1635–8.

Levanidov I.P. (1950) Chemical composition of the flesh of herring from the west coast of Sakhalin. *Rybnoe Khozyaistvo*, **2**, 37–40.

Lisac H. (1963) Variations in the fat content of Adriatic sardines during 1961. In *Proceedings and Technical Papers, General Fisheries Council for the Mediterranean*, **7**, 237–9.

Liston J. (1980) Microbiology in fishery science. In *Advances in Fish Science and Technology* (Ed. by J.J. Connell), pp. 138–57. Fishing News Books, Oxford.

Love R.M. (1970) *The Chemical Biology of Fishes*. Academic Press, London.

Love R.M. (1980) *The Chemical Biology of Fishes*. Vol. 2. Academic Press, London.

Love R.M. (1988) *The Food Fishes: their intrinsic variation and practical implications*. Farrand Press, London.

Lovern J.A. & Olley J. (1962) Inhibition and promotion of post mortem lipid hydrolysis in the flesh of fish. *Journal of Food Science*, **27**, 551–9.

Lovern J.A. & Wood H. (1937) Variations in the chemical composition of herring. *Journal of the Marine Biological Association of the UK*, **22**, 281–93.

Lyubavina L.A., Bakaneva G.F. & Dubintskaya G.M. (1975) Chemical composition and properties of the fat of some species of fish. *Trudy PNIRO* **36**, 160–70.

McBride J.R., MacLeod R.A. & Idler D.R. (1959) Proximate analysis of Pacific herring (*Clupea pallasii*) and an evaluation of Tester's 'Fat factor'. *Journal of the Fisheries Research Board of Canada*, **16** (5), 679–84.

MacCallum W.A., Adams D.R., Ackman R.G., Ke P.J., Dyer W.J., Fraser D.I. & Punjamapirom S. (1969) Newfoundland capelin: proximate composition. *Journal of the Fisheries Research Board of Canada*, **26**, 2027–35.

Mackie I. (1980) A review of some recent applications of electrophoresis and iso-electric focusing in the identification of species of fish in fish and fish products. In *Advances in Fish Science and Technology* (Ed. by J.J. Connell), pp. 444–50. Fishing News Books, Oxford.

Mackie I.M. (1982) General review of fish protein hydrolysates. *Animal Food Science and Technology*, **7**, 113–24.

Mackie I.M. (1983) New approaches in the use of fish proteins. In *Developments in Food Proteins* (Ed. by B.J.F. Hudson), pp. 215–62. Applied Science Publishers, London.

Mackie I.M. (1984) Advances in analytical techniques for evaluating the effects of freezing on the properties and characteristics of fish proteins. In *Proceedings of the IUFoST International Symposium, Chemical changes during food processing*, **II**, 91–106.

Mackie I.M., Hardy R. & Hobbs G. (1971) *Fermented fish products*. FAO Fisheries Report 100, FAO, Rome.

McLay R. (1982) Canning. In *Fish Handling and Processing* (Ed. by A. Aitken, I.M. Mackie, J.H. Merritt & M.L. Windsor), pp. 115–25. HMSO, Edinburgh.

Maciejczyk J. (1974) Technological value of Peruvian fish. *Przemysl Spozywczy*, **28**, 305–8.

Madhavan P., Balachandran K.K. & Choudhuri D.R. (1970) Suitability of iced stored mackerel and sardine for canning. *Fishery Technology*, **7** (1) 67–72.

Madhavan P., Nair T.S.U. & Balachandran K.K. (1974) A review on oil sardine. 1, Distribution, preservation and transportation. *Fishery Technology*, **11**, 88–92.

Magnusson H. & Moller A. (1985) Ropiness in the brine of sugar-salted herring. *Int. Journal of Food Microbiology*, **1**, 253–61.

Maslennikova N.V. (1976) Content of fat and protein in mackerel of varying size. *Rybnoe Khozyaistvo*, **5**, 62–3.

Milroy T.H. (1906) *The food value of the herring*, pp. 83–107. 24th Annual Report of the Fishery Board for Scotland.

Milroy T.H. (1909) Changes in the chemical composition of the herring during the reproductive period. *Biochemical Journal*, **3**, 366–90.

Minder L.P. & Gritskevich K.I. (1975) Food and technical value of blue whiting. *Trudy PNIRO* **36**, 91–106.

Miwa K., Sato Y. & Kinoshita E. (1958) Studies on the utilization of migratory fish. II, On the constituent parts of body of saury and their chemical components. *Bulletin of the Hokkaido Regional Fisheries Research Laboratory*, **19**, 66–71.

Miyauchi D. & Steinberg M. (1970) Machine separation of edible flesh from fish. *Fishery Industrial Research*, **6**, 165–71.

Moon N.J., Beuchat L.R., Kinkaid D.T. & Hayes E.R. (1982) Evaluation of lactic acid bacteria for extending the shelf life of shrimp. *Journal of Food Science*, **47**, 897–900.

Mukundan M.K., James M.A., Radhakrishnan A.G. & Antony P.D. (1979) Red and white meat of tuna (*Euthynnus affinis*): their biochemical role and nutritional quality. *Fishery Technology*, **16**, 77–82.

Mukundan M.K., Radhakrishnan A.G., James M.A. & Nair M.R. (1981) Comparative study of the nutrient content of fish and shellfish. *Fishery Technology*, **18**, 129–32.

Nandanie W.M.C., Jayasinghe C. & Fonseka T.S.G. (1985) *A preliminary study on the seasonal variation in the storage pattern and proximate composition of iced spotted sardine (Sardinella sirm)*. FAO Fisheries Report 317, supplement, Spoilage of tropical fish and product development, pp. 82–6.

Nazir D.J. & Magar N.G. (1963) Chemical composition of mackerel (*Rastrelliger kanagurta*) and changes in the nutritive value during storage. *Indian Journal of Technology*, **1**, 247–9.

Olley J. & Ratkowsky D.A. (1973) The role of temperature function integration in the monitoring of fish spoilage. *Food Technology New Zealand*, **8**, 13–17.

Ono T., Nagayama F. & Yamada T. (1959) Studies on the fat metabolism of fish. II, Contents of fat and cholic acid in the liver of Pacific mackerel. *Bulletin of the Japanese Society of Scientific Fisheries*, **24**, 862–4.

Paetow A., Schober B. & Papenfuss H.J. (1966) The chemical composition and organoleptic quality of fish from West African fishing grounds. *Fischerei-Forschung*, Wissenschaftliche Schriftenreihe. **4**, 99–101.

Pazos M., Wakao A. & Vicetti R. (1984) Lipid and fatty acid composition of Peruvian sardines. *Boletin de Investigacion, Instituto Tecnologico Pesquero*, **2** (1) 1–24.

Perova L.I., Odintsov A.B. & Semenov B.N. (1980) Technological characteristics of small tuna fish. *Rybnoe Khozyaistvo*, **8**, 66–9.

Pickston L., Czochanska Z. & Smith J.M. (1982) The nutritional composition of some New Zealand marine fish. *New Zealand Journal of Science*, **25**, 19–26.

Pillai V.K. (1974) Utilization of sardinella resources in India. In *Fishery Products* (Ed. by R. Kreuzer), pp. 212–15. Fishing News Books, Oxford.

Pirazzoli P., Ambroggi F. & Incerti I. (1980) Canned tuna in oil: changes in composition as related to cooking method and effect of storage temperature on ripening. *Industria Conserve*, **55**, 279–85.

Plete C., Ovejana F.M., Pulanco E., Arafiles L.M. & Alabastro E. (1985) Histamine content of smoke-dried frigate mackerel (*Auxis thazard* L.) In *Spoilage of Tropical Fish and Product Development*. FAO Fisheries Report 317, supplement, pp. 348–54.

Plimmer R.H.A. (1921) *Analyses and energy values of foods*. HMSO, London.

Podeszewski Z. & Stodolnik L. (1976) Lipids in commercial Baltic sea fish. VII, Changes in the lipid fraction during an annual cycle. *Bromatologia i Chemia Toksykologiczna*, **9**, 341–7.

Podsevalov V.N. & Perova L.I. (1973) Technochemical characteristics of some Atlantic species of fish. *Trudy Atlant. NIRO*, **52**, 146–64.

Podsevalov V.N. & Perova L.A. (1975) Fat content of

mackerel and scad flesh in relation to month of catching. *Rybnoe Khozyaistvo* **3**, 72.

Podsevalov V.N., Pavlova J.L. & Utrobina A.A. (1966) Chemical composition of some Atlantic fish. *Trudy Atlant. NIRO*, **16**, 3–27.

Popa G., Popescu N., Brinzac V. & Feteanu S. (1968) Contributi la studiul si expertiza sanitar-veterinara a pestelui oceanic. *Revue de Zootehnie si Medicina Veterinara*, **18** (6) 87–90.

Poulter R.G., Curran C.A. & Disney J.G. (1981) Chill storage of tropical and temperate water fish-differences and similarities. In *Advances in Technology in the Chilling, Freezing, Processing, Storage and Transport of Fish, Especially Underutilised Species*, pp. 111–23. International Institute of Refrigeration, Paris.

Quagliariello G. (1934) *La valeur alimentaire du poisson*. Mediterranean. Meeting Report. **8**, 117–26, International Council for Exploration of the Sea.

Ravichander N. & Keay J.N. (1976) The production and properties of minced fish from several commercially important species. *Conference proceedings: The production and utilization of mechanically recovered fish flesh (minced fish)* (Ed. by J.N. Keay), pp. 18–24. Ministry of Agriculture, Fisheries and Food, UK.

Reay G.A., Cutting C.L. & Shewan J.M. (1943) The nation's food. VI, Fish as food. II, The chemical composition of fish. *Journal of the Society of Chemical Industry*, **62**, 77–85.

Regier L.W., Ernst R.C. & Hale M.B. (1985) Processing requirements for the preparation of minced intermediates or surimi from menhaden. In *Proceedings of the International Symposium on Engineered Seafood Including Surimi* (Ed. by R.E. Martin), pp. 129–39. National Fisheries Institute, Washington.

Revankar G.D., Keshava N., Naidu A.K. & Baliga B.L. (1981) Fish mince – preparation and composition. *Indian Food Packer*, **35** (4), 20–24.

Saeki K. & Kumagai H. (1979) Muscle components of wild and cultured yellowtail. *Journal of the Food Hygiene Society of Japan*, **20**, 101–105.

Saeki K. & Kumagai H. (1984) Chemical components in ten kinds of wild and cultured fishes. *Bulletin of the Japanese Society of Scientific Fisheries*, **50**, 1551–4.

Sakaguchi H. (1976) Changes of biochemical components in serum, hepatopancreas and muscle of yellowtail by starvation. *Bulletin of the Japanese Society of Scientific Fisheries*, **42**, 1267–72.

Sakaguchi M., Murata M. & Kawai A. (1984) Changes in free amino acid contents in juvenile mackerel muscle during ice storage. *Bulletin of the Japanese Society of Scientific Fisheries*, **50**, 323–9.

Santos F.O. & Ascalan S.J. (1931) Amount of nutrients in Philippine food materials. *Philippine Agriculturist*, **20**, 402–409.

Sasa S. (1958) Chemical studies on herring meat. *Bulletin of the Faculty of Fisheries, Hokkaido University*, **8**, 319–45.

Schmidt P.J. (1952). Analyses of British Columbia salt herring. *Progress Reports of the Pacific Coast Stations*, **92**, 3–6, Fisheries Research Board of Canada.

Sen D.P. & Chaluvaiah G.L. (1968) Seasonal variation in the amount and characteristics of the oil of oil-sardine (*Sardinella longiceps*) fish. *Journal of Food Science and Technology* (Mysore), **5**, 117–22.

Sen D.P. & Revankar G.D. (1972) Seasonal variation in the amount of oil of oil-sardine (*Sardinella longiceps*) fish. *Journal of Food Science and Technology* (Mysore), **9**, 93.

Setna S.B., Sarangdhar P.N. & Ganpule N.V. (1944) Nutritive values of some marine fishes of Bombay. *Indian Journal of Medical Research*, **32**, 171–6.

Shenoy A.V. & Pillai V.K. (1971) Freezing characteristics of tropical fishes. 1, Indian oil sardine. *Fishery Technology*, **8** (1) 37–41.

Shewan J.M. (1971) The microbiology of fish and fishery products – a progress report. *Journal of Applied Bacteriology*, **34**, 299–315.

Shewan J.M. (1977) The bacteriology of fresh and spoiling fish and the biochemical changes induced by bacterial action. In *Handling, Processing and Marketing of Tropical Fish* (Ed. by P. Sutcliffe & J. Disney), pp. 51–66. Ministry of Overseas Development, London.

Shimeno S., Hosokawa H., Takeda M., Kajiyama H. & Kaisho T. (1985) Effect of dietary lipid and carbohydrate on growth, feed conversion and body composition in young yellowtail. *Bulletin of the Japanese Society of Scientific Fisheries*, **51**, 1893–8.

Shimizu Y., Machida R. & Takenami S. (1981) Species variations in the gel-forming characteristics of fish meat paste. *Bulletin of the Japanese Society of Scientific Fisheries*, **47** (1) 95–104.

Shimizu Y., Tada M. & Endo K. (1973) Seasonal variations in chemical constituents of yellowtail muscle. I, Water, lipid and crude protein. *Bulletin of the Japanese Society of Scientific Fisheries*, **39**, 993–9.

Sikorski Z.E., Kolakowska A. & Burt J.R. (1990) Post-harvest biochemical and microbial changes. In *Marine Food: Resources, Nutritional Composition and Preservation*, pp. 53–75. CRC Press Inc, Boca Raton.

Sikorski Z., Olley J. & Kostuch S. (1976) Protein changes in frozen fish. CRC Critical Review. *Food Science and Nutrition*, **8**, 97–120.

Simmonds C.K. & Heydenrych C. (1984) Composition of South African commercial fish species. *Annual Report*, No. 38, 72. Fishing Industry Research Institute, Cape Town.

Simmonds C.K. & Seaman P.D. (1981) Composition of South African commercial fish species. *Annual Report*, No. 35, 53. Fishing Industry Research Institute, Cape Town.

Simmonds C.K. & Tanner K.R.O. (1980) Analysis of South African commercial fish species. *Annual Report*, No. 34, 51–3. Fishing Industry Research Institute, Cape Town.

Skachkov V.P. & Yudina O.P. (1975) Black Sea sprat products manufacture. *Rybnoe Khozyaistvo* **10**, 74–6.

Smith W.V. & Ewart R.A. (1948) Kinetics of emulsion polymerisation. *Journal of Chemical Physics*, **16**, 592–9.

Smith J.G.M., Hardy R., McDonald I. & Templeton J. (1980a) The storage of herring in ice, refrigerated sea water and at ambient temperature. Chemical and sensory assessment. *Journal of the Science of Food and Agriculture*, **31**, 375–85.

Smith J.G.M., Hardy R. & Young K.W. (1980b) A seasonal

study of the storage characteristics of mackerel stored at chill and ambient temperatures. In *Advances in Fish Science and Technology*, (Ed. by J.J. Connell), pp. 372–8. Fishing News Books, Oxford.

Smith J.G.M., McGill A.S., Thomson A.B. & Hardy R. (1980c) Preliminary investigation into the chill and frozen storage characteristics of scad (*Trachurus trachurus*) and its acceptability for human consumption. In *Advances in Fish Science and Technology*, (Ed. by J.J. Connell), pp. 303–307. Fishing News Books, Oxford.

Sohn B.I., Carver J.H. & Mangan G.F. (1961) Composition of commercially important fish from New England waters. Part 1, proximate analyses of cod, haddock, Atlantic Ocean perch, butterfish and mackerel. *Commercial Fisheries Review*, **23** (2) 7–10.

Stansby M.E. (1976) Chemical characteristics of fish caught in the north-east Pacific Ocean. *Marine Fisheries Review*, **38** (9) 1–11.

Stansby M.E. & Lemon J.M. (1941) *Studies on the handling of fresh mackerel.* Research Report 1, Fish and Wildlife Service, US Department of Interior, Washington.

Stoddard J.H. (1968) *Fat contents of Canadian Atlantic herring.* Technical Report 79, Fisheries Research Board of Canada.

Storey R.M. & Owen D. A simple time-temperature function integrator. Torry Research Station, Aberdeen. Unpublished data.

Sulit J.I., Navarro O.B., San Juan R.C. & Caldito E.B. (1953) Proximate chemical composition of various species of Philippine market fishes. *Philippine Journal of Fisheries*, **2**, 109–22.

Sumner J.L. & Gorczyca E. (1981) Effect of vacuum packaging on the shelf life of fish held either in ice or at 4–6°C. *Refrigeration Science and Technology*, **4**, 365–70.

Suzuki T. (1981). *Fish and Krill Protein: Processing Technology.* Applied Science Publishers, London.

Svensson S. (1980) Stabilization of fish mince from gadoid species by pre-treatment of the fish. In *Advances in Fish Science and Technology*, (Ed. by J.J. Connell), pp. 226–32. Fishing News Books, Oxford.

Sweet C.W. (1973) Activity of antioxidants in fresh fish. *Journal of Food Science*, **38**, 1260–1.

Taarland T., Mathiesen E., Ovsthus O. & Braekkan O.R. (1958) Nutritional values and vitamins of Norwegian fish and fish products. *Tidsskrift for hermetikindustri*, **44**, 405–12.

Tashiro I., Itoh S. & Tsuyuki H. (1981) Seasonal variation of total lipids of horse mackerel. *Nippon Shokuhin Kogyo Gakkaishi*, **28**, 309–17, 482–90.

Thompson M.H. (1966) Proximate composition of Gulf of Mexico industrial fish. *Fishery Industrial Research*, **3** (2) 29–67.

Thorisson S. (1987) *Antioxidant properties of ethoxyquin and some of its oxidation products.* PhD. thesis, University of St Andrews.

Toyomizu M., Hanaoka K. & Yamaguchi K. (1985) Effect of release of free fatty acid by enzymatic hydrolysis of phospholipids on lipid oxidation during storage at −5°C. *Bulletin of the Japanese Society of Scientific Fisheries*, **47**, 614–20.

Toyomizu M., Nakamura T. & Shono T. (1976) Fatty acid composition of lipid from horse mackerel muscle. *Bulletin of the Japanese Society of Scientific Fisheries*, **42**, 101–108.

TRS (1989). Torry Research Station, Aberdeen, Scotland. *FAO Fisheries Technical Paper 309.*

Tsao C.-Y., Chon K.-C. & Jiang S.-T. (1980) Studies on freezing preservation of raw material of minced fish products. I. *Journal of the Chinese Agricultural Chemical Society.* **18**, 77–85.

Tsuchiya Y., Hata M., Asano M., Takahashi I., Nomura T. & Suzuki Y. (1953) Biochemical studies on skipper (*Cololabis saira*). I, general component. *Bulletin of the Japanese Society of Scientific Fisheries*, **19**, 513–17.

Tulsner M. (1965) The technological properties and processing possibilities of important north-west African commercial fish species. *Fischerei-Forschung*, **3**, 55–63. Wissenschaftliche Schriftenreihe.

Turnbull P.C.B. & Gilbert R.J. (1982) In *Adverse Effects of Foods* (Ed. by E.F.F. Jelliffe & D.B. Jelliffe). Plenum Press, New York.

Ueda T. (1976) Changes in the fatty acid composition of mackerel lipid and probably related factors. I, Influence of the season, body length and lipid content. *Bulletin of the Japanese Society of Scientific Fisheries*, **42**, 479–84.

Vakhrusheva M.N. & Nikiforova G.A. (1979) Technochemical characteristics of frozen west Pacific sardine. *Rybnoe Khozyaistvo*, **3**, 51–4.

van Wyk G.F. (1944) South African fish products. Part VIII, Composition of the flesh of Cape fishes. *Journal of the Society of Chemical Industry* (Trans.), **63**, 367–71.

Vasavan A.G., Ganghadharan P. & Padmanabhan V. (1960) Studies on the seasonal variation in the fat content of Indian oil sardines (*Sardinella longiceps*). *The Indian Food Packer*, October 1960, 7–10.

Venkataraman R. & Chari S.T. (1951) Seasonal variation in the chemical composition of mackerel (*Rastrelliger kanagurta* Russel). *Proceedings of the Indian Academy of Sciences* (Section B), **33**, 126–34.

Venkataraman R. & Chari S.T. (1953). Studies on mackerel fat variations: correlation of plankton fat with fat of fish. *Proceedings of the Indian Academy of Sciences* (Section B), **37**, 224–7.

Viviani R., Cortesi P., Crisetig G., Mancini L., Poletti R. & Borgatti A.R. (1968) First observations on seasonal changes of the lipids in the tissues of some Adriatic clupeids. *Rivista Italiana delle Sostanze Grasse*, **45**, 779–90.

Vlieg P. (1982) Proximate and fatty acid composition of the flesh of New Zealand red cod, hoki and jack mackerel. *New Zealand Journal of Science*, **25**, 155–8.

Vlieg P. (1985) Proximate analysis of commercial New Zealand fish species, 3. *New Zealand Journal of Technology*, **1**, 181–5.

Vlieg P., Habib G. & Clement G.I.T. (1983) Proximate composition of skipjack tuna (*Katsuwonus pelamis*) from New Zealand and New Caledonian waters. *New Zealand Journal of Science*, **26**, 243–50.

Watanabe K. (1963) Variations in chemical composition in some commercial fishes from the south of Brazil. *Bulletin of the Japanese Society of Scientific Fisheries*, **29**, 469–74.

Wesley P. (1982) Glucose oxidase treatment prolongs shelf

life of fresh seafood. *Food Development*, 16 (1) 36–8.

Whittle K.J., Hardy R. & Hobbs G. (1990) Chilled fish and fishery products. In *Chilled Foods – The State of the Art* (Ed. by T.R. Gormley), pp. 87–116. Elsevier Applied Science, London.

Whittle K.J., Robertson I. & McDonald I. (1980) Seasonal variability in blue whiting (*Micromesistius poutassou*) and its influence on processing. In *Advances in Fish Science and Technology* (Ed. by J.J. Connell), pp. 378–87. Fishing News Books, Oxford.

Wu C.S. & Toyomizu M. (1974) Enzymatic hydrolysis of triglyceride during cold storage of fish. *Kyushi Daigaku Nogakubu Fuzoku Suissan Jikkensho Gyosekki* 2, 33–9.

Yamada J. (1981) Content and distribution of lipid in sardine flesh. *Bulletin of the Tokai Regional Fisheries Research Laboratory*, **104**, 103–109.

Yamada K., Harada K., Kawahara T., Itoh R. & Tsukamoto S. (1983) Comparative rates of ATP degradation in rested and exhausted yellowtail in relation to rigor mortis. *Journal of the Shimonoseki University of Fisheries*, **31**, 59–66.

Yaroslavtseva L.D. (1966) Technochemical properties of some Indian Ocean fishes. *Rybnoe Khozyaistvo*, **42** (1) 60–64.

Young F.V.K. (1986) The refining and hydrogenation of fish oil. *Fish Oil Bulletin*, 17. The chemical and physical properties of crude fish oils for refiners and hydrogenators. *Fish Oil Bulletin*, 18. International Association of Fish Meal Manufacturers, Potters Bar.

CHAPTER 10

Total lipids and omega-3 fatty acids from seven species of pelagic fish

R. Pozo, B. Pérez-Villarreal and E. Saitua

INTRODUCTION

Nutritional factors affecting health, quality of life and longevity are important criteria used by consumers to select food. The principal nutrition-related health problems in the western industrialised nations arise from the overconsumption of fat, saturated fatty acids, cholesterol and sodium. Polyunsaturated fatty acids (PUFA) have been recommended as desirable dietary components since the 1950s (Nelson & Douglas 1965). The interests of the food industry have, however, been focussed on promoting vegetable oils (soybean and corn oils) as sources of PUFA. These oils are high in omega-6 fatty acids whereas fish oils are high in omega-3 fatty acids.

Epidemiological, clinical and biochemical studies have linked the dietary intake of PUFA from fish oils with beneficial effects in reducing the risk of coronary disease (Dyerberg *et al.* 1978, Stansby 1984, Dyerberg 1986, Kinsella 1986). Phillipson *et al.* (1985) found that diets containing fish oils led to decreases in both plasma cholesterol and triglyceride, as compared with a control diet without fish oils, in patients with elevated blood lipid levels. This beneficial effect is considered to be primarily due to omega-3 fatty acids and particularly to eicosapentaenoic acid (EPA) and less to docosahexaenoic acid (DHA). EPA and DHA have an anti-aggregatory effect that results from competitive inhibition of the synthesis of thrombotic eicosaenoid thromboxane or from increased production of prostacyclins which are anti-aggregatory agents (Kinsella 1988).

Many reports have been found dealing with the fat and fatty acid content of fish and shellfish (Gruger *et al.* 1964, Gooch *et al.* 1987, Hearn *et al.* 1987, Ackman and McLeod 1988). The fatty acid composition varies, between and within species, with a number of factors including diet, sex, size, season and area of catch (Ackman 1982). Pelagic (fatty) species seem to be better sources of EPA and DHA than groundfish (lean) species.

Although fat content data are available, there are no general guidelines on the fatty acid composition of fish from around the Spanish coast. As part of a national programme, the Research Institute for Fish Science and Technology (AZTI) carried out research to give a brief survey of the fatty acid composition, total fat, cholesterol and triglyceride content of commercially available pelagic species off the Spanish coast.

MATERIALS AND METHODS

Sample preparation

All fish analysed in this study were purchased fresh, wholesale in the local market. As soon as the fish were collected they were kept in ice and analysed within a few hours of landing. A sample for analysis was taken, from each fish, of the liver, belly, dark and white muscle and, when possible, roe tissue. The skin and bones were always discarded.

Samples were taken at monthly intervals for a year of anchovy (*Engraulis encrasicolus*), mackerel (*Scomber scombrus*), horse mackerel (*Trachurus trachurus*), Mediterranean horse mackerel (*Trachurus mediterraneus*) and pilchard (*Sardina pilchardus*). Each monthly sample was divided into different size groups: anchovy, two size groups (<17 cm and >17 cm); mackerel, three size groups (<30 cm, 30–40 cm and >40 cm); horse mackerel, three size groups (<25 cm, 25–35 cm and >35 cm); Mediterranean horse mackerel, two size groups (<35 cm and >35 cm); pilchard, two size groups (<18 cm and >18 cm). Four fish were sampled at random from each group. Three batches of albacore (*Thunnus alalunga*) were analysed, corresponding to the beginning (July), the middle (September) and the end (November) of the albacore fishing season. Each batch was separated into three size groups (<5 kg, 5–8 kg and >8 kg) and four fish were also sampled at random from each group. A single batch of bluefin tuna (*Thunnus thynnus*) was taken and divided into three size groups (<6 kg, 6–20 kg and >20 kg) and four fish were sampled at random from each group.

Lipid extraction/fatty acid methyl esters analysis

Fish oil was extracted from 20 g fish tissue by the method of Bligh and Dyer (1959), as modified by Hanson & Olley (1963), using the antioxidant butylated hydroxytoluene (BHT). Fish oil (100 mg) was saponified with 0.5 N methanolic-KOH and methylated for 1 hour with 14% BF_3-MeOH (Metcalfe *et al.* 1966). Fatty acid methyl esters were separated on a Hewlett-Packard 5890A gas chromatograph equipped with a flame ionization detector using a flexible fused silica column 25 m × 0.2 mm i.d. coated with 5% phenylmethyl silicone gum. Operating conditions were: temperature was programmed to be 45°C at splitless injection, rising by 25°C/min to 140°C, then by 4°C/min to 240°C and finally by 1°C/min to 260°C; helium carrier gas at 10 psi pressure. The gas chromatograph was equipped with a Hewlett-Packard 3393A integrator.

Fatty acid methyl esters were identified by comparison with known standards verified by mass spectrometry. Peak areas were used to determine the concentrations of fatty acid methyl esters in the samples, by reference to standard curves obtained by chromatographing standards under identical conditions.

Cholesterol/triglycerides analysis

Neutral lipid classes were separated by thin layer chromatography (TLC) on silica gel 60 plates (0.25 mm thick) using hexane: diethylether: acetic acid (80:20:2) as developing solvent. Samples were sprayed on with nitrogen using a TLC spotter (Camag-Linomat IV). Lipid classes were detected by the method of Fewster *et al.* (1969). Spots were identified by their migration characteristics relative to standards chromatographed simultaneously alongside the samples. Optical densities of the spots were measured with a TLC-scanner densitometer (Shimadzu TLC scanner CS-930).

Statistical analysis

Data were stored on computer files and processed by the statistical package SPSS-X V2.2 (SPSS Inc. 1986). Analyses of variance were carried out to test for between-batch and within batch effects. The significance was determined from F ratio of the between-batch and within-batch sums of squares.

RESULTS AND DISCUSSION

The total fat, EPA, DHA, cholesterol and triglyceride content in the white muscle of the seven species of fish studied are listed in Table 10.1. Size and seasonal variations are shown in Tables 10.2 and 10.3. The oil concentration varies widely among species, from 1.82% for anchovy to 8.51% for mackerel. Analysis of variance indicates that there are significant differences ($P < 0.05$) for size, tissue or month of analysis for total fat content in the species studied. Generally, the oil content of fish caught in spring was lower than that of those caught in autumn; fall fish usually contain increased fat in the muscle (Love 1980). Liver and belly tissue of the species studied had a fat content significantly ($P < 0.05$) greater than that of any other tissue examined.

EPA, DHA, cholesterol and triglyceride content vary between and within species with size, season

Table 10.1 Mean values of total fat, eicosapentaenoic acid (EPA), docosahexaenoic acid (DHA), cholesterol and triglyceride content in the white muscle of seven species caught off the Spanish coast. The values are pooled over all size groups and months.

Species	Total fat (g/100 g)	EPA (g/100 g)	DHA (g/100 g)	Cholesterol (mg/100 g)	Triglycerides (g/100 g)
Albacore	3.71	0.32	1.08	75.2	2.77
Anchovy	1.82	0.17	0.57	49.1	0.77
Bluefin	2.1	0.15	0.65	69.3	0.63
Horse mackerel	3.15	0.19	0.41	85.4	1.26
Mackerel	8.51	0.62	1.39	148.9	5.19
Mediterranean horse mackerel	3.84	0.24	0.47	93.7	1.73
Pilchard	5.71	0.51	0.52	70.8	3.51

Table 10.2 Seasonal variations of total fat, eicosapentaenoic acid (EPA), docosahexaenoic acid (DHA) and cholesterol content in the white muscle of mackerel, horse mackerel and anchovy. The values are pooled over all size groups.

	Mackerel				Anchovy				Horse mackerel			
	Fat g/100 g	EPA g/100 g	DHA g/100 g	Cholesterol mg/100 g	Fat g/100 g	EPA g/100 g	DHA g/100 g	Cholesterol mg/100 g	Fat g/100 g	EPA g/100 g	DHA g/100 g	Cholesterol mg/100 g
Jan.	12.00	0.89	1.86	177.6					1.34	0.06	0.23	38.1
Feb.	11.05	0.76	1.73	171.3					2.53	0.09	0.31	68.6
Mar.	8.26	0.59	1.43	166.1	0.94	0.08	0.36	31.7	2.10	0.11	0.22	71.8
Apr.	6.25	0.44	1.18	171.9	1.76	0.19	0.62	52.4	2.14	0.12	0.24	65.7
May	3.80	0.31	1.02	138.7	1.77	0.15	0.51	40.4	7.25	0.44	0.64	134.1
Jun.	2.56	0.26	0.90	109.3	1.29	0.12	0.41	34.8	5.26	0.31	0.51	133.1
Jul.	4.23	0.47	1.20	85.9	1.07	0.10	0.39	36.1	4.26	0.20	0.38	150.0
Aug.									3.62	0.19	0.44	81.5
Sep.	8.35	0.53	0.84	55.9	1.35	0.11	0.41	37.1	2.40	0.10	0.28	79.4
Oct.	11.71	0.64	1.87	137.2	3.85	0.36	1.07	83.5	2.50	0.11	0.27	61.3
Nov.	13.67	0.96	1.94	274.8	2.45	0.17	0.71	77.2	2.11	0.11	0.25	85.5
Dec.									2.31	0.10	0.29	55.7

Table 10.3 Size variations of total fat, cholesterol, EPA and DHA content in the white muscle of albacore and bluefin. Values are pooled over all months.

	Albacore			Bluefin		
	w < 5 kg	5–8 kg	w > 8 kg	w < 6 kg	6–20 kg	w > 20 kg
Total fat (g/100 g)	2.60	4.01	4.49	1.29	2.34	2.61
EPA (g/100 g)	0.23	0.33	0.38	0.08	0.16	0.18
DHA (g/100 g)	0.80	1.17	1.27	0.47	0.70	0.80
Cholesterol (mg/100 g)	56.4	70.6	98.8	45.0	60.8	101.3

Table 10.4 Fatty acid composition (weight %) from white muscle lipids of seven species of pelagic fish. The values are pooled over all size groups and months.

Fatty acid	Albacore	Anchovy	Bluefin	Horse mackerel	Mackerel	Mediterranean horse mackerel	Pilchard
14:0	2.55	3.65	2.15	5.81	4.10	5.51	7.61
15:0	0.57	0.61	0.95	0.74	0.56	0.70	0.74
16:0	21.42	22.69	24.04	26.20	18.92	26.52	26.26
16:1ω7	3.42	3.67	2.65	4.82	5.86	6.30	7.87
17:0	0.84	0.62	0.88	0.82	0.58	0.83	0.64
18:0	6.17	3.98	7.18	6.38	4.56	6.96	5.03
18:1ω9 + ω7	19.13	11.81	17.41	16.54	20.41	17.39	17.46
18:2ω6	1.07	1.37	1.70	1.26	1.28	0.92	1.09
18:3ω3	0.97	0.86	0.30	0.92	1.30	1.19	1.71
18:4ω3	0.16	0.34	0.21	0.32	0.29	0.29	0.27
20:0	tr	tr	0.20	0.30	0.55	0.36	0.62
20:1ω9	1.60	2.44	1.14	5.90	6.25	4.85	5.59
20:4ω3	0.87	0.43	0.37	tr	0.74	0.63	0.42
20:5ω3	7.95	8.58	6.40	5.07	8.79	6.78	8.37
22:1ω11	tr	4.25	tr	8.93	6.12	5.19	3.45
22:5ω3	2.15	3.41	2.84	1.36	2.45	1.24	2.22
22:6ω3	28.77	30.83	29.79	11.91	15.97	11.98	8.42
TOTAL ω3	40.87	44.45	39.91	19.58	29.54	22.11	21.41

tr – trace amounts

and tissue analysed. Bigger fish had an EPA and DHA concentration significantly ($P < 0.05$) greater than that of smaller ones and spring caught fish had lower levels than those caught in autumn. The total fat content showed the same size and seasonal variations. Belly tissue usually shows higher levels of EPA and DHA than dark and white muscles.

Triglycerides constitute the major lipid class in all species studied, except for anchovy and bluefin where the amount of this lipid class was less than 40% of total lipids. The levels of triglycerides are generally higher in bigger fish, in autumn caught fish and in the belly tissue. Cholesterol is present in the edible muscle of all species as free cholesterol; only in liver and fish roes were cholesterol esters found. Cholesterol levels showed the same variation as those of triglycerides and total fat.

Table 10.4 shows the fatty acid composition of total lipids from the white muscle of the seven species of pelagic fish studied. There were differences in total saturated, total mono-unsaturated fatty acids and total PUFA between species. The principal difference among species was for the fatty acid 22:1, which was present in trace amounts in albacore and bluefin and ranged from 3.45 to 8.93% in other species. Analysis of variance indicates that there are significant differences for either tissue analysed or season for total PUFA in the species studied. The major omega-3 fatty acids present were 22:6 and 20:5, respectively. The omega-3 PUFA constituted more than 35% of the total fatty acids of albacore, anchovy and bluefin. Mackerel, pilchard, Mediterranean horse mackerel

Table 10.5 Consumption of fish in Spain during 1988. (Ministry of Agriculture, Fisheries and Food, Madrid, Spain.)

	Annual consumption (millions kg)			kg/inhabitant/year		
	Fresh	Frozen	Canned	Fresh	Frozen	Canned
Anchovy	3.920	—	8.712	0.11	—	0.24
Albacore	11.576	0.987	12.429	0.33	0.03	0.35
Bluefin	9.350	0.553	18.758	0.26	0.01	0.53
Horse mackerel	15.051	—	7.360	0.42	—	0.21
Mackerel	1.025	—	—	0.03	—	—
Pilchard	110.376	2.258	11.482	3.31	0.06	0.32
Hake	145.274	175.272	—	3.73	4.46	—
Others	266.278	75.726	17.230	6.18	1.93	0.63
Crustaceans & molluscs	134.865	83.436	20.067	3.99	2.65	0.55

and horse mackerel were lowest in percentage of omega-3 PUFA. However, omega-3/omega-6 PUFA ratio for all the species studied was high, exceeding 10.

The statistics of the consumption of fish in Spain for 1988 (Table 10.5), showed that albacore, bluefin, anchovy and, to a less extent, horse mackerel are consumed more when canned in vegetable oil than fresh or frozen. The canning process involves various steps in which vegetable oil, usually soybean, is added in much quantities than the natural fish oil. In these conditions the fatty acid characteristics of the fish oil are probably hidden. Only pilchard and, especially, hake are consumed fresh or frozen in considerable amounts.

The fat content of the hake is low, approximately 1% (Pérez & Howgate 1987, Ackman 1988). In addition they are not good sources of omega-3 PUFA, the concentrations of EPA and DHA being very low (Ackman 1988). As a consequence, pelagic fish are of most interest for increasing the omega-3 fatty acids in the Spanish diet. In Spain, the use of fish fatty acids with beneficial effects on human health should be focussed on the exploitation and consumption of fish, fresh or frozen, more than canned, and on the pelagic species stated above. However, the consumption of groundfish like hake cannot be rejected because the diet would be more attractive and varied if they are included.

All fish have been considered to have similar nutritional value and selection was made on the basis of availability, traditional habits, freshness, flavour and other sensory characteristics. Species like mackerel, horse mackerel and sometimes pilchard have been rejected and considered as poor food. In addition, during the 1950s and 1960s a consumer survey led to surprising results: 63% of Spaniards did not consume mackerel (Anon 1985).

Recent data from the local government of the Basque country reports specific changes in consumer eating habits because of concerns about health and nutrition. Consumers now perceive seafood as better for health than red meats, although many factors affect purchase decisions. Taste continues to have the most influence on seafood purchases. Quality has become a close second to taste as a factor affecting seafood purchases, followed by nutrition and a number of other factors. When fish is intended to be a means of improving health, the fat content and the content of PUFA, specially EPA, must be considered. Mackerel and pilchard are potentially low cost valuable sources of EPA and contain relatively little cholesterol.

The results of this study show that albacore, pilchard and, notably, mackerel are good sources of omega-3 PUFA, and in particular of EPA and DHA. The EPA and DHA content of mackerel caught off the Spanish coast was, respectively, 0.62 g and 1.39 g/100 g of flesh, whereas bluefin had only 0.15 g and 0.65 g/100 g of flesh. In Japan, with a traditionally rich fish diet and low cardiovascular health risk, the intake of EPA has been estimated to be up to 0.5 g/day (Kagawa *et al.* 1982). Unfortunately, a similar figure has not been attainable in Spain. Society is, however, greatly concerned about the influence of the diet on health and a tendency to consume a more fish-oriented diet has been observed.

It is hoped that this study, along with others of similar nature (Gooch *et al.* 1987, Hearn *et al.* 1987, Ackman & McLeod 1988), may provide valuable information on selecting fish and fish oils for nutritional purposes.

BIBLIOGRAPHY

Ackman R.G. (1982) In *Nutritional evaluation of long-chain fatty acids in fish* (Ed. by S.M. Barlow & M. Stansby), pp. 25–88. Academic Press, New York.

Ackman R.G. (1988) Concerns for utilization of marine lipids and oils. *Food Technology*, **42**, 151–5.

Ackman R.G. & McLeod C. (1988) Total lipids and nutritionally important fatty acids of some Nova Scotia fish and shellfish food products. *Can. Inst. Food Sci. Technol. J.* **21** (4) 390–8.

Anon. (1985) *Consumo Alimentario en España 1985*. Ministerio de Agricultura, Pesca y Alimentación, Madrid.

Bligh E.G. & Dyer W.J. (1959) A rapid method of total lipid extraction and purification. *Can. J. Biochem. Physiol.* **37**, 911–17.

Dyerberg J. (1986) Linoleate-derived polyunsaturated fatty acids and prevention of atherosclerosis. *Nut. Rev.* **44**, 125–31.

Dyerberg J., Bang H.O., Stoffersen E., Moncada S. & Vane J.R. (1978) Eicosapentaenoic acid and prevention of thrombosis and atherosclerosis. *Lancet.* **2**, 117–21.

Fewster M.E., Burns B.J. & Mead J.F. (1969) Quantitative densitometric thin layer chromatography of lipids using copper acetate reagent. *J. Chromatogr.* **43**, 120–6.

Gooch J.A., Hale M.B., Brown Jr T., Bonnet J.C., Brand C.G. & Regier L.W. (1987) Proximate and fatty acid composition of 40 south-eastern US finfish species. In *NOAA Technical Report NMFS 54*.

Gruger E.H., Nelson Jr R.W. & Stansby M.E. (1964) Fatty acid composition of oils from 21 species of marine fish, freshwater and shellfish. *J. Amer. Oil Chem. Soc.*, **41**, 662–7.

Hanson S.W. & Olley J. (1963) Application of the Bligh and Dyer method for lipid extraction to tissue homogenates. *Biochem. J.* **89**, 101–2.

Hearn T.L., Sgoutas S.A., Heran J.A. & Sgoutas D.S. (1987) Polyunsaturated fatty acids and fat in fish flesh for selecting species for health benefits. *J. Food Science.* **52** (5) 1209–11.

Kagawa Y., Niskizama M., Suzuki M., Miyataki T, Hamamoto T., Goto K., Motoonoya E., Izumikkawa H., Hirata A. & Ebihara, A. (1982) Eicosapolyenoic acids of serum lipids of Japanese islanders with low incidence of cardiovascular diseases. *J. Nutr. Sci. Vitaminol*, **28**, 441–5.

Kinsella J.E. (1986) *Seafoods and fish oils in human health and disease.* Marcel Dekker Inc., New York and Basel.

Kinsella J.E. (1988) Food lipids and fatty acids importance in food quality, nutrition and health. *Food Technology*, **42** (10) 124–45.

Love R.M. (1980) *The Chemical Biology of Fishes* **2**. Academic Press, London.

Metcalfe L.D., Schimtz A.A. & Pelka J.R. (1966) A rapid preparation of fatty acid esters for gas chromatographic analysis. *Anal. Chem.*, **33**, 363–4.

Nelson A. & Douglas C. (1965) The effect of fat controlled diet on patients with coronary disease. *Med. Welt*, **36**, 2063–8.

Pérez B. & Howgate P. (1987) Composition of European hake, *Merluccius merluccius*. *J. Sci. Food Agric.*, **40**, 347–56.

Phillipson B.E., Rothrock D.W., Connor W.E., Harris W.S. & Illingworth D.R. (1985) Reduction of plasma lipids, lipoproteins and apoproteins by dietary fish oils in patients with hypertriglyceridemia. *N. Engl. J. Med.*, **312**, 1210–16.

SPSS Inc. (1986) *SPSS-X User Guide*, Chicago.

Stansby M.E. (1984) Fish or fish oil in the diet and heart attacks. *Mar. Fish. Rev.* **46** (2) 60–3.

CHAPTER 11

Oxidation of frozen sardine mince: effect of washing and antioxidants addition

F. Mestiri, V. Verrez, L. Han-Ching, M. Cardinal, and M. Jerome

INTRODUCTION

At present, surimi (refined washed mince) from fatty fishes is being extensively and intensively studied in the USA, Japan, Norway and other countries.

It is desirable for a number of reasons, such as yield, price and texture, to prevent oxidation in unrefined minces derived from such fish. The range of potential uses for this kind of raw material is wide; many new products may be developed from unrefined mince or mince mixed with different surimis, different fish minces or other materials (polysaccharides, proteins, etc.).

The results of some studies have been published (Nair *et al.* 1978, Pizzocaro *et al.* 1979), and Tejada *et al.* (1985) have shown that a storage life of 4 to 8 months at −20°C can be attained by vacuum packaging.

Antioxidants can also be used to increase the shelf life and improve stability of washed mince obtained from fatty fish (Hultin 1988). The efficiency of antioxidants on lipids varies widely according to their conditions of use, and depends on the method of preparation of the mince and the composition of the substrate.

In order to find the most satisfactory antioxidants for sardine mince, several additives have been tested on sardine lipids and minces prepared in a pilot plant, and the following is an account of the results obtained.

MATERIALS AND METHODS

Mince preparation and processing

In November 1988 60 kg of freshly caught sardines from the Bay of Biscay were purchased already filleted. The fillets were washed before being passed through a Baader 694 deboner equipped with a drum with 3 mm perforations. Part of the deboned flesh was washed in a pilot plant (1 part flesh : 5 parts water) for 15 minutes at 0–4°C with constant stirring and then dehydrated through a decanter centrifuge (Sharples P600) to a final moisture of 77.5%.

Washed and unwashed minces were mixed with additives in a kneading-trough blender, packaged in polyethylene/polyamide bags, frozen at −40°C and then stored at −10°C for 2 months.

Tables 11.1 and 11.2 summarise the percentages of antioxidants and additives and their method of addition used in the different experiments. The peroxide value (PV) and the thiobarbituric acid value (TBA) were determined periodically.

Sardine oil extraction and processing

Sardine oil was obtained from the fat recovered from the washing of the mince. The fat was centrifuged in order to eliminate water and cellular fragments, then decanted and passed through activated carbon. Two experiments were carried out: in the first, antioxidants were added to the oil alone, and in the second they were added to oil emulsified with water soluble muscle extract (Tables 11.1 and 11.2).

The water soluble muscle extract was obtained by washing the sardine mince (1 part flesh : 2 parts water), centrifugation (3500 rpm, 4°C) and freeze-drying the supernatant. Emulsification was carried out in a colloid mill (Stephan) under vacuum at constant speed (1500 rpm for 10 minutes). One litre of sardine oil was emulsified with 150 ml of a concentrated solution of the water soluble extract (60 g freeze dried extract dissolved in 150 ml water).

Antioxidants were added to the emulsion. Samples were stored at 30°C in stoppered flasks for a period of 11 weeks. Peroxide value, TBA value, carbonyl compounds (CC) content, and free fatty acids (FFA) content were determined periodically.

Table 11.1 Additives used in sardine oil and mince.

Antioxidants	Addition method in mince	Addition method in sardine oil	Concentration used in sardine mince or oil (%)
Butylated hydroxyanisole (BHA) (Sigma)	Dissolution in 30 ml soya oil	Direct addition to oil	0.02 (L)
2–6 di tert-butyl-4-methylphenol (BHT) (Merck)	Dissolution in 30 ml soya oil	Direct addition to oil	0.02 (L)
Tert-butyl hydroquinone (TBHQ) (Janssen chemica)	Dissolution in 30 ml soya oil	Direct addition to oil	0.02 (L)
D-tocopherols (TOCO) (Grindsted)	Dissolution in 30 ml soya oil	Direct addition to oil	0.05 (L)
Ascorbyl palmitate (Asc p) (Grindsted)	Dissolution in 30 ml soya oil	Dissolution in 1 ml ethanol	0.05 (L)
Citric acid (Cit A) (Sigma)	Dissolution in 30 ml water	Dissolution in 1 ml ethanol	0.04 (M)
Polyphosphates (PP) (BK Ladenburg)	Direct addition in mince	Dissolution in 1 ml water	0.3 (M)
Sorbitol (S) (Roquette)	Direct addition in mince		4 (M)

(L) % according to weight of lipids in mince
(M) % according to weight of mince

Oxidation evaluation

The thiobarbituric acid value was determined according to Vyncke (1970, 1975) as adapted by Smith & Whittle (1984). The peroxide value was evaluated according to the AOAC method. The carbonyl compounds content was determined according to Lawrence (1965) modified by Keller & Kinsella (1973). The percentage of free fatty acid was determined according to Lowry & Tinsley (1976).

The effect of washing on the fatty acids composition of mince

An analysis of fatty acids was carried out using gas chromatography, on both the unwashed and washed mince.

Fat extraction

Fat extraction was carried out at low temperatures in two steps: first, the neutral lipids were extracted with chloroform, and then the polar lipids were extracted with Folch solution (chloroform:methanol, 2:1).

Extraction 1 150 g mince and 100 ml $CHCl_3$ were mixed in a beaker, covered with aluminium for 15 minutes and then decanted. The lower layer was filtered through anhydrous Na_2SO_4. The same extraction was repeated on the upper layer.
Extraction 2 100 ml Folch solution and the upper layer of the second extraction were mixed together, decanted and then the lower layer was filtered.

The lower layers of extraction 1 and 2 were combined but could also have been treated separately. The solvents were evaporated under vacuum at 30°C and the remaining oil dried under N_2.

Sample preparation

The fatty acids were esterified according to the French standard (AFNOR NF T60-233). The methyl esters of the fatty acids were dissolved in heptane before injection.

Gas chromatography

The gas chromatograph (GIRDEL 30) was equipped with a capillary column, flame ionisation detector and a split injection port. The column (25 m × 250 μm id) was coated with CARBOWAX 20M (FS- WCOT) of 0.3 μm thickness.

The operating conditions were: injection, 220°C; column, isothermal at 190°C; detector, 250°C; carrier gas, hydrogen at 0.7 bar pressure; adjustment of the split leak at 50 ml/min.

Proximate composition of mince

The sardine minces were analysed as summarised in Table 11.3.

Table 11.2 Composition of samples in the different experiments: additives used.

Sardine oil alone	Emulsified sardine oil	Sardine mince	
		Washed mince (W)	Unwashed mince (UW)
Control (none)	Control (none)	Sorbitol alone (= control)	Sorbitol alone (= control)
BHA	BHA	Sorbitol + polyphosphates	Sorbitol + water (= control + water)
BHT	TBHQ	Sorbitol + D-tocopherols	Sorbitol + soya oil (= control + oil)
TBHQ	D-tocopherols	Sorbitol + polyphosphates + tocopherols	Sorbitol + BHA + soya oil
D-tocopherols	Polyphosphates		Sorbitol + citric acid + water
Polyphosphates	Citric acid		Sorbitol + polyphosphates + tocopherols
Citric acid			
Ascorbyl palmitate			

RESULTS AND DISCUSSION

The effect of antioxidants on oil

The reduction in free fatty acid content (Fig. 11.1) from the beginning of storage could be due to rapid oxidation. After this period the FFA content remained constant. It was found that BHA reduces the FFA content in oil, compared to the control, but it is difficult to interpret the results.

The presence of BHA, TBHQ, TOCO and PP caused a significant reduction in the quantity of malonaldehyde produced (Fig. 11.2). A significant reduction in the peroxide value was also observed for oils containing BHA, TOCO, PP and citric acid (Fig. 11.3). BHA, BHT, TOCO and citric acid seem to be more effective than the other antioxidants in reducing the production of carbonyl compounds (Fig. 11.4).

The data show that BHT and ascorbyl palmitate are ineffective in sardine oil. On the other hand, BHA and the other antioxidants reduce the oxidation of sardine oil slightly.

The effect of antioxidants on an oil emulsion of water soluble muscle extract

No significant difference in free fatty acid content was observed between the control and the samples with antioxidant (Fig. 11.5). While the FFA in sardine oil decreased during storage at 30°C, in sardine oil emulsified with soluble protein the FFA content increased slightly. This difference is almost certainly due to lipolytic enzymes present in the muscle extract.

Tsukuda (1976) found that in skipjack muscle the increase of FFA was due to hydrolysis of phospholipids and triglycerides and that the quantity of triglycerides decreased rapidly in the early period of frozen storage. For ordinary muscle, more than 90% of the initial amount had been broken down at −10°C, −20°C and −30°C. After 40 days at 10°C, the loss of phosphatidyl choline and phosphatidylethanolamine was observed to be from 56–70%.

However, the increase in FFA, in an oil emulsion of muscle extract, does not provide a measure of the rate of oxidation. The relationship between the release of FFA and lipid oxidation during frozen storage of fish has been discussed in various reports, and conflicting experimental results have been obtained (Castell *et al.* 1966, Shono & Toyomizu 1973, Mazeaud & Bilinski 1976). Nair *et al.* (1978) have stated that for sardines the FFA are more prone to oxidation than intact triglycerides and therefore the increased rate of hydrolysis of the glycerides in samples containing a higher percentage of fats may also contribute to a faster rate of oxidation. Furthermore Toyomizu *et al.* (1981) observed that for Pacific mackerel and carp the release of FFA did not lead to an inhibition of lipid oxidation.

In our experiments malonaldehyde seems to be produced to the same extent in sardine oil with or without water soluble muscle extract (Fig. 11.6). The rate of oxidation seems to be nearly the same in both these experiments. However, BHA and TBHQ would seem to have a slight effect on the rate of production of malonaldehyde.

Table 11.3 Chemical analysis of sardine mince.

Component	Method
Dry substance	Gravimetrically after drying at 105°C for 24 hours
Fat	Soxhlet technique
Protein	Kjeldahl technique
Ash	Gravimetrically after drying at 550°C for 24 hours

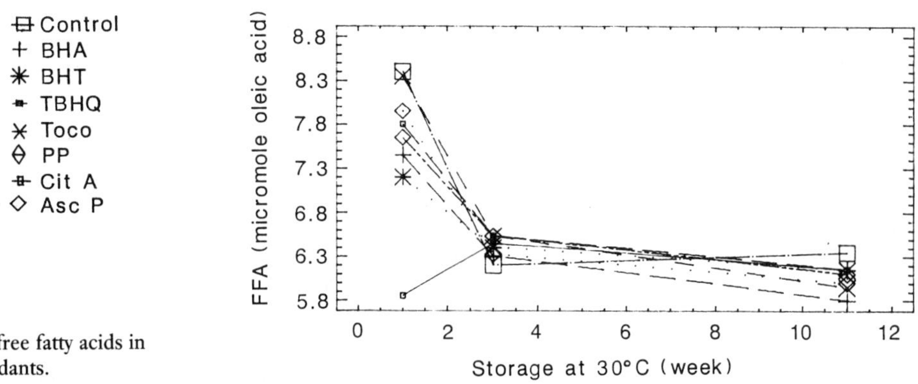

Fig. 11.1 Changes in free fatty acids in sardine oil with antioxidants.

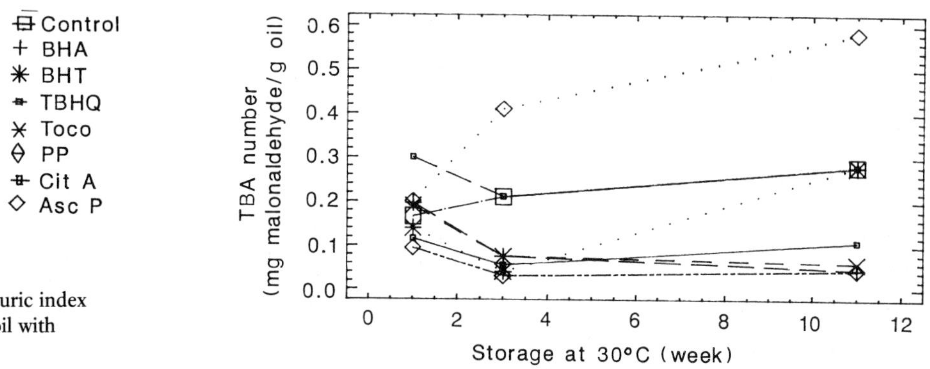

Fig. 11.2 Thiobarbituric index evolution in sardine oil with antioxidants.

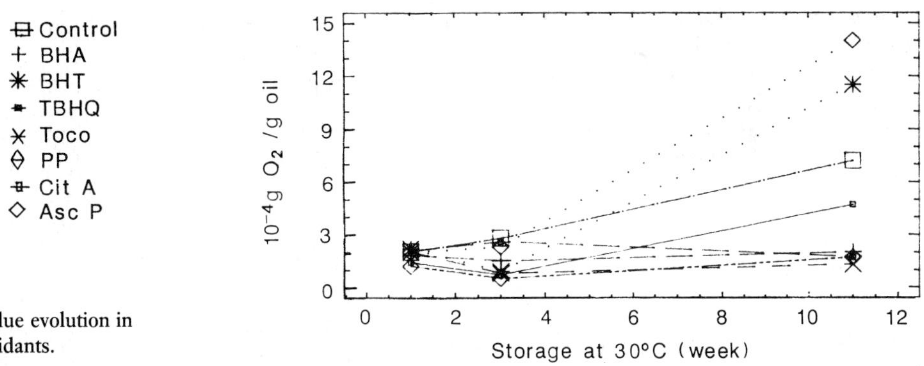

Fig. 11.3 Peroxide value evolution in sardine oil with antioxidants.

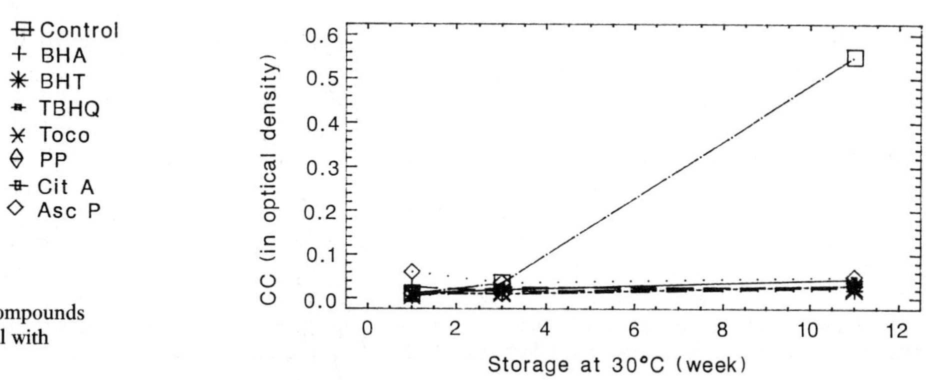

Fig. 11.4 Carbonyl compounds evolution in sardine oil with antioxidants.

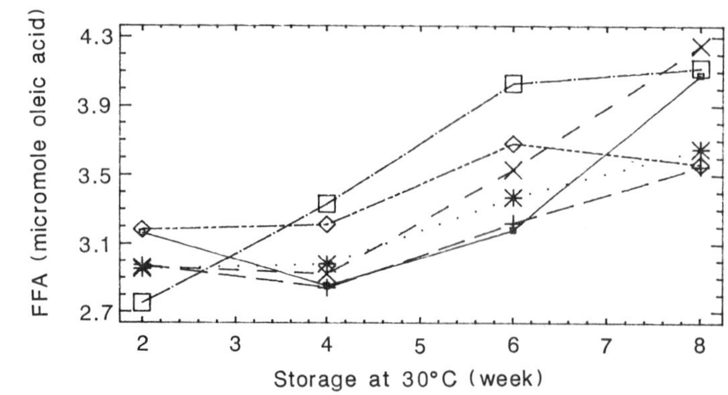

Fig. 11.5 Changes in FFA contents in sardine oil with muscle extract.

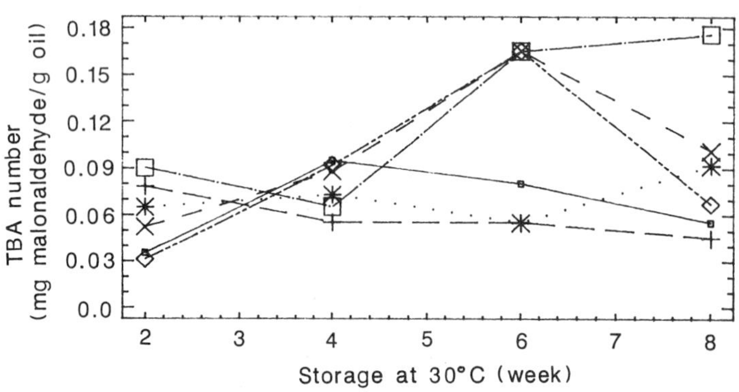

Fig. 11.6 Changes in TBA number in sardine oil with muscle extract.

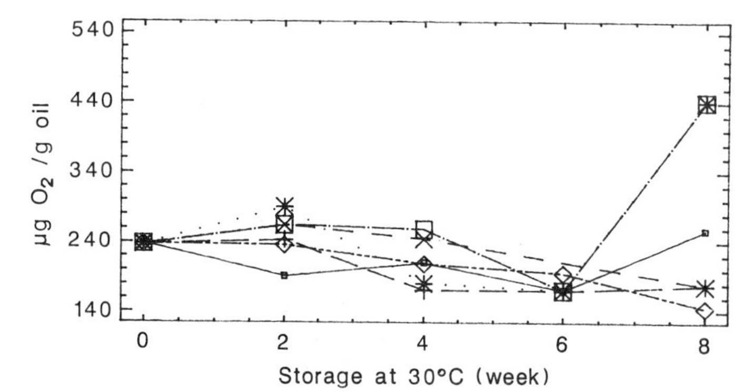

Fig. 11.7 Changes in peroxide value in sardine oil with muscle extract.

The peroxide value in the experiment remained relatively constant until the sixth week; at the eighth week an increase was observed in the control and the emulsions containing PP (Fig. 11.7). BHA, TBHQ and D-tocopherols, to a lesser extent, reduce sardine oil oxidation in the presence of water soluble muscle extract.

Ke *et al.* (1977) noticed that for mackerel skin lipids, TBHQ provides a greater protection against auto-oxidation of marine lipids than BHA, BHT and α-tocopherol. The authors reported that TBHQ has been recognised as one of the most effective antioxidants for lipids rich in polyunsaturated fatty acids.

CONCLUSIONS

BHA and TOCO seem to be effective against oxidation in sardine oil both with or without water soluble

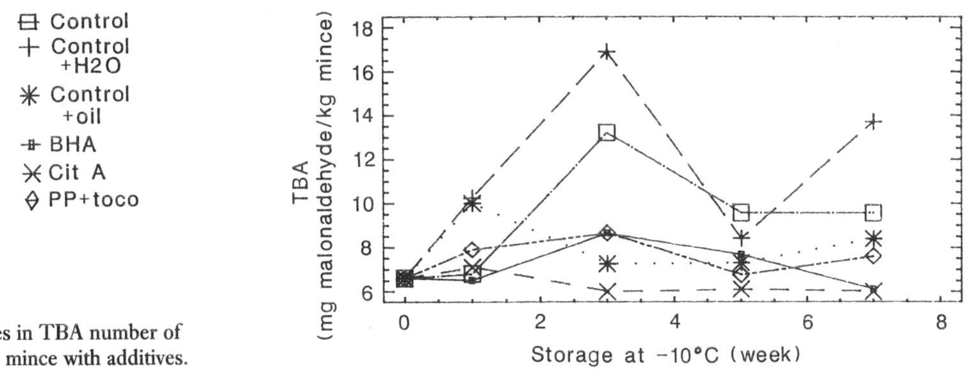

Fig. 11.8 Changes in TBA number of unwashed sardine mince with additives.

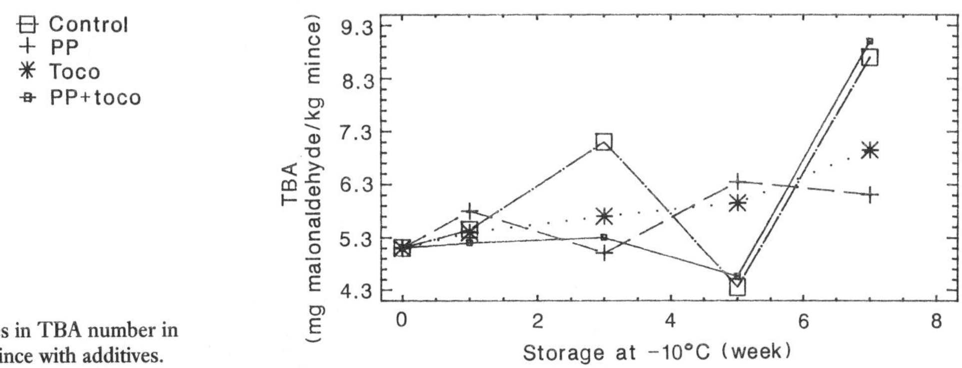

Fig. 11.9 Changes in TBA number in washed sardine mince with additives.

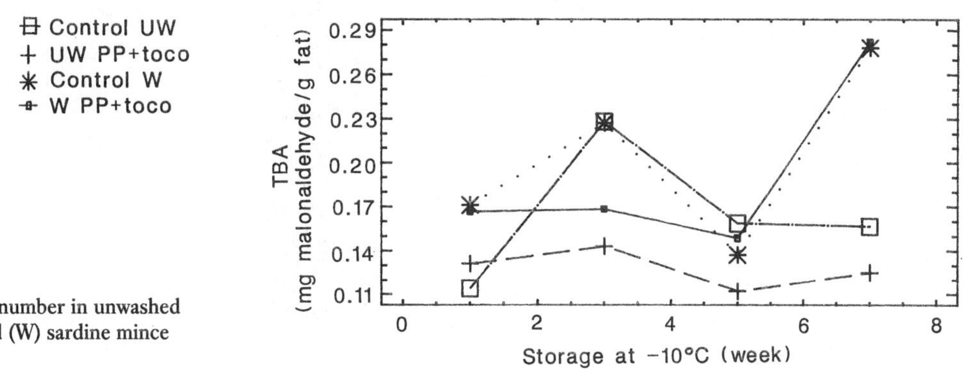

Fig. 11.10 TBA number in unwashed (UW) and washed (W) sardine mince with additives.

Table 11.4 Composition of washed/unwashed sardine mince.

	Moisture %	Protein % (6.25 × N)	Lipid %	Ash %
Unwashed mince	72.3	18.1	7.8	1.4
		(65.3)	(28.2)	(5.1)
Washed mince	77.5	16.8	3.1	0.7
		(74.7)	(15.8)	(3.1)

The numbers ( ) are expressed in % of dry matter

muscle extract, although TBHQ is effective only in the case of sardine oil mixed with the water soluble extract. It is possible that TBHQ has an effect on the lipolytic enzymes. PP and citric acid however are more effective in oil alone than in emulsified oil.

Washed and unwashed mince

The samples with sorbitol only, sorbitol and water, and sorbitol and soya oil were used as controls. It is

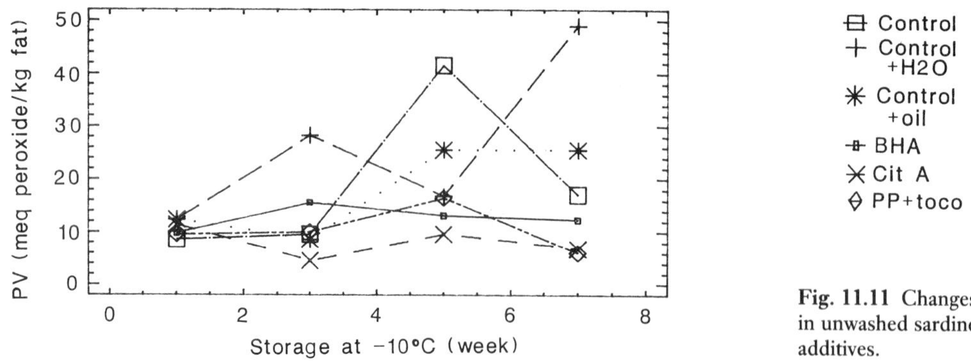

Fig. 11.11 Changes in peroxide value in unwashed sardine mince with additives.

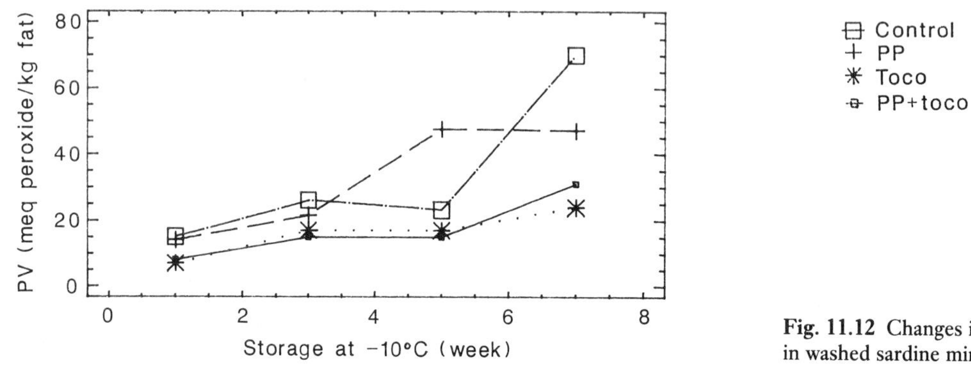

Fig. 11.12 Changes in peroxide value in washed sardine mince with additives.

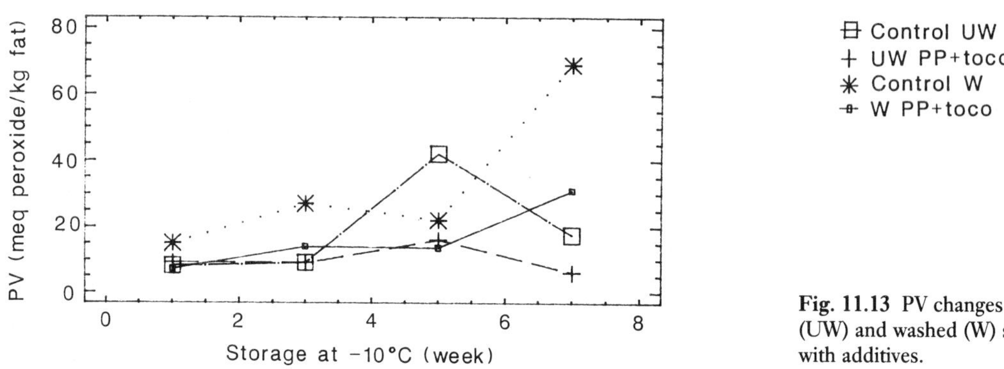

Fig. 11.13 PV changes in unwashed (UW) and washed (W) sardine mince with additives.

necessary to take these controls into consideration in a discussion of the effects of BHA which was dissolved in soya oil and citric acid, where the solvent was water.

Proximate composition of washed and unwashed sardine mince is indicated in Table 11.4.

Samples of unwashed fish mince with antioxidants reduced malonaldehyde formation (Fig. 11.8). The control with soya oil shows the same changes in TBA number as the samples with antioxidants; soya oil, purchased at a local market, is most probably already stabilised with additives and/or with natural antioxidants. These antioxidants could have had an inhibiting effect on lipid oxidation in sardine mince.

No significant differences in changes in TBA number (Fig. 11.9) in the different samples were observed in the case of washed mince (the variation factor of the TBA value is almost equal to 10%). If malonaldehyde content is expressed in mg of malonaldehyde per g of fat, the comparison between washed and unwashed mince is easier (Fig. 11.10).

Table 11.5 Composition of fatty acids (expressed in % of methyl esters) in unwashed and washed sardine fillet mince.

Fatty acids	Unwashed mince	Washed mince
C14	8.6	7.6
C14:1	0.2	0.2
C15	0.5	0.5
iso C16	0.1	0.1
C16	17.6	18.0
C16:1ω7	8.4	7.4
C17	0.4	0.4
C17:1ω8	0.9	0.8
iso C18	0.2	0.2
C18	3.6	3.9
C18:1ω9	9.9	9.1
C18:1ω7	2.5	2.5
C18:2ω6	0.8	0.8
C18:3ω6	0.2	0.2
C18:3ω3	0.7	0.6
C18:4ω3	2.3	1.9
C20	0.9	0.9
C20:1ω9 + ω11	3.7	3.5
C20:2ω6	0.1	0.1
C20:3ω6	0.2	0.2
C20:4ω6	0.6	0.7
C20:4ω3	0.8	0.8
C20:5ω3	13.6	12.7
C21:5ω3	0.5	0.4
C22:1ω11 + ω13	4.2	3.8
C22:4ω6	0.05	0.2
C22:5ω6	0.1	0.2
C22:5ω3	1.7	1.7
C22:6ω3	7.1	11.8
Total saturated	31.7	31.6
Total polyunsaturated	28.7	32.1
P/S ratio	0.9	1.0

The samples of washed mince have a higher TBA number, particularly at the end of storage.

Only samples with antioxidants added would appear to maintain a low peroxide value in both washed and unwashed mince (Figs. 11.11 and 11.12). In the case of washed mince with polyphosphates, there is no difference when compared with the control. When washed and unwashed fish mince are compared (Fig. 11.13), there is only a slight difference between the two controls at the end of storage. The peroxide value as expressed in fat content is also slightly higher in washed mince than in unwashed mince.

The analysis of the fatty acid composition of both washed and unwashed mince (Table 11.5) indicates a predominance of palmitic (C16:0), oleic (C18:1ω9), eicosapentaenoic (C20:5ω3) and docosahexaenoic (C22:6ω3) acids in both cases. Few differences were observed in the qualitative and quantitative composition of the fatty acids, though there was a slightly higher polyunsaturated acids/saturated acids ratio in the case of washed mince. It is possible that this difference might partly explain the higher degree of oxidation (as expressed in percentage of lipids) in the washed mince, the washing producing a slight reduction in the concentration of saturated acids.

Toyomizu *et al.* (1980) did not report a linear relationship between lipid content and lipid oxidation in their study of the pattern of lipid oxidation in the various tissues during the storage of sardines at −5°C, but they did point out that skin lipids are more prone to oxidation because they are composed of triglycerides alone and contain fewer polar lipids inhibiting oxidation. In our study, we have not determined the content of pieces of skin remaining in mince before and after washing, but it is probable that the great majority are left in the mince after washing.

Oxidation remains slightly lower in washed mince because of the smaller fat content due to the washing process. The washing of mince eliminated a large part of the fat (51.1%), minerals (39.2%) and water soluble components. Tejada *et al.* (1985) also noticed in sardine mince prepared in similar conditions, a significant loss of K, Mg, Ca, Fe and Cu.

In conclusion, it would seem to be beneficial before the frozen storage to wash the sardine mince once followed by dehydration in a centrifuge decanter or screw press before the addition of antioxidants.

Our preliminary results suggest that the samples mixed with antioxidants are more efficient than the controls. In unwashed fish mince, BHA is more effective than a combination of tocopherols-polyphosphates, whereas in washed mince this combination gives the best result compared to the control, but no synergetic effect was observed for this combination compared with tocopherols or polyphosphates alone.

The procedure used for selection of additives permits the rapid selection of efficient antioxidants in washed mince (tocopherols and a mixture of tocopherols and polyphosphates) but some caution must be exercised. For instance, the fact that BHT and ascorbyl palmitate are not efficient in sardine oil does not mean they will be ineffective in mince, because the two systems are not the same.

In addition, in combination with the experimental conditions used in our study, it would be useful to investigate the consequences of storage at lower temperatures, the type of packaging used etc. by undertaking a sensorial analysis during storage.

BIBLIOGRAPHY

Castell C.H., Moore B.A., Jangaard P.M. & Neal W.E. (1966) Oxidative rancidity in frozen stored cod fillets. *J. Fish. Res. Board Can.* **23**, 1385–401.

Hultin H.O. (1988) Potential lipid oxidation problems in fatty fish processing. In Fatty Fish Utilization: Upgrading from Feed to Food. In *Proceedings of a National Technical Conference in Raleigh* (Ed. by N. Davis), pp. 185–223. UNC Sea Grant College Publication, Raleigh.

Ke P.J., Nash D.M. & Ackman R.G. (1977) Mackerel skin lipids as an unsaturated fat model system for the determination of antioxidative potency of TBHQ and other antioxidant compounds. *J. American Oil Chemists' Society*, **54**, 417–20.

Keller J.D. & Kinsella J.E. (1973) Phospholipid changes and lipid oxidation during cooking and frozen storage of raw ground beef. *J. Food Science*, **38**, 1200–1204.

Lawrence R.C. (1965) Use of 2–4 dinitrophenyl hydrazine for the estimation of micro amounts of carbonyls. *Nature*, **205**, 1313.

Lowry R.R. & Tinsley I.J. (1976) Rapid colorimetric determination of free fatty acids. *J. American Oil Chemists' Society*, **53**, 470–72.

Mazeaud F. & Bilinksi E. (1976) Free fatty acids and the onset of rancidity in rainbow trout (*Salmo gairdneri*) flesh. Effect of phospholipase A. *J. Fish. Res. Board Can.*, **33**, 1297–302.

Nair P.G.V., Anthony P.D., Gopakumar K. & Nair M.R. (1978) Lipid breakdown in oil sardine (*Sardinella longiceps*) during frozen storage. *Fish Technol.*, **15**, 81–4.

Pizzocaro F., Fedeli E. & Gasparoli A. (1979) Studies on the autooxidation of sardine oil. In *Advances in Fish Science and Technology* (Ed. by J.J. Connell), pp. 258–63. Fishing News Books, Oxford.

Shono T. & Toyomizu M. (1973) Lipid alteration in fish muscle during cold storage. II, Lipid alteration pattern in jack mackerel muscle. *Bull. Japan. Soc. Sci. Fish.*, **39**, 417–21.

Smith J.G.M. & Whittle K.J. (1984) Comparative frozen storage characteristics of minced herring, mackerel, scad and pilchard and effects of packaging. Unpublished paper presented at a meeting of West European Fish Technologists' Association.

Tejada M., Moral A. & Borderias A.J. (1985) Conservation de hachis de sardine en entreposage congelé a −20°C: hachis obtenus a partir de sardine entière, et de sardine étêtée et éviscérée. *Int. J. Refrig.*, **8**, Sept, 294–304.

Toyomizu M., Hanaoka K. & Nakamura T. (1980) Lipid oxidation in the skin during storage of fish in the round at −5°C and the susceptibility predictable for lipid oxidation. *Bull. Japan. Soc. Sci. Fish.*, **46**, 1011–17.

Toyomizu M., Hanaoka K. & Yamaguchi K. (1981) Effect of release of free fatty acids by enzymatic hydrolysis of phospholipids on lipid oxidation during storage of fish muscle at −5°C. *Bull. Japan. Soc. Sci. Fish.*, **47**, 615–20.

Tsukuda N. (1976) Changes in the lipids of frozen fish. I, Changes in the lipids of skipjack during frozen storage. *Bull. Tokai Reg. Fish. Res. Lab.*, **84**, Feb. 31–41.

Vyncke W. (1970) *Fette Seifen Anstrichmittel*, **72**, 1084.

Vyncke W. (1975) *Fette Seifen Anstrichmittel*, **77**, 239.

CHAPTER 12

Use of a portable microwave fat meter with herring (*Clupea harengus*)

M. Kent, A. Lees and R.H. Christie

INTRODUCTION

The instrument described in this work (Kent 1989) is designed to solve a prevailing problem in the fish industry: the rapid determination of water and fat contents of products. Such tests as previously existed were destructive and time consuming. In whole fish flesh, such as herring and other pelagic species, the problem is to a certain extent simplified by the fact that a linear relationship exists between the fractions of water and fat (Iles & Wood 1965) (Fig. 12.1) and measurement of one serves to determine the other within the limits of validity of this relationship. Such a convenient relationship arises from the physiological need of the fish to maintain a slightly negative buoyancy even while the increasing fat content, which is of low density, tends to create the opposite effect. Variation in fat content arises from the seasonal effects of food availability and breeding cycle.

In the microwave frequency region (100 MHz–100 GHz) the dielectric properties of foodstuffs are dominated by those of water, especially when the proportion of water is large (see Mudgett *et al.* 1974 for example). The relationship between fat/water ratios and the dielectric properties of meat emulsions has been studied by Ohlsson *et al.* (1974) and some interesting results were obtained, notably that fat and water content could be determined from various dielectric measurements in the microwave range. However, in those data no correlation existed between the fat and water contents. The dielectric properties of herring flesh have been studied by Kent (1990) and it has been shown that there is an excellent correlation between the water content and the real part of the dielectric permittivity at all frequencies in the microwave range. Typical data obtained is shown in Figure 12.2 where permittivity is plotted against frequency, with water content as a parameter.

The dielectric properties of substances containing polar molecules such as water, exhibit dispersion in certain characteristic frequency bands. In those regions of dispersion the dielectric properties are complex, that is to say they have both real and imaginary components. The real part represents the capability of the material to store electromagnetic energy, and the imaginary part its ability to dissipate it. Many methods exist for the measurement of these properties but the one which has been used in this work depends on the dielectric sensitive properties of a form of transmission line known as microstrip or stripline.

A sensor based on this principle was developed earlier at Torry for the measurement of moisture content of a variety of materials (Kent 1972, 1973, 1974, Steele & Kent 1978, Kent & Price 1979) but in one publication (Steele & Kent 1978) its use was proposed for the measurement of the water content of herring and other fish.

STRIPLINE SENSOR

The evolution of stripline which is the basis of the sensor is shown in Figure 12.3. Its structure may be considered as that of a coaxial line which has been split longitudinally and opened out to expose the centre conductor. The outer conductor of the coaxial line becomes a flat conductor known as the 'ground plane', the dielectric filling the coaxial line becomes a flat 'substrate' and the centre conductor is flattened into a thin planar strip on its surface. Any dielectric material now placed in direct contact with this stripline will modify the transmission properties in a way which will be dependent on the dielectric properties of this material. As has already been stated, the dielectric properties of foodstuffs depend very much on their water content (see Mudgett *et al.* 1974), so these changes in transmission properties can be calibrated

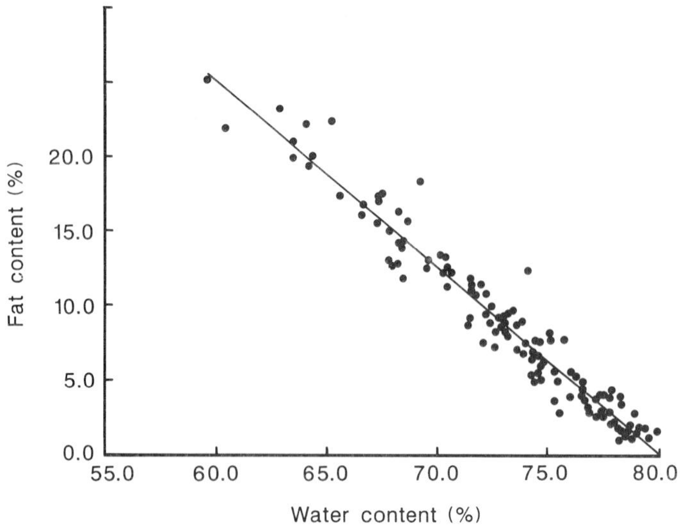

Fig. 12.1 Correlation between fat content and water content of herring both expressed as a total weight percentage fraction.

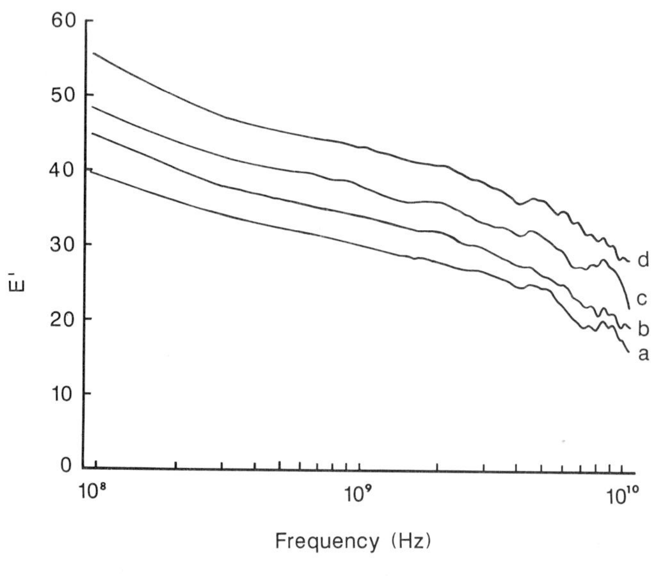

Fig. 12.2 Dielectric permittivity of herring flesh as a function of frequency with water content as parameter (a) 63.9%; (b) 67.09%; (c) 71.9%; (d) 78.2%.

against water content. In general, the amount of a particular constituent in any reasonably homogeneous material whose dielectric properties are affected by that constituent, can be measured.

The construction of the actual sensor is shown in Figure 12.4. A protective layer of polytetrafluoroethylene (ptfe) 0.1 mm thick is fixed directly over the strip; this protects it from erosion or damage and at the same time reduces the sensitivity to a level which can be easily handled. Without this layer nearly all of the energy can be coupled out of the strip by the superimposed material to be measured and the resultant power loss is too large to measure. The substrate material is a ptfe and ceramic mixture (Epsilam 10–3 Ms Co. Ltd) with a permittivity of ~10 and very low loss.

The substrate is 1.27 mm in thickness and the strip itself, as used in this work, is 1.7 mm in width and 53 mm long. The sensitivity of the sensor depends not only on its length but also very much on the dimensions and properties of the structure. In general the stripline does not have the same impedance as the coaxial line feeding it, which is characteristically 50Ω. This results in some impedance mismatching at

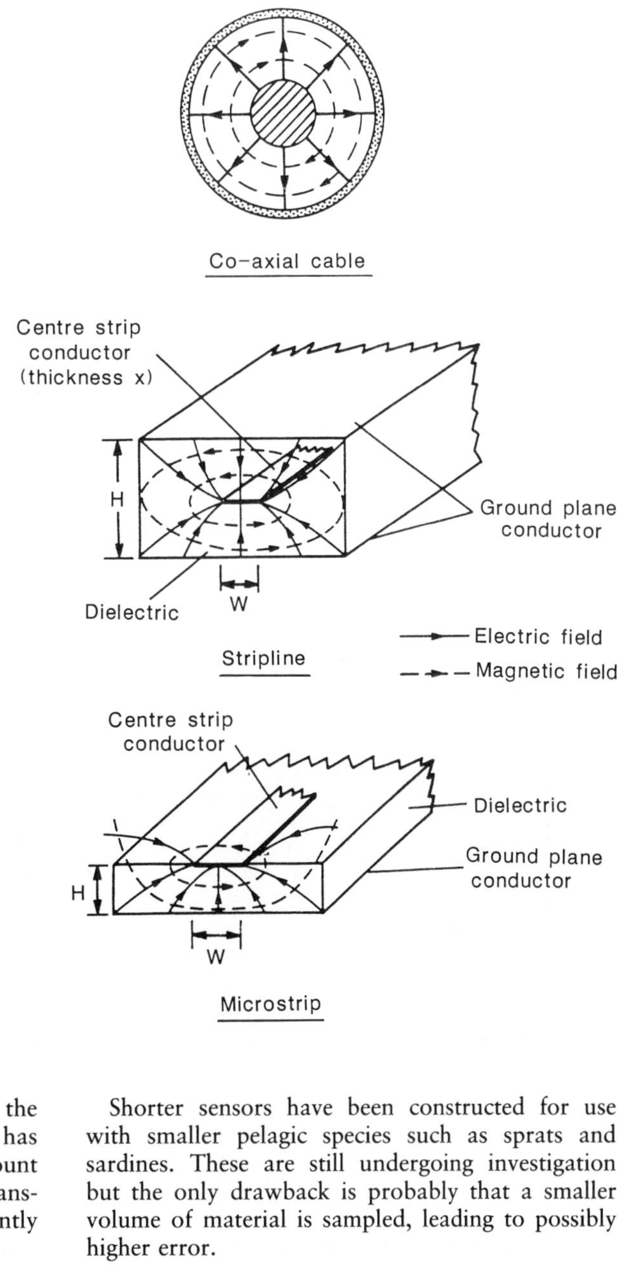

Fig. 12.3 Stripline construction compared to coaxial line.

the coaxial-to-stripline transition as well as at the boundaries of the loaded portion of the line. It has been shown however that such mismatches account for only a small proportion of the observed transmission losses in the line, which are predominantly due to the material of interest (Kent 1973).

To enable the transition from coaxial line to stripline to be made, the coaxial line is fed through the substrate and soldered to the surface of the strip at points A. In order that a flat surface can be presented to the fish these transition points are recessed slightly as shown. This makes negligible differences to the impedance of the line at those points. As mentioned, the whole structure is covered in thin ptfe (0.1 mm), which is uniformly flat.

Shorter sensors have been constructed for use with smaller pelagic species such as sprats and sardines. These are still undergoing investigation but the only drawback is probably that a smaller volume of material is sampled, leading to possibly higher error.

Microwave circuit

The transmission property of the sensor that is customarily measured is the insertion loss or attenuation of power in the line. This is measured by the microwave circuit shown in Figure 12.5. The boxed area is that part of the instrument housed in the hand-held

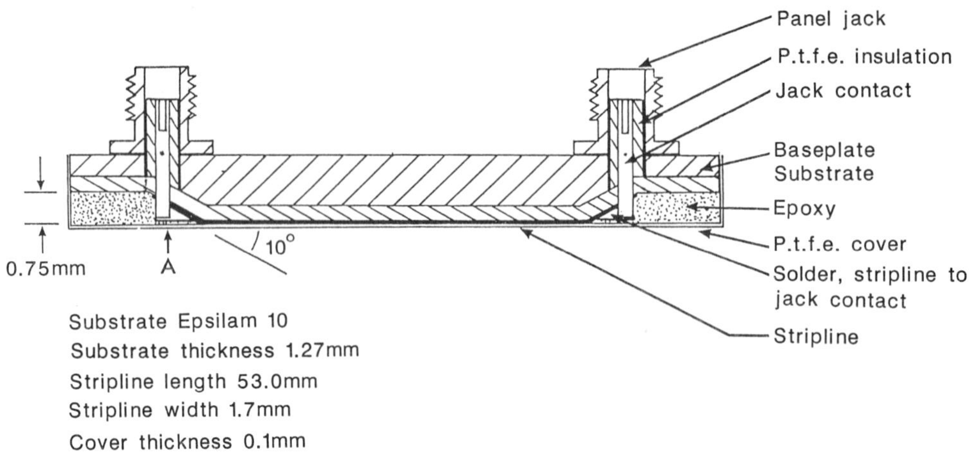

Fig. 12.4 Construction of stripline sensor used in the fat meter (not to scale).

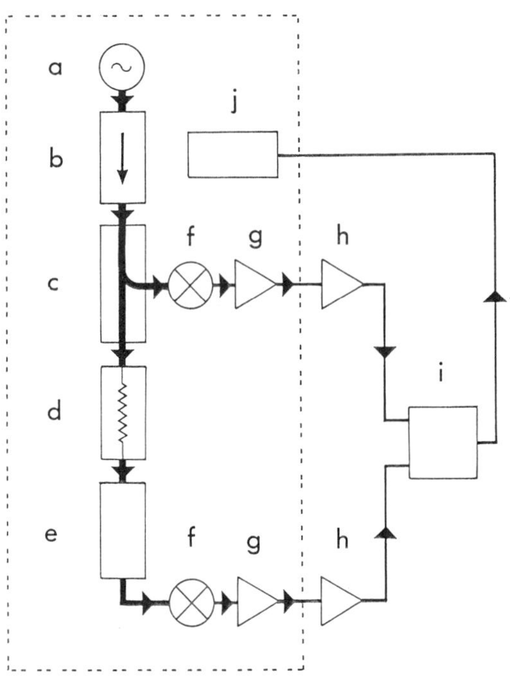

Fig. 12.5 Microwave circuit and associated electronics. (a) Gunn diode; (b) isolator; (c) 10 dB directional coupler; (d) 10 dB attenuator; (e) stripline sensor; (f) detectors; (g) preamplifiers; (h) trimming amplifiers; (i) log-ratio amplifier; (j) output display and/or microprocessor. Dotted line indicates components housed in handheld portion of fat meter.

probe. The rest of the electronics plus batteries and switches are in a separate unit connected to the probe by multicore cable (Fig. 12.5). Power from a microwave source such as a Gunn diode (a) is fed via an isolator (b) to a 10 dB directional coupler (c). The purpose of the isolator is to prevent any reflected power from interfering with the operation of the Gunn diode. The directional coupler feeds 10 dB of the power to a reference detector (f). The major part of the power is fed through a 10 dB attenuator (d) to the stripline sensor. The attenuator reduces the power to a level comparable with that in the reference arm. This level is also low enough to ensure square law operation of the detectors, given the input power of a few milliwatts. It also reduces the effect on both the Gunn diode and the directional coupler of reflected power from the various mismatches in the sensor. Apart from the obvious mismatch of the sensor impedance there are also reflections from the coaxial-to-stripline transitions which are difficult to eliminate entirely. After passing through the sensor (e) the remaining power is detected by another crystal detector (f).

The Gunn diode oscillates at ~10 GHz with a supply voltage of 7.5 V and delivers about 10 mW of power. The current drawn from the power supply for this is around 200 mA, so in order to conserve battery energy and life the Gunn diode is switched on only for the duration of the measurement. Since it never achieves thermal equilibrium it is therefore most probable that both the frequency and the power output are to some extent time-dependent. This is where the value of the reference system is demonstrated, since the ratio of the power in the signal detector to that in

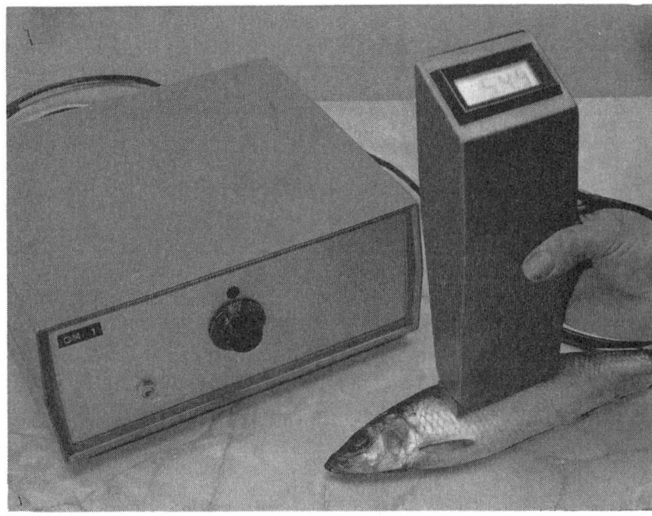

Fig. 12.6 General view of fat meter.

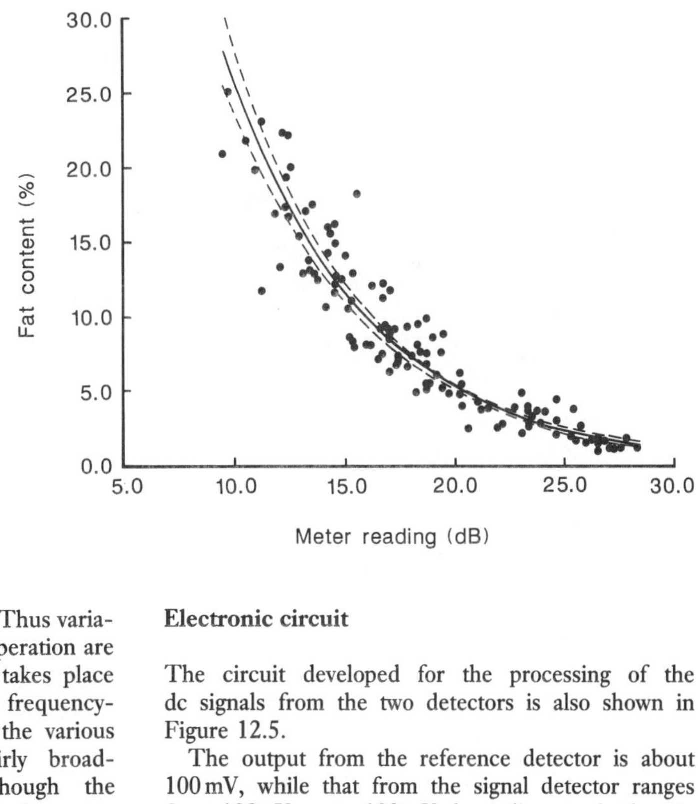

Fig. 12.7 Results of meter reading versus fat content for whole herring. The dashed lines represent the 95% confidence limits.

the reference detector is always measured. Thus variations in power level during this transient operation are not important. The frequency shift that takes place is small enough to render negligible any frequency-dependent changes that might occur in the various components. All components have fairly broadband characteristics (8–12 GHz). Although the system described operates in this particular frequency range (X-band) it can be constructed to function at any microwave frequency. For some applications it might in fact be better to make the measurements at 1–2 GHz.

Electronic circuit

The circuit developed for the processing of the dc signals from the two detectors is also shown in Figure 12.5.

The output from the reference detector is about 100 mV, while that from the signal detector ranges from 100 μV up to 100 mV depending on the loss in the sensor. These two dc signals are each fed to low gain (gain × 8) voltage follower type pre-amplifiers (g). The amplified signals are then fed through a further stage of amplification (h) to a logarithmic/ratio

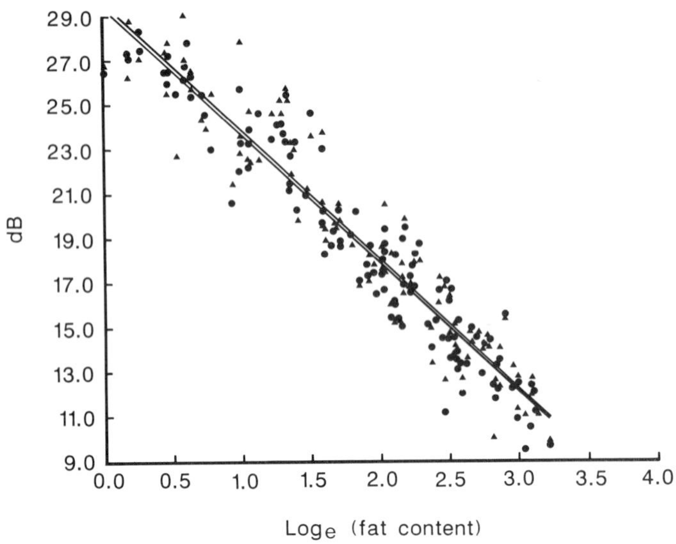

Fig. 12.8 Regressions of meter reading (dB) versus natural logarithm of the fat content for both left and right sides of the herring. There is no significant difference between the two lines.

device (i) which provides an output proportional to the logarithm of the ratio of the two signals. Such a ratio is readily expressed in decibels (dB) and is calibrated against fat or water content of the fish. The instrument is capable of measuring up to 30 dB of attenuation. In the present prototype instrument this signal is fed directly to a digital voltmeter in the hand-held part of the instrument (Fig. 12.6).

In a final version, the instrument will contain a microprocessor to perform a small number of operations. First it will store the calibration equations found from experiment to be necessary for each species, and secondly will provide a facility for averaging a predetermined number of readings. This latter function will allow the mean fat content of a batch of fish to be estimated, a desirable function as will be seen later.

MATERIAL AND METHODS

The results were obtained from 124 herring of known provenance from various fishing grounds on both sides of the Atlantic, for example, Bay of Fundy, Norway and the North Sea grounds. Much of the fish obtained had been frozen prior to the experiments but there seemed to be no significant difference between fresh of frozen and thawed fish. The exception to this was a batch which had obviously suffered some dehydration in the cold store (freezer burn) and for which the natural relationship between fat and water content no longer existed. The results for these were discarded.

Fat contents were determined on fillets with skin intact by the Foss-let method, and water contents were determined by drying in an air oven at 105°C for 48 hours. This period of drying, rather than 24 hours, has been shown by Christie *et al.* (1985) to yield better agreement between replicates.

RESULTS

After wiping dry the surface of the herring the sensor was placed on the middle portion of the fish parallel to and above the lateral line. As can be seen from Figure 12.7, the response of the instrument is not linear, the sensitivity in terms of dB per % fat decreasing with increasing fat content. This is probably due to a greater proportion of the water molecules in the flesh becoming rotationally hindered due to binding forces as the water content decreases. The water that remains is that which is more tightly bound to other molecular species in the tissue, such as proteins, sugars, ionic salts, etc., and has a lower permittivity than pure water in the microwave region. The sensor is known to respond to permittivity (Kent 1973). The calibration curve can be approximated very well by a logarithmic function relating attenuation A and fat content F:

$$\log_e(F) = a - bA$$

where $a = 4.85$
$b = 0.159$

Both left and right sides of the fish were measured but as shown by Figure 12.8 there was no significant difference in calibration between the two. It is therefore

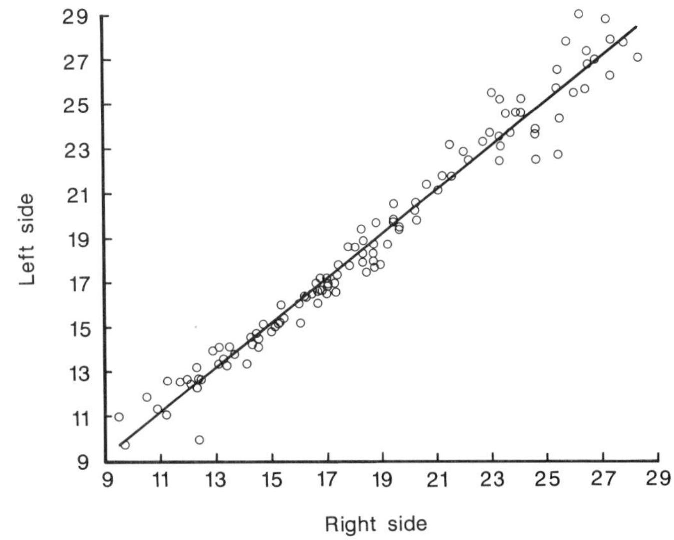

Fig. 12.9 Correlation between meter readings (dB) from left and right sides of herring showing increased scatter at high readings.

Table 12.1 Regression parameters for fat (F) and water (W) in herring. S_F is the standard deviation for F about the regression line (ii).

	This work	Iles & Wood
Number of samples	124	895
(i) W	79.66−0.755F	79.12−0.862F
(ii) F	100.4−1.254W	90.45−1.139W
r	0.973	0.991
S_F	1.353	0.614

permissible to measure on only one side of the fish and the choice of side is irrelevant.

It is interesting to note however that the correlation between sides becomes poorer as the fat content decreases to very low values and the meter reading increases (see Fig. 12.9). This may be a result of greater likelihood of surface moisture and greater variability in this from one side to another.

The consequences of drying the surface were investigated by making measurements on 50 fish and simply observing the effect of wiping dry the surface. There was a significant decrease in the measured value of attenuation, corresponding on the calibration to an increase of +1−2% in fat content for the region of 12 dB. At the same time there was an increase in variation between left and right sides but as this was only just significant at the 5% level, we can safely conclude that only a cursory wipe is needed to free the fish of excess surface moisture. Even that may not be necessary if the sensor is pressed firmly into the surface, thus extruding all free moisture.

Variability of results

The likely sources of variation in the measurements are:

(1) Variation in skin and/or scale thickness between the sensor and the flesh. The skin is known to have a relatively high moisture content compared with the underlying tissue.
(2) Small variations in residual surface moisture.
(3) Variations in distribution of fat below the skin and in light and dark muscle.
(4) Variation in the relationship between water content and fat content (Fig. 12.1). In this work the regression equation was found to differ from that of Iles & Wood (1965) (Table 12.1). This is due to major differences in the method of fat determination which in this work was on fillets alone but in the work of Iles and Wood had been on whole fish minus gonads. In addition the wider geographical sampling differs from the earlier work which was on North Sea herring alone.

Despite these perturbing influences the calibration data show a quite narrow 95% confidence interval, as can be seen in Figure 12.6 where the dashed curves represent these limits. For a fat content of around 28%, the 95% confidence interval is ±2.4% fat. At 10% fat this is reduced to ±0.5% fat and at even lower fat contents around 5% it becomes ±0.2%.

That is to say the 95% confidence interval in relative terms ranges from a value of around ±9% at high fat contents to around ±4% at low fat. The confidence interval refers of course to a number of independent measurements made on a representative sample of fish. For single fish the prediction interval must be invoked and the figures for that are much worse ranging from +16.6% or −10.4% at the high fat content to +2.9% or −1.8% at the low end of the range. Due to the logarithmic nature of the fitted curve, these prediction limits are asymetrical and in fact are unduly pessimistic at high fat contents. They show however that the instrument must be used essentially to measure the mean fat content of a batch. In its final form, it will contain the necessary electronics to perform this task in direct analogy with the Torry fish quality meter (Jason & Richards 1975).

As a final point, however, it has been observed that fish from a single batch, when such a quantity is the result of the by-catch, can vary in fat content from as little as 2% up to around 20%. One must ask the question then how useful or meaningful a batch measurement would be.

CONCLUSIONS

From the scatter of the data from individual fish about the calibration line it is concluded that such measurements could yield a value for average fat content of a batch within confidence limits of ±2.4% at worst (high fat content) and typically much less. Thus for the estimation of the fat content of batches the instrument seems useful, especially when it is noted that the technique is non-destructive. It must be remembered that part of this error arises from the method of calibration. Better fat analysis in the calibration procedure could result in a better performance of the instrument. It must be stressed however that a more accurate calibration would be against water content since the correlation with the fat content is inferred from the known relationship between fat and water.

BIBLIOGRAPHY

Christie R., Kent M. & Lees A. (1985) Microwave and infrared drying versus conventional oven drying in methods for moisture determination in fish flesh. *J. Food Tech.*, **20**, 117–27.

Iles T.D. & Wood R.J. (1965) The fat/water relationship in North Sea herring (*Clupea harengus*) and its possibile significance. *J. Mar. Biol. Ass. UK*, **45**, 353–66.

Jason A.C. & Richards J.C.S. (1975) The development of an electronic fish freshness meter. *J. PhysE. Sci. Inst.*, **8**, 826–30.

Kent M. (1972) The use of strip-line configurations in microwave moisture measurement. *J. Microwave Power*, **7**, 185–93.

Kent M. (1973) The use of strip-line configurations in microwave moisture measurement II. *J. Microwave Power*, **8**, 189–94.

Kent M. (1974) Improvements in or relating to measurement of the dielectric properties of materials. *UK Patent 1,354,474*.

Kent M. (1989) Microwave fat meter. *UK Patent Application 8915323.3*.

Kent M. (1990). Measurement of dielectric properties of herring flesh using transmission time domain spectroscopy. *Int. J. Food Sci. & Tech.*, **25**, 26–38.

Kent M. & Price T.E. (1979) Compact microstrip sensor for high moisture content measurement. *J. Microwave Power*, **14**, 363–65.

Mudgett R.E., Wang D.I.C. & Goldblith S.A. (1974) Prediction of dielectric properties in oil-water and alcohol-water mixtures at 3000 MHz, 25°C based on pure component properties. *J. Food Sci.*, **39**, 632–5.

Ohlsson T., Henriques M. & Bengtsson N.E. (1974) Dielectric properties of model meat emulsions at 900 and 2800 MHz in relation to their composition. *J. Food Sci.*, **39**, 1153–6.

Steele D.J. & Kent M. (1978) Microwave stripline techniques applied to moisture measurement in food materials. *Proceedings of the 1978 IMPI Symposium on Microwave Power*, pp. 31–6. Ottawa.

CHAPTER 13

Effect of frying on the nutritional quality of dried salted Indian mackerel (*Rastrelliger kanagurta*)

F.W. Ma'ruf, D.A. Ledward, R.J. Neale and R.G. Poulter

INTRODUCTION

Dried-salted fish have been consumed for generations in many parts of the world and world-wide it has been estimated that about 12 million tonnes (wet weight equivalent) of dried fish are produced annually (FAO 1981). Although several studies have been carried out on temperate water species, little research on tropical species, processed by traditional methods, has been reported, especially with regard to their nutritional quality.

Fish body oils are known to be very susceptible to oxidation by atmospheric oxygen and when dried such fats have little protection against oxidation (Wolfe 1975). Such products may develop distinctive 'oxidised' flavours, but these are not necessarily unacceptable (Waterman 1976). It has been claimed that as well as modifying flavour some compounds formed by peroxidising lipids are toxic (Sanders 1987), and that lipid peroxides can destroy vitamins A and E (Olcott 1962). Dried fish may also develop a yellow or brown coloration due to non-enzymic browning reactions involving either lipid oxidation products (Karel 1973) and/or reducing sugars (Jones 1962).

As well as affecting the colour and flavour of the product, peroxidising lipids (and reducing sugars) may decrease the nutritional value of the protein by reacting with one or more of the essential amino acids present. However this aspect has received little attention, although a few reports suggest that salting and drying *per se* have little effect on the nutritional value of fish protein (Cutting 1962).

Excessive heat treatment is known to impair the nutritional value of fish proteins (Aitken & Connell 1979). Opstvedt (1988) has recently reviewed current views on the effects of smoking and drying on the protein quality of fish muscle and concluded that at drying temperatures of less than 100°C losses are moderate; our previous results would support this conclusion (Ma'ruf *et al.* 1990). He also concluded that, although the results are not uniform, normal cooking procedures at temperatures up to 100°C have little effect. However frying, especially in oxidised oil, has been shown to reduce the digestibility of fish protein (Motohiro 1988). In view of the importance of these types of products to the developing countries a study of well-characterised samples made by a widely used technology was carried out to evaluate some of the nutritional changes undergone during processing and subsequent cooking (frying).

Results reporting the effect of different salt contents on both the chemical and nutritional quality are reported elsewhere (Ma'ruf *et al.* 1990).

MATERIALS AND METHODS

Materials

Indian mackerel (*Rastrelliger kanagurta*), usually about 50–60 g wet weight, were used throughout this work. The fish, caught by gill net, were purchased from three different landing sites and processed within one day of catching in Indonesia. About 80 kg of fish were used, of which about 10 kg were frozen to act as controls. The salt used for brining, prepared locally by the evaporation of sea water, contained significant concentrations of impurities including about 0.2 ppm copper and 30 ppm iron (Ma'ruf *et al.* 1990). All other chemicals were AR grade (Fisons, Loughborough). Most measurements were made on at least three replicates taken from 15 to 20 ground fish. In the feeding trials 1 kg of sample was used for each treatment.

Processing

The fresh mackerel were dressed, split, washed and immersed in 5% (by weight) salt solutions for about 16 hours. On removal from the brine these samples were rinsed and sun-dried on interwoven bamboo mats raised from the ground. At 16.00 hours every day the fish were removed to store and replaced on the drying mats at 08.00 hours the following day. Drying, as judged by experienced processors, was complete in 2–3 days at which point the fish were placed in covered, woven baskets and stored at the processor's for 2–4 days prior to packing. On departure for the UK the dried samples were sealed into loose fitting polythene bags, while the frozen fish were mixed with dry ice and placed in an insulated container. On arrival at Nottingham about 48 hours later the dried fish, still in polythene bags, were stored at 30°C and analysed at intervals over 20 weeks; the polythene-wrapped frozen fish were stored at −25°C and freeze-dried prior to analysis.

It is usual to fry dried-salted fish prior to consumption and so selected samples were fried in coconut oil by two methods. In the first the samples were cooked until 'done' as judged by an Indonesian housewife (household frying). In later experiments the frying was for 2, 5 or 10 minutes at 150°C. In these later experiments fresh oil was used for every sample. Shallow frying was employed so that one surface was open to the atmosphere, and samples were turned at regular intervals.

Analytical techniques

Proximate analysis

Protein was determined by the Kjeldahl procedure, fat by extraction with petroleum ether (BP 40–60) for 16 hours and moisture by oven drying at 105°C for 16 hours (Egan *et al.* 1981).

Chemical tests

TBA value was estimated by the distillation method of Tarladgis *et al.* (1960) except that 5 drops of antioxidant solution (0.3 g BHA + 0.3 g BHT + 5.4 g propylene glycol + 1 ml 0.2% EDTA + 4 g Tween 40/100 ml) were added prior to distillation.

Fluorescent pigment content was determined by the method of Hanson & Normington (1985), as modified by Ma'ruf *et al.* (1990).

Subjective assessment indicated that glacial acetic acid extraction did not remove all the colour from the fried samples and thus the degree of browning was estimated from the reflectance of the finely-ground, compressed tissue at 420 nm using a Perkin-Elmer Model 124 reflectance spectrophotometer (Ma'ruf 1989).

Amino acid analysis of the samples was performed as described by Obanu *et al.* (1975).

Table 13.1 Composition of the basal casein diet.

Ingredient	g/kg
Casein	150
Maize starch	489
Glucose	150
Oil (maize)	150
Salt mix*	50
Vitamin mix*	10
D.L. methionine	1

* Composition as described by Payne & Stewart 1972.

Nutritional evaluation

The protein quality of the fish was assessed by rat feeding trials in which net protein utilisation (NPU), biological value (BV) and true digestibility (TD) were measured (Miller & Bender 1955).

The fish were pulverised in a hammer mill and stored at 0°C in water impermeable bags until required. The basal diet is shown in Table 13.1. The fish sample under test replaced all the casein and methionine and part of the oil in the basal diet, so that each diet contained 10% fish protein (N × 6.25) and 15% fat (some from maize oil and some from fish).

The rats, males and females of the Wistar strain, were weaned at 21 days and fed on the basal diet *ad libitum* for 7 days. After weighing (range 68–80 g) each rat was placed in a stainless steel cage with a wire mesh back and floor, underneath which a metal tray to collect faeces was placed. Food and water were fed *ad lib.* and six rats were fed on each diet. Six rats were also fed a protein-free diet. The dietary intake of each rat was recorded over seven days at the end of which time the rats were killed by CO_2 inhalation and cervical dislocation and reweighed. The nitrogen content of the carcasses was estimated from their moisture contents (Miller & Bender 1955), determined by oven drying at 105°C for 48 hours, and that of the faeces (and diets) by Kjeldahl analysis.

RESULTS

Chemical quality

The chemical composition of the dried fish is given in Table 13.2. During storage for 20 weeks at 30°C the

Table 13.2 Proximate composition (wet weight basis) of the dried, salted fish during storage at 30°C and the effect of frying thereafter. Means of triplicates.

Storage time (weeks)	Frying regime at 150°C	Protein	Fat	Moisture
0	None	55.0	3.4	26.7
0	'household'*	52.2	18.5	18.4
10	None	56.0	2.7	24.6
10	2 min	60.1	17.6	14.5
20	None	61.6	4.3	18.4
20	2 min	56.7	16.7	12.3
20	5 min	55.6	20.1	8.9
20	10 min	54.1	25.6	6.9
Freeze dried control		76.7	6.5	6.5

* Cooked in reused oil until 'acceptable'.

Table 13.3 Effect of storage at 30°C on the chemical quality of dried salted fish and the subsequent effect of frying.

Storage time (weeks)	Frying regime	TBA[x] value	Fluorescence[y]	Browning[z]
0	None	27.3[a]	12.2[b]	21.2[b]
0	'household'	7.7[b]	15.2[a]	12.6[ah]
10	None	7.8[b]	14.7[a]	18.2[d]
10	2 min	7.7[b]	18.9[c]	11.1[ch]
20	None	8.5[bc]	18.8[c]	17.0[f]
20	2 min	9.8[c]	18.4[c]	10.9[ch]
20	5 min	8.8[bc]	18.4[c]	10.4[c]
20	10 min	7.4[b]	21.5[d]	8.0[e]
Freeze dried control		—	6.3[e]	22.9[b]

[x] Expressed as mg malonaldehyde/kg sample (mean of 3).
[y] Expressed as relative emission units/g sample/ml (mean of 3).
[z] Expressed as % reflectance at 420 nm (mean of 5).
Values in the same column bearing the same letter are not significantly different ($P > 0.05$).

only significant change in composition was due to some loss of moisture (Table 13.2). Frying led to the expected increase in fat and decrease in moisture (Table 13.2).

As found previously, storage in air at 30°C caused an increase in both browning and the content of fluorescent compounds (Ma'ruf *et al.* 1990). On frying, all samples browned still further and most exhibited significant increases in the content of fluorescent compounds in the tissue (Table 13.3). TBA values however exhibited no clear trend, confirming that this index is of little value in estimating the quality of the lipids present in these products (Ma'ruf *et al.* 1990). Even two minutes' frying in fresh oil led to a significant ($P < 0.05$) increase in browning (loss of reflectance) and in most cases a significant increase in fluorescent product content was also seen following frying (Table 13.3). The exceptions were the samples stored for 20 weeks at 30°C which increased in fluorescent product content only after 10 minutes' frying.

The apparent amino acid content of selected samples was also determined (Table 13.4) and it can be seen that frying in fresh oil caused an apparent loss of lysine and glutamic acid in addition to the losses already induced by storage at 30°C (Ma'ruf *et al.* 1990, Table 13.4). It is interesting to note that 'household' frying appeared to lead to a loss of all amino acids.

Nutritional quality

During storage at 30°C there were significant decreases in NPU, BV and TD of the fish (Table 13.5). It is also seen from Table 13.5 that prolonged frying led to further decreases in NPU and BV.

Table 13.4 Amino acid content of dried fish previously brined in 5% salt before and after frying at 150°C (g/100 g protein). Results of single analysis.

	0 weeks at 30°C		20 weeks at 30°C			
Amino acid	control	'household' frying	control	2 min frying	5 min frying	10 min frying
Asp	10.6	8.9	10.6	11.0	11.3	10.4
Thr	5.1	4.1	5.0	5.2	5.4	4.9
Ser	4.1	3.3	3.9	4.1	4.3	3.9
Glu	18.4	16.4	17.4	20.0	20.1	13.8
Pro	4.1	3.9	5.2	6.2	6.6	6.1
Gly	6.2	5.0	5.9	7.9	8.4	6.1
Ala	8.0	6.1	8.2	8.8	9.3	8.5
Val	7.5	6.0	7.8	7.2	7.3	7.6
Ileu	7.1	5.3	7.2	6.6	6.6	7.0
Leu	10.9	8.4	11.2	11.0	11.1	10.9
Tyr	5.2	4.0	5.3	5.7	4.8	5.0
Phe	6.5	4.8	6.7	6.4	6.3	6.2
His	3.6	3.0	3.2	3.3	3.4	3.3
Lys	11.7	9.2	10.3	10.7	11.1	9.4
Arg	5.8	5.2	5.2	5.8	6.2	5.5
$\frac{1}{2}$ Cyst	1.2	0.9	1.0	1.1	1.0	1.8
Met	3.7	3.1	3.4	3.6	3.6	3.9

Table 13.5 Protein quality of dried salted fish during storage at 30°C and the effect of subsequent frying.

Storage time (weeks)	Frying regime	NPU	BV	TD	Feed intake g/7 days	Growth rate g/7 days
0	none	75.6[a]	77.4[b]	97.7[a]	72.9[bcd]	26.0[be]
0	'household'	42.2[b]	43.3[e]	97.9[a]	59.1[a]	5.6[ad]
10	none	67.8[c]	69.9[b]	97.0[a]	66.9[acd]	31.4[b]
10	2 min	66.6[c]	70.3[b]	94.6[b]	80.0[b]	40.6[c]
20	none	53.7[ef]	56.8[c]	94.6[b]	68.3[ad]	13.6[df]
20	2 min	51.5[df]	55.0[c]	93.5[b]	69.1[d]	11.8[df]
20	5 min	43.9[b]	46.4[a]	94.3[b]	66.2[ad]	8.5[d]
20	10 min	40.6[b]	43.2[a]	93.9[b]	63.5[ad]	7.3[d]
Freeze dried control		80.0[a]	81.6[b]	98.0[a]	82.3[b]	43.0[bc]

Each value is the mean of six determinations.
Values in the same column bearing the same superscript are not significantly different ($P > 0.05$)

DISCUSSION

As discussed elsewhere, the storage of dried-salted fish leads to the formation of both fluorescent compounds and brown pigments and both these parameters correlate well with the decrease in NPU of the fish (Ma'ruf *et al.* 1990). The present study indicates that a typical frying regime causes further decreases in NPU and concomitant increases in fluorescence and browning. In the 20 week stored samples an increase in fluorescent pigment content was only seen after 10 minutes' frying. These samples did, though, possess high contents of fluorescent compounds even before frying and thus these results suggest that there is a limit to the amount of such compounds that can be generated. There was a significant correlation between browning and fluorescent content ($r = -0.86$, $p < 0.01$). There was no significant correlation between TBA value and protein quality but this is not surprising since it is well established that malonaldehyde can react with proteins and thus is not a reliable indicator of lipid oxidation in these products.

The correlation coefficient between browning and NPU ($r = -0.83$, $p < 0.01$) was superior to that found between NPU and fluorescence production ($r = 0.75$, $p < 0.05$). In an earlier study not involving frying, Ma'ruf *et al.* (1990) also found that browning, measured on acetic acid extracts of dried, unfried fish, was a better indicator of nutritional quality than fluorescence production. Although massive losses in NPU were found on frying, the decreases in amino acid contents (especially the limiting S-amino acids) do not seem sufficient to explain this loss in nutritional quality. It is possible that the large decreases observed are due to the formation of complexes that render the amino acids biologically unavailable yet are labile to acid (such as might occur in the Maillard reaction with lysine). 'Household' frying appeared to lead to a loss of all amino acids (Table 13.5) and Obanu *et al.* (1975) also observed an apparent loss of all amino acids on prolonged storage at 38°C of some intermediate moisture meat products. They attributed this to the extensive formation of covalent, acid resistant linkages between the proteins that effectively rendered them 'non-available'.

The cause of the decrease in nutritional quality may well not be solely due to loss of amino acid availability since the reduced food intake and growth rate observed, in most cases on storage and/or frying (Table 13.5), suggest toxic compounds may also form (Sanders 1987, Ma'ruf *et al.* 1990). But whatever the cause the importance of these results to populations obtaining a significant proportion of their protein intake from dried-salted fish cannot be ignored.

BIBLIOGRAPHY

Aitken A. & Connell J.J. (1979) Fish. In *Effects of heating on foodstuffs* (Ed. by R.J. Priestley), pp. 219–54. Applied Science Publishers, London.

Cutting C.L. (1962) The influence of drying, salting and smoking on the nutritive value of fish. In *Fish in Nutrition* (Ed. by E. Heen & R. Kreuzer), pp. 161–79. Fishing News Books, Oxford.

Egan H., Kirk R.S. & Sawyer R. (1981) *Pearsons Chemical Analysis of Food*, 8th edn., Churchill Livingstone, London.

FAO (1981). *The prevention of losses in cured fish.* FAO Fisheries Technical Paper 219, FAO, Rome.

Hanson S.W. & Normington G. (1985) *Potentially harmful compounds in cured fish.* Report 7, ODA Project R3693, Humberside College of Higher Education, Grimsby.

Jones N.R. (1962) Browning reaction in dried fish products. In *Recent Advances in Food Science*, 2 (Ed. by J. Hawthorn & J.M. Leitch), pp. 74–80. Butterworths, London.

Karel M. (1973) Protein interactions in biosystems. Protein-lipid interactions. *J. Fd. Sci.*, **38**, 756–63.

Ma'ruf F.W. (1989) *Chemical and nutritional evaluation of Indonesian dried-salted fish.* PhD thesis, University of Nottingham.

Ma'ruf F.W., Ledward D.A., Neale R.J. & Poulter R.G. (1990) Chemical & nutritional quality of Indonesian dried-salted fish (*Rastrelliger kanagurta*). *Int. J. of Fd. Sci. & Tech.*, **25**, 66–77.

Miller D.S. & Bender A.E. (1955) The determination of the net utilization of proteins by a shortened method. *British J. Nutr.*, **9**, 383–8.

Motohiro T. (1988) Effect of smoking and drying on the nutritive value of fish. A review of Japanese Studies. In *Fish Smoking & Drying* (Ed. by J.R. Burt), pp. 91–120. Elsevier Applied Science, London.

Obanu Z.A., Ledward D.A. & Lawrie R.A. (1975) Storage of intermediate moisture beef at tropical temperature. II, Effect on eating quality. *J. Fd. Tech.*, **10**, 667–74.

Olcott H.S. (1962) Oxidation of fish lipids. In *Fish in Nutrition* (Ed. by E. Heen & R. Kreuzer), pp. 112–16. Fishing News Books, London.

Opstvedt J. (1988) Influence of drying and smoking on protein quality. In *Fish Smoking and Drying* (Ed. by J.R. Burt), pp. 23–36. Elsevier Applied Science, London.

Payne P.R. & Stewart R.J.C. (1972) Cubed diets of high and low protein value. *Laboratory Animals*, **6**, 135–40.

Sanders T.A.B. (1987) Toxicological considerations in oxidative rancidity of animal fats. *Fd. Sci. and Tech. Today*, **1**, 162–4.

Tarladgis B.G., Watts B.M., Younathan M.T. & Dugan L.R. (1960) A distillation method for the quantitative determination of malonaldehyde in rancid foods. *J. American Oil Chemists Society*, **37**, 44–8.

Waterman J.J. (1976) *The production of dried fish.* FAO Fish Technical Paper 160, FAO, Rome.

Wolfe M.L. (1975) The effect of smoking and drying on the lipids of West African herring (*Sardinella* spp.). *J. Fd. Tech.*, **10**, 515–22.

CHAPTER 14

Fishing gear influence on the spoilage of albacore (*Thunnus alalunga*) in ice

B. Pérez-Villarreal and R. Pozo

INTRODUCTION

A variety of different types of gear are used in tuna fishing. The most common are purse seines, pole and line, longliners, trollers and handliners (Goodrick 1987, Lawlor & Adams 1988). The fishing method chosen depends on the species, environmental conditions, type of fleet and economics.

Albacore or white tuna, *Thunnus alalunga*, has traditionally been caught in the north-east Atlantic by two fishing techniques: trolling or pole and line with live bait. Recently, new catching methods have been introduced for albacore using pelagic trawls and tanglenets as the type of gear, with the purpose of making this fishery more efficient (Prado 1988). As a consequence, the question of the differences in the quality of albacore caught by the new and traditional methods was raised.

The influence of the seasonal, environmental and physiological factors and handling procedures on the quality of fish has been studied in a wide variety of species (Stansby & Lemon 1941, Lemon & Regier 1977, Bullard & Collins 1978, Smith *et al.* 1980, Vyncke 1980, Reppond & Collins 1983, Samuels *et al.* 1984, Barile *et al.* 1985, Ravesi *et al.* 1985) including tunas (Crawford & Finch 1968, Wilson 1982, Gibson 1984, Otwell 1988).

Although some reports have been found dealing with the effect of method of catching on fish quality (Shewan 1949, Nickelson & Vanderzant, 1976, Botta *et al.* 1987), only a few refer to tuna species (Cramer *et al.* 1981, Goodrick 1987).

Albacore belongs to the family Scombridae. Fish of this family have high levels of free histidine in the muscle tissue, which can be transformed by histidine decarboxylating bacteria to histamine (Ienistea 1973). Histamine formation depends on fish species, bacterial load, method of catching and handling, temperature and processing techniques (Martin *et al.* 1978).

The aim of the present study was to establish the influence of fishing gear and of method of on-board handling on the spoilage of fresh albacore stored in ice, making special reference to differences in histamine formation and shelf-life between line-caught and net-caught albacore.

MATERIAL

Two batches of albacore were selected for the purpose of this study during the 1988 albacore fishing season. As a traditional method, the trolling technique was chosen for the line-caught albacore group. Standard pelagic trawls were used for the net-caught albacore group. Both batches were obtained from commercial fishing vessels working in waters of the north Atlantic. Fish were stowed and covered with ice in refrigerated holds until landed.

Line-caught albacore was unloaded at Bermeo port. Details of when the fish had been caught and homogeneity of the batch were obtained. Net-caught tuna was obtained from Lorient market. Fish from this group had been gutted and the gills removed at sea. It was known that the fish had been caught 2–3 days before purchase. Fish from both groups were transported to the laboratory in boxes, covered with ice and stored in a cold room at approximately 2–3°C. Ice was added as required.

As soon as the fish arrived, six samples were analysed to assess the initial quality of each batch. To study the influence of the method of catching on the spoilage of albacore and determine possible differences on the shelf-life, three or four fish from each batch were removed for analysis at intervals of 3 or 4 days during 40 days of storage in ice.

METHODS

Sensory assessment

The fish were examined in the raw and cooked state by a trained panel of eight assessors experienced in evaluating the freshness of fish, following scales constructed in previous work (Pérez-Villarreal & Pozo 1990). Scales for raw fish have a maximum of 9 and a minimum of 3, which describes when the fish was considered completely spoiled. On the raw fish, the appearance of skin, eyes, white and red muscle and belly flaps, and the texture of skin in net-caught albacore were evaluated. These parameters, together with appearance and odour of the gills, were assessed in line-caught albacore.

A steak of approximately 3 cm thick was cut from the centre region of each fish and cooked in a closed casserole over boiling water without addition of any water or condiments. The same panel of assessors evaluated the odours, flavour and texture of the cooked fish. Scales for cooked fish have a maximum of 5, and a minimum of 1 representing fish which would definitely be unfit for consumption.

Chemical analyses

A protein-free extract was prepared from the white muscle by blending 100 g of white muscle in 300 ml of 0.6 M perchloric acid, and filtering.

Total volatile bases were determined by making aliquots of the extract alkaline with 20% NaOH and distilling into boric acid solution. The distilled bases were titrated with standard hydrochloric acid solution. Results were expressed as mg of nitrogen/100 g.

Trimethylamine was determined by the Dyer picrate method (1945) as modified by Murray & Gibson (1972). Nucleotides and related compounds were measured by the ion pair, reverse phase HPLC procedure of Murray & Thomson (1983). Histamine was determined by HPLC with fluorescence detection following derivatisation with o-phthaldialdehyde (OPA) (Pozo & Saitua 1988).

Microbiological methods

Samples were prepared from the white muscle by homogenising 100 g of white muscle in 400 ml of chilled sterile 0.1% peptone phosphate buffer, pH 7.0–7.2, in a Stomacher (Seward Medical, London). Decimal dilutions of the homogenate were prepared in the same buffer. Total viable counts (TVC) were determined by plating 0.1 ml of suspension on to Bacto Plate Count Agar (DIFCO 0479) and incubating at 20°C for 5 days for the psychrophilic count and at 37°C for 3 days for the mesophilic count.

A pseudomonad count was made using Pseudomonas Agar Base (OXOID CM559) with CFC supplement (OXOID SR103). Plates were incubated at 25°C for 48 hours.

Statistical analyses

Data was stored on computer files and processed by the statistical package SPSS-X V2.2 (SPSS 1986). Statistical evaluation followed standard procedures.

RESULTS AND DISCUSSION

Sensory assessment

Sensory scores for general appearance (freshness) on the raw fish, and odour and flavour on the cooked fish, were related to storage times expressed as days in ice. The linear regressions calculated by the least squares procedure gave the equations:

Line-caught

$$\text{Score (general appearance)} = 9.12 - 0.162 \text{ days}$$

$$n = 28 \qquad r^2 = 0.935$$

$$\text{Score (cooked odour)} = 4.93 - 0.088 \text{ days}$$

$$n = 28 \qquad r^2 = 0.918$$

$$\text{Score (cooked flavour)} = 4.94 - 0.082 \text{ days}$$

$$n = 28 \qquad r^2 = 0.897$$

Net-caught

$$\text{Score (General appearance)} = 7.28 - 0.137 \text{ days}$$

$$n = 38 \qquad r^2 = 0.736$$

$$\text{Score (cooked odour)} = 5.11 - 0.115 \text{ days}$$

$$n = 38 \qquad r^2 = 0.935$$

$$\text{Score (cooked flavour)} = 5.71 - 0.154 \text{ days}$$

$$n = 30 \qquad r^2 = 0.889$$

The coefficients of determination (r^2) for these equations indicate that there is a good correlation of sensory scores with storage time in ice.

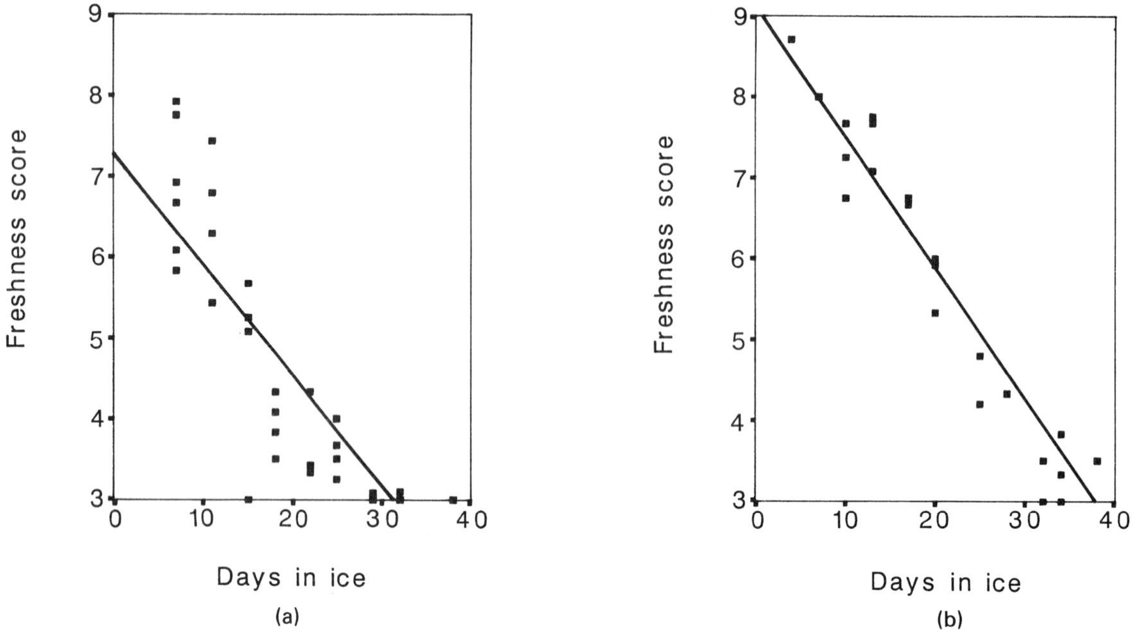

Fig. 14.1 Changes in freshness score of raw fish during storage in ice. (a) Net-caught. (b) Line-caught.

The scatter plots and linear regressions for general appearance of the raw fish are shown in Figure 14.1. After comparison of both regressions, slopes were not significantly different ($P > 0.1$). Significant differences ($P < 0.01$) were found between methods of catching, in general appearance at the initial stages of the storage (as soon as the fish arrived at the laboratory). Net-caught albacore showed a big variety of physical damage: scales had been ripped off in wide areas, skin had been raised, they presented numerous bruises and signs of compaction of the fish against the net in the skin and the muscular bundles had ruptured in some samples. A large proportion (63%) of the fish showed two or more of these types of physical damage. Line-caught albacore did not show such physical damage, but presented blood bruises caused by the use of gaffs that go through the muscle for landing fish.

In the cooked fish, significant differences ($P < 0.01$) between methods of catching were found in the coefficients of the linear regressions for odour and flavour. Spoilage in line-caught albacore happened more slowly than in net-caught albacore.

If we establish as the limits of good quality fish a score of 5.5 for raw fish and 3.5 for cooked fish, these are reached in approximately 20 days for line-caught and 13 days for net-caught albacore.

Differences in sensory quality can be explained by the differences in the catching methods. The use of pelagic trawls and tangle-nets results in a very long struggle by the fish inside the net. The fish are highly stressed because of the struggle, and death can result from the high pressure exerted on the tuna inside the net and from the lack of oxygen. Substantial amounts of physical damage inside and outside tuna can occur in these circumstances. Fish metabolism is also accelerated (Love 1980), producing for example an accumulation of lactic acid in the muscle tissue that can cause death or effect changes in the fish quality during storage, thus shortening shelf-life.

Chemical analyses

The initial concentrations of total volatile bases (TVB) and their changes during storage time did not show significant differences ($P > 0.1$) between methods of catching. Initial concentrations were around 30 mgN/100 g, similar to the values found by Aldrin & Grandgirard (1971) and Pérez-

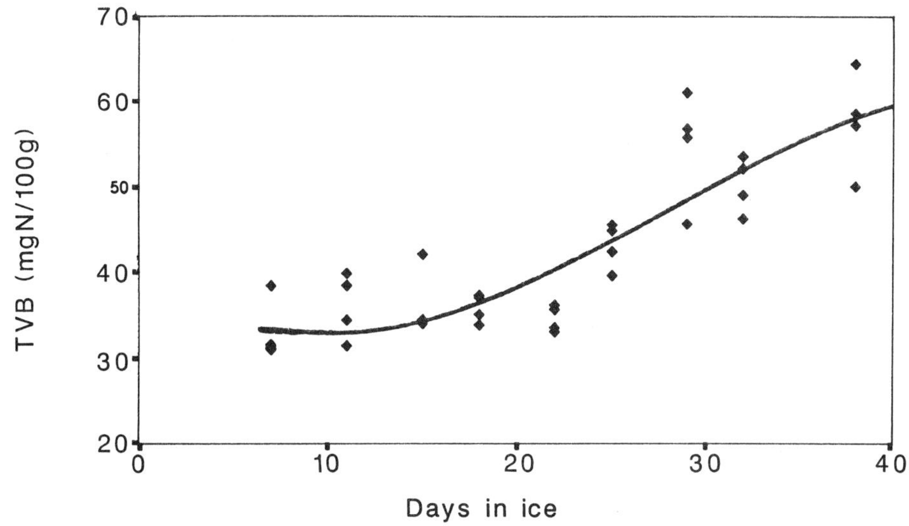

Fig. 14.2 Change in total volatile base content during storage in ice (net-caught).

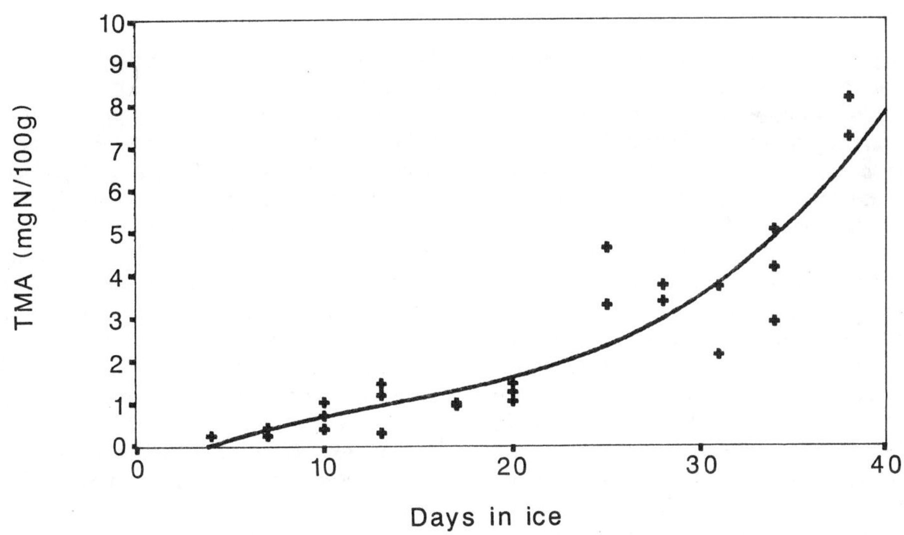

Fig. 14.3 Change in trimethylamine content during storage in ice (line-caught).

Villarreal & Pozo (1990), and the increase during the first 30 days in ice was very small for both methods. Data for net-caught albacore is plotted against days in ice in Figure 14.2.

Trimethylamine (TMA) for both methods of catching increased very slowly from values of nearly zero. After 25 days in ice, TMA concentration reached 9–10 mgN/100 g at the end of the storage trial (Fig. 14.3), a low value compared with other tuna species (Putro & Saleh 1985). Signifi-

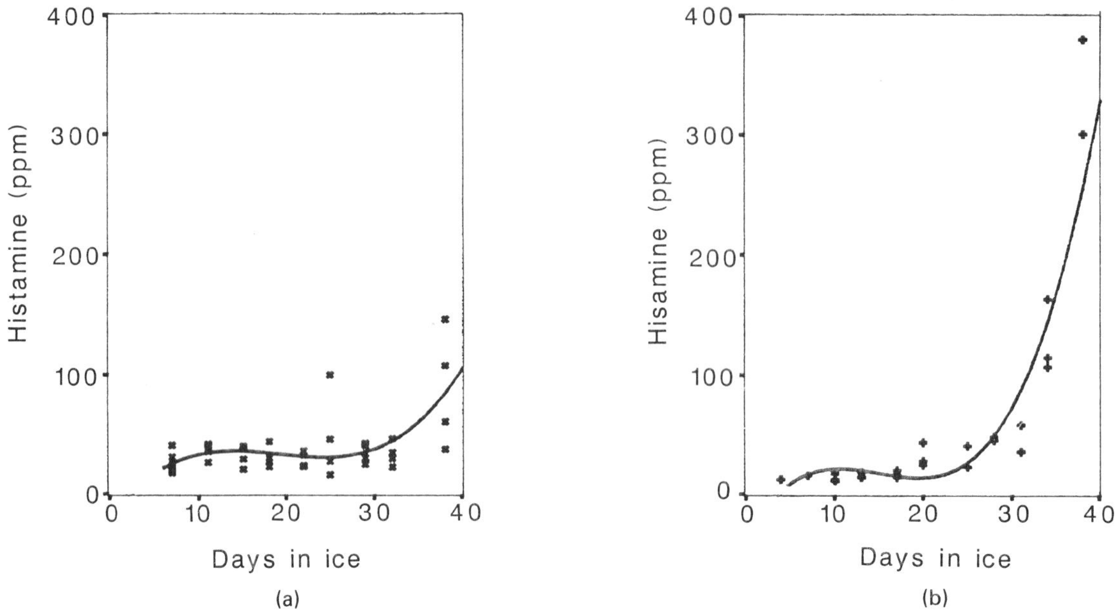

Fig. 14.4 Changes in histamine content during storage in ice. (a) Net-caught. (b) Line-caught.

cant differences ($P > 0.1$) between line-caught and net-caught fish were not shown.

The adenosine nucleotides and their catabolites have been used as indicators of freshness in several species. In general, changes in hypoxanthine (HYP) content have been shown to have a close relationship to freshness (Burt *et al.* 1976). The changes in HYP concentration during storage were similar for both methods of catching, showing an approximately linear increase. The increase was at a slower rate for line-caught albacore, reaching 90 mg/100 g at the end of the storage time. At this time, net-caught albacore gave values of nearly 140 mg/100 g (about 30% of the sum of all the degradation products).

The use of specific levels of histamine as indices of quality has not been considered a reliable index of spoilage. Nevertheless, histamine content was measured in this work to establish the influence of fishing gear on its initial concentration. It was found that the initial concentrations of histamine were very small (12–16 ppm) and no significant differences ($P > 0.1$) between line-caught and net-caught albacore were found. Under the conditions of this experiment (fish covered with ice and stored in a chillroom at 2–3°C), concentrations remained nearly constant during the first 30 days of storage (Fig. 14.4). Maximum values at the end of the storage trial (after the fish would be considered unfit for consumption) varied from 150 ppm (net-caught) to 400 ppm (line-caught).

In summary, it appears that different methods of catching and on-board handling do not influence the initial concentration and evolution pattern of the classical chemical parameters used as indices of spoilage (TVB, TMA, ATP degradation products), nor the initial histamine concentration.

Microbiological analyses

The normal microflora of fish caught in temperate waters consists mainly of gram-negative psychrophilic organisms. During storage in ice the low temperature favours growth of this group, so that it dominates the flora of spoiled fish. Within this flora, pseudomonads are important in the development of the typical odours and flavours of spoiled fish (Shaw & Shewan 1968, Shewan 1976, Hobbs & Hodgkiss 1982).

Total viable counts (TVC) of mesophilic micro-organisms were also made to establish the possible influence of different methods of catching and on-board handling on this type of flora. TVC at 37°C indicated an initial bacterial load at the beginning of

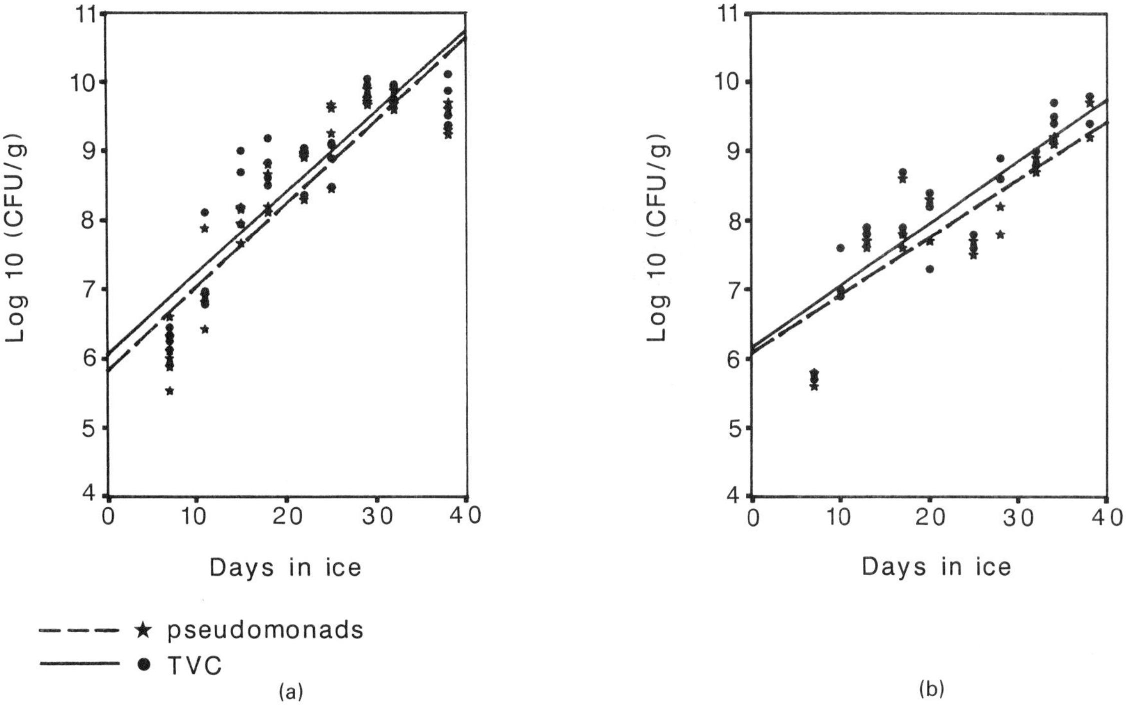

Fig. 14.5 Changes in microbiological counts during storage in ice. (a) Net-caught. (b) Line-caught.

the storage period of approximately 10^4 CFU/g; significant differences between methods of catching ($P > 0.1$) were not detected. During storage in ice this parameter did not show important changes with time in either group, there being a big variation within batches. TVC of mesophilic organisms would not be a good indicator of spoilage in albacore.

Figure 14.5 shows the scatter diagram of TVC of psychrophilic organisms and counts of pseudomonads during storage time for both methods of catching.The regression equations for the two counts are:

Line-caught

TVC, $\log_{10}$ (CFU/g) = 6.16 + 0.090 days

$n = 27 \quad r^2 = 0.787$

pseudomonads, $\log_{10}$ (CFU/g) = 6.08 + 0.084 days

$n = 23 \quad r^2 = 0.746$

Net-caught

TVC, $\log_{10}$ (CFU/g) = 6.05 + 0.117 days

$n = 38 \quad r^2 = 0.803$

pseudomonads, $\log_{10}$ (CFU/g) = 5.81 + 0.121 days

$n = 38 \quad r^2 = 0.810$

The coefficients in the regression equations for both parameters were found to be not significantly different from each other ($P > 0.1$) in either line-caught or net-caught albacore. The present data seems to show that the majority of the TVC flora in albacore are active spoilers of the pseudomonads and closely related groups.

Significant differences ($P > 0.1$) were not found in the inital load (10^6 CFU/g) between catching methods. The regression slopes of TVC on storage time for line-caught and net-caught albacore were slightly different ($P < 0.05$), reaching counts of 10^8 CFU/g after 25 days in ice for net-caught and 32 days in ice for line-caught.

It seems as if the spoilage of net-caught albacore is more rapid than of line-caught, even though the initial bacterial load is similar in both groups. Several reasons can explain this. Natural barriers in net-caught albacore have been destroyed by physical damage to the skin (scales ripped off, skin grazed, muscular bundles ruptured) helping the penetration of bacteria from external sources. Also, during the removal of guts and gills at sea, the epithelium that protects the internal cavities can be torn, making it easier for the intrinsic flora that normally lives in guts and gills to pass into the muscle tissue.

BIBLIOGRAPHY

Aldrin J.F. & Grandgirard A. (1971) Qualité du thon albacore (*Thunnus albacores*). Essais d'appréciation par voie chimique. *Rec. Méd. Vét.*, **147**, 583–606.

Barile L.E., Milla A.D., Reilly A. & Villadsen A. (1985) Spoilage patterns of mackerel (*Rastrelliger faughni Matsui*). 1, Delays in icing. *ASEAN Food J.* **1** (2) 70.

Botta J.R., Bonnell G. & Squires B.E. (1987) Effect of method of catching and time of season on sensory quality of fresh raw Atlantic cod (*Gadus morhua*). *J. Food Sci.*, **52** (4) 928–31.

Bullard F.A. & Collins J. (1978) Physical and chemical changes of pink shrimp, *Pandalus borealis*, held in carbon dioxide modified refrigerated seawater compared with pink shrimp held in ice. *Fish Bull.*, **76**, 73.

Burt J.R., Gibson D.M., Jason A.C. & Sanders H.R. (1976) Comparison of methods of freshness assessment of wet fish. II, Instrumental and chemical assessments of boxed experimental fish. *J. Fd. Technol.*, **11**, 73–89.

Cramer J.L., Nakamura R.M., Dizon A.E. & Ikehara W.N. (1981) Burnt tuna: conditions leading to rapid deterioration in quality of raw tuna. *Mar. Fish. Rev.*, **43**, 12–16.

Crawford L. & Finch R. (1968) Quality changes in albacore tuna during storage on ice and in refrigerated sea water. *Fd. Tech.*, **22**, 87–90.

Dyer W.J. (1945) Amines in fish muscle. *J. Fish. Res. Board Can.*, **6**, 351–67.

Gibson D.J.M. (1984) *Southern Bluefin tuna for the sashimi market.* Ministry of Fisheries and Agriculture, Wellington.

Goodrick B. (1987) Postharvest quality of tuna meat, a question of technique. *Fd. Tech. in Australia*, **39** (7), 343–5.

Hobbs G. & Hodgkiss W. (1982) The bacteriology of fish handling and processing. In *Developments in Food Microbiology* (1) (Ed. by R. Davies), pp. 71–117. Applied Science Publishers, London.

Ienistea C. (1973) Significance and detection of histamine in food. In *The Microbiological Safety of Foods* (Ed. by B.C. Hobbs & J.H. Christian), p. 327. Academic Press, London.

Lawlor F.J. & Adams C.M. (1988) Tuna fishing methods and gear in the Gulf of Mexico and south Atlantic. In *Yellowfin tuna: Fishing gear, production and quality.* Sea Grant Report 91, pp. 36–45. Washington.

Lemon D.W. & Regier L.W. (1977) Holding of Atlantic mackerel (*Scomber scombrus*) in refrigerated sea water. *J. Fish. Res. Board Can*, **34**, 439.

Love R.M. (1980) *The Chemical Biology of Fishes*, **2**. Academic Press, London.

Martin R.E., Gray R.J.H. & Pierson M.D. (1978) Quality assessment of fresh fish and the role of the naturally occurring microflora. *Food Technol.*, **32** (5) 188–92.

Murray C.K. & Gibson D.M. (1972) An investigation of the method of determining trimethylamine in fish muscle extracts by the formation of its picrate salt. *J. Fd. Technol.*, **7**, 35–46.

Murray J. & Thomson A.B. (1983) Reverse phase ion pair separation of nucleotides and related products in fish muscle. *J. High Resolution Chromatog. & Chromatog. Comms.* **4**, 209–210.

Nickelson R. & Vanderzant C.V. (1976) *Bacteriology of shrimp.* Paper presented at 1st Annual Tropical and Subtropical Fisheries Technology Conference. Texas A&M University, College Station.

Otwell W.S. (1988) Post-harvest handling considerations for fresh Yellowfin tuna. In *Yellowfin tuna: Fishing gear, production and quality.* Sea Grant Report 91, pp. 46–51. Washington.

Pérez-Villarreal B. & Pozo R. (1990) Chemical composition and ice storage of albacore (*Thunnus alalunga*). *J. Food Sci.*, **55**, 678–82.

Pozo R. & Saitúa E. (1988) Determinación de histamina en pescado y sus productos derivados por HPLC con detección fluorimétrica. *Alimentaria*, **196**, 27–9.

Prado J. (1988) Trawling for albacore. *Infofish Int.*, **4**, 50.

Putro S. & Saleh M. (1985) Effect of delayed icing. (a) Skipjack tuna. *FAO Fish. Tech. Pap.* **252**, 30.

Ravesi E.M., Licciardello J.J., Tuhkunen B.E. & Lundstrom R.C. (1985) The effect of handling or processing treatments on storage characteristics of fresh spiny dogfish, *Squalus acanthias. Mar. Fish. Rev.*, **47** (1) 48.

Reppond K.D. & Collins J. (1983) Pacific cod (*Gadus macrocephalus*): Change in sensory and chemical properties when held in ice and in CO_2 modified refrigerated seawater. *J. Food Sci.*, **40**, 1552.

Samuels R.D., DeFeo A., Flick G.J., Ward D.R., Rippen T., Riggins J.K., Coale C. & Smith C. (1984) *Demonstration of a quality maintenance program for fresh fish products.* Report to Mid-Atlantic Fisheries Development Foundation. Virginia Polytechnic Inst. and State University, Sea Grant, VPI-SG-84-04R, Blacksburg.

Shaw B.G. & Shewan J.M. (1968) Psychrophilic spoilage bacteria of fish. *J. Appl. Bact.*, **31**, 89–96.

Shewan J.M. (1949) Some bacteriological aspects of handling, processing and distribution of fish. *J. Royal Sanit. Inst.*, **59**, 394.

Shewan J.M. (1976) The bacteriology of fresh and spoiling fish and the biochemical changes induced by bacterial action. In *Handling, Processing and Marketing of Tropical Fish* (Ed., by P. Sutcliffe & J. Disney), pp. 51–66. Tropical Products Institute, London.

Smith J.G.M., Hardy R., McDonald I. & Templeton J. (1980) The storage of herring (*Clupea harengus*) in ice, refrigerated sea water and at ambient temperature. Chemical and sensory assessment. *J. Sci. Food Agric.*, **31**, 375–85.

SPSS Inc. (1986). *SPSSX user guide.* Chicago.

Stansby M.E. & Lemon J.M. (1941) *Studies on the handling of fresh mackerel* (*Scomber scombrus*). Res. Rept. 1, US Fish

and Widlife Serv., Washington.
Vyncke W. (1980) Quality assessment of gutted and ungutted gurnard (Trigla spp.) by organoleptic and objective methods. *Z. Lebensm. Unters. Forsch.*, **171**, 352–4.
Wilson N.D.C. (1982). *Some aspects of the handling and processing of tuna in New Zealand's South Island west coast fishery, a preliminary investigation*. Report No. 5, Palmerston North: Massey University Food Technology Research Center.

CHAPTER 15

Quantitative and qualitative determination of biogenic amines in fish

J.P. Gouygou, C. Sinquin, M. Etienne, A. Landrein and P. Durand

INTRODUCTION

Biogenic amines formed by decarboxylation of amino acids can be considered good indices for determining the freshness of seafood. Their presence in relatively small amounts indicates a spoilage stage, and their good thermal stability allows an evaluation of the raw material that had been used for canning.

Special attention is being paid to histamine because its presence beyond a certain amount has often been associated with cases of food (scombroid) poisoning. Histamine, especially that which is contained in seafood and sea products, can certainly be regarded as a sanitary problem (Massenot & Lariviere 1988) and an economic problem. As we have no adequate control of the parameters that may cause excess histamine formation, it is difficult to predict, from a given histamine concentration in a batch of raw material, what the final concentrations in the industrially processed end-product will be. For producers. notably those who export, the risk that their batches may be rejected by buyers and that they may lose market outlets is not negligible.

The risk of histamine formation is increased by the number of stages through which the raw materials as well as the finished product must pass, as required by international trade. For example, tuna is caught in the Indian Ocean, frozen, stored and conveyed in the refrigerated compartments of cargo ships, unloaded in France, stored, defrosted, canned, then marketed or exported; anchovies are fished off the coasts of Morocco, salted, canned, then exported to France where they are repackaged and marketed. As a consequence, the relatively high incidence of scombroid poisoning has encouraged public health services to focus on quality control and to set standards for the acceptance or rejection of commercial marine products.

For these different reasons and within the brief for the tasks assigned to the Institut Français de Recherché pour l'Exploitation de la Mer, in the field of upgrading, control and monitoring of marine foodstuffs we have undertaken specific studies on biogenic amines. Two main objectives were defined: the perfection of quantitative methods and the development of qualitative methods suitable for routine controls. This work was begun about four years ago and in this paper we summarise the different steps of the experimental work which has been conducted since then.

EXPERIMENTAL APPROACH

Quantitative methods

Current regulations are concerned exclusively with histamine and not with biogenic amines. Many countries have already established a maximum permitted concentration (MPC) of histamine in marine products. However few countries have defined official standards or introduced strict regulations. Most of them merely propose guidelines. This overall lack of regulation and accurate standards is an illustration of the difficulty in defining adequate maximum permitted concentrations.

In France an MPC of 10 mg of histamine/100 g (as applied to fresh tuna) was set on 4 July 1984 by the veterinary departments concerned with food hygiene. This standard is becoming the reference value for the acceptance of processed marine products such as canned or packaged fish.

In our 'control department' the method used for histamine measurement has been the fluorimetric method derived from that of Lerke & Bell (1976). This method has proved to be accurate and sensitive: its detection limit is approximately 1 mg of histamine/ 100 g of fish sample. However, it is rather slow (maximum 4–5 samples/hour) and requires strict precautions and care. Moreover if the pH requirements are

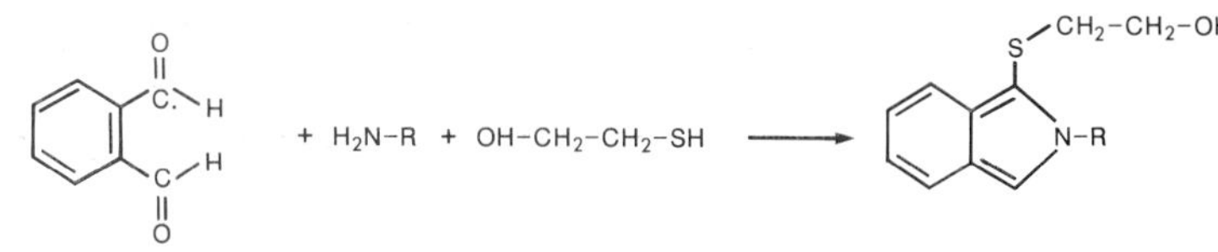

Fig. 15.1 The fluorimetric method for determining biogenic amines in fish.

not strictly met, histidine is liable to interfere with histamine measurements. In order to avoid these drawbacks and in order to be able to determine histamine and other biogenic amines simultaneously, we have started to work on the HPLC method. HPLC techniques are now widely employed in many laboratories. They are efficient, rapid and can be automated.

Isocratic HPLC histamine determination

The first part of our study was concerned with histamine measurement (Gouygou *et al.* 1987) using an isocratic HPLC system, in trichloroacetic acid extracts after derivatisation with o-phthalaldehyde (OPA). This HPLC method proved to be a sensitive, simple and selective method for the analysis of histamine, where no specific extraction procedure is necessary. However under these chromatographic conditions biogenic amines such as cadaverine, putrescine and spermidine were not detected. Putrescine and cadaverine can notably increase the toxicity of histamine, therefore there might be some interest in simultaneously determining these biogenic amines.

Histamine and biogenic amines determination in a HPLC gradient system

The fluorimetric method for determining biogenic amines in fish is based on the derivatisation of amines with OPA in the presence of mercaptoethanol, which forms strongly fluorescent adducts that are much stronger than the histamine fluorophore (Fig. 15.1).

The OPA complexes obtained from fish TCA extracts were separated by reversed-phase HPLC under gradient elution conditions.

The amine-OPA complex was prepared by mixing a 0.5 ml standard solution or sample with 1 ml OPA reagent (90 mg OPA dissolved in 10 ml borate buffer, 1 ml methanol and 0.2 ml of 2-mercaptoethanol) and leaving it for four minutes in a tube protected from light. Immediately afterwards 4 ml of ethylacetate were added. The contents were mixed on a vortex mixer for exactly one minute and the two phases were allowed to separate. The supernatant was then injected into the HPLC system.

The sample was prepared by blending 50 g of fish with 100 ml 10% TCA solution, and a clear extract obtained by centrifuging and filtering through a 0.45 μm filter.

Reversed-phase HPLC under gradient elution conditions was found to give a satisfactory separation of biogenic amines in 25 minutes. The gradient was obtained from solvent A and solvent B at a constant flow rate of 1.5 ml/min.

Solvent A was 10% acetonitrile in water with monosodium phosphate (75 mmoles/l) and B was 60% acetonitrile in water with monosodium phosphate (75 mmoles/l). Fluorescence was monitored at 350 nm excitation and 450 nm emission wavelengths. OPA-derivatised standard amines produced five peaks at 7.6 minutes retention time for histamine, 19.2 minutes for tyramine, 20.2 minutes for spermidine, 23.0 minutes for putrescine and 24.1 minutes for cadaverine (Fig. 15.2). The peak areas were proportional to the amine concentration and a linear relationship was observed (Fig. 15.3).

A chromatogram illustrating the HPLC separation of tuna extract is shown in Figure 15.4.

Our experiment showed that the HPLC method allows the simultaneous analysis of biogenic amines and their quantitative determination in 25 minutes. This method does not require a tedious extraction procedure provided the operating method is correctly followed.

Qualitative methods

This type of analysis meets the requirements of routine analysis for control purposes in a laboratory situation with limited equipment and semi-quantitative results can be obtained fairly rapidly.

In this field enzymatic methods offer special advantages which are linked to enzyme characteristics, chiefly with regard to sensitivity, due to high catalytic activity, and specificity. References to only a few papers were obtained from a bibliographic survey on enzymatic determination of biogenic amines. The measurement of histamine in biological fluids was described more than ten years ago (Beaven 1975). However, this method, based on methylation of histamine by a histamine-N methyl tranferase (HMT), used a radioactive labelled substrate which greatly limits its utilisation.

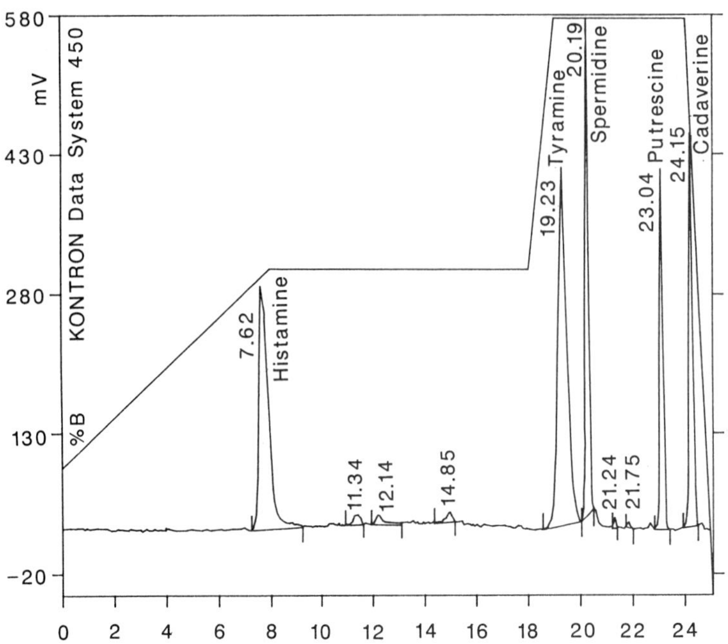

Fig. 15.2 HPLC separation for amines standard (histamine, tyramine, spermidine 0.1 mg/ml; putrescine, cadaverine 0.025 mg/ml).

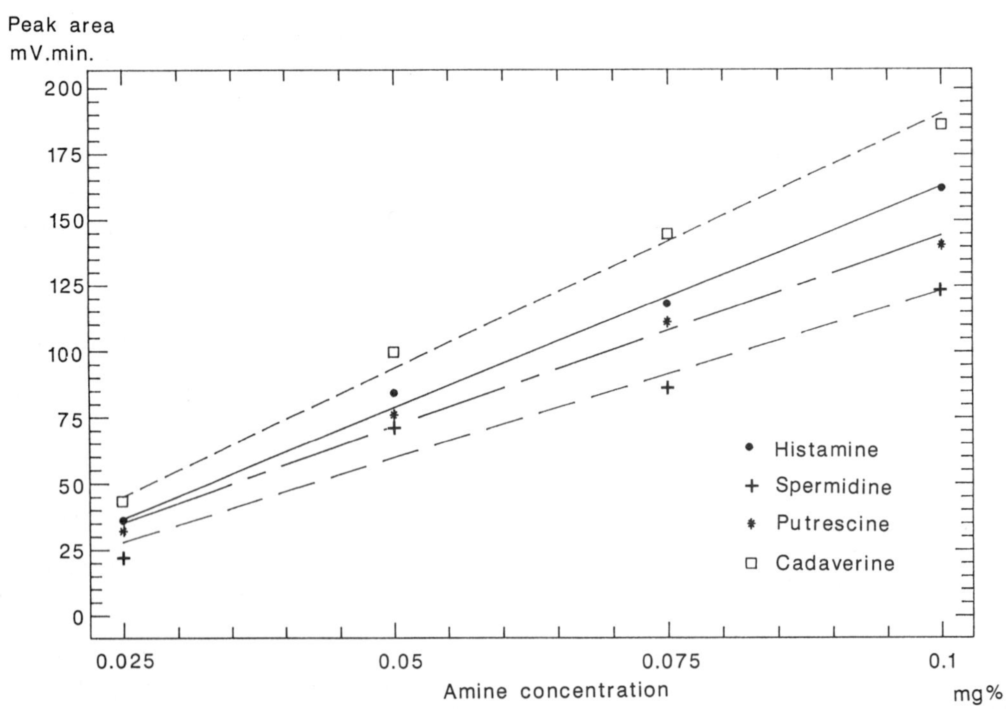

Fig. 15.3 Linearity of peak area for increasing amine concentration.

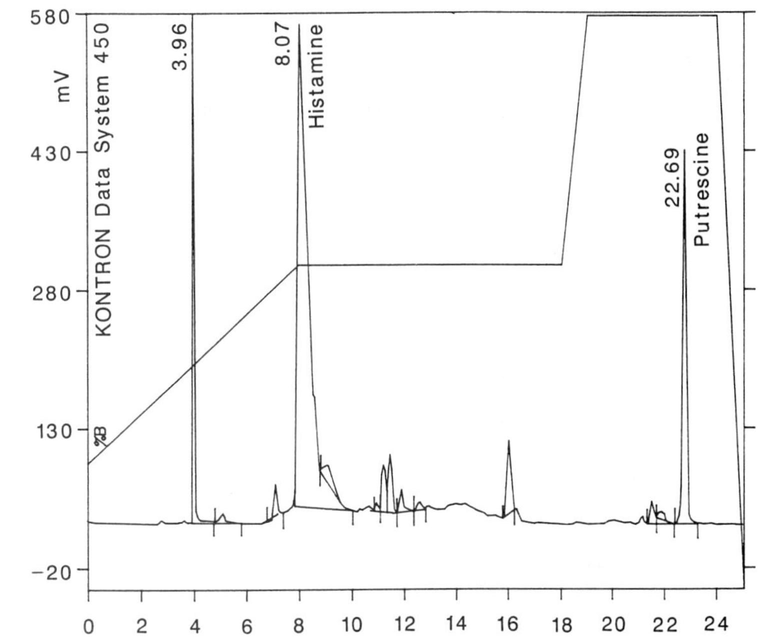

Fig. 15.4 HPLC separation of tuna extract.

More recently, Lerke *et al.* (1983) have described a rapid test for histamine measurement in fish. This qualitative evaluation was made using a bienzymatic system involving, first, diamine oxidase (DAO) and then peroxidase (HRP – horse radish peroxidase). This second enzyme reveals hydrogen peroxide formed during the first enzymatic reaction by oxidising the leuco form of a synthetic substrate (LCV – Leuco Crystal Violet) to form a spectrophotometrically measurable compound as shown in the following equation:

$$(1)\ R{-}CH_2{-}NH_2 + O_2 + H_2O \xrightarrow[\text{oxidase}]{\text{amine}} R{-}CHO + NH_3 + H_2O_2$$

$$(2)\ H_2O_2 + LVC \xrightarrow[\text{HRP}]{\text{peroxidase}} H_2O + \text{chromophore (crystal violet)}\ \lambda = 596\,\text{nm}$$

This method was also of particular interest to us because of the use of an oxidase. This class of enzymes has been widely used in the preparation of enzyme electrodes also called biosensors.

An attempt was made to use such a device for the measurement of biogenic amines.

The enzyme electrode

Numerous works deal with biosensors and many papers have been published during the last ten years on this topic. Enzyme electrodes are formed by the joining of an electrochemical sensor and an enzymatic layer. In the case of oxidases, enzyme electrodes are obtained by coupling an enzymatic film of immobilised oxidase with an oxygen electrode, currently named the Clark electrode.

Oxidase catalysed general reaction:

$$XH_2 + O_2 \rightarrow X + H_2O_2$$

where XH_2 is the reduced form of the substrate and X its oxidised form.

The decrease in oxygen concentration, due to enzymatic reaction occurring in the enzymatic film, is monitored amperometrically by the Clark electrode (Romette *et al.* 1982) in a fully computerised system (Kernevez 1983). On this basis a device has been developed in France at the University of Technology of Compiègne in conjunction with a private company (Seres).

We have tested this system for its ability to measure biogenic amines. The first part of our experimental work was to produce amine oxidase membranes and the first step consisted of the extraction of specific amine oxidases. These enzymes are widely found in the animal and vegetable kingdoms and in bacteria. They are classified in terms of the substrate they oxidise: mono amine oxidase (MAO), diamine oxidase (DAO), polyamine oxidase (PAO) or in terms of the

Table 15.1 Enzyme characteristics of amine oxidase in plants, bacteria and mammals (Gerard *et al.* 1985).

Enzyme source	Name	Specific activity	Substrate specificity
Lens esculenta	DAO	70	cadaverine, putrescine, spermidine
Lathyrus sativus	DAO	2	cadaverine, putrescine, spermidine, spermine, tyramine, tryptamine
Pisum sativum	DAO	70	cadaverine, putrescine, lysine, ornithine, tyramine, histamine
Ulcia faba	AO	8	cadaverine, putrescine, spermidine, histamine, spermine, agmatine
Arachis hypogea	AO	0.6	cadaverine, putrescine, spermidine, spermine
Soybean	AO	40	cadaverine, putrescine, spermidine, spermine, histamine, tyramine
Aspergillus niger	AO	9.5	cadaverine, putrescine, spermidine, histamine, tyramine, spermine
Micrococcus rubens	Put O	9.2	putrescine, spermidine
Candida boldini	Benz O	4.5	cadaverine, putrescine, histamine, tyramine, benzylamine
Sarcina lutea	Tyr O	40	tyramine, dopamine
Serratia marcescens	Spdm O	10	spermidine
Penicillium chrysogenum	Agm O	7.4	agmatine
Human placenta	DAO	7.0	cadaverine, putrescine, spermidine, histamine
Human placenta	MAO	1.6	cadaverine, putrescine, spermidine, histamine, tyramine, spermine
Dog intestine	DAO	0.1	cadaverine, putrescine, histamine, benzylamine
Pig kidney	DAO	1.6	cadaverine, putrescine, spermidine, histamine
Rabbit serum	MAO	0.2	histamine, tyramine, tryptamine, mescaline
Pig plasma	AO	200	histamine, tyramine, tryptamine, benzylamine

best oxidised substrate: putrescine oxidase, spermidine oxidase etc. No 'histamine oxidase' has been discovered so far (Table 15.1) although some of the enzymes do oxidise histamine.

At present the only DAO commercially available is that of the pig kidney (P DAO) but it has a weak specific activity. We were in consequence obliged to purify a DAO from lentil seeds (L DAO) which has a strong specificity for the biogenic amines found in spoiled fish; and a DAO from human placenta (H DAO) having a broad specificity and reacting on histamine (Gerard *et al.* 1985). The kinetic parameters of these enzymes are given in Table 15.2.

Contrary to the data given by Floris *et al.* (1983), the extracted L DAO was shown to be active on histamine. Because of this activity and its relatively easy preparation we have worked with L DAO and tested the possibility of utilisation in the analysis of the freshness of marine products (Durand *et al.* 1987).

Using the automated electrode system, the biogenic amine pool was evaluated in fish stored in ice (Fig. 15.5). The experimental results show the increase, during storage, of biogenic amines measured by the electrode signal and the very good correlation existing between the enzyme electrode method and total volatile base (TVB-N) and trimethylamine (TMA) measurements.

These results clearly show the capacity of the automated enzyme system to evaluate fish quality. However the routine utilisation of this device was found to be very delicate. Moreover it can be used only in a laboratory situation with skilled people and the apparatus is rather costly.

The know-how acquired in the production of immobilised L DAO films which are used in the automated electrode system led to the idea of making test strips.

Test strips

The test strips were produced from an immobilised enzyme film containing both L DAO and HRP. They were made in the same way as the L DAO film produced for the enzyme electrode, as previously described (Durand *et al.* 1987). They were obtained by co-entrapment of the enzyme molecules and an inactive matrix protein (gelatin or bovine serum albumin) with a bifunctional reagent (glutaraldehyde): these films of about 40 cm^2 can be cut into strips of about 2 cm^2.

By taking the experimental conditions given by Lerke *et al.* (1983), it was possible to detect the presence of histamine or biogenic amine as shown in Figure 15.6.

The main interest of strip tests, beside their ease

Table 15.2 Kinetic parameters of L DAO and H DAO.

	L DAO	H DAO
Putrescine	K_m = 0.23 mM V_m = 280 mM mn^{-1}	K_m = 2 mM V_m = 115 mM mn^{-1}
Cadaverine	K_m = 0.12 mM V_m = 260 mM mn^{-1}	K_m = 2.5 mM V_m = 85 mM mn^{-1}
Spermidine	K_m = 0.17 mM V_m = 90 mM mn^{-1}	K_m = 2.7 mM V_m = 60 mM mn^{-1}
Histamine	K_m = 0.71 mM V_m = 10 mM mn^{-1}	K_m = 1.2 mM V_m = 95 mM mn^{-1}

of use, consists mainly in the possibility of their reutilisation. Unfortunately in the experimental conditions used, a rapid absorption of the dye (crystal violet) was observed on the film which made it impossible to re-utilise without running the risk of obtaining false readings.

The initial Lerke idea was then taken up using the free enzyme which, in our case was L DAO, and an attempt was made to implement it using a fast plate analyser (Galgani & Bocquene, in press).

Fast plate reading

The recent development of fast plate readers, used frequently in immunology for the ELISA test, gives the possibility of performing simultaneously a large number of kinetic measurements. It was possible to screen for the presence of biogenic amines in a large number of fish extracts and to reveal the presence of histamine in these samples. In this case the preparation of the extracts was found to be the main limiting factor.

The method involves a two step sequential enzyme system (Lerke *et al.* 1983). First diamine oxidase catalyses the breakdown of biogenic amines leading to the formation of hydrogen peroxide. This can then be detected by the formation of crystal violet from the leuco base in the presence of peroxidase. This compound is detected at λ = 540 nm.

Standard amines and HRP (20 U/ml) were prepared in a 0.15 M phosphate buffer (pH = 7). Leuco crystal violet (0.5 mg/ml) was dissolved in HCl (0.5%) and stored under refrigeration and protected from light. The sample was obtained by blending 20 g of fish with 100 ml of phosphate buffer for one minute. The sample was then centrifuged and filtered.

150 µl of sample, 50 µl of HRP and 75 µl of L DAO were introduced into the holes of the microplates. After 10 minutes at 23°C (pH = 7) LCV was added. Readings of optical density were made after

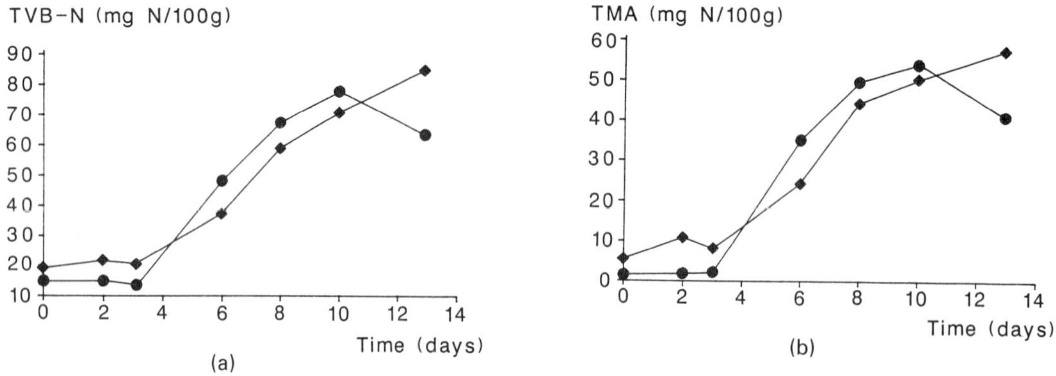

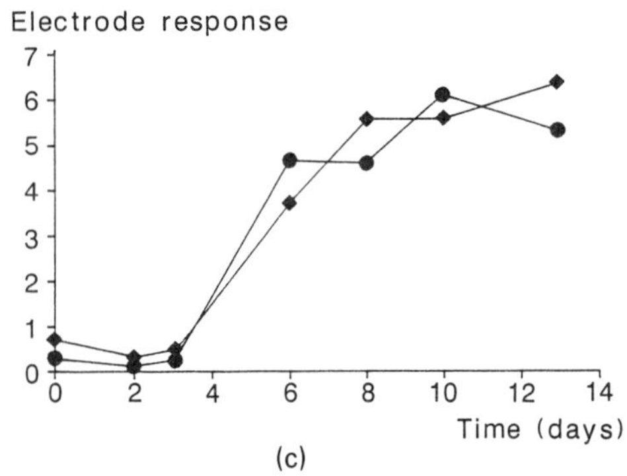

Fig. 15.5 Evaluation of fish freshness of whiting and European hake. (a) TVB-N in mg N/100 g; (b) TMA in mg N/100 g; (c) enzymic method, extract in phosphate buffer.

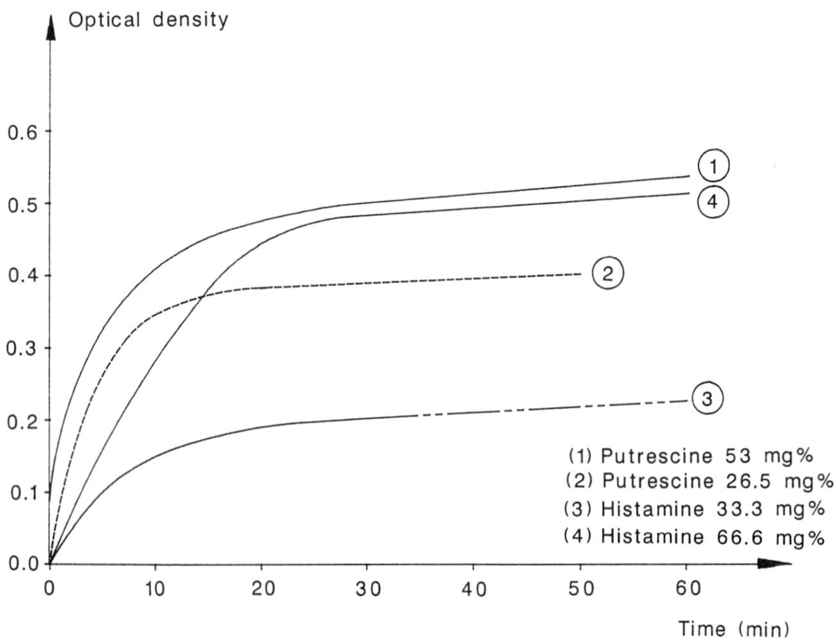

Fig. 15.6 Kinetics of colorimetric detection of amines in presence of L DAO/0.4 U + HRP/2 U.

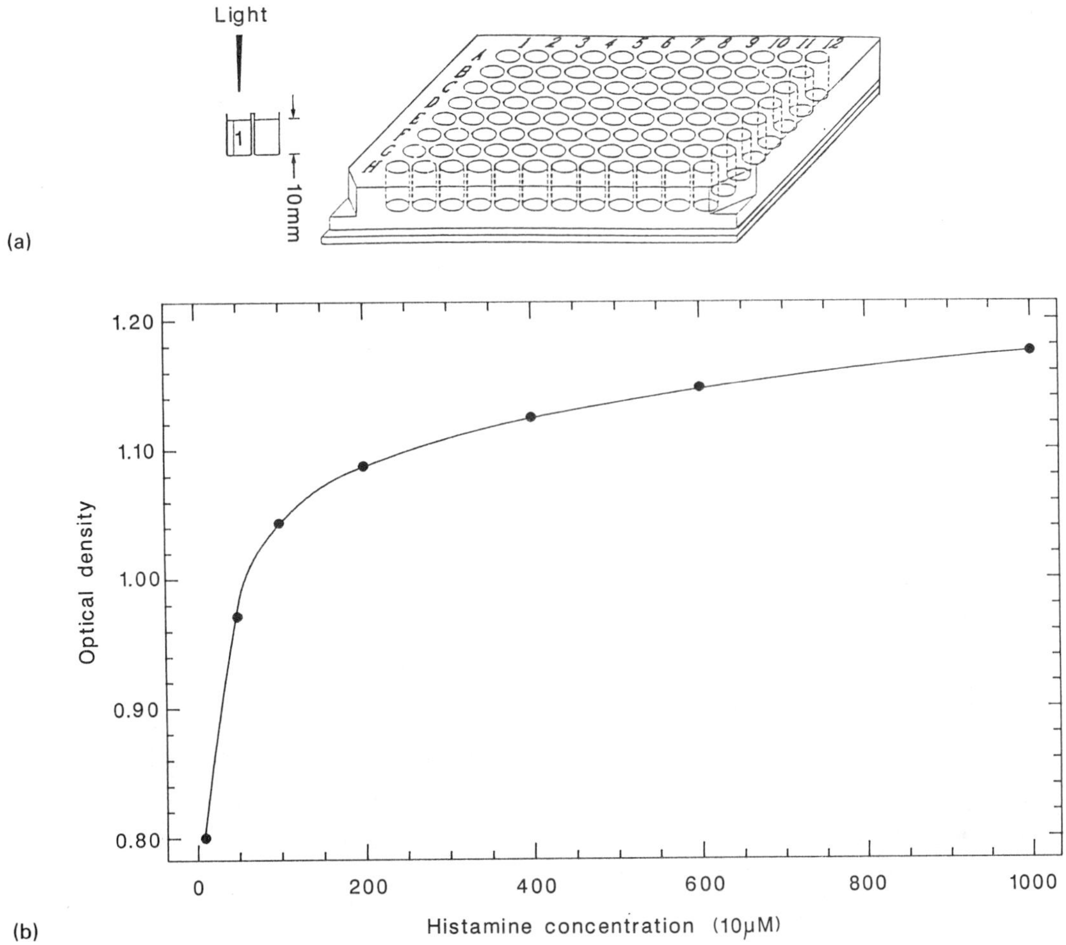

Fig. 15.7 (a) Fast plate reader; (b) optical density for increasing concentrations of histamine.

two minutes using the microplate reader at $\lambda = 540$ nm. (Fig. 15.7).

This test gave a strong positive response at low levels of putrescine and cadaverine (0.D: 2.5 and 1.5 for 2 mg/100 g fish). The detection limit was found to be 2 mg/100 g fish for histamine.

Samples of tuna (*Thunnus alalunga*) from the 'control department' were analysed using the DAO enzyme method (fast plate reader) and the HPLC method. The results are given in Table 15.3.

The DAO enzyme method allows a semi-quantitative estimation of the presence of biogenic amines. A lack of coloration indicates an absence of biogenic amines and so is an indicator of fish freshness. A low degree of coloration seems to indicate the presence of histamine. In this event, the level has to be evaluated by the HPLC method. A high degree of coloration indicates either a large amount of histamine or the presence of cadaverine and/or putrescine.

CONCLUSION

The enzymatic method using L DAO offers new perspectives in the field of evaluation of the freshness of fish via the semi-quantitative estimation of the biogenic amine pool. However a more exact determination would require the supplementary step of separation by ion-exchange chromatography. This technique adds to the range of chromatographic methods, in particular HPLC methods and could lead, in the near future, to the fabrication of kits which could be used outside the

Table 15.3 Analytical results of biogenic amines in tuna by HPLC and enzymatic methods.

		HPLC method	
Samples of tuna	Fast plate reader O.D.	Histamine mg/100 g fish	Cadaverine mg/100 g fish
A	0.98	58	4.5
B	1.04	66	5.4
C	0.50	30	2.4
D	1.30	144	4.8
E	0.19	37	0.9

laboratory. The discovery of new enzymes, especially one with a specific activity for histamine, could hasten the development of such techniques. The development of the use of such enzymes in fast plate readers could turn out to be of particular interest.

At present HPLC offers the best means of performing a simultaneous analysis of biogenic amines. After 25 minutes and without employing a tedious extraction procedure, HPLC allows the quantitative determination of amines to be carried out. The method is sensitive, simple and selective and the analysis can be performed rapidly in an automated system.

BIBLIOGRAPHY

Beaven M.A. (1975) Radiochemical assay procedures for drugs and transmitters. *Handbook of Psychopharmacology* (Ed. by L.L. Ivesen & S.H. Snyder), **I**, pp. 253–90. Plenum Press, New York.

Durand P., Gouygou J.P. & Sinquin C. (1987) *New methods for fish quality determination.* Unpublished paper presented at a meeting of the West European Fish Technologists' Association.

Floris G., Giartosio A. & Rinaldi A. (1983) Diamine oxidase from *Lens esculenta* seedlings: purification and properties. *Phytochemistry* 22 (9), 1871–4.

Galgani F. & Bocquene G. (in press) Semi-automated colorimetric and enzymatic assays for aquatic organisms using microplate readers. *Wat. Res.*

Gerard H., Romette J.L. & Durand P. (1985) *Prospect of using immobilized enzymes in biochemical evaluation of seafood.* Unpublished paper presented at a meeting of the West European Fish Technologists' Association.

Gouygou J.P., Sinquin C. & Durand P. (1987) High pressure liquid chromatography determination of histamine in fish. *J. Fd. Sci.*, **52** (4) 925–7.

Kernevez J., Konate L. & Romette J.L. (1983) Determination of substrate concentrations by a computerized enzyme electrode. *Biotechnol. Bioeng.*, **XXV**, 845–55.

Lerke P.A. & Bell L.D. (1976) A rapid fluorimetric method for the determination of histamine in canned tuna. *J. Fd. Sci.*, **41**, 1282–4.

Lerke P.A., Porcuna M.N. & Chin H.B. (1983) Screening test for histamine in fish. *J. Fd Sci.*, **48**, 155–7.

Massenot C. & Lariviere L. (1988) Rapport de synthèse sur les toxi-infections alimentaires collectives déclarées aux directeurs des services vétérinaires en 1987. *Bulletin d'Information et de Documentation de la Direction Générale de la Concurrence, de la Consommation et de la Repression des Fraudes*, **11**.

Romette J.L., Tran N.D., Durand P., Boitieux J.L. & Navarro, J.L. (1982) In *Enzyme Engineering* (Ed. by I. Chibata), **6**. Plenum Publishing Corporation, New York.

CHAPTER 16

Colour changes in the skin in relation to the formation of histamine in dorado (Mahi-mahi, *Coryphaena hippurus*)

C.D. Wood, N. Camba and M. Grijalva

INTRODUCTION

Dorado or dolphin fish (Mahi-mahi, *Coryphaena hippurus*) is a fish popular for consumption in Ecuador and for export, particularly to the USA. It is a medium-sized pelagic species with a predominantly white flesh. The majority of the catch comes from the small-scale sector which uses small wooden or fibre-glass boats. Traditionally the fish was not chilled on board the boats and the early batches sent for export experienced a series of problems due to poor quality, especially high levels of histamine.

Histamine formation in fish has been reviewed extensively by Arnold & Brown (1978). In brief, histamine is formed by the action of certain types of bacteria as the fish spoils. Fish which contain high levels of the amino acid histidine in its free form can accumulate very high levels of histamine. This can occur with many pelagic species such as tuna, mackerel and sardine. The precise mechanisms of toxicity are not well understood but fish containing high levels of histamine have been implicated in incidents of food poisoning (so called scombroid poisoning). Although other factors may be involved (Clifford *et al.* 1989) the presence of high levels of histamine is taken as indicative of fish which can cause this type of poisoning. The symptoms of poisoning vary from case to case but include diarrhoea, nausea, severe headaches and cardiac palpitations.

Bostock *et al.* (1985) have shown that the formation of high histamine levels in dorado can take place in 9 hours at ambient Ecuadorian temperatures (25–30°C), but at temperatures of 20°C or less histamine is not accumulated until after the fish has become obviously spoiled. It appears that the bacteria responsible for histamine formation are not very active at chill temperatures (Eitenmiller *et al.* 1982). Therefore the use of ice on the fishing boats to obtain a good quality product is essential for an artisanal fishery under tropical conditions, unless the fishing trips are of very short duration.

In recent years the Ecuadorian fishing industry has considerably improved the handling of dorado for export. The use of ice in the fishing boats is now normal in the handling of fish for export. Quality problems do, however, still exist. The fish can die on the fishing lines and start to spoil hours before being iced. The fishermen do not always use the ice well; it may be used in large chunks and inadequately mixed with the catch in the bottom of the boat. Even when wooden containers are used to ice the catch it is rare that any thermal insulation is used, so all the ice may have melted before the fish is landed. Poor quality dorado can reach the fish traders and in turn the processing plants which export the product.

The chemical analysis for histamine is a complicated process which needs a laboratory and certain equipment and chemical reagents. A rapid chemical/enzymic method has been developed suitable for use without laboratory facilities (Barratt & Camba 1985) but the industry needs a simpler, non-chemical, non-destructive method of inspection to evaluate its products. Many sectors of the industry use the skin colour change from yellow to white of the fish's ventral surface as the most important indicator of quality. However there has apparently been no information in scientific literature about this change in relation to histamine formation. Therefore the present study was conducted to investigate the colour change, analysing samples for histamine and classifying the quality of the fish using conventional sensory methods during the spoilage of dorado at actual or simulated Ecuadorian ambient temperatures.

MATERIALS AND METHODS

Sources of dorado

For the spoilage trial at actual ambient temperatures fresh fish was obtained during a fishing voyage of the training ship *Sirius* of the Fisheries School (Escuela de Pesca), Manta, Ecuador. Long lines were used for fishing and normally, but not always, the fish was brought aboard alive.

For the two trials conducted in the laboratory at simulated ambient temperatures, fish was obtained from the exporter *Cepromar*, Guayaquil. The fish had been kept on ice for about 24 hours from capture to the start of the trial, but without knowing the exact handling conditions.

Fish handling

Under ambient conditions on the 'Sirius' (Experiment 1)

Starting early in the night, three recently captured fish were left at ambient temperatures (23–26°C), fish number 1 being dead and numbers 2 and 3 alive when landed on the boat. Every 1–3 hours they were inspected and their state of freshness was classified using the system of Connell (1990) and the colour of the skin was noted. The colour was classified visually using a scale from 0–100% in which 0% was for a fish with a completely white ventral surface and 100% for a fish with the maximum possible yellow coloration. For this classification the extension and intensity of yellow coloration of the skin was considered.

After the first 9 hours, in addition to the inspection, samples were taken (at intervals of 2–4 hours) of the muscle from one side of the fish, starting near the head. The samples were taken as parallel portions, leaving at least 10 mm between each. The samples obtained were frozen immediately after being taken. These were analysed for their histamine content. The fish were left until their organoleptic characteristics indicated that they were unfit for human consumption (bad odours, poor appearance and texture).

In addition to this experiment a comparative study of the colour changes of the skin of dorado at ambient temperatures and on ice was conducted.

Under simulated ambient conditions in the laboratory (Experiments 2 and 3)

Experiment 2 was conducted using an electric kiln (Afos mini kiln, England) to maintain the fish at temperatures between 30°C and 35°C. In experiment 3 the fish was submerged in water between 30°C and 35°C to control its temperature. Six fish were used in each experiment, which were each conducted using the same methods of sensory inspection, classification of colour and quality, and histamine analysis that were used in experiment 1. The fish were inspected every 2–4 hours and the taking of samples for histamine analysis was started after 8 hours at intervals of 2–4 hours.

Histamine analyses

The fish samples were skinned and blended in a Hobart mixer/blender. Analyses were performed by the flourometric method of Taylor *et al.* (1978).

RESULTS

Figures 16.1–16.6 show the changes of skin colour, quality and histamine content of the fish in relation to storage time under three different sets of conditions. Figures 16.1, 16.3 and 16.5 show colour and quality,

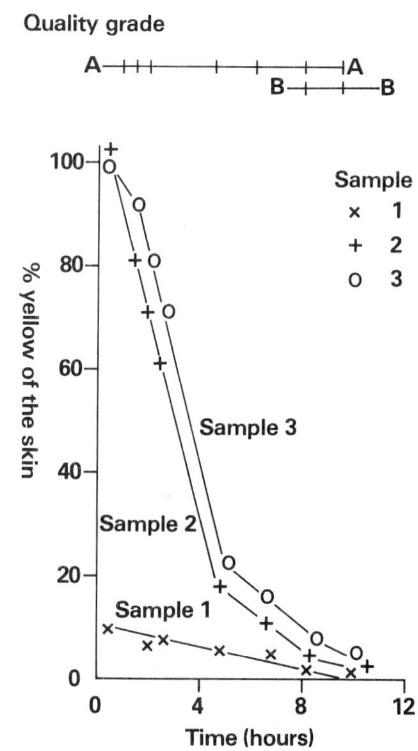

Fig. 16.1 Loss of skin colour of dorado at 23–26°C (experiment 1).

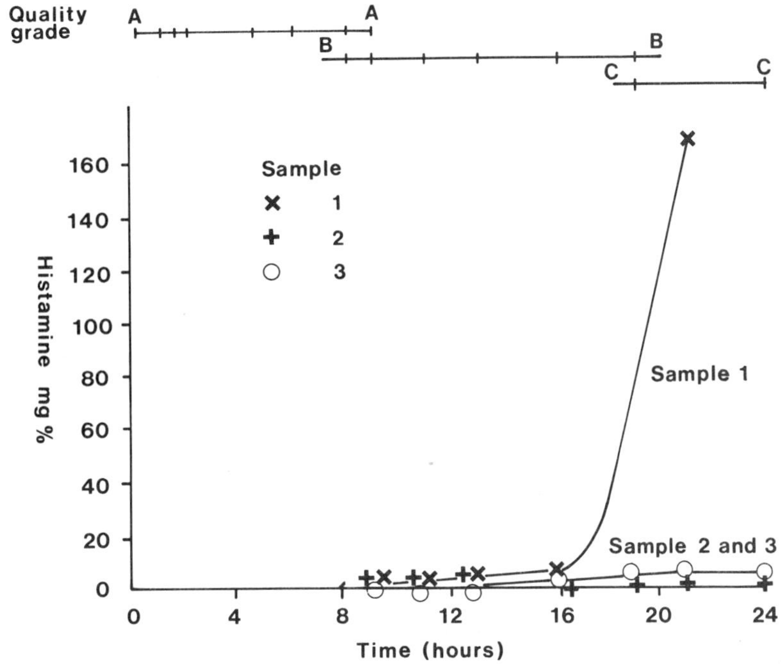

Fig. 16.2 Accumulation of histamine in dorado at 23–26°C (experiment 1).

Figures 16.2, 16.4 and 16.6 histamine content and quality. Due to the presence of coincident data, for the preparation of the figures lines were not drawn for each sample. Lines were selected which show the tendencies of the data and which indicate the differences between the fish. The horizontal lines above the graphs represent the quality grade of the fish, the short vertical lines indicating the time of inspection. Overlapping lines mean that there was fish of both grades.

In experiment 1 (Figs. 16.1 and 16.2) under ambient conditions (temperatures of 23°C to 26°C), the loss of the yellow colour was more rapid than the accumulation of histamine. Only the first fish, which was landed dead on the fishing line, had accumulated significant levels of histamine during 24 hours of storage. However, it had lost much of its colour before being landed. The second and third fish were in rigor mortis between 4.5 and 6 hours after dying. The first fish had passed this stage before it was landed. A fourth fish captured the following day was in rigor mortis between 1.5 and 3 hours after being killed, losing the yellow colour of its skin in 7 to 8 hours as had the second and third fish (Fig. 16.1).

In comparison, dorado stored in ice pass through rigor mortis after 40 hours and lose very little coloration during this period. During experiment 1 dorado with yellow skin was of grade A, without significant histamine levels. The change from yellow to white coincided more or less with the change from grade A to B, which took place after 8 hours at ambient temperatures. There was no formation of significant levels of histamine until the fish was of grade C (not acceptable due to its sensory characteristics). The colour change of the skin was not useful in distinguishing between fish of grade B and grade C.

The two experiments conducted in the laboratory indicated that under the conditions used, the change in the colour of the skin can be much slower. In experiment 2, in the kiln, the fish skin dried out and as Figure 16.3 indicates, under these conditions there was little loss of the yellow coloration. Deterioration and histamine formation, however, continued in the normal way (Fig. 16.4) and it was possible to obtain fish with strongly coloured yellow skin which was of grade C quality and with a very high histamine content (over 100 mg%). Figure 16.5 indicates that in the fish in water at temperatures between 30°C and 35°C (experiment 3) the loss of yellow colour from the skin was slower than in the first experiment (under ambient conditions). After 12 hours there were fish (especially number 5) which retained something of the yellow

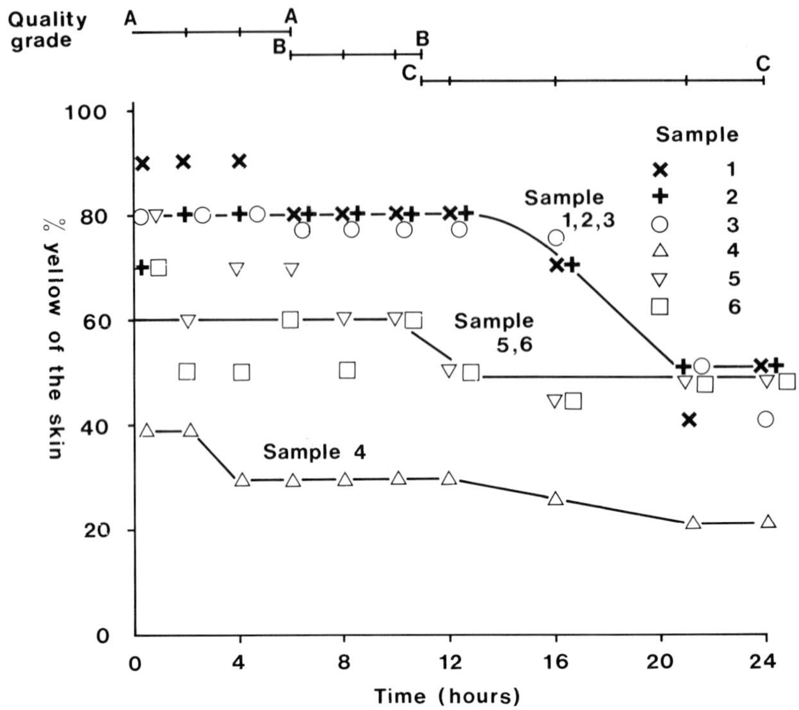

Fig. 16.3 Loss of skin colour of dorado at 30–35°C under dry conditions (experiment 2).

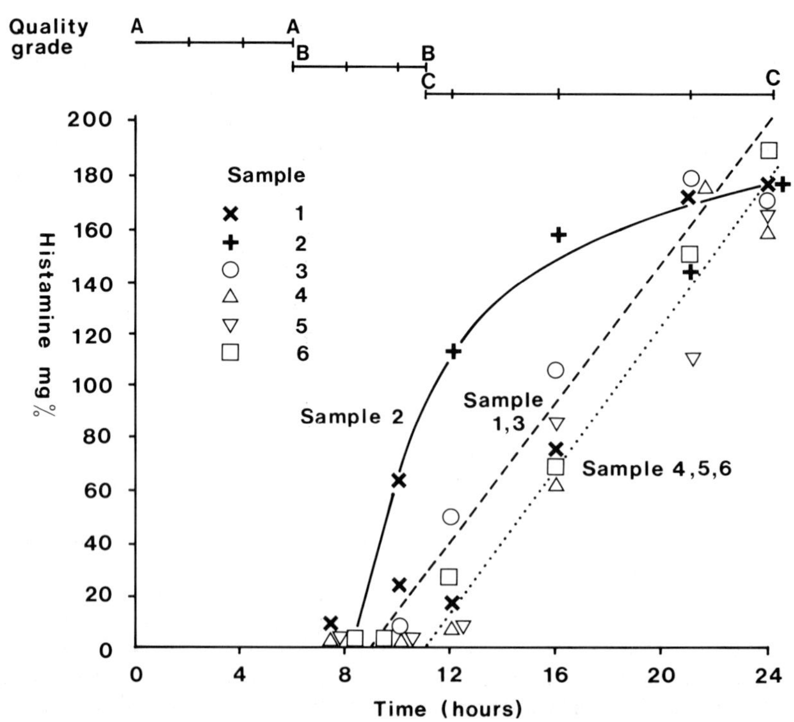

Fig. 16.4 Accumulation of histamine in dorado at 30–35°C under dry conditions (experiment 2).

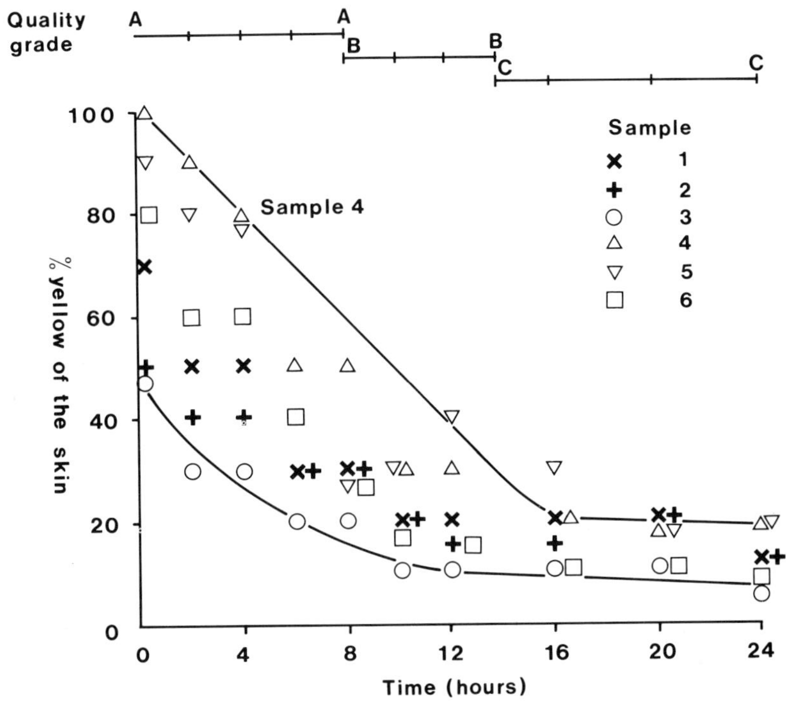

Fig. 16.5 Loss of skin colour at 30–35°C under wet conditions (experiment 3).

coloration of the skin but had very high levels of histamine.

In experiments 2 and 3 it is worth noting that none of the samples taken from fish classified as grade A contained significant levels of histamine. However certain fish classified as grade B (normally considered as fit for human consumption) contained high levels of histamine and in various cases in experiment 3 they contained more than 100 mg%, a level considered to be associated with scombrotoxicity.

DISCUSSION AND CONCLUSIONS

Relationship between the yellow coloration and quality of dorado

Recently captured dorado has a yellow ventral surface. There are, however, significant differences in the intensity and extension of coloration between fish. During the deterioration of the fish there is a loss of colour. The three experiments have shown that the speed of colour loss depends not simply on the rate of deterioration of the fish but on the handling conditions used after capture. It is notable that if the skin can dry out during handling the loss of yellow coloration is much slower than under moist conditions, while general deterioration is not significantly affected.

It can be concluded, therefore, that there is no fixed relationship between the quality or freshness of the fish and the intensity or extent of the yellow coloration of the skin. The first experiment (on board the fishing boat) indicated, however, that under those 'normal' conditions fish with an intense yellow colour was of good quality and 'white' fish was of inferior quality. Therefore this change in colour is useful as a quality indicator but reliable only in combination with other quality indicators such as smell, appearance of the eyes etc.

Yellow coloration and histamine content

As with general quality, there was no fixed relationship between skin colour and histamine content. The data indicates that the fish classified as grade A due to their sensory characteristics did not contain histamine and usually had a yellow coloured skin. The reverse, how-

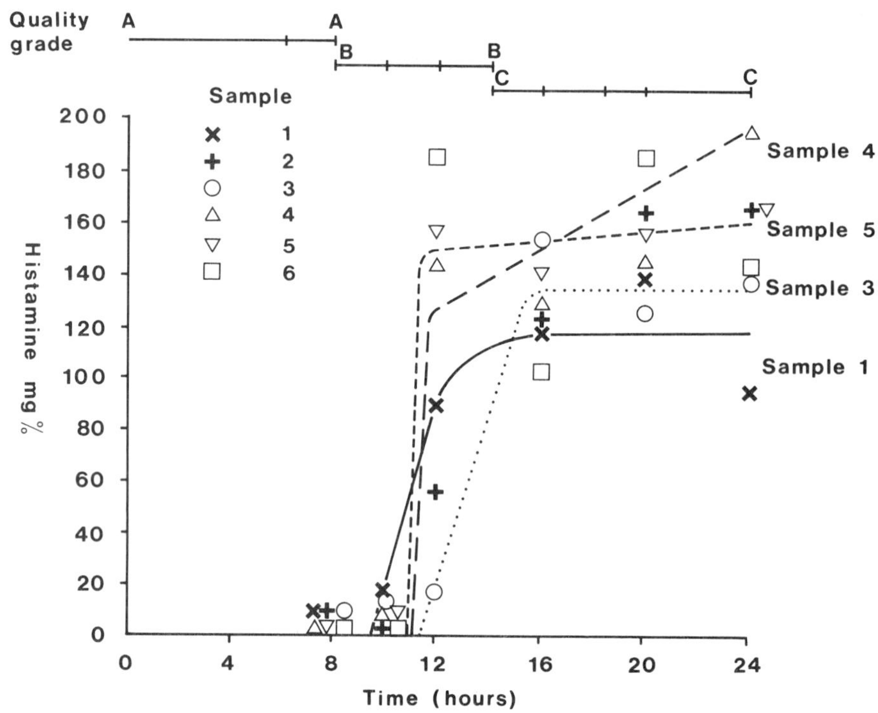

Fig. 16.6 Accumulation of histamine in dorado at 30–35°C under wet conditions (experiment 3).

ever, is not true; under certain conditions yellow skinned fish may be of grade B or C and could contain histamine.

The change of colour is not useful in distinguishing between fish of inferior quality (grade B) and fish not fit for human consumption (grade C) because the two grades had the same skin colour under the conditions used. Smell and texture of the fish are the most important indicators for distinguishing the two grades, but neither the colour of the skin nor the smell and texture can distinguish between inferior quality fish which are apparently fit for human consumption but which in reality contain excessively high levels of histamine. This finding is in agreement with Bostock *et al.* (1985) who concluded that organoleptic analysis cannot reliably identify fish with potentially toxic levels of histamine.

Temperature and histamine formation

The data obtained in this study supports the findings of Bostock *et al.* (1985) that histamine formation in dorado takes place at relatively high temperatures (above approximately 20°C). This is presumably because the prevailing histamine-producing organisms prefer higher temperatures, so that at 30°C to 35°C histamine formation is sudden and rapid after 8 to 10 hours. Cooling the fish even to temperatures such as 15°C to 20°C prevents histamine formation for over 40 hours by which time they become obviously spoiled (Bostock *et al.* 1985).

BIBLIOGRAPHY

Arnold S.H. & Brown D. (1978) Histamine toxicity from fish products. *Adv. Food Res.*, **24**, 113–54.

Barratt A. & Camba N. (1985) *Modificacion del metodo rapido enzimatico colorimetrico para detectar la presencia de histamina en productos pesqueros. Boletin Cientifico y Tecnico*, **8** (2) 29–36. Instituto Nacional de Pesca, Guayaquil, Ecuador.

Bostock T., Barratt A. & Camba N. (1985) A study of histamine in dorado (Mahi-mahi, *Coryphaena hippurus*) and its relationship with the quality of the product from the Ecuadorian fishery. *Boletin Cientifico y Tecnico*. **8** (2) 1–24. Instituto Nacional de Pesca, Guayaquil, Ecuador.

Clifford M.N., Walker R., Wright J., Hardy R. & Murray C.K. (1989) Studies on volunteers on the role of histamine in suspected Scombrotoxicosis. *J. Sci. Food Agric.*, **47**, 365–75.

Connell J.J. (1990) *Control of Fish Quality*. Fishing News Books, Oxford.

Eitenmiller R.R., Orr J.H. & Wallis W.W. (1982) Histamine formation in fish: microbiological and biochemical conditions. In *Chemistry and Biochemistry of Marine Food Products*, pp. 39–50. AVI Publishing Co., Westport, Connecticut.

Taylor S.L., Lieber E.R. and Leatherwood M. (1978) A simplified method for histamine analysis of foods. *J. Food Sci.*, **43**, 247–50.

CHAPTER 17

Further volunteer studies on scombrotoxicosis

P. Ijomah, M.N. Clifford, R. Walker, J. Wright, R. Hardy and C.K. Murray

INTRODUCTION

Scombrotoxic fish poisoning, or scombrotoxicosis, are terms coined to describe adverse reactions which sometimes occur following consumption of fish such as tuna, bonito or mackerel (*Scomberesocidae* and *Scombridae* families), collectively referred to as scombroid fish. Occasional reports have also implicated non-scombroid fish, such as sardine, anchovies and herring, but from the UK point of view the most important and commonly implicated fish is mackerel. A common feature of all these fish is that they contain red muscle, rich in histidine, varying from 1 g/kg in herring to as much as 15 g/kg in tuna. During storage under adverse high-temperature conditions, histidine may be decarboxylated to histamine by microbial decarboxylases.

The symptoms of scombrotoxicosis are shown in Table 17.1 and may vary in severity from a mere tingling of the lips and mouth with flushing, through mild gastric discomfort and nausea, to the most extreme situation with nausea, vomiting and diarrhoea. It is rare for an individual to present with all the symptoms; most of the incidents in the UK have involved gastrointestinal symptoms with nausea and diarrhoea but often without vomiting. The onset is usually rapid – sometimes within minutes, often within 2–3 hours of ingesting the fish.

Table 17.1 Symptoms associated with scombrotoxin poisoning from canned fish: data from 26 incidents affecting 71 people. (Murray *et al.* 1982.)

Symptoms	Number of incidents
Rash – bright red	14
Hot flush and sweating	13
Burning in mouth – peppery taste	12
Nausea	8
Vomiting	8
Headache	4
Diarrhoea	4
Stomach pain	4
Swelling or soreness of the tongue	3
Dizziness	2
Swelling of face	1

The fact that the toxin is stable, that it appears to be produced on storage of scombroid fish at relatively high ambient temperatures (10–20°C) and that scombroid fish subject to such temperature abuse commonly contain relatively high levels of histamine, led to the suggestion that scombrotoxicosis might be a manifestation of histamine intoxication. Early studies, therefore, centred on the concentrations of histamine in fish associated with scombrotoxic incidents (Table 17.2).

From this table it will be noted that, although in many cases samples of either the incriminated fish or fish from the same batch contained high histamine levels, sometimes exceeding 100 mg/100 g, many incidents appeared to result from fish with much lower concentrations of histamine, at least in the sample analysed. In the years 1976 to 1986, less than half the incidents were associated with histamine concentrations in excess of 5 mg/100 g. Conversely, in many cases higher levels of histamine have been found in fish which did not elicit symptoms. This suggests that histamine may not be the causal agent.

Studies on scombrotoxicosis have been confused by the possible interacting variables involving both the sample and the consumer, listed in Table 17.3, and most investigations have been limited in their ability to disentangle these. First, the amounts of histamine consumed by individuals involved in incidents of scombrotoxicosis are usually unknown; often the fish analysed has been from a different fish from the same batch as that causing toxicosis, more rarely from 'plate waste' representing portions of fish actually consumed.

Table 17.2(a) Histamine content of fish associated with incidents of scrombotoxicosis.

	No. of samples		
Histamine concentration (mg/100 g fish)	Incriminated fish	Fish from same batch	All fish (%)
>100	30	23	53 (22)
50–100	12	4	16 (7)
20–50	9	2	11 (4)
5–20	9	12	21 (9)
<5	93	46	139 (58)
Total	153	87	240 (100)

Table 17.2(b) Incidents of scombrotoxic fish poisoning. (Bartholomew *et al.* 1987.)

Year	Number of suspected incidents	Number of cases	Number of incidents with >5 mg histamine/100 g fish
1976	3	9	2
1977	1	1	—
1978	3	9	1
1979	44	178	29
1980	28	79	16
1981	24	41	14
1982	33	120	14
1983	29	31	10
1984	37	62	6
1985	29	58	5
1986	27	50	4
Total	258	638	101

Table 17.3 Potentially confusing factors.

Samples	Consumers
Histamine	Wrong diagnosis
Microbial toxins	Weight
Other components/contaminants	Sex
or metabolites	Blood group
	Medication (MAOI)
Wrong sample seized	Idiosyncratic intolerance
Lack of sample history	True allergy

The variation in histamine concentrations in different parts of the same fillet, or between different fish from the same batch, has not been systematically studied in relation to the ability of the sample to elicit symptoms. Hence simple comparisons of histamine concentrations in sound fish and fish associated (with varying degrees of closeness) with incidents, has led to very variable results.

Secondly, as the histamine is produced by microbial decarboxylation of histidine, it follows that samples with elevated concentrations are also microbially contaminated and the possibility must be considered that the gastrointestinal symptoms result from the presence of a microbial toxin. In this case, the presence of histamine may be coincidental but not causal. Alternatively, other microbial metabolites may be involved by potentiating, directly or indirectly, the effects of histamine, which would also be consistent with the apparent poor correlation between symptoms and histamine levels. In this context, monoamine oxidase inhibitors might be involved, for instance other microbially produced amines could inhibit detoxication of histamine either competitively or non-competitively.

A further complication arises from the fact that, in actual incidents, a detailed sample history is lacking; such a history would ideally include the location where the fish were caught and recent feeding habits (to allow for 'relay toxicity' from other organisms), whether or not the fish were parasitised (which could speed up proteolysis), and a detailed thermal history etc.

In the case of possible variations in consumer sensitivity to the putative toxin, the high frequency of people affected in some of the incidents would suggest that we are not dealing with an allergy to the fish – although clearly true allergy or other idiosyncratic intolerance may cloud the picture further – possibly leading to false positives as far as true scombrotoxicosis is concerned. If we are not dealing with an allergen, it would be expected that any response would be dose related, with dose expressed on a bodyweight basis. In this case, it is obvious why the concentration of histamine (or other toxin) in the fish may correlate poorly with the effect; it is the amount consumed per unit of bodyweight which describes the effective dose. There may also be sex-dependent differences in sensitivity (some sex-dependent differences in drug detoxication mechanisms and sensitivity to drugs are known) and this may be compounded by differences in lean body mass between males and females.

Finally, if histamine is the causal factor, medication with monoamine oxidase (MAO) inhibitors, or histamine antagonists (H_1 or H_2 antagonists) could affect the outcome.

With all these potentially confusing factors, and the fact that there is no suitable animal model for man, it is not surprising that scombrotoxicosis has proved extremely difficult to study, or that inadequately controlled studies have been very variable. In one recent study in our laboratories (Clifford *et al.* 1989) we attempted a single blind study in which volunteers of both sexes received on different occasions smoked mackerel associated with an incident in Glenrothes which contained 270 mg histamine/100 g; experimentally spoiled mackerel containing 114 mg/100 g; or sound mackerel spiked with histamine at levels of up to 600 mg/100 g. In this study, even when volunteers ingested 300 mg histamine in 50 g fish, no significant symptoms were elicited other than perioral tingling. This underlines the difficulty of studying scombrotoxicosis experimentally and eliciting symptoms. A very expensive study, including material associated with an actual incident, failed to yield a single clearly positive response. However, it did add weight to the argument that histamine *per se* cannot be the scombrotoxin; some other material must be involved either as the toxin or as a potentiator of histamine.

Recently a sizeable amount of smoked mackerel, which was associated with an outbreak of scombrotoxicosis on an oil rig, became available and it was decided to set up another controlled volunteer study using this material, and controlling or monitoring as many of the variables as possible. The purpose of this chapter is to present the preliminary results of this study, which is still in progress.

EXPERIMENTAL DESIGN

Volunteers

The studies involved healthy, medically-screened, informed volunteers who were not receiving any medication. There were a total of 26 volunteers: 15 males (bodyweight 52–105 kg) and 11 females (bodyweight 53–64 kg).

Fish

Smoked mackerel fillets associated with an incident of scombrotoxicosis on an oil rig, or material from the same consignment, were stored frozen until required. In order to ensure that the histamine content of each fillet used was known, the fillets were quartered and samples for analysis were taken from diagonal quarters. This also gave an indication of fish-to-fish and batch-to-batch variability of the material used in the study.

Samples for volunteer testing (50 g) were taken from diagonal quarters, and where the size of the fillet permitted, after removal of the material for analysis, two volunteer samples were prepared from the same fillet.

Protocol

The detailed conduct of the tests was similar to that previously described (Clifford *et al.* 1989). The subjects attended at 0900h having fasted overnight and the studies were conducted in a single-blind fashion under medical supervision. After ingestion of a 50 g sample of fish with toast and tea or orange juice *ad libitum*, the volunteers completed a questionnaire at regular intervals over a 24 hour period and recorded the following symptoms: headache, flushing, oral tingling, visual disturbances, abdominal discomfort, nausea, palpitations, wheeziness, diarrhoea and flatulence, as well as any other symptoms.

RESULTS AND DISCUSSION

The histamine concentrations in the various batches of mackerel studied are shown in Table 17.4. It can be seen that inter-fish sample variation within the same batch was as much as five-fold (Batch B6/E) and most batches proved highly variable. This illustrates the difficulty in retrospective analytical studies on samples from batches implicated in scombrotoxic incidents; there may be great uncertainty about the amount of histamine actually consumed. In the present study, since each fillet was analysed separately this uncertainty has been minimised.

For the first time in our laboratories we were able to elicit the range of symptoms normally associated with scombrotoxicosis. Because of time limitations not all batches have yet been tested in all individuals, and in some cases the analytical figure for histamine in the fillet consumed is still awaited, although the mean figure for each batch is known. The amount of histamine ingested has been calculated on a bodyweight basis from the concentration in the fillet where available, or from the batch mean value.

Taking nausea/vomiting and/or diarrhoea as the criteria for a positive response, the results available to date are summarised in Table 17.5; again the variability in response is evident. When the doses of histamine ingested are calculated on a bodyweight basis (Table 17.6), the lowest dose eliciting a response in 'naturally spoiled' fish was 0.3 mg/kg bodyweight whereas the highest dose *not* giving the symptoms of nausea/vomiting and diarrhoea was almost ten times this figure at 2.8 mg/kg bodyweight. In parallel studies using sound fish 'spiked' with histamine, doses of 3–4 mg/kg bodyweight did not elicit the symptoms of scombrotoxicosis, but only mild subjective symptoms such as perioral tingling.

Again these results would seem to suggest that hista-

Table 17.4 Variation in histamine content of fish samples used in study.

Batch no.	No. of samples	Histamine content (range) mg/kg	Mean	Standard deviation
B6/C	18	740–2160	1436	424.9
B6/A	6	510–1700	848	430.3
PH7	6	1000–1800	1448	303.0
PH8	6	960–1750	1470	304.9
PH9	6	1700–2560	1933	315.6
B6/B	20	540–2200	1485	448.1
B6/E	17	470–2800	1421	531.0
B6/F	10	870–1960	1431	329.6
BP2/B	17	440–2660	1132	598.0

Table 17.5 Effect on normal volunteers of fish implicated in scombrotoxicosis. Total number of volunteers: 26 (15 males, 11 females). Percentage of volunteers responding: 20% males, 36% females.

Sample	No. of tests (n)	No. of volunteers responding (diarrhoea and/ or vomiting)	% response
B6/E	19	6	32
B6/B	21	5	24
BP2/B	17	4	24
B6/C	15	1	7
B6/F	9	0	0
B6/A	4	0	0
PH7	7	2	29
PH8	5	0	0
PH9	4	0	0

mine *per se* is not the scombrotoxin; some other toxin or potentiator must be involved. However, one has to consider that there may be substantial differences in response among the volunteers. At the moment the opportunity to compare the effect of samples *from the same fillet* in two individuals is limited to eight pairs (Table 17.7), so any conclusions at the moment must be tentative since only tests on the same fillet will duplicate all the possible sample variables. In these paired comparisons there were substantially different responses between each pair in seven of the eight comparisons and only in comparison number 6 were the symptoms similar in both volunteers. In some of the other cases, the differences might have been due to differences in sex or the fact that the non-responder was significantly heavier. However, there do appear to be genuine differences in sensitivity unrelated to sex or body weight in that, in comparisons 2, 3 and 8

Table 17.6 Lack of correlation between histamine content in mackerel and symptoms of scombrotoxicosis in volunteers.

	Variation in response: Histamine mg consumed/kg body weight
Spoiled fish	
Lowest value giving vomiting and/or diarrhoea	0.3
Highest value not giving these symptoms	2.8
Spiked fresh fish	
Highest value not giving these symptoms	>3–4

involving pairs of the same sex, the volunteer with the lighter bodyweight presented the milder symptoms. With regard to female volunteers SD and CM, the difference in sensitivity was confirmed using two different batches of fish (paired comparisons 2 and 5); these comparisons were not complicated by pregnancy in either volunteer.

The data does not indicate that scombrotoxicosis results solely from allergy or idiosyncratic intolerance since all but one of the volunteers who have responded positively on at least one occasion have reacted negatively on others (Table 17.8). Although most subjects who developed diarrhoea also complained of nausea, only one subject (NM) vomited when she received each of the spoiled batches B6/B, B6/E, BP2/B and PH7; in this instance, an idiosyncratic intolerance cannot be excluded.

The data summarised in Table 17.7 might appear to suggest an inter-sex difference in sensitivity, since the frequency of positive responses was twice as high in females, and two males (BJ and DC) have not responded to any of the samples, even from batches producing a high frequency of positive responses. However, this may be misleading since these are the two heaviest volunteers (105 and 89 kg respectively) and in paired comparison number 7, the heavier male volunteer presented with more severe symptoms than the lighter female. Too little data is available yet from lightweight males and comparable bodyweight females to clarify whether there is a real inter-sex difference in sensitivity.

Returning to the question of other factors in the fish which might cause scombrotoxicosis or might modulate the response to the toxin, this preliminary data does suggest that, even corrected for bodyweight, the correlation between histamine dose and response is

Table 17.7 Cross presentation studies on the same fillet.

Paired comparison No.	Batch	Histamine consumed (mg)	Volunteers/body weight, sex* and symptoms			
1	B6/E	140 mg	*KP (67 kg) (m)* Headache Flushing **Diarrhoea**		*DC (89 kg) (m)* No Symptoms	
2	B6/E	80 mg	*SD (64 kg) (f)* Oral tingling Headache Flushing	 Nausea Flatulence **Diarrhoea**	*CM (57 kg) (f)* Oral tingling Nausea	
3	B6/E	74 mg	*FC (64 kg) (m)* Flushing Shivering Wheeziness	 Flatulence Nausea **Diarrhoea**	*CD (52 kg) (m)* No Symptoms	
4	B6/B	99 mg	*NM (57 kg) (f)* Headache Stomach ache Nausea	 Vomiting **Diarrhoea**	*BJ (105 kg) (m)* No Symptoms	
5	B6/B	110 mg	*SD (64 kg) (f)* Oral tingling Flushing Headache	 Nausea **Diarrhoea**	*(57 kg) (f)* No Symptoms	
6	B6/B	100 mg	*JL (60 kg) (f)* Oral tingling Flushing Headache	 Flatulence Nausea	*SD (64 kg) (f)* Oral tingling Flushing Headache	 Flatulence Nausea
7	B6/B	27 mg	*RM (57 kg) (f)* Flushing Wheeziness		*KP (67 kg) (m)* Nausea **Diarrhoea**	
8	B6/C	76 mg	*SD (64 kg) (f)* Oral tingling Headache Flushing	 Flatulence Nausea	*NM (57 kg) (f)* Headache Shivering Nausea	 **Diarrhoea** **Vomiting**

* m = male, f = female

too poor to explain scombrotoxicosis simply as histamine intoxication. Now that we have duplicate samples from fish which we know can elicit a response in an individual volunteer, more detailed studies are planned similar to our earlier attempts using H_1 and H_2 antagonists (Clifford *et al.* 1989). These should clarify whether some other agent is potentiating histamine or whether histamine is involved at all, other than as a non-effective covariant; a lack of effect of histamine antagonists would be strongly indicative of a totally different causal mechanism. It is also planned to study the effects of MAO inhibitors as potentiators and to investigate the feasibility of developing an animal model for scombrotoxicosis.

In summary, the results of earlier experiments in our laboratories and of the preliminary results in the present series indicate that:

(1) Histamine alone is not the causative agent of scombrotoxicosis.
(2) There are marked differences between individuals in sensitivity to scombrotoxin(s).

Table 17.8 Number of positive/negative responses to mackerel samples* by individual volunteers showing at least one positive response⁺.

Volunteers		No. of positive response	No. of negative response
SD	(F)	3	3
JL	(F)	2	3
NM	(F)	4	0
SA	(F)	1	3
KP	(M)	4	1
FC	(M)	1	4
JW	(M)	1	2

* All samples from batches implicated in outbreak of scombrotoxicosis
⁺ Positive response as defined using the criteria outlined in the text; 19 other volunteers have not responded to date.

(3) The effect is probably not an allergic response so some other (metabolic?) explanation must be sought.

BIBLIOGRAPHY

Bartholomew B.A., Berry P.R., Rodhouse J.C., Murray C.K. & Gilbert R.J. (1987) Scombrotoxic fish poisoning in Britain: features of over 250 suspected incidents from 1976 to 1986. *Epidem. Inf.* **99**, 775–82.

Clifford M.N., Walker R., Wright J., Hardy R. & Murray C.K. (1989) Studies with volunteers on the role of histamine in suspected scombrotoxicosis. *J. Sci. Food Agric.* **47**, 365–75.

Murray C.K., Hobbs G. & Gilbert R.J. (1982) Scombrotoxin and scombrotoxin-like poisoning from canned fish. *J. Hyg. Camb.* **88**, 215–19.

CHAPTER 18

The application of electrophoretic techniques to the identification of species of canned tuna and bonito

I.M. Mackie, M. Chalmers, P. Reece, A.E. Scobbie and A.H. Ritchie

INTRODUCTION

The problems of identifying and differentiating between closely related species of fish such as those of the tuna and salmon families, when canned, are long-standing. Although the various species are differentiated commercially and are named on labels, often to comply with labelling regulations (The Food Labelling Regulations 1984 and The Food Labelling (Scotland) Regulations 1984) there is at present no certain method of differentiating one from another when only a portion of flesh is available for examination. Generally, canned tuna is free of skin and bones – often a useful indicator of species. As differences between species are largely those of colour, taste and texture – all sensory attributes – differentiation by one or all of these criteria would at best be uncertain. Nonetheless, species are differentiated in canned products and different import duties are applied – for example, bonito is subject to a lower duty than tuna – even when the inadequacy of the means of establishing the authenticity of a species is fully acknowledged.

When flesh is raw there is no difficulty in differentiating even the closely related species of tuna and bonito on the basis of the separation profiles of the water-soluble, or sarcoplasmic, proteins obtained on electrophoresis or isoelectric focusing (Mackie 1980, Laird *et al.* 1982). As each species has a characteristic profile, the identity is established by comparison with those of authentic species examined under the same experimental conditions.

Establishing the identity of the species of cooked, but not autoclaved, flesh is somewhat more difficult as the constituent proteins are denatured by heat, and those of interest for species identification of raw flesh are no longer extractable with water. However, by using more powerful solvents such as 6M urea (Mackie 1980) or 2% sodium dodecyl sulphate (SDS) (Scobbie & Mackie 1988), it is possible to dissolve all the flesh proteins and separate them into species-characteristic profiles on electrophoresis. The separation profiles show a greater overall similarity between species than those of aqueous extracts of raw flesh, largely because of the contribution from the myofibrillar structural proteins, e.g. myosin, actin, tropomyosin and others which show few, if any, species differences on electrophoresis. Of the two analytical procedures, that using SDS is preferred because SDS is easier to handle and it gives better resolution of the proteins of interest for species identification.

When flesh has been subjected to more severe conditions of heating, as in autoclaving in the canning process, the proteins undergo further denaturation depending on the time and temperature of the treatment. In many instances the electrophoretic profiles obtained for commercially canned fish are too indistinct and too undifferentiated to be of any value in species identification, presumably because of the heat damage that such samples received during the autoclaving stage. For such products it has been found that more satisfactory results have been obtained if species identification is based on profiles obtained on isoelectric focusing of the peptides, produced on reaction with the protein-splitting agent cyanogen bromide (Gross 1967). This reagent splits proteins at the methionine residues only and releases water-soluble protein fragments which have been shown to give species-characteristic profiles (Mackie & Taylor 1972, Mackie, 1980).

For species which are well separated genetically, the profiles of the cyanogen bromide peptides of canned flesh can be readily distinguished one from another (Mackie 1980) but as the relationship becomes closer, the species-characteristic zones can be reduced to such an extent that species differentiation is not possible. Five species of Pacific salmon, for example, cannot be sub-divided beyond two groups using this procedure (Mackie 1980) but the situation regarding differentiation of species of the tuna/bonito family after canning is not known. Accordingly the opportunity was taken to evaluate the current species identification procedures in a recent collaborative trial between Torry Research Station and the Instituto de Investigaciones Marinas, Vigo, Spain. The analytical results obtained by Torry Research Station form the basis of this chapter.

EXPERIMENTAL

Raw material

Authenticated whole specimens of the following seven species of tuna and bonito were obtained and canned by the Instituto de Investigaciones Marinas in Vigo:

Bonito	*Sarda sarda*	(Sarda)
Yellowfin tuna	*Thunnus albacares*	(Rabil)
Bigeye tuna	*Thunnus obesus*	(Patudo)
Skipjack tuna	*Katsuwonus pelamis*	(Listado)
Bluefin tuna	*Thunnus thynus*	(Atun)
Little tunny	*Euthynnus alletteratus*	(Bacoreta)
Albacore	*Thunnus alalunga*	(Albacora)

The fish were either processed immediately or held frozen at −30°C until required.

Canning procedure

In preparation for canning, the fish were eviscerated and steam cooked at 102–103°C until the backbone temperature reached 70°C. After cleaning (removal of skin, bone and red muscle) 82 g of flesh was transferred to oval cans (OL-120) 150.1 × 64.7 × 28.8 mm, total volume 120 ml. Approximately 35 ml of vegetable oil and 2.0 g salt were added. The cans were then vacuum steamed and heat processed in a retort under one of the following conditions:

(1) 115°C for 55 min
(2) 118°C for 35 min
(3) 118°C for 55 min

Preparation of SDS extracts of canned flesh

A portion of the canned flesh was freed of as much oil as possible by blotting with a paper towel. A 0.25 g portion of the blotted flesh was then taken and suspended in 1.0 ml n-hexane in an Eppendorf tube. The mixture was then separated by centrifugation at 11 600 × *g* and the hexane removed. The hexane washed flesh, blotted free of hexane, was then broken up and extracted in 1.0 ml (2%/w/v) aqueous SDS according to the procedure of Porzio & Pearson (1977). The suspension was heated for 35 minutes at 60°C and the solution of proteins obtained clarified by centrifugation at 11 600 × *g* and diluted 8.5 times (20 μl + 150 μl) using Laemmli sample buffer (Laemmli 1970). The diluted extracts were then heated in a boiling water bath for two minutes, and cooled and clarified by centrifugation at 11 600 × *g*. The extracts were either analysed immediately or stored frozen at −30°C until required for analysis. A sample loading of 6–10 μl was normally used.

Preparation of cyanogen bromide peptides from canned flesh

A 1.0 g portion of the canned flesh, freed of oil as far as possible by blotting with a paper towel, was broken up and transferred to a 100 ml round bottom flask. Formic acid (20 ml; 70%) solution and 0.5 g cyanogen bromide were added, and the mixture was allowed to stand overnight at room temperature, then diluted with 20 ml water and taken to dryness on a rotary evaporator. The residue was suspended in 20 ml water, re-dried and finally resuspended in water (10 ml). The pH of the mixture was then adjusted to 6.5 with 2M sodium hydroxide solution. After dialysis against distilled water for 24 hours the mixture was then centrifuged at 10 000 × *g* for 10 minutes at 4°C. The supernatant solution obtained was recentrifuged and finally freeze-dried and stored at −20°C until required for isoelectric focusing.

Procedure for SDS polyacrylamide gel electrophoresis

The protein solutions were analysed by SDS-PAGE on 12.5% total acrylamide (T) concentration (2.6% Crosslinker, C) separating gels (140 × 120 × 1.5 mm) incorporating a 5.3% T (2.6%C) stacking gel as described by Laemmli (1970). Electrophoresis was carried out at 40 mA per slab until the tracking dye reached the bottom of the gel in a BioRad Protean Cell (BioRad Laboratories, Watford) with cooling to approximately 15°C. Following electrophoresis the gels were fixed with several charges of 50% (v/v) aqueous methane containing 0.1% formaldehyde solution (38% w/v). After fixing, the gels were silver stained as described by Wray *et al.* (1981) but with the inclusion of 10% (v/v) acetic acid in the final wash (Scobbie & Mackie 1988).

Standard molecular weight markers were obtained from Pharmacia (phosphorylase b, 94 000; bovine serum albumin, 67 000; ovalbumin, 43 000; carbonic anhydrase, 30 000; trypsin inhibitor, 20 100 and α-lactalbumin, 14 000).

Procedures for isoelectric focusing

Preparation of gels

The polymerising solution was prepared by combining 10 ml 29.1% acrylamide monomer, 10 ml 0.9% bis-acrylamide, 3.6 ml ampholine (pH 3.5–9.5) (Pharmacia LKB) and water to give a total volume of 60 ml. The solution was degassed and, after the addition of 2 ml 1% ammonium persulphate and 0.06 ml N,N,N′,N′-tetramethyl ethylenediamine (TEMED), it was transferred to the gel mould assembly of two glass plates (260 × 125 mm) separated by a 2 mm thick silicone gasket and containing gel-bond in contact with the inner surface of one of the plates. After allowing polymerisation to take place for 2 hours at room temperature, the gel-bond with the gel attached was carefully removed from the assembly, covered with a clean plastic sheet, sealed in a polythene bag and stored at 4°C until required.

Isoelectric focusing

Isoelectric focusing was carried out on an LKB Multiphor using an LKB Power Supply (Pharmacia LKB) set at 1500 V, 50 mA and 30 W for a total run time of 150 minutes. The solutions of cyanogen bromide peptides were applied on filter paper strips at the anode end of the gel. Cathode and anode electrolytes were 1.0M sodium hydroxide and 1.0M orthophosphoric acid respectively. After isoelectric focusing the gels were fixed in 20% trichloroacetic acid (TCA), and stained with silver nitrate after the method of Warlaw *et al.* (1988).

RESULTS AND DISCUSSION

It is of interest in the context of this chapter to compare the separation profiles of the water-soluble or sarcoplasmic proteins of the raw flesh of the seven species (Fig. 18.1). As would be expected for species within a family, in this case the Scombridae, there is an overall similarity in the profiles. Those species most closely related – bluefin tuna, yellowfin tuna and albacore – show the greatest correspondence of protein zones, whereas the more distantly related bonito and skipjack tuna show greater divergence from the tuna profiles. Each of the seven species, however, gives a profile which is readily distinguishable from the others and which enables unequivocal identification of the seven species to be made.

Effect of heat on fish flesh

The nature of the changes which take place on heat denaturation of flesh proteins is poorly understood but it is generally believed that the bonds, largely H-bonds, which are involved in the secondary and tertiary structures are broken, resulting in an unfolding of the native configuration. These changes which take place between 30°C and 60°C are largely physical rather than chemical and are accompanied by an association, often random, of the unfolded chains through hydrophobic and H—bonds, resulting in coagulation and loss of enzyme activity.

On the other hand, collagen, the main connective tissue protein, is broken down into water soluble subunits (gelatin) on heat denaturation, usually between 50°C and 60°C. It is only at temperatures above 90°C that chemical changes are involved. For example, oxidation of SH groups is known to occur and to lead to the formation of covalent S—S—bonds. As the temperature and time of heating are further increased, chemical changes, as yet poorly understood, occur and associated with them is some loss in the nutritional value of the flesh. The digestibility of lysine is diminished, possibly as a result of reaction of the ε amino group (Opstvedt *et al.* 1984). These changes are seen on electrophoresis as a reduction in the intensity of stained protein bands, possibly associated with a loss of basic dye-binding groups and an overall increase in the background staining of the gels (Hoffmann 1977).

As the extent to which these chemical changes affect the tuna species on canning has not previously been established, a systematic study was carried out on all seven species using both the SDS electrophoresis of SDS extracts and isoelectric focusing of cyanogen bromide peptides. At the same time a study was made on the effect of the three heat treatments on the separation profiles obtained by both systems. Alternative methods of producing species-specific fragments from canned flesh, e.g. cleaving with pepsin or N-bromo-succinimide, were also investigated but without much success and are not discussed in this chapter.

SDS electrophoresis

On SDS electrophoresis in high concentration acrylamide gels (12.5%), those proteins of molecular weight less than 60 000 are well resolved (Fig. 18.2).

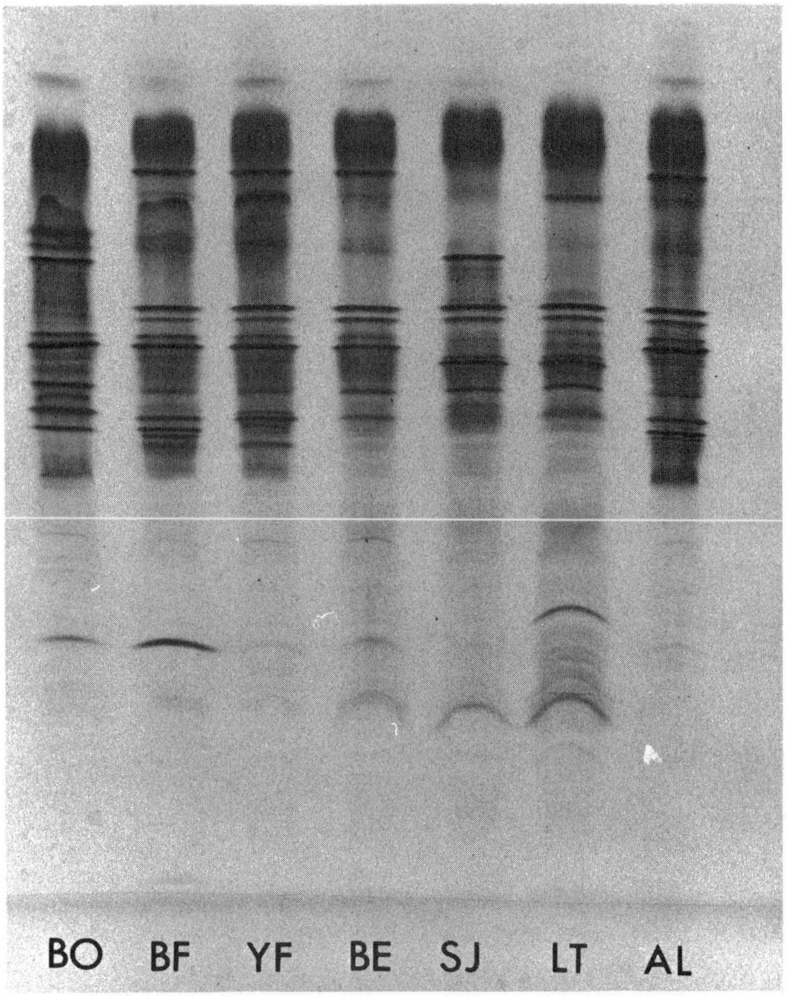

Fig. 18.1 Isoelectric focusing of water-soluble or sarcoplasmic proteins of raw flesh. Left to right, bonito, bluefin tuna, yellowfin tuna, bigeye tuna, skipjack tuna, little tunny and albacore (Coomassie Blue staining).

As some of these proteins or sub-units are derived from the species-specific sarcoplasmic proteins, some differences would be expected in the separation profiles for the different species, but on the other hand those zones corresponding to the myofibrillar proteins would be expected to show few if any differences in molecular weight profiles between species. Actin and the tropomyosin sub-unit, for example, fall within this molecular weight range but would not be expected to show any species differences. It has previously been shown (Scobbie & Mackie 1988) that SDS electrophoresis of SDS extracts of both cooked and raw fish gave identical separation profiles and that even for closely related species such as those of the gadoid family – cod, haddock and Alaska pollack – significant differences in the mobility of only one or two out of over 20 zones could be used to differentiate the species. Although the effect of autoclaving has been to reduce somewhat the intensity of staining for a given concentration of protein, the profiles are sufficiently distinct to allow differentiation of all seven species on close examination. The most divergent profile is that of skipjack tuna, while the profiles showing the greatest coincidence of zones are those of albacore and little tunny. Thus, contrary to expectations, the effect of heat even under the severest conditions used in the study has not been sufficient to affect the profile to any significant or consistent extent (Fig. 18.3).

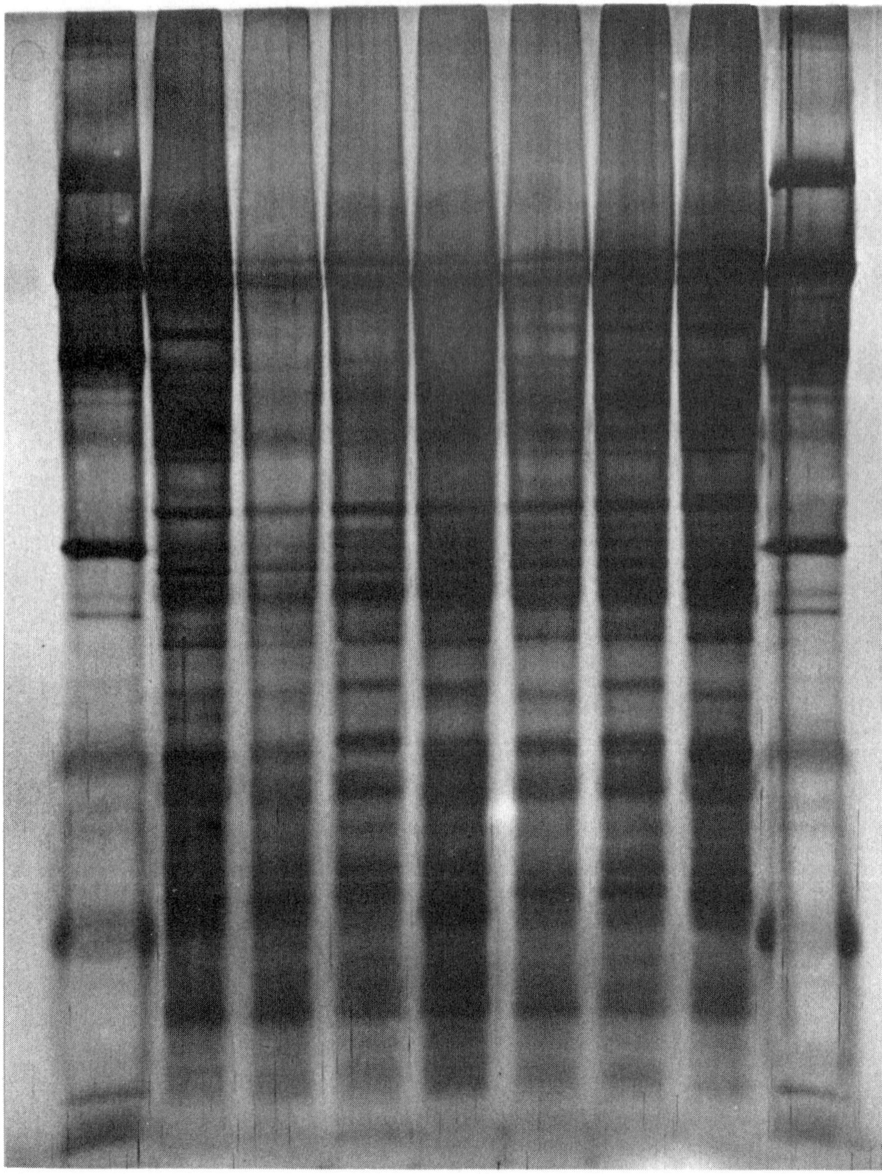

Fig. 18.2 SDS electrophoresis of SDS extracts of canned flesh. Left to right, standard molecular weight markers, albacore tuna, little tunny, bluefin tuna, skipjack tuna, bigeye tuna, yellowfin tuna and bonito, standard molecular weight markers.

Isoelectric focusing of cyanogen bromide peptides

When the cyanogen bromide peptides of the seven species canned under 'normal' conditions were separated by isoelectric focusing, the profiles obtained showed an overall similarity but some distinct differences between the seven species. The increased sensitivity obtained with silver staining (Fig. 18.4) over that obtained with Coomassie Blue, allows the analyses to be performed at concentrations giving improved resolution of the protein zones and facilitates differentiation of the species. Comparison of the profiles shows that those for the closely related species – alba-

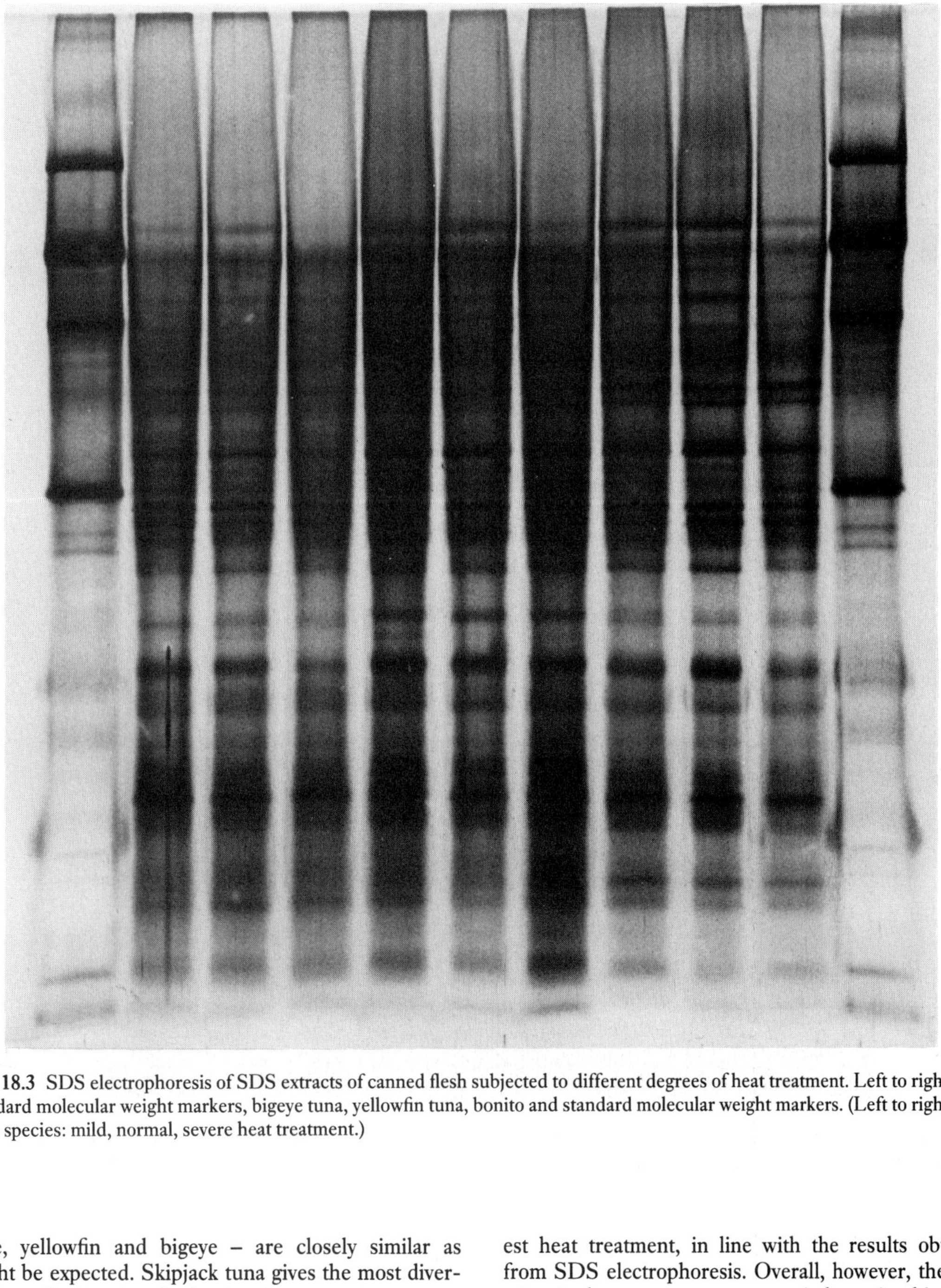

Fig. 18.3 SDS electrophoresis of SDS extracts of canned flesh subjected to different degrees of heat treatment. Left to right, standard molecular weight markers, bigeye tuna, yellowfin tuna, bonito and standard molecular weight markers. (Left to right for each species: mild, normal, severe heat treatment.)

core, yellowfin and bigeye – are closely similar as might be expected. Skipjack tuna gives the most divergent profile but that for bonito shows a greater correspondence with the tunas than might be expected.

In other work it was also observed that for some species there was a reduction in staining intensity of protein zones when the fish was subjected to the highest heat treatment, in line with the results obtained from SDS electrophoresis. Overall, however, the variation in heat treatments as carried out in this study had little effect on the separation profiles and it would seem that isoelectric focusing of the cyanogen bromide peptides offers one way of identifying the species of canned tunas and bonitos.

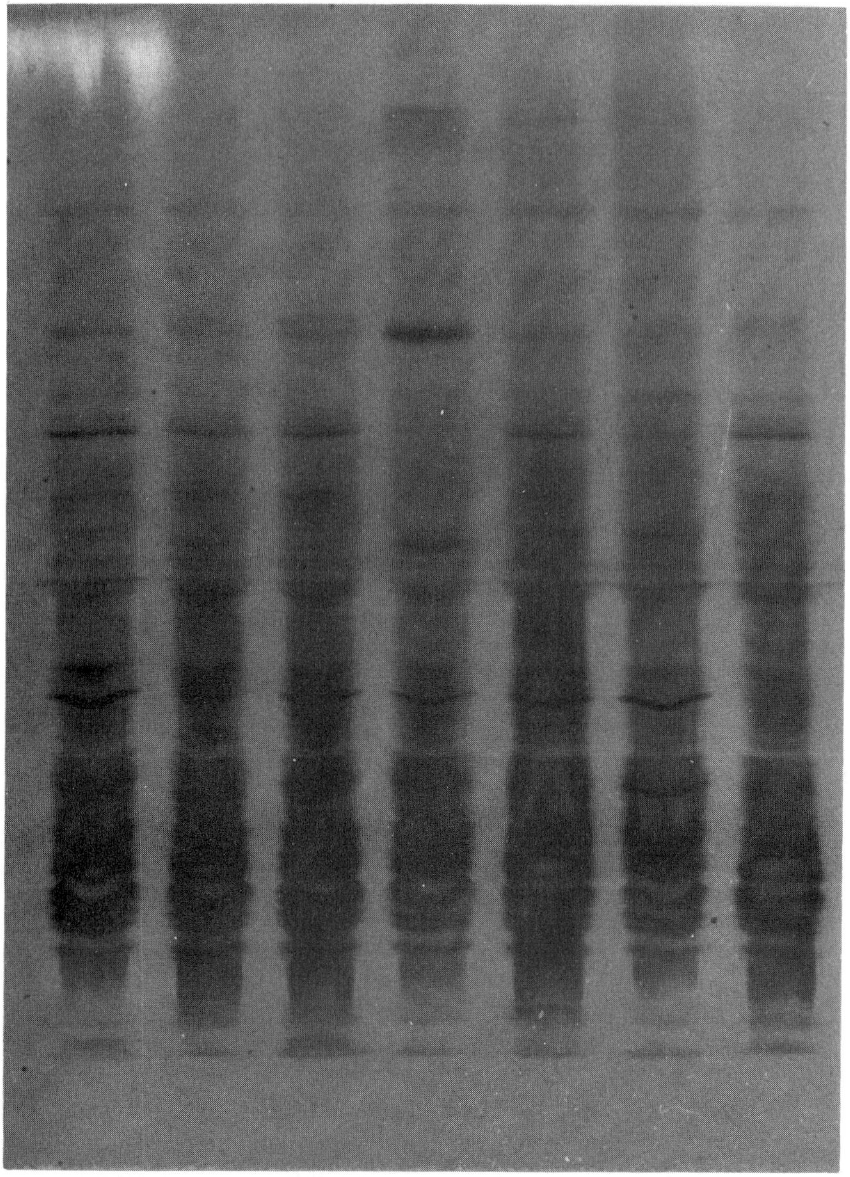

Fig. 18.4 Isoelectric focusing of cyanogen bromide peptides of canned flesh. Left to right, bonito, yellowfin tuna, bigeye tuna, skipjack tuna, bluefin tuna, little tunny and albacore (silver staining).

CONCLUSIONS

From the results obtained in this study it would seem that SDS electrophoresis of SDS extracts of canned fish and isoelectric focusing of cyanogen bromide peptides could be used as complementary procedures for the identification of the closely related species of tuna and bonito.

Further studies are required on the effect of the heat treatment on the separation profiles, with particular reference to conditions found in commercial practice.

ACKNOWLEDGEMENTS

The authors wish to thank Dr Gallardo Abuin and Dr Ricardo Martin for their collaboration and for obtaining and preparing the samples of canned tuna and bonito. The work was supported by a travel and subsistence grant (1988–89) from the British Council and the Consejo Superior de Investigaciones Cientificas of Spain under the Acciones Integradas Hispano-Britanicas (1988–89) programme.

BIBLIOGRAPHY

Gross E. (1967) The cyanogen bromide reaction. In *Methods in Enzymology*, (Ed. by C.H.W. Hirs) **XI**, pp. 238–42. Academic Press, New York.

Hoffman K. (1977) The influence of heat on meat proteins, studied by SDS electrophoresis. In *Physical, Chemical and Biological Changes in Food Caused by Thermal Processing* (Ed. by T. Høyem & O. Kvåle.) pp. 311–27. Applied Science Publishers, London.

Laemmli U.K. (1970) Cleavage of structural proteins during the assembly of the head of bacteriophage T4. *Nature*, **227**, 680–85.

Laird W.M., Mackie I.M. & Ritchie A.H. (1982) Differentiation of species of fish by isoelectric focusing on agarose and polyacrylamide gels – a comparison. *J. Assoc. Publ. Analysis*, **20**, 125–35.

Mackie I.M. (1980) A review of some recent applications of electrophoresis and isoelectric focusing in the identification of species of fish in fish and fish products. In *Advances in Fish Science and Technology* (Ed. by J.J. Connell) pp. 444–51. Fishing News Books, Oxford.

Mackie I.M. & Taylor T. (1972) Identification of species of heat sterilized canned fish by polyacrylamide disc electrophoresis. *Analyst*, **97**, 609–11.

Opstvedt J., Miller R., Hardy R.W. & Spinelli J. (1984) Heat-induced changes in sulphydryl groups and disulphide bonds in fish protein and their effect on protein and amino acid digestibility in rainbow trout (*Salmo gairdneri*). *J. Agric. Food Chem.*, **32**, 929–35.

Porzio M.A. & Pearson A.M. (1977) Improved resolution of myofibrillar proteins with sodium dodecylsulphate-polyacrylamide gel electrophoresis. *Biochim. Biophys. Acta.*, **490**, 27–34.

Scobbie A.E. & Mackie I.M. (1988) The use of sodium dodecylsulphate polyacrylamide gel electrophoresis in fish species identification – a procedure suitable for cooked and raw fish. *J. Sci. Fd. Agric.*, **44**, 343–51.

The Food Labelling Regulations (1984) *Statutory Instruments* 1984 No. 1305.

The Food Labelling (Scotland) Regulations (1984). *Statutory Instruments* 1984 No. 1519 (S128).

Warlow R.S., Morgan J., Nicola N. & Bernard C.C.A. (1988) A nondenaturing vertical isoelectric focusing polyacrylamide slab gel system suitable for silver staining and electrophoretic blotting. *Analyt. Biochem.*, **175**, 474–81.

Wray W., Boulekas T., Wray V.P. & Hancock R. (1981) Silver staining of proteins in polyacrylamide gels. *Analyt. Biochem.*, **118**, 474–81.

CHAPTER 19

Influence of iced storage and washing on the protein composition of sardine (*Sardina pilchardus*) minces

I. Filipi, M.L. Nunes, N. Vrucinic-Filipi and D. Roseg

INTRODUCTION

Sardine *Sardina pilchardus* lately has become a very interesting source for mince production. As mentioned by Suzuki (1981), only 30% of sardines caught have been used for human consumption. This has roused scientists to make more efforts to find better ways of utilising sardine. Using previous experience in mince and surimi from Alaska pollack, they started to investigate small pelagic fish as material for the production of kamaboko related products (Nunes 1981, Babbitt 1986, Keay 1986, Regenstein, 1986, Rhodes, 1986).

Sardine meat is characterised by a high fat content, a significant percentage of dark meat, and a large quantity of sarcoplasmic proteins which inhibit gel formation. Washing of mince is very effective in removing these undesirable components, as well as in increasing the concentration of myofibrillar proteins, thereby improving gel strength and elasticity (Babbitt 1986).

This study was carried out to investigate the effect of iced storage and three washing cycles on the protein composition of sardine mince.

MATERIALS AND METHODS

Sardines were caught south-west of the Portuguese port of Peniche and transported in insulated boxes mixed with ice to (INIP) Lisbon. Immediately after reaching the laboratory, the fish were divided into four batches. The first batch was used as fresh sardine and the others were well mixed with ice and maintained at 1–3°C for one, two and three days.

The fish were headed and gutted manually, washed and rinsed, and finally deboned through a Baader 694 fish deboner at maximum belt tension, using a stainless steel drum with openings of 3 mm diameter. The raw minces were washed three times, using 3 parts of cooled water (8–10°C) to 1 part of fish, followed by filtration through cheesecloth. A solution of $NaHCO_3$ (0.5% w/w) was used for the first washing, followed by a tap water rinse. Finally the mince was washed with a 0.15% (w/v) NaCl solution. The prepared material was packed into freezer trays and quickly frozen at −40°C. The frozen samples were packaged under vacuum in PVC bags and stored at −30°C until analysed.

Protein composition (sarcoplasmic, myofibrillar, alkali-soluble and stroma) and also non-protein nitrogen were determined according to the method of Hashimoto *et al.* (1979). The total nitrogen content of each fraction was measured by the Kjeldhal method (AOAC 1984).

Sodium dodecylsulphate–polyacrylamide gel electrophoresis (SDS–PAGE) was performed according to Laemlli (1970). Myofibrillar and alkali-soluble fractions were dialysed against 25 mM tris/HCl buffer (pH 7.4) containing 0.2% SDS and 0.1% β-mercaptoethanol. Dialysed samples as well as untreated samples of sarcoplasmic fractions were subjected to electrophoresis on gradient gel (7.5–15.0%). The loaded protein concentrations in each run are indicated in Table 19.1. Protein molecular weight standards ranging from 14 300 to 200 000 daltons were purchased from BRL (Basel, Swiss). After electrophoresis, the gels were stained with Coomassie Blue R-250 for two hours on a shaker, then destained with destain I (10% HAc, 25% isopropanol) and destain II (10% HAc, 10% isopropanol). The gels were dried under vacuum.

RESULTS AND DISCUSSION

Effect of iced storage

Sardine, like other small pelagics, is quite fragile and spoils quickly but its shelf life can be extended by

Table 19.1 Protein concentrations loaded in electrophoretic runs.

	Protein concentration (μg)															
	Lanes															
Protein fractions	1	a	b	c	5	a	b	c	9	a	b	c	13	a	b	c
Sarcoplasmic	73	89	95	67	65	56	44	44	59	46	55	41	54	53	57	46
Myofibrillar	132	132	138	139	131	141	117	109	134	128	109	50	127	126	124	113
Alkali-soluble	34	33	36	84	53	98	94	207	61	98	67	176	108	123	104	220

Table 19.2 Influence of iced storage on nitrogen distribution of sardine minces.

Storage time days	Sarcoplasmic fraction %	Myofibrillar fraction %	Alkali-soluble fraction %	Stroma fraction %	Non protein nitrogen %	Total protein mg/g
0	29.6	53.3	2.8	1.4	12.9	25.72
1	26.6	53.9	4.5	1.5	13.5	26.62
2	24.5	56.0	5.2	1.8	12.5	24.91
3	22.7	53.0	9.3	2.5	12.5	24.84

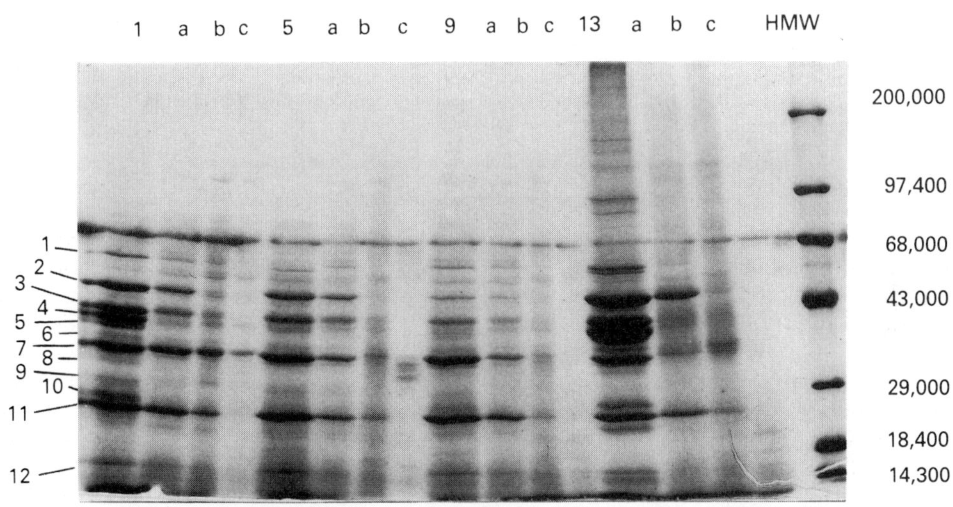

Fig. 19.1 Gradient SDS–PAGE patterns of sarcoplasmic fractions. Unwashed minces: lane 1 – fresh sardine; lane 5 – one day iced; lane 9 – two days iced; lane 13 – three days iced. Washed minces: lane a – first washing; lane b – second washing; lane c – third washing.

using ice on board. An alternative to this practice is the application of ice just after landing, and storage in a refrigerated plant. Fast mechanised systems can also contribute to its optimal utilisation.

The changes in protein composition influenced by iced storage are presented in Table 19.2. These results show that storage on ice leads to a reduction in sarcoplasmic proteins. In contrast, myofibrillar proteins remained unchanged while alkali-soluble and stroma proteins increased.

Changes caused by iced storage were followed by SDS–PAGE. The influence on the sarcoplasmic protein fraction can be seen in Figure 19.1. Lane 1 shows several very dense bands which represent components whose molecular weight was estimated to be 56 k (56 000 D), 46 k, 41 k, 39 k, 37 k, 35 k, 29.8 k, 25.8 k,

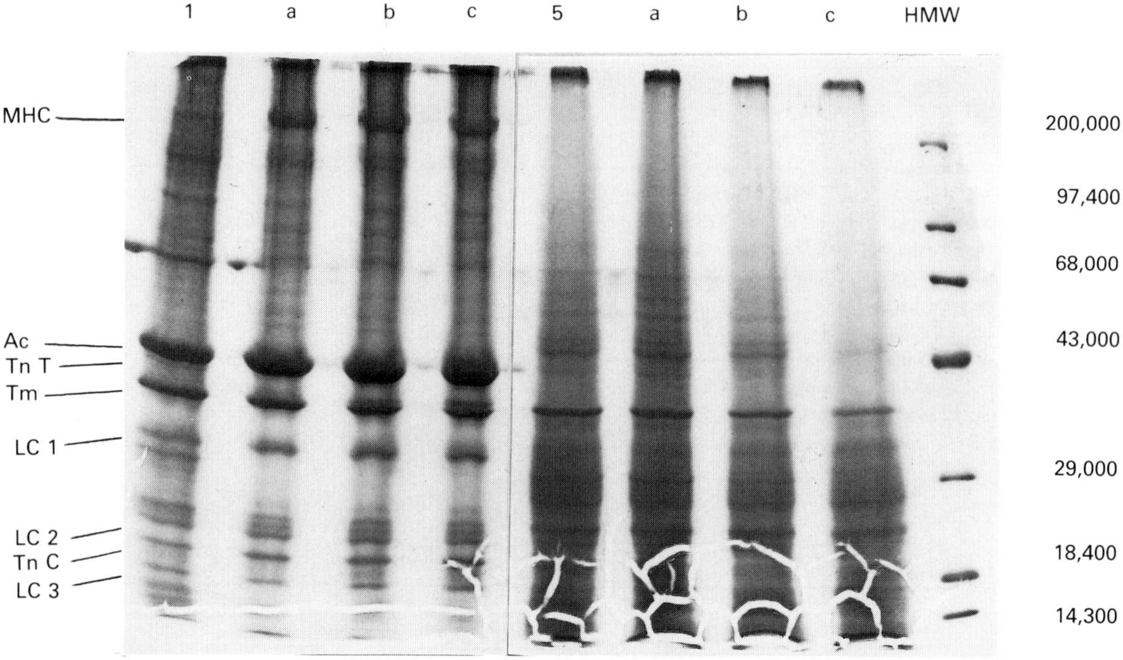

Fig. 19.2 Gradient SDS–PAGE patterns of myofibrillar fractions. Unwashed minces: lane 1 – fresh sardine; lane 5 – one day iced. Washed minces: lane a – first washing; lane b – second washing; lane c – third washing. MHC = myosin heavy chain; Ac = actin; Tn T = troponin T; Tn C = troponin C; Tm = tropomyosin; LC 1, LC 2, LC 3 = myosin light chains.

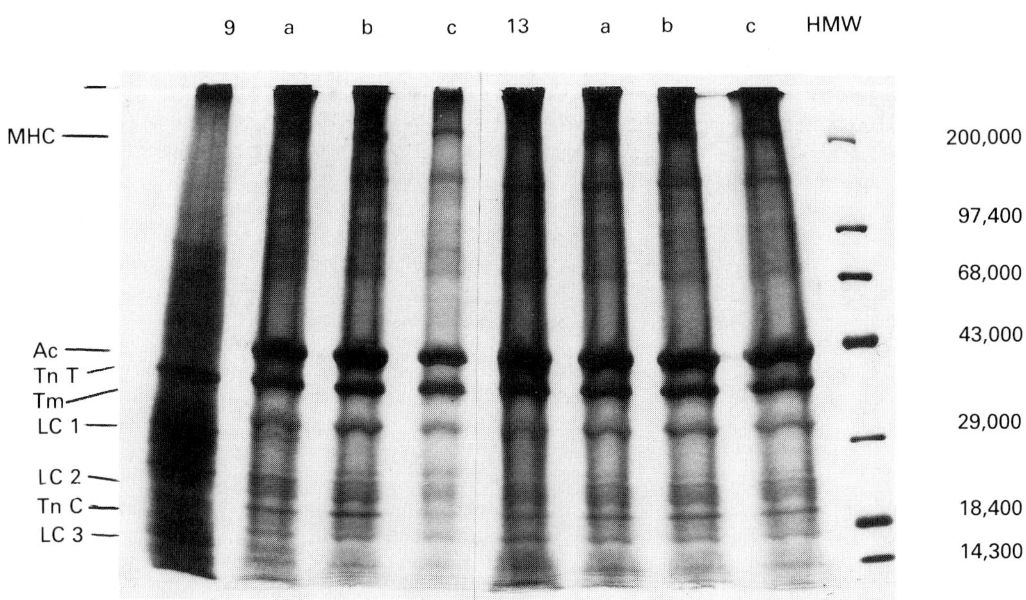

Fig. 19.3 Gradient SDS–PAGE patterns of myofibrillar fractions. Unwashed minces: lane 9 – two days iced; lane 13 – three days iced. Washed minces: lane a – first washing; lane b – second washing; lane c – third washing. MHC = myosin heavy chain; Ac = actin; Tn T = troponin T; Tn C = troponin C; Tm = tropomyosin; LC 1, LC 2, LC 3 = myosin light chains.

25 k, and 16 k. By comparing lanes 1, 5 and 9 it is noticeable that the major bands become fainter with longer iced storage. This is caused by leaching out of water soluble proteins. Lane 13 does not follow these changes, the bands being denser and with background staining, which is usually associated with protein denaturation. These results are contrary to those expected and require a further study which must be supported by the identification of each protein band.

The results given in Table 19.2 do not exhibit significant differences in the quantity of myofibrillar proteins, but the electrophoretic patterns shown in Figures 19.2 and 19.3 display different components for each day during iced storage. Fresh sardine mince (Fig. 19.2 lane 1) exhibits all the major bands of myofibrillar proteins including myosin heavy chain (MHC), actin (Ac), troponin T (Tn T), tropomyosin (Tm), myosin light chains (LC 1), (LC 2), (LC 3), and troponin C (Tn C).

Figure 19.2, lane 5 and Figure 19.3, lane 9 represent the myofibrillar fraction after one and two days on ice and display patterns different from those in Figure 19.2, lane 1. In both cases proteolysis had taken place. There was also the complete disappearance of MHC and actin while other bands became very faint. The gradual disappearance of these heavy chain sub-units can be due to a highly-active enzyme system in sardine during iced storage. High molecular weight protein aggregates which are too large to enter the gel during electrophoresis were also observed, as referred to by Ohnishi & Rodger (1980). Changes in the preparation parameters, such as higher temperature, pH, ionic strength etc., can be possible reasons for this effect.

The influence of particular parameters on protein composition will be examined further. The myofibrillar fraction of sardine stored for three days in ice (Fig. 19.3, lane 13) exhibits major bands, but proteolysis and aggregation had also taken place.

According to the method of preparation, the alkali-soluble fraction was prepared by overnight extraction with 0.2 M NaOH. Under these conditions a greater part of the remaining proteins was solubilised while the stroma fraction remained insoluble. As shown in Figure 19.4, lanes 1, 5, 9, and 13, actin is the major band detected. On the top of the gel high molecular weight aggregates can be seen.

Effect of washing

Washing is an indispensable operation in mince stabilisation and in surimi production. The aim of washing has been described by a number of authors (Suzuki 1981, Grantham, 1981, Adu *et al.* 1983, Lee 1986). The main purpose of washing is to eliminate water soluble proteins and to concentrate myofibrillar proteins which are essential for gelatinisation of surimi based products.

The influence of washing on protein composition was followed both by measuring the protein of each fraction and by their SDS-PAGE patterns. Each step of the washings was examined. Results are presented in Figure 19.5. With regard to the sarcoplasmic fraction, the highest decrease was found after the first washing (approx. 35%) and this slowly decreased until the last washing. In Figure 19.1 these results can be observed. Components estimated to be 37 k, 32 k and 29.8 k disappear completely after the first washing. Furthermore, 46 k, 41 k, 34 k and 25 k components remained until the second washing (in mince from fresh sardine) and only the 34 k component was found after complete washing. Moreover, other components became fainter or disappeared completely during washing. Simultaneously, the washing also removes flavour compounds, pigments and other substances that can affect mince quality.

The first and second washings have a favourable effect on the solubilisation of myofibrillar proteins. The decrease found after the third washing can be attributed to some solubilisation of the myofibrillar fraction as a consequence of the presence of salt in the last washing water, and also to the fact that pH and hardness of washing water have not been controlled. The highest decrease was observed in mince from two days' iced sardine. On the other hand, alkali-soluble and stroma proteins increased as the myofibrillar fraction decreased. Stroma proteins resulted mostly as alkali-insoluble proteins. Based on these results it can be suggested that the salt content in the last washing water of sardine minces must be lower than 0.15% in order to avoid certain difficulties in the preparation of surimi and surimi-based products, namely a premature setting of the protein.

SDS-PAGE patterns of myofibrillar fractions, Figures 19.2 and 19.3, show the beneficial effect of washing on the resultant mince. The principal component of myofibrillar proteins, MHC, in minces produced from fresh sardine became more abundant after the first washing and remain unchanged until the end of the process, while the 29 k component was removed. In the minces prepared from iced fish, the myofibrillar fraction (Figs 19.2 and 19.3) was not improved by washing, presumably because of irreversible changes, compared with the unwashed mince. SDS-PAGE of the alkali-soluble fraction displayed many bands which are supposed to be proteolytic breakdown products of myofibrillar proteins.

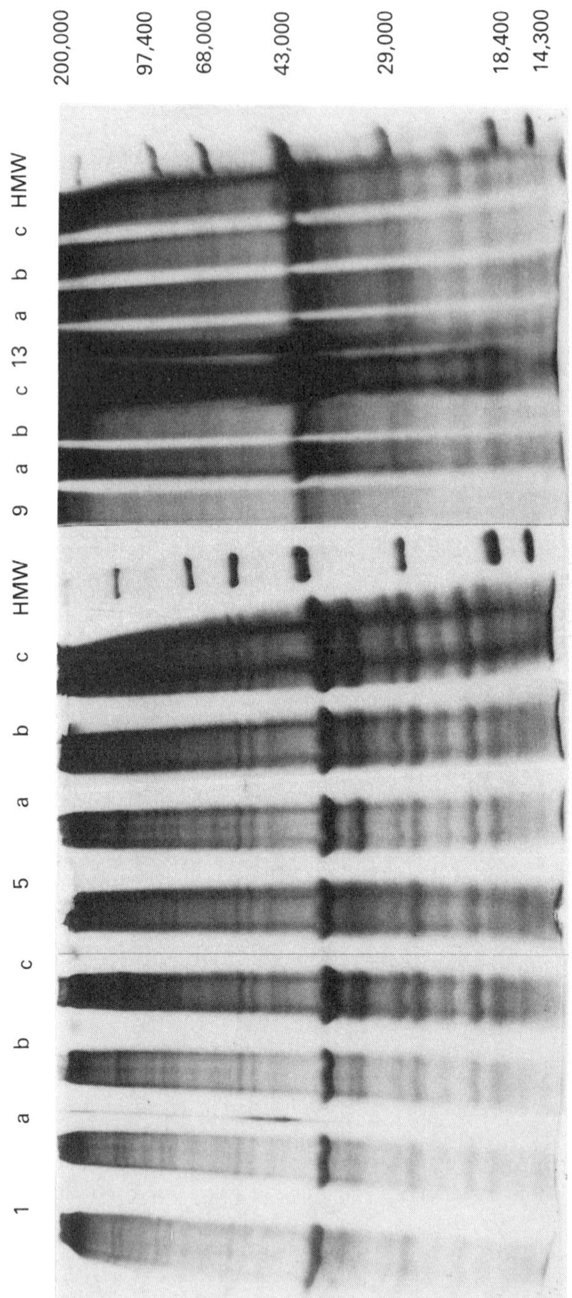

Fig. 19.4 Gradient SDS–PAGE patterns of alkali-soluble fractions. Unwashed minces: lane 1 – fresh sardine; lane 5 – one day iced; lane 9 – two days iced; lane 13 – three days iced. Washed minces: lane a – first washing; lane b – second washing; lane c – third washing.

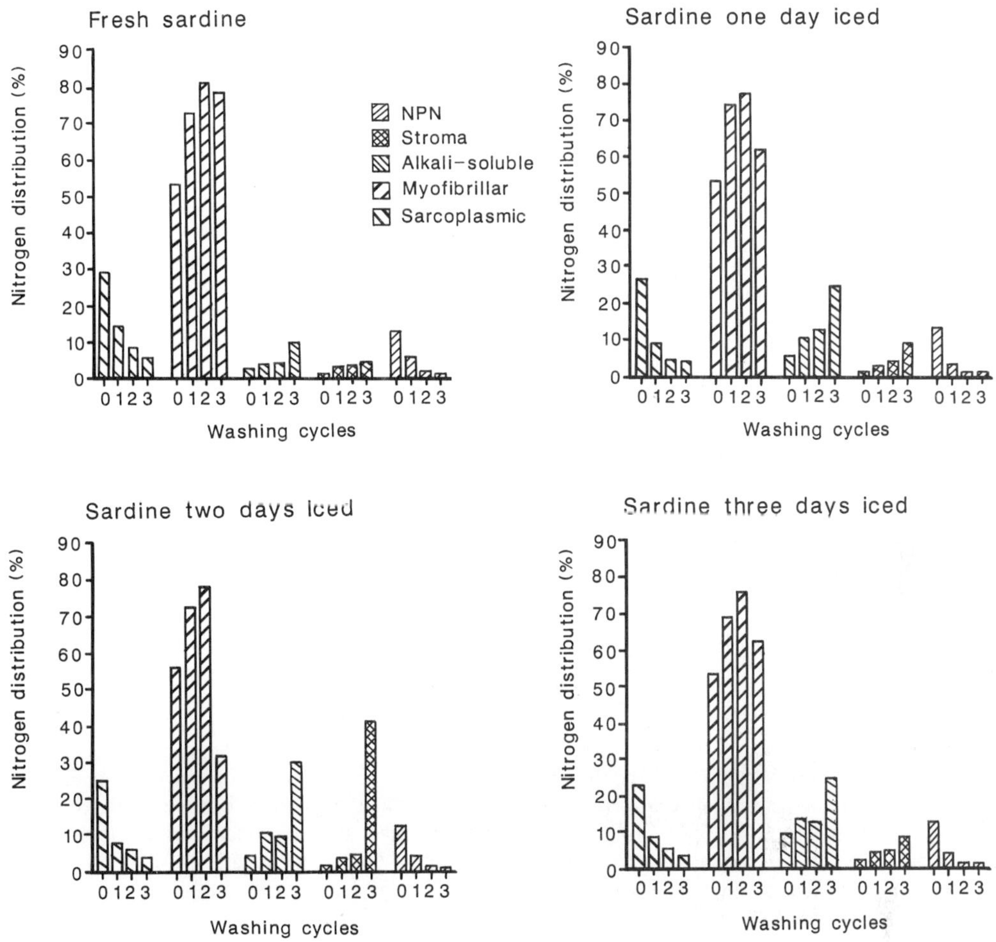

Fig. 19.5 Effect of iced storage and washing cycles on nitrogen distribution of sardine minces.

BIBLIOGRAPHY

Adu G.K., Babbitt J.K. & Crawford D.L. (1983) Effect of washing on the nutritional and quality characteristics of dried minced rockfish flesh. *J. Food Sci.*, **48**, 1053–5.

AOAC (1984) *Official Methods of Analysis*, 14th edn. Association of Official Analytical Chemists, Washington.

Babbitt J.K. (1986) Suitability of seafood species as raw materials. *Food Technol.*, **40**, 97–100.

Grantham G.J. (1981) Minced fish: A review. *FAO Fish. Tech. Pap.* 216.

Hashimoto K., Watabe S., Kono M. & Shiro K. (1979) Muscle protein composition of sardine and mackerel. *Bull. Jap. Soc. Sci. Fish.*, **45**, 1435–41.

Keay J.N. (1986) Surimi – the European perspective. *Infofish Marketing Digest*, **6**, 29–32.

Laemlli U.K. (1970) Cleavage of structural proteins during the assembly of the head of bacteriophage T4. *Nature*, **227**, 680–5.

Lee C.M. (1986) Surimi manufacturing and fabrication of surimi-based products. *Food Technol.*, **40**, 115–24.

Nunes M.L. (1981) Minced fish from two small pelagic species. *Bol. Inst. Nac. Invest. Pescas.*, **9**, 69–79.

Ohnishi M. & Rodger G.W. (1980) Analysis of the salt-soluble protein fraction of cod muscle by gel filtration. In *Advances in Fish Science and Technology* (Ed. by J.J. Connell), pp. 422–8. Fishing News Books, Oxford.

Regenstein J.M. (1986) The potential for minced fish. *Food Technol*, **40**, 101–106.

Rhodes R.J. (1986) The evolving market for surimi-based products in the United States. *Infofish Marketing Digest*. **6**, 17–19.

Suzuki T. (1981) *Fish and Kril Protein: Processing Technology*. Applied Science Publishers, London.

CHAPTER 20

Influence of chilled and frozen storage of sardines on the gelling properties of surimi

A. Leinot, C. Ozenne and J.C. Cheftel

INTRODUCTION

The surimi industry in France is still in a developmental stage. Two firms, Scoma (Lorient) and Comapêche (St Malo) are producing surimi from underutilised species including sardine (*Sardina pilchardus*), horse mackerel (*Trachurus trachurus*), pout (*Gadus luscus*) and blue whiting (*Micromesistius poutassou*).

Roussel & Cheftel (1988) and Roussel (1988) investigated the influence of various heat processing temperatures on the final texture of kamaboko gels prepared from sardine stored for 1–3 days in ice. The textural characteristics of kamaboko gels from very fresh sardine were markedly enhanced by a gel-setting step (37°C for 30 minutes) while incubation at 60°C for 60 minutes led to soft gels ('modori' effect). The texture of kamaboko gels from less fresh sardine was not improved after a gel-setting step and became softer as a result of the modori effect.

Ishikawa *et al.* (1977), Ishikawa (1978) and Ishikawa *et al.* (1979) studied the influence of fish handling and freshness, and of harvest season, on the textural characteristics of kamaboko gels from Japanese sardine (*Sardinops melanosticta*). Gels prepared from sardine stored for 3 days in ice indicated a severe drop in textural characteristics (gel strength and folding score). These textural characteristics were inferior in gels prepared from sardine caught in the summer compared with gels from sardine caught during the winter. The initial pH of sardine flesh appeared to influence markedly the quality of the final gel. Sardine surimi was found to respond to the gel setting and modori phenomena.

Since variations in the quality characteristics of sardine surimi were observed, it was considered of interest to further investigate the effects of sardine freshness (varying length of chilled or frozen storage) on the gelling properties of surimi.

MATERIALS AND METHODS

Material

Sardines (*Sardina pilchardus*) were caught near Sète in December 1988. Their length was 18 ± 3 cm, their weight 35 ± 6 g and their fat content 7.5 ± 0.1%. The sardines were stored in ice on board for about 4 hours, then stored on land in ice (for a maximum period of 90 hours), or at 10°C (70 hours maximum) or 20°C (50 hours maximum).

Sardines kept 20 or 60 hours in ice on land were also frozen at −40°C by spraying liquid nitrogen, stored at −20°C for up to 3 months and then thawed in tap water.

Sardines used for solubility experiments were caught in December 1989. Their length was 15 ± 1 cm, their weight 33 ± 6 g and their fat content 3.8 ± 0.1%. Ten of these sardines were kept alive for 4 hours by placing them in a bin containing sea water with air bubbling.

Preparation of surimi

Fish were beheaded, gutted, washed in tap water, filleted by hand, and minced using an A 707/A Kenwood mincer (Havant, UK) fitted with a perforated plate with 4 mm diameter holes. Fish and fillets were maintained below 10°C at all times. The mince was washed three times according to Suzuki (1981), using five parts of solution to one part mince (temperature <10°C). Gentle manual agitation of the mince for 5 minutes was followed by a period of 5 minutes without agitation to allow for fat decantation and removal. The mince was partially dewatered between washes and the final dewatering was carried out using a Calor centrifugal strainer (Lyon, France). The mince was then mixed with sucrose and sorbitol (each to a concentration of 4 g per 100 g of washed mince) for 5 minutes in a Magimix 1800 (Robot-Coupe, Bagnolet, France), keeping the temperature below 10°C. The resulting surimi was packed in polyethylene bags (with the open end folded over and unsealed) and kept at −40°C for between 1 day and 4 months.

Preparation of kamaboko gels

Frozen surimi (100 g, water content around 77%) was thawed at 4°C (air) for 15 hours and mixed for 2 minutes in a cold room at 4°C with 3 g NaCl in the Magimix mixer. The resulting paste was introduced into PVDC casings (31.5 mm in diameter, 40 μm thick; Krehalon, Deventer, the Netherlands) using a manual stuffer. Kamaboko gels (water content around 75%) were subjected to a 30 minutes setting in a 37°C water bath, cooked for 50 minutes in a 90°C water bath, cooled for 30 minutes in running water and finally kept for 15 hours in a refrigerator at 4°C before analysis.

In the case of surimi from frozen sardine, 5 g potato starch per 100 g surimi were also added (in order to avoid water drip after cooking). The resulting paste was mixed again for 4 minutes while the final water content was adjusted to 75% with tap water.

Texture evaluation of kamaboko gels

The penetration test was carried out with a Stevens LFRA (St Albans, UK) texture analyser. Gel samples (31.5 mm diameter; 20 mm height) were penetrated with a 5 mm diameter flat cylindric probe at 60 mm/min. Gel strength was measured as the product of breaking force at rupture (newton) and distance at rupture (millimetre); 10 to 15 measurements were carried out on 3–4 slices for each kamaboko gel.

The folding test was carried out on carefully prepared gel slices (31.5 mm diameter; 3 mm height). Each slice was double folded and quality assessed according to the method given by the White Fish Authority (Anon. 1978). The folding score was expressed using a seven-point scale. The maximum value of 7 corresponded to the absence of cracking when the double folded sample was pressed strongly with the fingers, while the value of 6 corresponded to a cracking under the same conditions (but to an absence of cracking when the single folded sample was pressed). Measurements were carried out on at least five samples of the same gel.

Analytical methods

Water content was determined by drying 2 g samples at 103°C for 24 hours. Fat content was measured by petroleum ether extraction (Osborne & Voogt 1978), except that a Kumagawa extraction unit was used. Protein content was determined using the Lowry method; bovine serum albumin was used as a standard. All measurements were carried out in duplicate.

Determination of the solubility of muscle proteins at various ionic strengths

The distribution of sardine muscle proteins (mince from fillets) was determined according to their solubility at various ionic strengths, and used as another quality index. The water and salt extraction procedure was similar to that of Webb *et al.* (1976): 5 g of mince were placed into a glass jar and 100 ml of 0.05M $NaHCO_3$ (pH 7.0, μ = 0.05, 4°C) were added. After homogenisation in an Ultra-Turrax mixer for 30 seconds, the mixture was centrifuged at 8000 *g* for 20 minutes. The supernatant containing water soluble proteins and non-protein nitrogen was collected. The residue was extracted with 100 ml of 0.05M $NaHCO_3$ (pH 7.0, 4°C) containing 0.55M NaCl (final μ = 0.60). The supernatant contained salt soluble proteins, while the final residue contained salt insoluble proteins. Temperature was maintained below 10°C in a cold room throughout the extraction procedure. The protein content was determined in the water soluble, salt soluble and salt insoluble fractions. All measurements were carried out in duplicate. Solubility of muscle proteins at the start of storage was obtained from 10 sardines kept alive for 4 hours after fishing.

Sensorical assessment of the freshness index

The sensorial assessment of fresh (and in some cases frozen) sardines was carried out using the EC grading scheme based on appearance, odour and texture of fish (EC 1976). The EC index varied between 0 (not fresh) and 3 (very fresh). The fish were classified into four grades: 3 > extra > 2.7; 2.7 > A > 2; 2 > B > 1; 1 > C > 0. All the measurements were carried out on at least 20 fish.

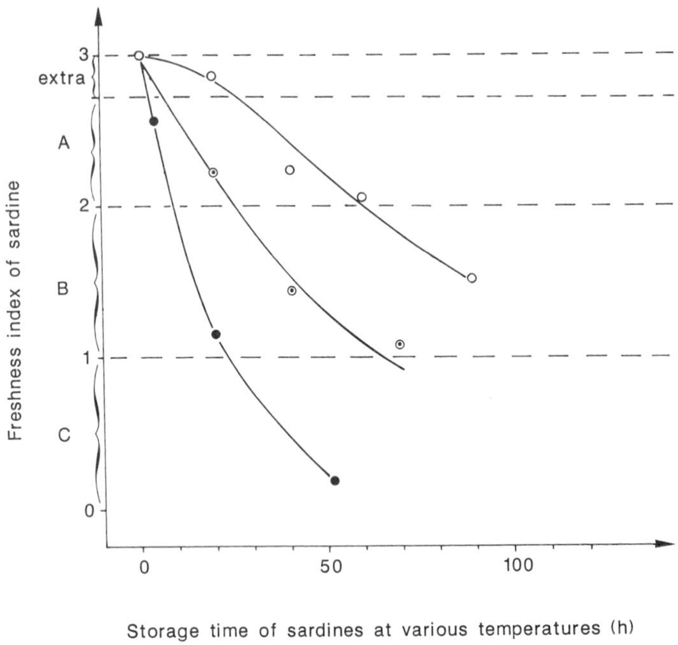

Fig. 20.1 Influence of storage time on the freshness index of sardine. Storage at 0°C (○); 10°C (⊙) or 20°C (●). Storage time after landing the fish.

RESULTS AND DISCUSSION

Freshness index of sardines

The change in the freshness index of sardine as a function of the length of storage at various temperatures (0°C, 10°C or 20°C) is shown in Figure 20.1 (the storage time is expressed as hours after landing the fish). As expected, the freshness index decreased faster at higher storage temperatures. The freshness index of sardine stored at 20°C decreased to almost 0 after 50 hours, to 1.1 after 70 hours at 10°C and to 1.5 after 90 hours at 0°C.

The freshness index of sardine kept in ice for 20 h (extra grade), then stored at −20°C for 7 days, and finally thawed in tap water, decreased to 1.8 (not shown in Fig 20.1). A longer frozen storage period of 4 months resulted in a slightly lower freshness index (1.5).

Gel strength and folding test of kamaboko gels

The resistance to the folding test and the gel strength of kamaboko gels prepared from sardine stored for various periods of time at 0°C, 10°C or 20°C are illustrated in Figures 20.2 and 20.3. It can be seen from Figure 20.2, where the folding score is plotted against gel strength, that the latter represents a more sensitive criterion than the folding score for assessing the gelling properties of surimi.

The change in the gel strength of the above mentioned kamaboko gels as a function of storage time and temperature is illustrated in Figure 20.3. Sardine stored at 20°C displayed a rapid decrease in kamaboko gel strength, and lost gel forming ability within 40 hours of storage. Sardine stored for 40 hours at 10°C or 60 hours at 0°C gave kamaboko gels with a 43–53% reduction in gel strength but with a folding score of 6 or above.

The resistance to the folding test and the gel strength of kamaboko gels prepared from extra grade or A/B grade sardine that was frozen and stored at −20°C for up to 3 months are shown in Figures 20.4 and 20.5. As for gels without starch (Figs. 20.2 and 20.3), gel strength represents a more sensitive criterion than the folding score for quality assessment.

It is not possible to compare directly the kamaboko gels from Figures 20.3 and 20.5 since the latter contain 5% potato starch (in order to avoid the drip that normally occurs on cooking of gels from frozen sardine). Indeed, the addition of 5% potato starch reduced markedly the gel strength of kamaboko gels, as can be seen from a comparison of control gels (from sardine stored for 20 and 60 hours in ice) in Figures 20.3 and 20.5. This may be due to the simultaneous decrease in protein concentration (from 11.5% to 8%).

Figure 20.5 shows that the gel strength of kamaboko

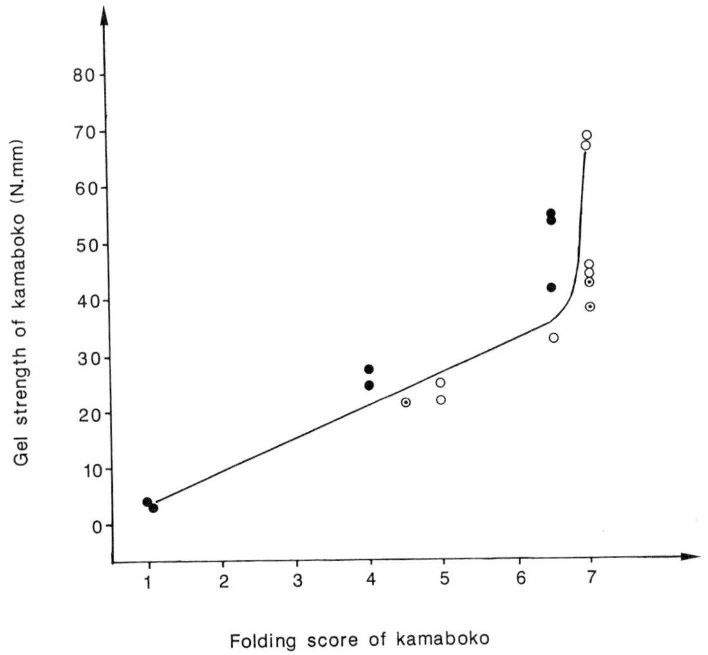

Fig. 20.2 Correlation between the gel strength and the folding score of kamaboko. Storage at 0°C (○); 10°C (⊙) or 20°C (●). All gels prepared without potato starch.

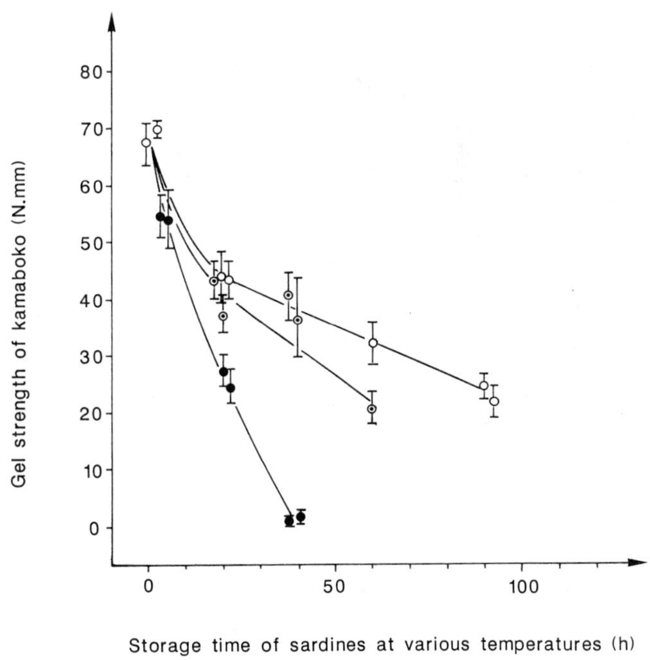

Fig. 20.3 Influence of sardine storage time on the gel strength of kamaboko. Storage at 0°C (○); 10°C (⊙) or 20°C (●). Storage time after landing the fish. All gels prepared without potato starch.

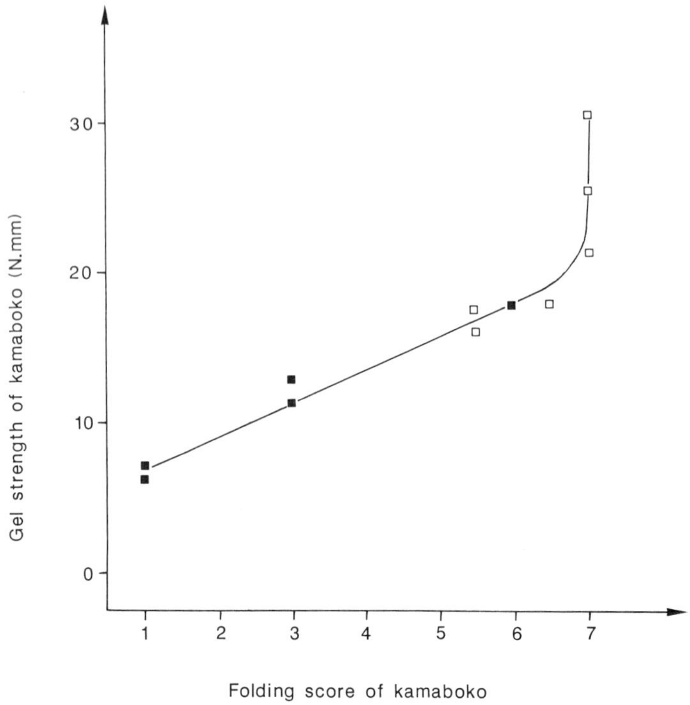

Fig. 20.4 Correlation between the gel strength and the folding score of kamaboko. Extra grade (□) or A/B grade (■) sardine stored at −20°C. All gels prepared with 5 g of potato starch per 100 g of surimi.

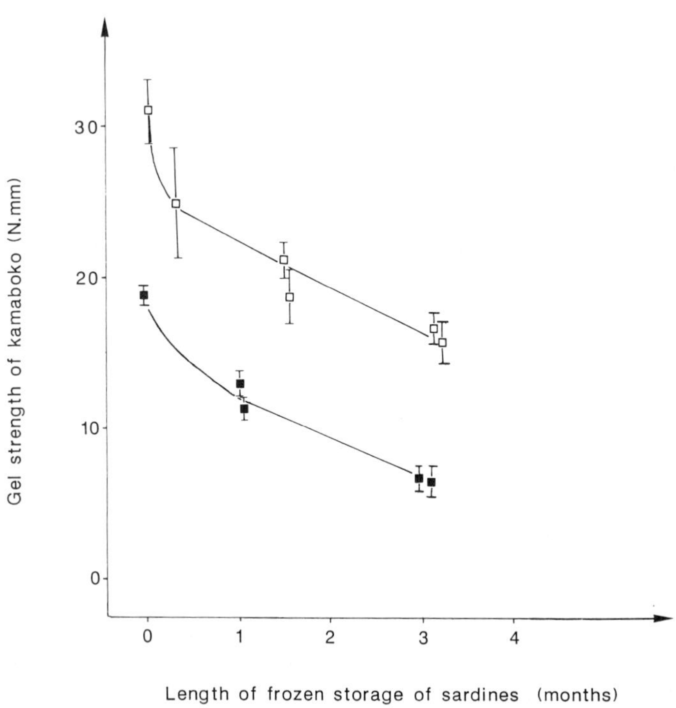

Fig. 20.5 Influence of the length of frozen storage of sardine on the gel strength of kamaboko. Extra grade (□) or A/B grade (■) sardine stored at −20°C. The control 0 time sardines were not frozen. All gels prepared with 5 g of potato starch per 100 g of surimi.

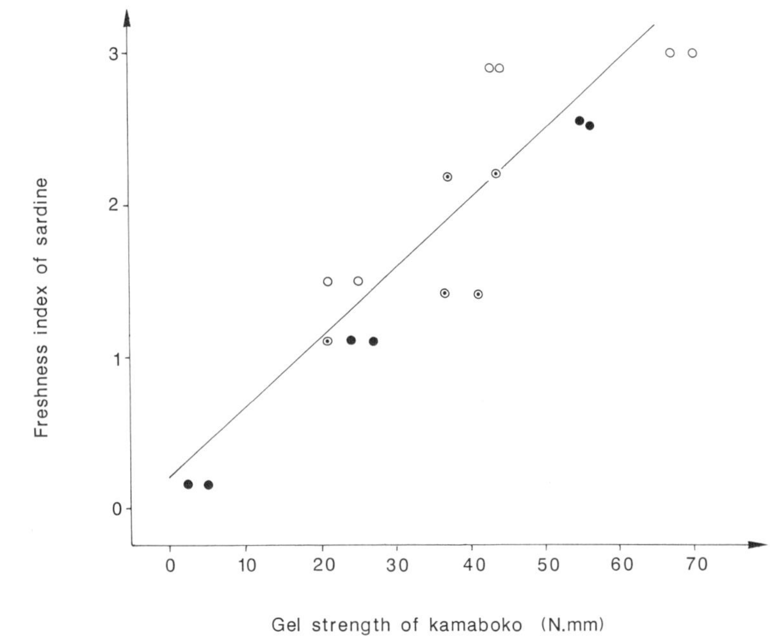

Fig. 20.6 Correlation between the freshness index of chilled sardine and the gel strength of the resulting kamaboko. Storage at 0°C (○); 10°C (⊙) or 20°C (●). All gels prepared without potato starch.

from extra grade sardine decreased by 15% when the sardine was kept frozen at −20°C for 3 days, and by 45% when the frozen storage reached 3 months. The gel strength of kamaboko from A/B grade sardine decreased by 62% after 3 months of frozen storage. The score at the folding test remained 6 or above until extra grade sardine was kept frozen for about 45 days, while A/B grade sardine kept frozen for 30 days already gave very poor quality gels (folding score = 3).

These results appear to indicate that it remains possible to prepare surimi of acceptable quality from frozen sardine provided extra grade sardines are used for freezing, and provided frozen storage (−20°C) does not exceed 45 days. Such conclusions should be checked with sardine caught at other seasons. While the use of frozen sardine may be advantageous from the standpoint of price control, the necessity to add starch and the risk of lipid oxidation represent serious drawbacks.

Correlation between the freshness index of sardine and the gel strength of kamaboko gels

A correlation can be observed between the freshness index of sardine and the gel strength of kamaboko gels, for sardine stored at 0°C, 10°C or 20°C (Fig. 20.6). The correlation factor is close to 0.90. Such a correlation may be useful to predict the gelling properties of surimi from the sensorial evaluation of a batch of purchased sardine.

Arrhenius relationship between the rate constant (K) of variation of kamaboko gel strength and sardine storage temperature

The five curves from Figures 20.3 and 20.5 can be plotted on semi-logarithmic paper as GS = exp (−K . t), where GS is gel strength and t the storage time of sardine.

The rate constants (K) of variation of kamaboko gel strength were calculated as 29.6×10^{-7}, 45.1×10^{-7} and $192 \times 10^{-7}\ s^{-1}$ at 0°C, 10°C and 20°C respectively; (the correlation factor was higher than 0.90 in all cases).

The rate constants (K) of variation of kamaboko gel strength as a function of the frozen storage of extra grade and A/B grade sardine at −20°C were calculated as above and found to be equal to $0.732 \times 10^{-7}\ s^{-1}$ and $1.183 \times 10^{-7}\ s^{-1}$ respectively; (the correlation factor was higher than 0.95 in both cases).

The logarithm of the rate constant (K) of variation of kamaboko gel strength can be plotted against the reciprocal of the absolute temperature of sardine storage according to the Arrhenius equation (Fig.

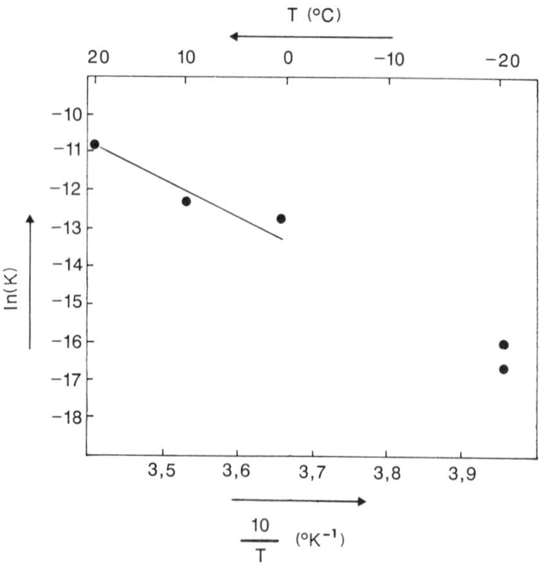

Fig. 20.7 Arrhenius plot of the rate constant (K) of variation of kamaboko gel strength versus sardine storage temperature (°K). Gels prepared with or without potato starch.

20.7). An activation energy of 14.6 kcal/mole can be calculated between 0°C and +20°C; (the correlation factor is close to 0.94).

Solubility of muscle proteins at various ionic strengths

The change in solubility of muscle proteins as a function of the length of storage at 0°C is shown in Figure 20.8. Water soluble and, especially, salt soluble protein fractions were high immediately after death (as measured on sardines kept living for 4 hours after fishing) and decreased markedly within 20 hours of storage.

The water soluble fraction then decreased slowly during longer storage, while the salt soluble fraction increased slowly. The increase in the salt insoluble fraction after 20 hours of storage at 0°C accounts for the decrease in the salt soluble protein fractions.

These rapid changes in protein solubility immediately after death could be due to aggregation phenomena related to the onset of rigor mortis. Rigor mortis was observed to take place within 4 hours of iced storage after fishing.

Half the proteins were found to be salt insoluble in 'fresh' sardines (*Sardinops melanosticta*) by Ehira & Uchiyama (1987), but measurements were not done immediately after death. In a previous study carried out in this laboratory, Roussel (1988) observed little change in the pH value of sardine muscle (from pH 6.4 to 6.2) during the first 24 hours of iced storage. This suggests that the rapid protein solubility variations observed in Figure 20.8 are not due to pH changes, but rather to rigor mortis.

CONCLUSIONS

The following conclusions can be drawn from the data reported above.

Sardine freshness markedly influences the gelling properties of surimi. A significant correlation is observed (r = 0.9) between the freshness index of sardine (stored at 0°C or above) and the gel strength of kamaboko. This makes it possible to predict the potential gelling properties of surimi from the freshness evaluation of whole sardine.

Keeping sardine in ice for 20 hours caused a 35% reduction in kamaboko gel strength. The score in the folding test, however, remained at 6 or above until the sardine was kept for more than 60 hours in ice. Gel strength therefore represents a more sensitive criterion than the folding score for assessing the quality of kamaboko gels.

When extra grade sardine (20 hours in ice) was stored at −20°C for 45 days before preparation of surimi, the gel strength of the resulting kamaboko was found to decrease by 35%. The folding score remained at 6 or above, but decreased progressively during longer periods of frozen storage. It therefore remains possible to prepare kamaboko gels of acceptable quality from frozen sardine provided it is extra grade at the time of freezing, and provided frozen storage (−20°C) does not exceed 45 days.

The rate constants (K) of variation of kamaboko gel strength as a function of the length of sardine storage at 20°C, 10°C, 0°C and −20°C were calculated and found to decrease markedly when the sardine storage temperature decreased. An Arrhenius plot of K as a function of storage temperature gave an activation energy of 14.6 kcal/mole in the temperature range 0°C to +20°C.

The effects of seasonal variation on the quality of kamaboko gels from chilled or frozen sardine are being investigated.

ACKNOWLEDGEMENTS

This study has benefited from a grant, No. 88/5571032, given by IFREMER, Nantes. We are grateful to Dr T. Akahane for his advice and comments,

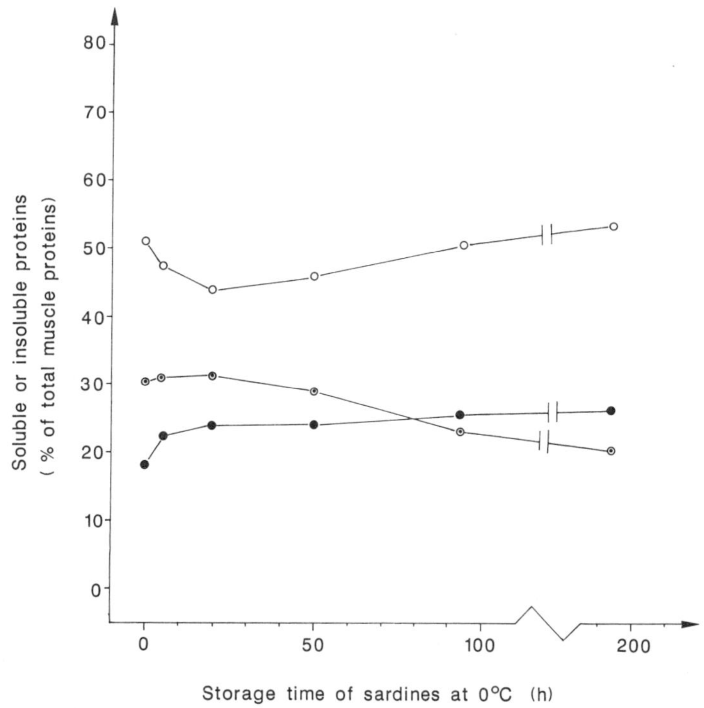

Fig. 20.8 Influence of sardine storage time on the solubility of muscle proteins at different ionic strengths. Water soluble proteins (⊙); salt soluble proteins (○); salt insoluble proteins (●). Total muscle proteins for each storage time = water soluble protein + salt soluble + salt insoluble protein. Sardine stored at 0°C or processed immediately after death. Sardine were caught in December 1989 (fat content: 3.8%).

and to the Sathoan fish producers Cooperative (Sète) for providing sardine.

BIBLIOGRAPHY

Anon. (1978) *Quality Control Test Procedure for Surimi Production.* Field report 625 of White Fish Authority Industrial Development Unit, Kingston upon Hull.

EC (1976) EC Regulation No 103176 (Council of 19 January, 1976).

Ehira S. & Uchiyama H. (1987) Determination of fish freshness using the K value and comments on some other biochemical changes in relation to freshness. In *Seafood Quality Determination* (Ed. by D. E. Kramer & J. Liston), pp. 137–52. Elsevier Science Publishers, Amsterdam.

Ishikawa S. (1978) Fish jelly product (kamaboko) and frozen minced meat (frozen surimi) made of sardine. II, Influence of the temperature during manufacturing process on the jelly strength of kamaboko. *Bull. Tokai Regional Fisheries Res. Lab.*, **94**, 37–44.

Ishikawa S., Nakamura K. & Fujii Y. (1977) Fish jelly product (kamaboko) and frozen minced meat (frozen surimi) made of sardine. I, Freshness and handling of the fish material affecting its kamaboko forming ability. *Bull. Tokai Regional Fisheries. Res. Lab.*, **90**, 59–66.

Ishikawa S., Nakamura K., Fujii Y., Yamano G., Sugiyama T., Shinozaki K., Tobita K. & Yamaguchi Y. (1979) Fish jelly product (kamaboko) and frozen minced meat (frozen surimi) made of sardine. III, Influence of the treatment methods for materials just after catch on the kamaboko forming ability of sardine meat. *Bull. Tokai Regional Fisheries Res. Lab.*, **99**, 31–42.

Osborne D.R. & Voogt P. (1978) *The Analysis of Nutrients in Foods*, pp. 155–65. Academic Press, London.

Roussel H. & Cheftel J.C. (1988) Characteristics of surimi and kamaboko from sardines. *Intern. J. Food Sci. Technol.*, **23**, 607–23.

Roussel H. (1988) *Préparation et propriétés gélifiantes du surimi de sardine. Influence de la fraîcheur du poisson, du procédé technologique et de divers additifs chimiques sur les caractéristiques du surimi et des gels protéiques.* Thèse, pp. 74–6. Université des Sciences et Techniques du Languedoc, Montpellier.

Suzuki T. (1981) *Fish and Krill Protein Processing Technology*, pp. 101–103 and 134–7. Applied Science Publishers, London.

Webb N.B., Hardy E.R., Giddings C.G. & Howell A.J. (1976) Influence of mechanical separation upon proximate composition, functional properties and textural characteristics of frozen Atlantic croaker muscle tissue. *J. Food Sci.*, **41**, 1277–81.

CHAPTER 21

By-products from pelagic fish

N. Urdahl

INTRODUCTION

This subject has been dealt with by many authors, including Lovern (1950), Lovern (1955), Stansby & Dassow (1963), Stansby (1967), Windsor & Barlow (1981) and Ledward *et al.* (1983).

For an industrial company the principal products are those that give the highest contribution to the income of the company, while by-products are those of less importance. In the fishing industry the situation is somewhat different, since principal products are often defined as products for human consumption while everything else is a by-product. Using this definition this chapter deals with the following utilisations of pelagic fish:

- fish as a fertiliser
- fish meal and fish oil
- fish as feed
- fish silage
- fish hydrolysate
- fish as raw material for chemical and pharmaceutical products.

Different forms of fish protein concentrate (FPC), surimi and even minced fish have been listed as by-products, but these groups of products have been left out here.

FISH AS A FERTILISER

From ancient times fish offal and even dried fish have been used as fertilisers on farms in coastal districts. Fish solubles, containing water soluble compounds of the fish tissue, have also been used for this purpose (Stansby 1963). Today some people (those who can afford it) prefer so-called biodynamically grown foods. In such production, only natural organic fertilisers of biological origin are used. Perhaps this may again open up a market for specialised fish products as a fertiliser, but the volume will probably not be high.

FISH MEAL AND FISH OIL

Among the by-products, fish meal and fish oil are the most important. It is difficult to fix a time when these products first came into use, but in a description of a certain district on the west coast of Norway from 1762 we can read (Strøm 1762):

> 'People around here cook heads and guts of herring to obtain an oil. Used as fuel in lamps, this oil is much better than fish liver oils'.

A footnote to the paper tells how in 1760 enormous quantities of herring were caught, but the price was low. Some people started to cook the whole herring to extract the oil. 'From 6–8 barrels of herring they may obtain one barrel of beautiful oil, 'according to the author.

The manufacture of fish meal and fish oil developed into an industry during the last decades of the nineteenth century and is still developing.

Manufacturing process for fish meal and oil

In Figure 21.1 the manufacturing process is shown as a flow sheet. From a storage tank the raw material is fed to the cooker for coagulation of the proteins. Via a strainer the cooked fish is fed to the press and out come presscake and pressliquor. The pressliquor is treated in decanters and separators to take out suspended solids (sludge) and oil. The water (stickwater) contains water soluble proteins, minerals, vitamins etc. and after concentration in a multi-stage evaporator it is mixed with

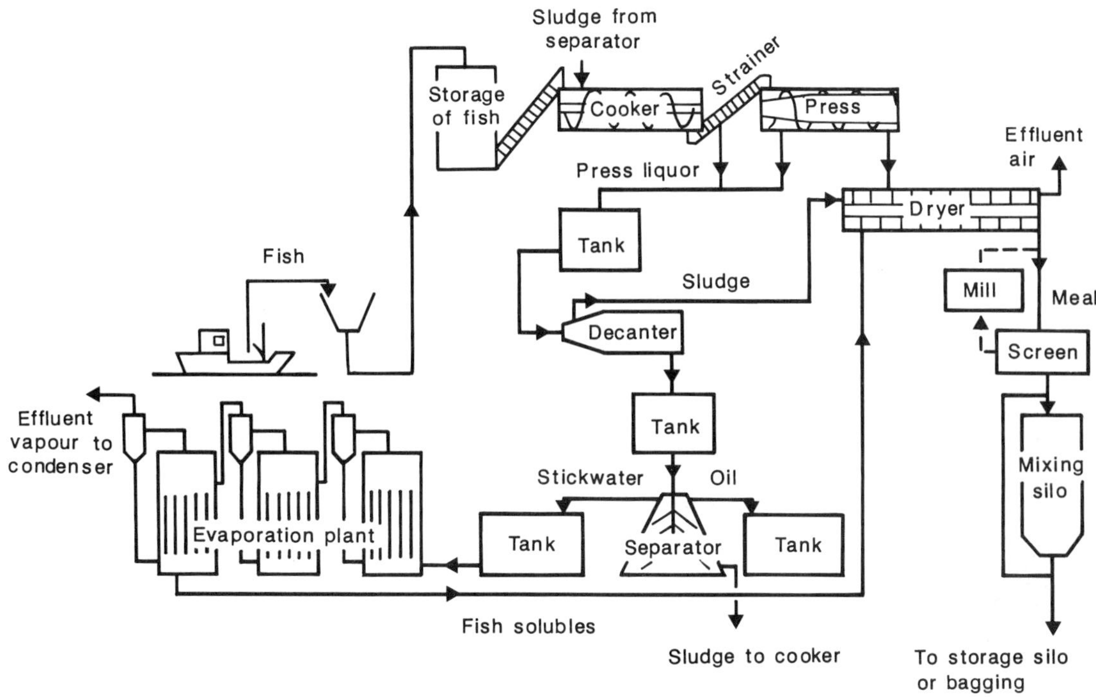

Fig. 21.1 Simplified flow sheet describing the fish meal process.

the presscake and dried. Usually 20–30% of the dry matter of the fish is found in the solubles. After drying, the fish meal passes through a screen and a mill, and the meal is mixed and often pelletised prior to storage.

The flow sheet shows the general principles of the process, but in practice there are several modifications. For instance the effluent gases from the dryer may be used as a source of energy in a waste heat evaporator and substantial energy savings are achieved. On paper the process seems to be quite simple, but most of the unit operations in chemical engineering are included. In modern factories the pieces of equipment are supplied with process regulators and the different unit operations are linked together by advanced control systems.

Composition and types of fish meal

Types of fish meal may refer to meal produced from different species of fish, but may also refer to differences in processing. Presscake meal is dried presscake where at least 20% of the dry matter of the fish is left out. This will be reflected in the composition of the meal. Full meal or whole meal means that all dry matter of the fish is found in the meal. Today most fish meals in international trade are full meal. The expression brown fish meal usually refers to meal from pelagic fish while white fish meal is produced from lean fish or offal from lean fish. White fish meal is often presscake meal, especially if produced on board factory ships.

The gross composition of some types of fish meal is shown in Table 21.1 (Barlow & Windsor 1984). The upper section of the table is meal from pelagic species while the lower section deals with white fish meal. There is no marked difference in composition, but normally the content of both protein and fat are lower in white fish meal and this is compensated by a higher content of ash. This may also be influenced by the content of fish solubles, however, and the incorporation of fish offal.

Moisture contents between 5% and 10% are quite normal. The salt content in the body fluids of all fish is nearly the same, but the water content of the fish differs from species to species and from season to season. Consequently the salt content of fish meals may differ even if they are full meals. Use of refrigerated sea water to preserve the catch may also increase the salt content of the meal.

Table 21.1 Average figures for protein, oil, moisture and salt in fish meals from various origins (Windsor & Barlow 1981).

Type of meal	Origin	Protein, % (N × 6.25)	Oil, % (diethyl ether)	Moisture %	Salt %
Herring type	UK	72.0	9.7	7.7	1.4
	Norway	71.9	7.5	8.4	1.3
	Denmark	—	—	—	1.6
Capelin	Iceland	66.4	11.3	8.5	2.1
Anchovy/pilchard	South Africa	66.1	9.7	10.1	2.9
	South W. Africa	65.0	9.3	9.4	3.1
Anchovy	Peru	66.4	9.7	8.6	—
White fish	South Africa	64.0	6.3	7.6	2.4
	Iceland	65.8	3.4	7.0	1.9
	UK	64.5	4.5	10.0	—
Fish	Germany	60.4	8.5	10.5	—
	Belgium	65.0	6.8	—	3.5

Table 21.2 Mean value total amino acid composition (g/16g N) of fish meals (determined mainly by ion-exchange chromatography) (Windsor & Barlow 1981).

	Herring meals	Anchovy meals	Pilchard and Maasbanker meals	Tuna (mixed species) offal meals	Menhaden meals	White fish meals
Lysine	7.73	7.75	7.94	7.30	7.56	6.90
Methionine	2.86	2.95	2.71	2.75	2.82	2.60
Cystine	0.97	0.94	0.95	0.79	0.90	0.93
Tryptophan	1.15	1.20	1.02	1.05	1.07	0.94
Histidine	2.41	2.43	3.02	3.41	2.32	2.01
Arginine	5.84	5.82	5.95	6.43	6.04	6.37
Threonine	4.26	4.31	4.38	4.34	3.97	3.85
Valine	5.41	5.29	5.41	5.31	5.10	4.47
Isoleucine	4.49	4.68	4.48	4.46	4.40	3.70
Leucine	7.50	7.62	7.30	7.20	7.14	6.48
Phenylalanine	3.91	4.21	3.91	4.10	3.95	3.29
Tyrosine	3.13	3.40	3.23	3.28	3.22	2.60
Aspartic acid	9.10	9.49	9.37	9.30	9.07	8.54
Serine	3.82	3.84	4.27	4.18	3.61	4.75
Glutamic acid	12.77	12.96	12.92	11.93	12.70	12.79
Proline	4.15	4.17	4.52	5.43	4.58	5.34
Glycine	5.97	5.62	6.92	8.15	6.78	9.92
Alanine	6.25	6.31	6.17	6.76	5.94	6.31
Crude prot. %	73.6	65.4	65.4	53.24	62.01	65.01
Moisture %	6.93	8.01	9.0	6.20	8.25	8.49

Fish meal is mainly used as a protein source in compound feeds and the amino acid composition is very important. Table 21.2 shows the amino acid composition in different fish meals (Windsor & Barlow 1981). We may conclude that the content of the different amino acids calculated as a percentage of protein is nearly the same in all fish meals, irrespective of species.

Table 21.3 shows the content of some of the most important minerals in fish meal (Windsor & Barlow 1981). Another survey is given by Julshamn *et al.* (1978). From Table 21.3 it seems as though a high

Table 21.3 Content of essential minerals in fish meal (Windsor & Barlow 1981).

Element	Peruvian anchovy	Norwegian herring	White fish	Menhaden
Ash, %	15.4	10.1	20.0	18.0
Calcium, %	3.95	1.95	8.0	5.26
Phosphorus, %	2.60	1.50	4.8	2.98
Sodium, %	0.87	0.42	n.a.	0.34
Magnesium, %	0.25	0.11	0.15	0.14
Potassium, %	0.65	1.20	0.9	0.72
Iron, ppm	246	150	300	438
Copper, ppm	10.6	5.4	7	11.4
Zinc, ppm	111	120	100	151
Manganese, ppm	9.7	2.4	10	35.6
Iodine, ppm	n.a.	2.4	n.a.	n.a.
Selenium, ppm	1.39	2.78	1.50	2.22

content of ash gives a high content of calcium and phosphorus. When looking at potassium and selenium the situation seems to be the opposite, but this may just reflect differences in the inclusion of fish solubles.

Utilisation of fish meal

The traditional utilisation of fish meal is as a protein source in compound feeds, mainly for poultry, pigs and to some extent ruminants. In this context fish meal competes with vegetable proteins and crystalline amino acids. Today most feed mills use least cost programs for the formulation of compound feeds, whereby they can take the value of all nutrients in fish meal and other ingredients into account. Up to 10% of fish meal may be included in compound feeds for domestic animals, depending on recipes and prices. Further details are available from literature on this subject (FAO 1970; IAFMM 1972; Pike 1975a and b; Barlow & Pike 1977; Miller 1978; Miller *et al.* 1981; Gulbrandsen 1984; Opstvedt 1985).

Fish farming has grown as an industry during recent years and fish meal is one of the principal ingredients in fish feeds, especially dry feeds. For this purpose fish meal can only partly be replaced by vegetable proteins (Pike 1989).

Today drainage of nitrogen to fresh water and sea water recipients is a problem in all countries with intensive agriculture and animal husbandry. This may lead to restrictions on how to make up compound feeds; for instance, it may require well-balanced proteins of high digestibility. This may strengthen the position of fish meal as a protein source in animal feeds.

Table 21.4 Fish meal – World production (1000 t).

	1983	1984	1985	1986	1987
Chile	828	1023	1125	1282	1081
Denmark	322	319	289	308	264
Ecuador	22	148	256	206	117
Iceland	68	172	175	178	159
Japan	1133	1263	1167	1179	950
Mexico	64	56	78	85	104
Norway	347	287	238	194	186
Peru	252	568	717	973	821
South Africa	172	115	115	124	260
Thailand	205	223	220	203	217
USA	490	455	474	407	490
USSR	605	674	657	747	729
Other	775	777	768	772	n.a.
Total world	5283	6080	6279	6658	n.a.

World production of fish meal and some of the main producing countries are listed in Table 21.4 (Bowman 1988). On average the world production is 6–7 million tonnes, which requires 35–40 million tonnes of fish as raw material for fish meal. Some of the main catching areas are shown in Figure 21.2.

Composition and utilisation of fish oils

In some countries this type of industry is still called the fish oil industry, indicating oil as the principal product, but in fact today it is a by-product of the production of fish meal.

Some important technical and physical characteristics of fish oils are listed in Table 21.5 (Urdahl & Nygård 1982). A typical fish oil contains less than 0.5% water and sludge and normally less than 2% unsaponifiable substances. Usually 95% of the fish oil consists of triglycerides. As shown in Table 21.6, fish oils may contain a broad spectrum of fatty acids; small amounts of branched fatty acids and fatty acids with uneven numbers of carbon atoms are special for fish oils (Young 1986a).

The approximate content of some principal fatty acids of typical fish oils are listed in Table 21.7 (Young 1986a). The first 6 oils may be looked on as one group originating from fish caught in the Atlantic Ocean, the North Sea and the Barents Sea. The second group (the other oils in the table) are produced from raw materials caught in the Mexican Gulf and in waters outside South Africa, Japan and Latin America. The first group is characterised by a high content of monoenic fatty acids and a moderate content of polyunsaturated fatty acids. The second group is charac-

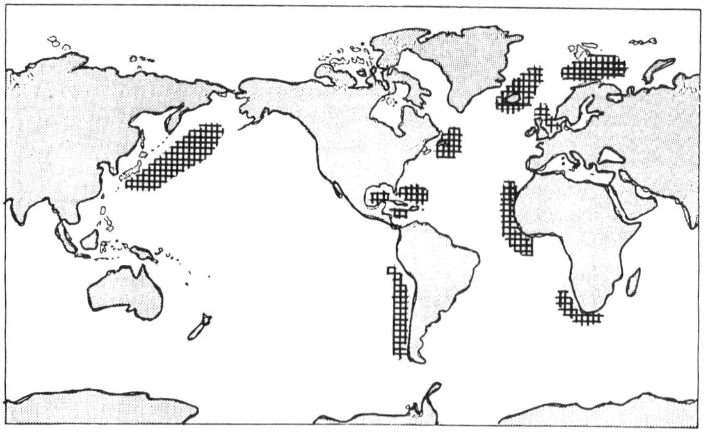

Fig. 21.2 Important areas for catching of fish as raw material for fish meal and fish oil.

Table 21.5 Physical characteristics of fish oils.

Specific heat, cal/g	0.50–0.55
Heat of fusion, cal/g	ca.54
Calorific value, cal/g	ca.9500
Slip melting point, °C	10–15
Flash point, °C	
as triglycerides	ca.360
as fatty acids	ca.220
Boiling point, °C	>250
Specific gravity:	
15°C	ca.0.92
30°C	ca.0.91
45°C	ca.0.90
Viscosity, cp:	
20°C	60–90
50°C	20–30
90°C	ca.10

Table 21.6 Fatty acids found in fish oils.

12:0	18:branched	21:0
14:0	18:0	21:5 n−2
14:1	18:1	22:0
15:branched	18:2 n−9	22:1 n−11
15:0	18:2 n−6	22:2
16:0	18:2 n−4	22:3 n−3
16:1	18:3 n−6	22:4 n−3
16:2 n−7	18:3 n−3	22:5 n−3
16:2 n−4	18:4 n−3	22:6 n−3
16:3 n−4	19:branched	23:0
16:3 n−3	19:0	24:0
16:4 n−4	19:1	24:1
16:4 n−1	20:0	
17:branched	20:1	
17:0	20:2 n−9	
17:1	20:2 n−6	
	20:3 n−6	
	20:3 n−3	
	20:4 n−6	
	20:4 n−3	
	20:5 n−3	

terised by high contents of polyunsaturated fatty acids, especially C20:5 (n−3) (EPA), and a high iodine number. In general, the fatty acid composition of fish oils reflects the fatty acid composition in the feed of the fish. Consequently the catching area and season is more important for determining the fatty acid composition than the fish species.

The world production of fish body oils is listed in Table 21.8 (Bowman 1988). The yearly production is approximately 1.5 million tonnes, which means 1.5–2% of the total world production of fats and oils, which is dominated by soybean oil and palm oil. Most of the fish oils are used as hydrogenated marine fat (Young 1986b) in margarine and shortenings. In some countries partially hydrogenated fish oils are winterized and used as salad oils. In all foodstuffs fish oils may be substituted with vegetable oils and consequently the price of fish oils is determined by the prices of soybean oil and palm oil. Fish oils may also be used as an energy source in feedstuffs both in crude and refined form (Opstvedt 1985). Fish oil is an essential energy source in fish feeds and this market has increased in recent years (Pike 1989).

The positive effects on health of polyunsaturated fatty acids from fish are well documented (Simopoulos *et al.* 1986), and this has already had a marked effect on the image of fish oils. To include unhydrogenated fish oils in staple foods is a real challenge. If we succeed, it may have a marked effect on the price of fish oils. The increasing recognition of the health benefits of fish oils has opened up a market for this product as a health food. Fish oils or concentrates of

Table 21.7 Approximate content of principical fatty acids typical of fish oils, %.

	Capelin	Herring	Sprat	Norway Pout	Mackerel	Sand eel	Menhaden	Sardine Pilchard	Horse Mackerel	Anchovy
C14:0	7	7	ns	6	8	7	9	8	8	9
C16:0	10	16	16	13	14	15	20	18	18	19
C16:1	10	6	7	5	7	8	12	10	8	9
C18:1	14	13	16	14	13	9	11	13	11	13[1]
C20:1	17	13	10	11	12	15	1	4	5	5[2]
C22:1	14	20	14	12	15	16	0.2	3	8	2
C20:5	8	5	6	8	7	9	14	18	13	17
C22:6	6	6	9	13	8	9	8	9	10	9
Iodine value	94/164	95/160	125/147	ca.140	136/167	144/173	150/200	141/212	158/193	178/200

[1] Contains element of C16:4.
[2] C20:1 and C18:4 combined.
ns = not stated.

Table 21.8 Fish body oils – World production (1000 t).

	1983	1984	1985	1986	1987
Chile	57	184	184	225	172
Denmark	89	82	77	88	85
Iceland	27	90	119	106	82
Japan	350	417	404	483	430
Norway	217	190	152	99	71
Peru	17	143	121	252	109
South Africa	31	18	28	29	83
USA	181	169	129	153	136
USSR	73	93	85	95	94
Other	83	122	170	132	n.a.
Total world	1125	1508	1469	1662	n.a.

n.a. = not available.

polyunsaturated fatty acids are sold in capsules under different trade names. It may become a very profitable business for those who produce and sell the capsules, but so far the impact on the prices of fish oils has been negligible.

Both in the form of oil and as fatty acids derivatives, fish oils are used for many technical purposes like tanning, detergents, ore flotation, water repellents, lubricants, plasticisers, corrosion inhibitors etc. A review of the utilisation of fish oils in these ways is given by Dyer (1967).

Fish oils are still used as fuel – unfortunate but a fact. Partly because of a temporary surplus of fats and oils and partly because of high taxation on ordinary fuel oils, some companies have found it more profitable to use fish oil as fuel in the fish meal factory than to sell it. We are then back to the basic utilisation mentioned by Strøm (1762) more than 200 years ago.

FISH AS FEED

Fish and fish offal have been used as feed for animals throughout the ages. On farms in coastal districts it was common to find both eggs and bacon tasting distinctly of fish. Even today fish and fish offal, both fresh and frozen, are used as feeds but mainly to fur animals, in fish feeds and in pet foods. After grinding, the fish is mixed with vitamins, minerals, binders, etc. and served as moist pellets or canned. The cost of processing may be low but the high water content increases the cost of storage and transportation, and shelf life is short except when canned.

FISH SILAGES

Fish silage of different kinds has been used as an animal feed for many decades, but mostly in districts with a fishing industry. Processing methods are illustrated in Figure 21.3.

As an initial step the fish or fish offal is minced. A very simple way of producing silage is to add acids to prevent bacterial growth, and to store the mixture. Instead of acids, carbohydrates and lactic acid bacteria may be added, which will prevent the growth of harmful bacteria.

During storage the fish is liquefied by the action of digestive and other enzymes in the fish. This may be stopped at a certain degree of degradation by heating to inactivate the enzymes. Oil may be removed to obtain de-oiled silage. To facilitate storage and transportation some of the water may be removed by evaporation. If it is undesirable to have any enzymatic degradation, the minced fish may be heated prior to the addition of acids or carbohydrates and lactic acid bacteria.

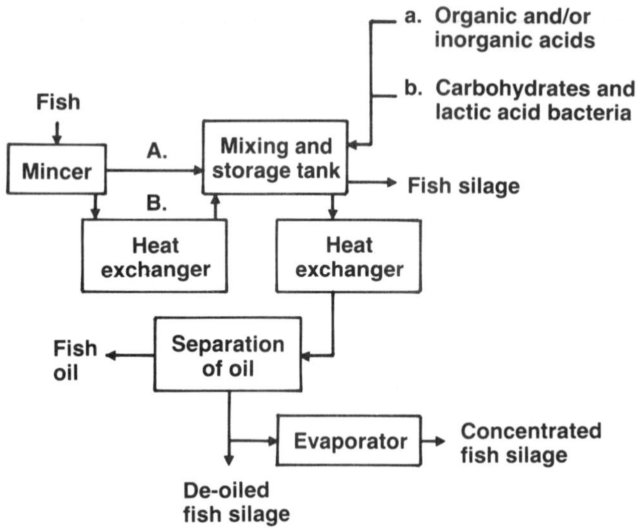

Fig. 21.3 Alternative sequences of silage production from pelagic fish.

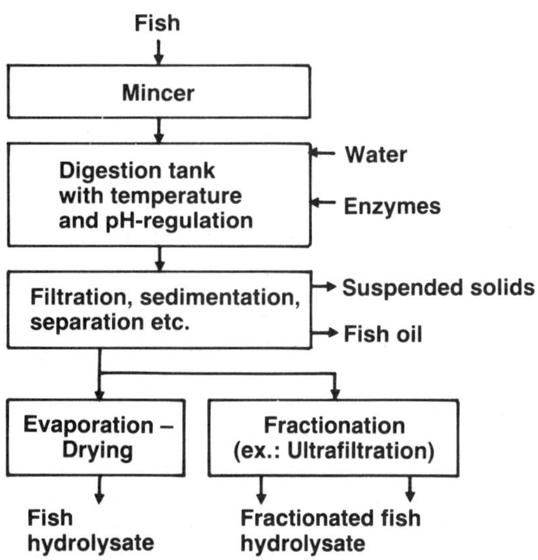

Fig. 21.4 Principal steps in the processing method for fish hydrolysate from pelagic fish or fish offals.

Fish silage is used as a feed ingredient and the composition will be influenced both by raw materials and processing procedures. The liquid form is a drawback, since the other ingredients in traditional animal feeds are in the form of powders. Today fish silage has found a new market in fish feeds as an ingredient in moist feed pellets.

Further details are given by Windsor & Barlow (1981) and Raa *et al.* (1983).

FISH HYDROLYSATES

A flow sheet illustrating the processing of fish hydrolysates is given in Figure 21.4. The process is similar to the ones used for the processing of fish silage. The fish is minced prior to the addition of water and compounds to prevent bacterial growth. The suspended solids are liquefied by enzymatic reactions caused by the addition of proteolytic enzymes. Suspended solids and fish oil are removed and the fish hydrolysate may

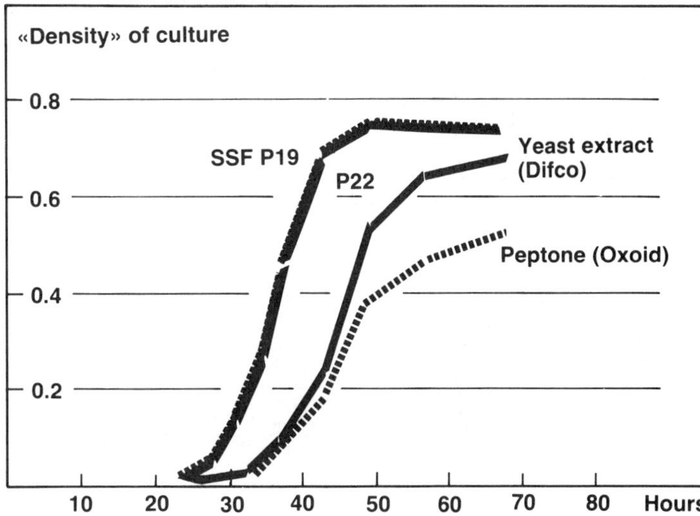

Fig. 21.5 Growth rate of lactic acid bacteria in media containing fish hydrolysate (SSF P19 and P22).

be concentrated to a viscous liquid with a high content of dry matter. For some purposes it may even be dried to a powder or granules. Hydrolysates may also be fractionated, for instance according to molecular weight.

Fish hydrolysate may be used as a protein or amino acid source in compound feeds in the same way as fish silage. In refined form it may be used as a flavour compound both for feeds and foods. It may also be used as a constituent of growth media for industrial fermentation processes; an example is shown in Figure 21.5 (SSF 1988). The market for such products seems to increase as a result of biotechnological research.

Important questions for all uses of fish hydrolysates are the variation in molecular weight of the different peptides and the amino acid sequence. This may be partly controlled by the type of enzymes used and the processing conditions during digestion.

For more details see Gilberg (1978), Aksnes (1988), Aksnes & Brekken (1988), Martinez & Gilberg (1988), Aksnes (1989) and Gilberg *et al.* (1989).

FISH AS A RAW MATERIAL FOR CHEMICAL AND PHARMACEUTICAL PRODUCTS

In this context we had better look at marine organisms in general and not only fish. The development of 'downstream technology' as a part of biotechnology has opened up new possibilities for the enrichment and purification of compounds.

The following substances are of interest:

- concentrates of specific fatty acids
- complex lipids
- enzymes and hormones
- nucleic acids and nucleotides
- natural organometallic compounds
- antibiotics with defined specificity
- others

Compounds like these may also be found in terrestrial plants and animals, but one important difference has to be considered. Marine mammals, fish, algae, bacteria, feed and faeces, particles of organic and inorganic matter, disolved nutrients etc. are all retained in the same medium – sea water. Such conditions lead to tough competition between species resulting in survival of only the fittest. As a consequence, some marine organisms have developed advanced biochemical defence systems which may be utilised both in pharmaceutics and industry. Further information is given in Grant & Mackie (1977), Aksnes (1988), Martinez *et al.* (1988), Olafsen (1988) and Gilberg *et al.* (1989).

PELAGIC FISH AS AN INDUSTRIAL RAW MATERIAL OF BIOLOGICAL ORIGIN

'Fish ought to be used as food because a lot of people are starving,' is often put forward as an argument. When looking at some of the disasters and tragedies presented to us on television, we can easily understand that argument. Lack of food in some parts of the world is not, however, caused by insufficient production capacity in the world as a whole. In my opinion the

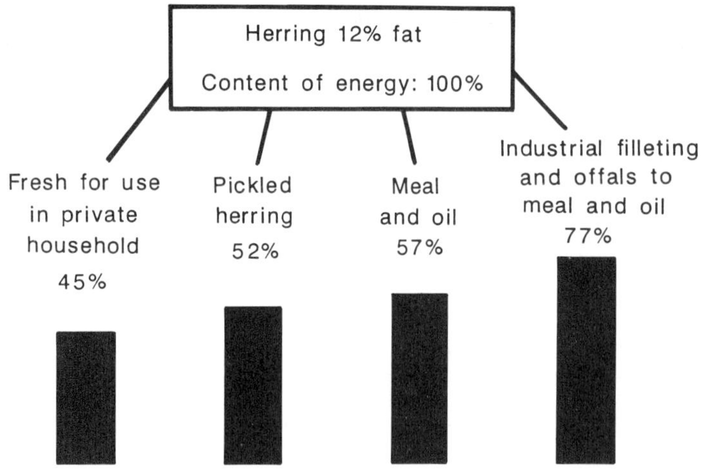

Fig. 21.6 Yield of energy in food products from herring.

reasons are rather to be found in politics, national economies, agricultural and industrial policies, traditions, lack of education, etc. For that reason, fish may as well be looked on as a general raw material for industrial purposes.

To illustrate how much food, in terms of energy, may be obtained from fish when it is utilised in different ways, I have made some calculations (Fig. 21.6) (Urdahl, 1984).

A given amount of herring with 12% fat has a content of energy equal to 100%. When consumed directly in the home, the head, guts, back bone etc. are discarded and the energy yield as food becomes 45%. In pickled herring, because of the salt content, the distribution of dry matter is different from fresh herring so it gives a 52% yield of energy as food.

If fish meal and fish oil are produced, the oil is used in food products (100% of the energy as food). Fish meal is used in feed for animals and of course the yield of energy may vary. I have used 20% but in some cases it may be higher or lower. When adding together the yields of energy in food products resulting from the original herring, we arrive at 57%.

Today, industrial filleting and the use of offal for fish meal and fish oil are more common and the yield of energy in food products may be increased to 77%.

When looking at fish resources, all nations and most fishermen agree that we should not take out more than the sustainable biological yield and we must utilise it in such a way that we do not destroy the environment. When these conditions are fulfilled I feel that pelagic fish may be looked on as a general raw material with many built-in challenges and opportunities. Many different products may be manufactured and the economics will define what is a principal product and what is a by-product. This may differ from one company to another.

BIBLIOGRAPHY

Aksnes A. (1988) Location of enzymes responsible for autolysis in bulk stored capelin (*Mallotus villosus*). *J. Sci. Fd. Agric.*, 263–71.

Aksnes A. (1989) Effect of proteinase inhibitors from potato on the quality of stored herring. *J. Sci. Fd. Agric.*, 225–34.

Aksnes A. & Brekken B. (1988) Tissue degradation, amino acid liberation and bacterial decomposition of bulk stored capelin. *J. Sci. Fd. Agric.*, 53–60.

Barlow S.M. & Pike I.H. (1977) *The role of fat in fish meal in pig and poultry nutrition*. Techn. Bull. 4, International Association of Fish Meal Manufacturers, Potters Bar.

Barlow S.M. & Windsor M.L. (1984) *Fishery by-products*. Techn. Bull. 19, International Association of Fish Meal Manufacturers, Potters Bar.

Bowman, S.J. (1988) *Digest of selected statistics*. 28th Annual Conference of International Association of Fish Meal Manufacturers, Potters Bar.

Dyer, J.A. (1967) General industrial and potential uses of fish oils. In *Fish oils* (Ed. by M.E. Stansby), pp. 270–79. The AVI Publishing Company Inc., Westport, Connecticut.

FAO (1970) Available amino acid content of fish meals. *Fish. Rep.* 92, FAO, Rome.

Gilberg A. (1978) Proteolytic activity and frequency of burst bellies in capelin. *J. Fd. Technol.*, 409–16.

Gilberg A., Batista I. & Strøm E. (1989) Preparation and characterization of peptones obtained by a two-step enzymatic hydrolysis of whole fish. *Biotechnol. and Appl. Biochem.* **11**, 413–23.

Grant P.T. & Mackie A.M. (1977) Drugs from the sea – fact or fantasy? *Nature*, **267**, 786–8.

Gulbrandsen K.E. (1984) *Fish meal in diets for young pigs*. Techn. Bull. 17, International Association of Fish Meal

Manufacturers, Potters Bar.

IAFMM (1972) *The value of fish meal in practical livestock diets*. Techn. Bull. 2, International Association of Fish Meal Manufacturers, Potters Bar.

Julshamn K., Haugenes J. & Utne F. (1978) The content of 14 major and minor elements (minerals) in Norwegian fish species and fish products, determined by atomic absorption spectrophotometry. *Fisk. Dir. Skr., Ser. Ernæring*, **1** (4) 117–35.

Ledward D.A., Taylor A.J. & Lawrie R.A. (1983) *Upgrading wastes for feed and foods*. Butterworths, London.

Lovern J.A. (1950) The possible use of fish guts as food for animals. *J. Sci. Fd. Agric.*, **10**, 314–16.

Lovern J.A. (1955) Recent developments in the fish by-products industry. *J. Sci. Fd. Agric.*, **5**, 233–9.

Martinez A. & Gilberg A. (1988) Autolytic degradation of belly tissue in anchovy (*Engraulis encrasicholus*). *Intern. J. Fd. Sci. and Technol.*, **23**, 185–94.

Martinez A., Olsen R.L. & Sérra J.L. (1988) Purification and characterization of two trypsin-like enzymes from the digestive tract of anchovy (*Engraulis encrasicholus*). *Comp. Biochem. Physiol.*, **91B**, 677–84.

Miller E.L. (1978) *Evaluating the protein contribution of feed-stuffs for ruminants*. Techn. Bull. 5, International Association of Fish Meal Manufacturers, Potters Bar.

Miller E.L., Galway N.W., Pike I.H. & Newman G. (1981) *Milk production response of dairy cows fed a supplement of fish meal on commercial farms*. Techn. Bull. 14, International Association of Fish Meal Manufacturers, Potters Bar.

Olafsen J. (1988) *Marine biokjemikalier*. Paper presented at open meeting of Norwegian Fisheries Research Council, Ålesund.

Opstvedt J. (1985) *Fish lipids in animal nutrition*. Techn. Bull. 22, International Association of Fish Meal Manufacturers, Potters Bar.

Pike I.H. (1975a) *The role of fish meal in diets for poultry*. Techn. Bull. 3, International Association of Fish Meal Manufacturers, Potters Bar.

Pike, I.H. (1975b) *The role of fish meal in diets for pigs*. Techn. Bull. 6, International Association of Fish Meal Manufacturers, Potters Bar.

Pike, I.H. (1989) The role of fish oil in feeds for farmed fish – estimated current and potential use. Techn. Bull. 25, International Association of Fish Meal Manufacturers, Potters Bar.

Raa J., Gilberg A. & Strøm T. (1983) Silage production – theory and practice. In *Upgrading wastes for feed and foods* (Ed. by D.A. Ledward, A.J. Taylor & R.A. Lawrie), pp. 117–32. Butterworths, London.

Simopoulos A.P., Kifer R.R. & Martin R.E. (1986) *Health Effects of Polyunsaturated Fatty Acids from Seafood*. Academic Press, Orlando.

SSF (1988) *Hydrolyserte fiskeproteiner*. In annual report of The Norwegian Herring Oil and Meal Industry Research Institute (SSF), Bergen.

Stansby M.E. (1963) Miscellaneous Industrial Products. In *Industrial Fishery Products* (Ed. by M.E. Stansby & J.A. Dassow, pp. 247–52. Reinhold Publishing Corporation, London.

Stansby M.E. (1967) *Fish Oils*. The AVI Publishing Company, Westport, Connecticut.

Stansby M.E. & Dassow J.A. (1963) *Industrial fishing technology*. Reinhold Publishing Corporation, London.

Strøm H. (1762) Sundmørs beskrivelse referred to in *En industri av verdensformat* (Ed. by K. Fasting), 1954, p. 14. Sildemelfabrikkenes Landsforening, Oslo.

Urdahl N. (1984) *Mel og olje – en fornuftig ressursutnyttelse*. Meldinger fra SSF **2**, 16–19.

Urdahl N. & Nygård, E. (1982) Marine oils – composition and available quantities. In *Proceedings of Lipidforum symposium on oleochemicals*, pp. 40–50. Gothenburg.

Windsor M. & Barlow S.M. (1981) *Introduction to fishery by-products*. Fishing News Books, Oxford.

Young, F.V.K. (1986a) The chemical & physical properties of crude fish oils for refiners & hydrogenators. *Fish Oil Bull.* 18, International Association of Fish Meal Manufacturers, Potters Bar.

Young F.V.K. (1986b) The refining and hydrogenation of fish oils. *Fish Oil Bull.* 17, International Association of Fish Meal Manufacturers, Potters Bar.

CHAPTER 22

Fish meal quality

G.P. Savage

INTRODUCTION

Fish meal is a valuable source of protein, lipids, minerals and vitamins (Windsor & Barlow 1981, Barlow & Windsor 1983). It is used in nearly every country, particularly for poultry, pig and aquaculture feeding rations where it is recognised as a high quality protein source. A large proportion of the world's fish catch is made into meal and oil. This proportion may well increase if underutilised species, which have little consumer appeal, are exploited further. Virtually any fish or shellfish can be used to make fish meal. Material used to make fish meal and oil falls into two categories:

(1) whole fish caught in very large quantities having unacceptable composition, size or quality for direct human use
(2) waste materials from filleting, canning and other fish processing (Hampton 1981).

The proximate composition of pelagic fish has been extensively investigated. Protein levels of 16 ± 3% are a feature of nearly all fish (Windsor & Barlow, 1981) and New Zealand species generally fall within this range (Vlieg, 1984a,b; 1985a,b). As fish meals are obtained by separating protein and ash from water and oil, meals with very similar compositions might be expected irrespective of the species being processed. That this is not always the case may be related to the quality and type of raw material used (FAO 1975). Meals made from scraps, for example, will contain a larger progortion of bones, thus lowering the crude protein content of the final meal (Table 22.1). Johnston & Savage (1987) showed that meals made from whole hoki (*Macruronus novaezelandiae*) had a higher crude protein content and biological value than meals made from orange roughy (*Hoplostethus atlanticus*) scraps. Processing factors such as cooking and drying temperatures, grinding and antioxidant stabilisation will also affect the quality of the meal.

Day-to-day variation in the quality of fish meal produced at one factory is quite common. Even though the processing parameters and raw material composition remained the same, Johnston and Savage (1987) noted that the biological value of fish meal produced at the same plant, but on a different day, had a 15% lower biological value.

Fish spoilage due to autolysis, lipolysis and microbial action will result in soft and degraded flesh which will coagulate poorly during cooking. This will cause difficulties during processing, leading to a lower protein and a higher oil content in the meal. Storage conditions will also affect the quality of the meal.

The crude protein levels and the quality of a meal may also be affected by the addition of condensed stick water back into the meals. Triple-effect evaporators which reach temperatures as high as 150°C are generally used to condense stick water and this may affect the overall protein quality of the resulting so called 'whole meal'.

FATS AND FATTY ACIDS

Fluctuations in oil and moisture levels occur both within species and in different seasons (Stansby & Olcott 1963). High oil contents are a bonus to meal manufacturers but processing difficulties can often be encountered. This has been shown in New Zealand with difficulties associated with processing orange roughy and oreo dory (*Pseudocyttus maculatus*). These meals often have an unacceptably high oil content unless processed carefully.

Carroll (1986) has summarised the recent considerable interest in the biomedical effect of fish oils. It is clear that polyunsaturated fish oils are very effective in lowering serum triglyceride and cholesterol levels (Phillipson *et al.* 1985, Herold & Kinsella 1986,

Table 22.1 Comparison of a meal manufactured from whole hoki fish with a meal made from orange roughy scraps (Johnston and Savage 1987).

	Hoki meal (mean of 7 meals)	Orange roughy meal (mean of 8 meals)
Moisture	9.5	7.1
Protein	67.2	59.6
Oil	5.9	8.1
Ash	15.5	21.3
True digestibility	87.0	87.8
Biological value	82.7	74.6
Utilisable protein	48.3	39.1

Norum & Drevon 1986). In addition it has been shown that diets containing fish oils have an inhibitory effect on blood clotting and this reduces the risk of thrombosis which is often a major factor in heart attacks and strokes (Dyerberg 1982 and 1986, Herold & Kinsella 1986, Norum & Drevon 1986).

The main polyunsaturated fatty acids in such marine fish oils are eicosapentaenoic acid (C20:5n–3) and docosahexaenoic acid (C22:6n–3). Until recently the only rich natural sources were products of marine origin, i.e. fish and shellfish, especially fatty fish (mackerel, herring, pilchard and salmon) found in colder oceans. However, substantial enrichment of these omega-3 fatty acids can occur in the tissues of broiler chickens fed either fish oils or fish meals (Hulan *et al.* 1988). The positive nutritive value of the residual omega-3 fatty acids in fish meals should not be overlooked and efforts should be made to prevent oxidation of these fatty acids during processing and later storage of the meal.

The effectiveness of the addition of the antioxidant ethoxyquin to anchovy meals to protect the nutritionally important polyunsaturated fatty acids such as eicosapentaenoic and docosahexaenoic acid from oxidation during storage of the processed dried meal has been clearly demonstrated by De Koning & Mol (1989). Their results also suggest that the addition of ethoxyquin during manufacture of the meal would prevent the rapid oxidation of these fatty acids which occurs during the normal meal drying process (30 minutes at an outlet temperature of 70–80°C). The addition of an antioxidant should be standard procedure in the manufacture of high quality fish meals.

It is interesting to note that the digestibility of the omega-3 fatty acids found in a New Zealand commercially extracted fish meal is considerably lower than the same fatty acids in the added marine oil when fed to chinook salmon (Table 22.2). This suggests that the unextracted omega-3 fatty acids remaining in fish meal are only partially available to chinook salmon, which is in contrast to the high availability of all the other fatty acids found in fish meal and fish oil.

Table 22.2 Digestibility coefficients of the ω3 essential fatty acids in chinook salmon (Bird & Savage 1989, and unpublished data).

	Added Marine Oil			
	Basal	5%	10%	15%
18:3	90	96	98	99
20:5	87	97	97	98
22:5	63	88	96	98
22:6	41	81	88	95

ASH CONTENT

The total ash content of meals can range from 10–20% depending on the proportion of bones and scrap used to manufacture the meal. High levels of ash are normally found in poorer quality meals. Fish meal contains a wide range of essential elements and heavy metals (Barlow & Windsor 1983). In certain circumstances some of these heavy metals may limit the use of a few fish meals (Johnston & Savage, 1989 and in press).

Minerals such as phosphorus and selenium are important resources for the compound industry. Phosphorus has a high biological availability of 88% in chicks, which appears to be unaffected by the condition of the fish or the method of processing (Huque & Frisjensen 1985). Selenium on the other hand had a biological availability of 77% in freshly processed fish meal, which fell to 51% when partly-decomposed fish was processed into fish meal (Huque & Frisjensen 1985).

Fish meal contains significant quantities of manganese but Halpin & Baker (1986) have observed that addition of fish meal to chick diets containing just adequate amounts of dietary manganese had an adverse effect on feed gain and caused an increased incidence of perosis. This suggests that the manganese in fish meal is largely unavailable and that some other component in the fish meal may be binding some of the remaining manganese in the diet.

PROTEIN AND AMINO ACIDS

It is well known that fish protein has a well-balanced amino acid profile and hence high protein quality.

A consistent feature of experiments where various protein sources are compared is the high apparent digestibility of the individual amino acids in fish meal (Green 1989). This suggests that fish meals would be more suitable than meat meal for inclusion into diets when emphasis is on provision of highly digestible nutrients. The overall quality of the meal, however, is affected by the processing techniques used, especially in the cooking and drying stages. It was believed that cooking temperatures close to 100°C were necessary to liberate oil, displace physiologically bound water and coagulate the protein. However, fat cells have weak walls and require only gentle heating or mechanical pressure to disrupt. Results obtained from cooking fresh winter capelin (a difficult raw material to process) at different temperatures gave best fat extractions at around 55°C. This led to the development of the low temperature process and to many established factories lowering their cooking temperatures (Sand 1977).

The residence time and the temperature used to dry the processed fish have a considerable effect on the nutritive value of the final meal. The different systems used to manufacture and dry meals have a significant effect on these parameters (Umemura 1982). The amino acids cysteine, histidine, tryptophan and lysine are particularly susceptible to destruction during heat treatment. It is for this reason that the amino acid profile does not always adequately predict the nutritional value of a meal. Unavailable amino acids which are subsequently destroyed by heat and oxidative reactions are released on acid hydrolysis and estimated in this profile.

Johnston & Savage (1987) clearly showed that meals made principally from hoki at one factory contained very similar levels of crude protein (Table 22.3). If the meal was dried at 78.5°C a highly digestible meal with a high biological value was produced. A similar meal was produced the following day using the same raw material dried at 125°C. The true digestibility of the meal was reduced by 73% and the biological value by 45%. The utilisable protein in the overheated meal was only a fraction of that found in the properly processed meal.

Drying meals at high temperatures not only reduces the overall quality of the meal but also predisposes chickens fed the meal to develop gizzard erosion. Wessels & Post (1989) showed that fish meals heated to 130°C or higher during drying caused considerable gizzard erosion when used in starter diets. This did not, however, affect the live weight gain during the experimental period. The fineness of the fish meal had no effect on this phenomenon but in some instances

Table 22.3 Mixed hoki meal manufactured at the same factory and dried at two different temperatures.

	78.5°C	125°C
Crude protein	63.9	68.4
True digestibility	87.0	23.4
Biological value	80.0	44.4
Utilisable protein	44.5	7.1

added dietary lysine reduced the incidence of gizzard erosion.

Okazaki *et al.* (1983) isolated a compound gizzerosine which they believe was formed during the raised temperature drying of their mackerel meal. They suggest that this compound is responsible for the gizzard erosion associated with high temperature dried meals. Gizzerosine appears to be formed from free histidine and histamine by reaction with protein in fish meal during the drying process. Meals prepared from sardine and mackerel can contain high levels of histamine formed from microbial decarboxylation of histidine. The level of histamine in fish meal is known to be elevated by the return of fish solubles to the meal (Toyama *et al.* 1981). Gizzerosine appears to mimic the physiological action of histamine in the body, leading to increased secretion of gastric acid resulting in gizzard erosion. Watanabe *et al.* (1987) showed that stomach erosion in rainbow trout (*Salmo gairdneri*) does occur when fed elevated levels of histidine and histamine. They showed that while low levels of dietary histamine promoted dietary protein digestion, higher levels led to stomach abnormalities and a darkening of the body colour. Watanabe *et al.* (1987) went on to show that a small increase in stomach erosion of rainbow trout was observed when heat treated (130°C for 3 hours) whole sardine meal was fed for eight weeks.

It is important that overprocessed meals should be identified quickly. Further attention should be given to the storage of fish and fish scraps prior to processing to reduce microbial decarboxylation of histidine, and attention should be given to the effect of returning fish solubles to make so called whole meals. Control of these adverse effects may be more important in the preparation of high quality meals intended for aquaculture use.

Johnston & Savage (1987) concluded that a low temperature rendering plant with careful control of the drying process could consistently produce high quality meal even from fish scraps. Other plants such as an Alfa Laval or Harris steam plant can produce good quality meals provided there is careful control of the operating conditions.

MEASUREMENT OF PROTEIN QUALITY OF FISH MEAL

Crude protein is the most valuable component of fish meal and the evaluation of its quality is a most important issue, particularly when the meal is intended for aquaculture use. The quality of any protein, including fish meal, is best assessed using biological tests on the animal of interest, e.g. poultry, pigs, fish etc. For reasons of cost and equipment it is often more practical to undertake this type of testing using the rat as an experimental model (Eggum 1970, Johnston & Savage 1987). Assays using animals can identify the quality of the different types of meal but they involve a considerable amount of experimental work, time delay and expense. This time delay is unacceptable in commercial practice.

The essential amino acid composition is a major determinant of the nutritional quality of a protein. The availability of these amino acids is, however, an equally important consideration. Fish meal is a significant source of lysine so the measurement of fluorodinitrobenzene (FDNB) available lysine (Carpenter 1960) as modified by Carpenter and Booth (1973) is an important chemical method used to evaluate the quality of fish meals. FDNB binds with the available e-amino group of lysine. Heat damaged protein will not react with FDNB (Booth 1971). In good quality fish meal over 85% of the total lysine should be available. More recently a new dye binding method (Hurrell *et al.* 1979) using the azo dye acid orange 12 has also been used to measure the extent of heat damage of a fish meal. Both methods are satisfactory for use in a single laboratory wishing to compare the quality of fish meals of known history, for example within a single fish meal factory. They are not, however, suitable for comparison of fish meals of unknown history by different laboratories (Barlow *et al.* 1984).

Several *in vitro* methods for the measurement of protein digestibility have been developed. The pepsin method (EEC 1972) with all its modifications is the most well known. It is frequently stated that while this method can discriminate between commodities with widely differing digestibilities, it cannot be relied on to identify more subtle differences between meals (Green 1989).

Other methods use enzyme mixtures such as the pepsin-pancreatin system (Akeson & Stahman 1964) or enzyme preparations e.g. pronase isolated from *Streptomyces griseus* (Ford & Salter 1966). Multi-enzyme systems (Hsu *et al.* 1977) have been developed to complete analyses in one hour with a high degree of sensitivity. The multi-enzyme system consists of trypsin, chymotrypsin and an intestinal peptidase. *In vitro* digestibility is estimated by measuring the pH drop of the protein suspension after 10 minutes of digestion. The results correlate well with *in vivo* values determined in rats. A modified 20 minute assay in which a fourth enzyme is added was described by Satterlee *et al.* (1979). Enzyme digestion at constant pH offers obvious advantages as the pH can be maintained at each enzyme's optimum. Pederson & Eggum (1983) have shown that good correlations can be obtained with *in vivo* values using an *in vitro* multi-enzyme digestion procedure in a pH stat. The uptake of titrant correlated with the true protein digestibility of the same protein determined in rats.

Recent experiments at Lincoln (Johnston & Savage, unpublished) have shown, using a cheaper commercial preparation of proteolytic enzymes (Alcalase 2.4 L) in a pH stat, that excellent correlations with *in vivo* rat results can be obtained. The advantage of this method is that the technique is both simple, quick, reliable and uses a cheap enzyme source. Results obtained so far suggest that, provided a blank titration (no enzyme present) is carried out on each sample, the method will accurately predict the digestibility of a range of meals without the need for fine grinding of the samples or pre-extraction to remove the fat content, which appears to be a prerequisite of other enzyme based methods.

AQUACULTURE FEEDS

Many cold and warm water fish have a high protein requirement and many require more than 30% crude protein in the diet (Cowey 1980). Fish meal is recognised as an excellent source of protein for most fish species. Its generally high quality results in its wide inclusion as the main protein source in diets for farmed fish. Attempts to substitute other proteins for fish meal, either wholly or in part, have frequently resulted in decreased growth rates (Dabrowski *et al.* 1989).

The increasing cost of high quality fish meal poses real problems for cost effective feed formulation. Feed represents a major cost in intensive aquaculture. In addition it is clear that the world production of fish meal will not meet the expected increasing demand. This will encourage more active research into the development of alternative protein sources, perhaps with appropriate amino acid supplementation. At the same time it will become increasingly important to identify the different qualities of fish meal being offered on the world market so that this useful resource can be put to best possible use.

CONCLUSIONS

Fish meal is a good source of protein and essential amino acids and has a high biological value in diets for animals. High quality fish meal is widely used in the pig and poultry industries and increasing amounts are being used in intensive aquaculture. In all these uses it is of paramount importance that the essential amino acids in fish meal are highly digestible and available to the animal consuming them. It is therefore necessary that feed compounders are able to identify each grade of fish meal available in the market so that it can be optimally formulated into the appropriate diet.

ACKNOWLEDGEMENTS

The author wishes to thank J.N. Johnston for assistance in the preparation of this paper.

BIBLIOGRAPHY

Akeson W.R. & Stahmann M.A. (1964) A pepsin-pancreatin digest index of protein quality. *J. of Nutrition*, **83**, 257–61.

Barlow S.M., Collier G.S., Jurtiz J.M., Burt J.R., Opstvedt J. & Miller E.L. (1984) Chemical and biological assay procedures for lysine in fish meals. *J. of Sci. of Fd. and Agri.*, **35**, 154–64.

Barlow S.M. & Windsor M.L. (1983) Fishery by-products. In *CRC Handbook of Nutritional Supplements*, **II** (Ed. by M. Rechcigl (Jr)). CRC Press, Boca Raton, Florida.

Bird J.N. & Savage G.P. (1989) The absorption of astaxanthin by Chinook salmon (*Oncorhynchus tshawytscha*). *Proceedings of the Nutrition Society of New Zealand.* **14**, 174–7.

Booth V.H. (1971) Problems in the determination of FDNB-available lysine. *J. Sci. of Fd. and Agri.*, **22**, 658–66.

Carpenter K.J. (1960) The estimation of available lysine in animal-protein foods. *Biochemical Journal*, **77**, 604–10.

Carpenter K.J. & Booth V.H. (1973) Damage to lysine in food processing; its measurement and significance. *Nutrition Abstracts and Reviews*, **43**, 423–51.

Carroll K.K. (1986) Biological effects of fish oils in relation to chronic diseases. *Lipids.* **21**, 731–2.

Cowey C.B. (1980) Protein metabolism in fish. In *Protein Deposition in Animals.* (Ed. by P.J. Buttery & D.B. Lindsay), pp. 271–88. Butterworth, London.

Dabrowski K., Poczyczynski P.G., Kock G. & Berger B. (1989) Effect of partially or totally replacing fish meal protein by soybean meal protein on growth, food utilization and proteolytic enzyme activities in rainbow trout (*Salmo gairdneri*). New *in vivo* test for exocrine pancreatic secretion. *Aquaculture*, **77**, 29–49.

De Koning A.J. & Mol T.H. (1989) Lipid determination in fish meal: an investigation of three standard methods applied to stabilised and non-stabilised anchovy meals at increasing stags of maturity. *J. Sci. of Fd. and Agri.*, **46**, 259–66.

Dyerberg J. (1982) In *Nutritional Evaluation of Long-Chain Fatty Acids in Fish Oil.* (Ed. by S.M. Barlow & M.E. Stansby), pp. 245–61. Academic Press, New York.

Dyerberg J. (1986) Linolenate-derived polyunsaturated fatty acids and prevention of atherosclerosis. *Nutritional Reviews*, **44**, 125–34.

Eggum B.O. (1970) In *The Production of Fish Meal and Oil.* FAO Fisheries Technical Paper 142, FIIM/T142, Rome.

European Economic Community (1972) Determination of the solubility of crude protein in pepsin and HCl. *Journal Officiel des Comminunautes Europeannes*, 3rd Directure No. L. 123/11.

FAO Fishing Industries Division (1975) In *The Production of Fish Meal and Oil.* FAO Fisheries Technical Paper 142, FIIM/T142, Rome.

Ford J.E. & Salter D.N. (1966) Analysis of enzymatically digested food proteins by Sephadex-gel filtration. *British Journal of Nutrition*, **20**, 843–60.

Green S. (1989) A note on the digestibilities of nitrogen and amino acids in meat, skimmed-milk and fish meals in young pigs. *Animal Production*, **48**, 237–40.

Halpin K.M. & Baker D.H. (1986) Long term effects of corn, soybean meal, wheat bran and fish meal on manganese utilization in the chick. *Poultry Science*, **65**, 1371–4.

Hampton M. (1981) Fish processing – past, present and future. *Food Flavourings, Ingredients, Processing, Packaging*, **3**, 21–2.

Herold P.M. & Kinsella J.E. (1986) Fish oil consumption and decreased risk of cardiovascular disease: a comparison of findings from animal and human feeding trials. *American Journal of Clinical Nutrition*, **43**, 566–98.

Hsu H.W., Vavak D.L., Satterlee L.D. & Miller G.A. (1977) A multi-enzyme technique for estimating protein digestibility. *J. Fd. Sci.*, **42**, 1269–73.

Hulan H.W., Ackman R.G., Ratnayake W.M.N. & Proudfoot F.G. (1988) Omega-3 fatty acid levels and performance of broiler chickens fed redfish meal as redfish oil. *Canadian Journal of Animal Science*, **68**, 533–47.

Huque Q.M.E. & Frisjensen J. (1985) Biological availability of selenium and phosphorus in fish meal as affected by condition of fish and type of meal. *British Poultry Science*, **26**, 289–97.

Hurrell R.F., Lerman P. & Carpenter K.J. (1979) Reactive lysine in foodstuffs as measured by a rapid dye-binding procedure. *J. Fd. Sci.*, **44**, 1221–31.

Johnston J.N. & Savage G.P. (1987) Protein quality of New Zealand fish meals. *New Zealand Journal of Technology*, **3**, 123–8.

Johnston J.N. & Savage G.P. (1989) Protein quality and mercury levels in New Zealand fish meals. In *Proceedings of the 30th Annual Poultry Convention* (Ed. by M.R. Patchell), pp. 109–32. Massey University, Palmerston North, New Zealand.

Johnston J.N. & Savage G.P. (in press) A review of mercury consumption and toxicity with special reference to fish and fish meal. *Nutrition Abstracts and Reviews Series A.* (in press).

Norum K.R. & Drevon C.A. (1986) Dietary n–3 fatty acids and cardiovascular diseases. *Arteriosclerosis*, **6**, 352–5.

Okazaki T., Noguchi T., Igarashi K., Sakagami Y., Seto H., Mori K., Naito H., Masumura T. & Sugahara H. (1983) Gizzerosine, a new toxic substance in fish meal, causes severe gizzard erosion in chicks. *Agricultural and Biological Chemistry*, **47**, 2949–52.

Pedersen B. & Eggum B.O. (1983) Prediction of protein digestibility by an *in vitro* enzymatic pH-stat procedure. *Zeitschrift fur Tierphysiologie, Tierernährung und Futtermittelkunde.* **49**, 265–77.

Phillipson B.E., Rothrock D.W., Connor W.E., Harris W.S. & Illingsworth D.R. (1985) Reduction of plasma lipids, lipoproteins and apoproteins by dietary fish oils in patients with hypertriglyceridemia. *New England Journal of Medicine*, **312**, 1210–16.

Sand G. (1977) *Advances in the production of fish meal.* Manufacturers' symposium on the production and use of fish meal. International Association of Fish Meal Manufacturers, Potters Bar.

Satterlee L.D., Marshall H.F. & Tennyson J.M. (1979) Measuring protein quality. *Journal of the American Oil Chemists' Society*, **56**, 103–9.

Stansby M.E. & Olcott H.S. (1963) Composition of fish. In *Industrial Fishing Technology* (Ed. by M.E. Stansby), Robert E. Krieger Publishing Co., Huntingdon, New York.

Toyama K., Okuzumi M., Yokoi T., Aoe H. (1981) Histamine content of fish meal. *Nippon Suisan Gakkaishi*, **47**, 415–19.

Umemura Y. (1982) Studies on gizzard erosion – inducing substance in fish meal. *Japan Agricultural Research Quarterly*, **16**, 125–30.

Vlieg P. (1984a) Proximate analysis of 10 commercial New Zealand fish species. *New Zealand J. of Sci.*, **27**, 99–104.

Vlieg P. (1984b) Proximate analysis of commercial New Zealand fish species 2. *New Zealand J. of Sci.*, **27**, 427–33.

Vlieg P. (1985a) Proximate analysis of commercial New Zealand fish species 3. *New Zealand J. of Tech.*, **1**, 181–9.

Vlieg P. (1985b) Proximate analysis of commercial New Zealand fish species 4. *New Zealand J. of Tech.*, **1**, 245–9.

Watanabe T., Takeuchi T., Satoh S., Toyama K. & Okuzumi M. (1987) Effect of dietary histidine or histamine on growth and development of stomach erosion in rainbow trout. *Nippon Suisan Gakkaishi*, 1207–14.

Wessels J.P.H. & Post B.J. (1989) Effect of heat treatment of fish meals, fines and the addition of lysine as related to gizzard erosion in chickens. *J. Sci. of Fd. and Agri.*, **46**, 393–406.

Windsor M. & Barlow S. (1981) *Introduction To Fishery By-products*. Fishing News Books, Oxford.

CHAPTER 23

Utilisation of pelagic fish for human consumption

K.J. Whittle and C.D. Wood

INTRODUCTION

The world catch of pelagic fish currently represents some 50% of the global catch of marine fish, and the resource remains substantial but variable (Chapter 1). However, in contrast with other major categories such as demersal fish and shellfish, only about 42% of the small pelagics are used directly for human consumption. The remainder is reduced to meal and oil for animal feeding stuffs, eventually becoming available for human consumption by conversion to other protein resources and food products. However, the conversion of fish meal protein to animal protein is less efficient than direct consumption. For example, broiler chickens fed a balanced diet containing 4–5% fish meal can typically convert about 60% of this protein into chicken protein; the conversion ratio is lower for pigs and cattle.

This seemingly wasteful use of the resource continues while the worldwide demand for fish and fish products for food remains buoyant (Table 23.1). Potentially the opportunities are substantial for diverting an increasingly greater proportion of the pelagic supplies directly to human consumption. The rationale for encouraging such moves in terms of world food supply and composition, as well as nutritional and health advantages, is inescapable and the benefits are likely to be very large. The trends shown in Table 23.1 towards greater use for human consumption are encouraging.

These benefits do not just involve the development of the fish industry. Malnutrition is a problem in many less developed countries (LDCs). The reasons are complex: lack of money to buy foods which may be available in sufficient quantities; social customs may inhibit the consumption of fish, meat or other products which could enhance diets weak in essential nutrients; many staple foods in LDCs, such as rice and maize, contain relatively little protein and its biological value is low. While it is essential to ensure that basic foods of this type are in sufficient supply, this will not in itself prevent malnutrition. Adequate supplies of protein, fats, vitamins and minerals are important particularly for small children and pregnant and lactating women whose requirements for body building nutrients are particularly high. Without doubt the debilitating effect of inadequate diet has been and continues to be a significant factor in restricting the development of LDCs. There is the potential not only to maximise the economic return by utilising pelagic fish for human consumption, but also the potential to increase the supplies of protein for human consumption to alleviate existing and future nutritional deficiencies. If food supplies fail to keep up with population growth the result can only be a decline in living standards.

The opportunities and the associated problems of utilisation, which occur in both LDCs and the developed countries have occupied fish technologists worldwide for many years with varying degrees of success. The problems vary according to species, geographical location and potential markets but can be related generally to the following factors:

- the sheer potential quantity of catches
- the small size and migratory patterns of many of the species
- the very seasonal nature of the fisheries
- large annual fluctuations in yield from the fisheries
- the wide variability in composition (particularly in oil content)
- the bony structures
- the dark flesh
- the stronger flavours
- rapid spoilage of the fresh fish
- development of rancidity in frozen storage, reducing shelf-life
- consumer prejudices against these species.

Table 23.1 Use of the world catch. (FAO Fishery and Fish Product Statistics.)

Year	1970	1978	1987
Total catch (10^6 tonnes)	68	72	93
Human consumption (%)	63.2	71.0	72.4
Fresh	26.3	28.1	21.8
Frozen	13.9	18.3	23.8
Cured	11.5	11.2	14.2
Canned	11.5	13.4	12.6
Non-human/industrial (%)	36.8	29.0	27.6

Various aspects of these factors are dealt with in the reviews and research papers presented in this volume. In this review, by selection and analysis of three examples from different parts of the world, we examine the changing patterns of utilisation of the pelagic resources. The aim is to provide a basis for improved strategies in the future.

SOUTH-WESTERN INDIA – INDIAN OIL SARDINE (*Sardinella longiceps*)

There is an important fishery for small pelagic fish, especially the oil sardine, off the west coast of India. The fish is caught close inshore mainly by the small-scale sector using small wooden canoes which operate from beaches, so it is generally landed in a fresh condition even though the fish is not chilled on board. The changes in utilisation of this resource have been an interesting example of how the general development of the economic infrastructure, coupled with a growing demand for food, has enabled the fisheries industry to develop.

Development of the fishery, supplies and the industry

Nair (1973) described how the abundant sardine catches were used for manure in the 19th century. By the start of the 20th century demand for fish oil in Europe had stimulated the extraction of sardine oil by very crude, small-scale methods; the meal was still used for manure. This industry was virtually brought to an end by the failure of the fishery in the 1940s. Its recovery in the late 1950s and 1960s coincided with the growth of good roads and increasing availability of lorries for transport. This, coupled to a much wider availability of ice, has meant that the possibilities of distributing fresh fish have increased massively over the last 30–40 years. Of the approximately 270 000 tonnes of sardine (of all species) caught in India in 1971, some 60% was probably distributed as fresh fish (Madhavan *et al.* 1974).

Markets, handling and processing

Catches of the oil sardine have been about 200 000 to 250 000 tonnes/year in recent years, and Table 23.2 shows the general figures for fish utilisation in India, indicating that about 80% of the catch is used as fresh fish.

Table 23.2 Utilisation of fish in India, 1982–6. (FAO statistics.)

Year	1982	1983	1984	1985	1986
Total landings (10^3 tonne)	2373	2504	2855	2824	2925
Used as products					
Frozen fish, excluding fillets (10^3 tonne)	7	19	20	10	12
Fresh, chilled and frozen fish, exported (10^3 tonne)	10	19	15	10	13
Cured fish (10^3 tonne)	124	137	138	168	146
Fish meal (10^3 tonne)	23	25	35	31	29
Canned fish			data not available		
Apparent use as fresh fish, derived from data above (10^3 tonne)	1901	1975	2281	2186	2359
Apparent use as fresh fish as % total landings	80.1	78.9	79.9	77.4	80.6

As well as being iced, small pelagic species are also salted. Calicut is a major centre for curing, although the increasing demand for fresh fish has led to competition for raw material and a decline in fish curing. Sardine from glut catches is still used to produce fish meal and oil in the Mangalore area, but the importance of this trade has declined considerably. The relatively few industrial plants still in existence are often idle due to lack of raw material. There has been a marked increase in the consumption of fresh sardines and a corresponding decline in cured sardine and meal production. Infrastructural development has enabled fresh fish to reach more consumers on a regular basis, the small scale entrepreneurial distribution system expanding to supply new markets as they have appeared.

New developments

Although considerable progress has been made towards using this resource to provide nutritionally and commercially valuable products for human consumption, scope for improvement and new initiatives still remains, notably in better handling methods to improve quality, and in the development of high quality processed products for higher income markets.

The development of flexible, retortable pouches as a possible alternative to canned products is of considerable interest. The cans themselves are very expensive in many LDCs because they are not produced locally or must be imported; transport costs of empty cans are high. Initial studies by ODNRI and Central Institute of Fisheries Technology, India (CIFT) in association with the Metal Box Company demonstrated that acceptable products could be made by pouching Indian oil sardines. Recent studies undertaken for the Overseas Development Administration (ODA) indicated that many of the materials needed for making the pouches will shortly be produced in India and that pouched products could be up to 5–8% cheaper than the equivalent canned product. Several product types have been developed although final product formulation and acceptability trials have still to be conducted. The products are intended for middle class consumers, particularly in north-eastern India where there is relatively limited access to fish. The products would also be suitable for institutional markets such as the military.

ECUADOR – SARDINE (*Sardinops sagax*) AND MACKEREL (*Scomber japonicus*)

Development of the fishery, supplies and the industry

The development of the fishery for small pelagic fish in Ecuador has been reviewed by Menz and French (1982). Purse seining was introduced as recently as 1962 to supply a modest canning and fish meal industry. The first large capacity fish meal plant was opened in 1967 and stimulated an increase in landings of small pelagics from 8500 tonnes to 21 000 tonnes. By 1971, it had risen to some 42 000 tonnes. In 1972–3 the Peruvian anchoveta fishery failed and some of

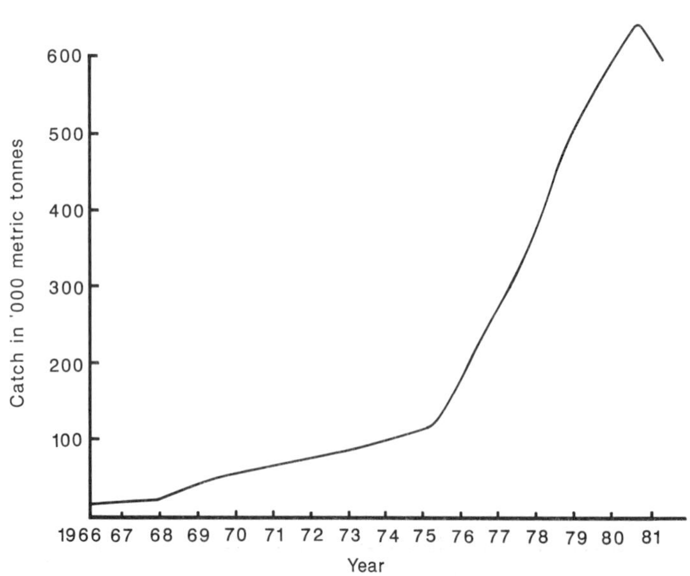

Fig. 23.1 Catches of small pelagic fish in Ecuador.

Table 23.3 Utilisation of fish landings in Ecuador 1982–6. (Wood *et al.* 1988.)

Year	1982	1983	1984	1985	1986
Landings used					
Total (10^3 tonnes)	665.7	227.8	896.2	1511.6	1019.3
Industrial sector					
Total (10^3 tonnes)	622.5	196.3	863.8	1479.6	991.9
Canned (%)	15.5	23.0	10.5	6.0	8.0
Fresh/frozen (%)	7.0	25.0	8.5	5.0	9.0
Meal (%)	71.0	38.0	77.0	87.0	80.0
Artisanal sector					
Total (10^3 tonnes)	43.2	31.6	32.4	33.0	27.4
All uses (%)	6.5	14.0	4.0	2.0	3.0

Table 23.4 Apparent internal consumption of fish in Ecuador 1982–6 (10^3 tonnes). (Derived from Wood *et al.* 1988.)

Year	1982	1983	1984	1985	1986
Fresh fish from artisanal sector	24.5	23.5	20.8	23.7	18.4
Fresh and frozen fish (mainly frozen) from industrial sector	0.48	0.25	0.24	0.42	0.04
Canned fish (tuna, sardine, mackerel; product weight)	12.8	13.3	14.5	18.2	19.7

the boats and meal plants were sold to Ecuador. This stimulated a rapid increase in activity, landings rising to 110 000 tonnes in 1974, 383 000 tonnes in 1977, and over 500 000 tonnes for the first time in 1979 (Fig. 23.1).

The 1980s have seen a much slower growth in the industry. Year to year fluctuations in catches due to climatic and oceanographical conditions in Ecuadorian coastal waters (El Niño phenomenon) have led to falls in catches in some years. However, landings of small pelagic fish now normally constitute some 90% of Ecuador's total landings of fish. For example, in 1986 small pelagics were nearly 880 000 tonnes of a total catch of just over 1 million tonnes (Table 23.3). The industry has been characterised by importing sophisticated fishing boats to catch the fish, and canning and fish meal plants to process it; the products are primarily intended for export. The massive expansion of this fishery has had only a relatively limited impact on fish consumption within Ecuador.

Markets, handling and processing

Ecuador has a very active small-scale fishery, comparable with that found in India, which supplies the bulk of the fresh fish consumed within the country. There does not appear to be any general shortage of good quality, small pelagic fish being landed. There is a strong traditional preference for white fleshed fish and for fresh rather than frozen products. Table 23.3 shows the scale of industrial landings (principally small pelagics but including tuna and shrimp) and those for the artisanal sector, and the principal methods of utilisation. Table 23.4 indicates the apparent internal consumption of fish in Ecuador. The data demonstrate clearly that the landings of small pelagic fish are much greater than the total fish consumption within the country.

The bulk (some 80–90%) of the small pelagic landings are used for fish meal and oil production. Some is used locally as animal feed but most is exported. Frozen sardine and mackerel are also produced, almost exclusively for export. Salted and dried mackerel are prepared in small quantities, and minor quantities of mackerel are also consumed fresh.

Sardine and mackerel are canned, this being their most popular form for consumption throughout Ecuador. In remote areas far from the coast it is the only form of fish product available on a regular basis. The major fish canners use purse seiners equipped with refrigerated sea water (RSW) to chill their catch, but they are expensive to buy and operate. Unrefrigerated seiners can land some fish in a reasonably fresh condition if they are operating near to their home base. Fish is selected from the final catch for canning in some of the smaller plants. However this is not recommended as histamine can accumulate if the fish is held at elevated temperatures for about 8 hours or more.

Under tropical conditions histamine formation is indicative of a very significant danger. NRI has been active in investigating and monitoring histamine formation in Ecuador, Peru and India. It has been shown that under tropical conditions sardine (*Sardinella longiceps*) and other species can accumulate potentially toxic levels while still appearing to be in acceptable condition. Outbreaks of food poisoning may well be a factor in the relative unpopularity of these species.

Apart from fish from unrefrigerated boats, canners do at times have surplus catches from the RSW boats. This may lead to the curtailing of fishing or even the diversion of acceptable quality fish to the meal plant. Additionally, the shoals often contain more than one species. The predominant type is canned but other species, plus any damaged fish, are sorted out by hand and used for meal. A large cannery may divert as much as 1 tonne per day of fresh fish of this type to the meal plant, a small proportion from a plant capable of processing over 100 tonnes of fish in a day but, nevertheless, a large amount for a fish trader in Ecuador used to working with the small-scale fishery sector.

Despite the fact that meat and white fleshed fish are preferred foods, it is perhaps surprising to find in Ecuador that fresh sardine and mackerel are not popular but canned products are. Eating habits are undoubtedly a significant factor; canned products are ready to eat, whereas fresh sardine and mackerel are unfamiliar and many householders probably do not know how to prepare them. Canning also softens the bones and alters the flavour, making the fish more attractive to eat. Even so, these fish are extremely popular in other countries so there does not appear to be any fundamental reason why fresh sardine should not become popular in Ecuador. It could be a relatively inexpensive product where distribution costs are low.

New developments

While there are several major problems to be solved, there does appear to be some potential in distributing small pelagics as fresh fish. A campaign to promote good nutrition by increased consumption of small pelagic species would be one positive step towards stimulating a market. Handling practices would have to be improved; sardine is soft textured and very susceptible to physical damage. Fish traders in Ecuador do not currently use boxes; fish is bulked with ice in the backs of lorries for distribution. Suitable boxes would have to be introduced to ensure that iced sardines arrived at their destinations in reasonable condition.

Nevertheless, in Ecuador it seems unrealistic to expect more than a modest proportion of the small pelagic catch to be used as fresh fish. In a good year for the fishery, the present population could have a supply of over 100 kg per person compared with a current annual per capita consumption of about 4 kg. Clearly if a major proportion of the fish is to be used for direct human consumption, it must be processed and preserved in some way.

There is considerable potential in introducing products, which are well established in some countries, to regions where they are unfamiliar or not generally popular. Some adaptation of the products may be required, and undoubtedly they would need careful marketing. ODNRI/ODA has been instrumental in a programme aimed at introducing salted mackerel to the rural communities in remote mountainous parts of Ecuador. These communities previously had very little access to fish and very little purchasing power. The product consisted of split mackerel which was salted and distributed in saturated brine held in plastic buckets. Nearly 5 tonnes of product were sold over the six month period of the programme, but while a level of acceptability was achieved, production was disrupted by the non-availability of fish due to particularly adverse fishing conditions. It was not possible to demonstrate the economic viability of the operation and so it appeared unlikely, at least in the short term, that the private sector would be interested in adopting the production and distribution of this product. Nevertheless, this could be a particularly effective method of improving the nutritional status in these rural areas, should a government or aid agency be prepared to provide the necessary incentives.

It is not only in Europe that smoking is used to make a more attractive product. Various fish, including mackerel (*Scomber japonicus*), have been smoked experimentally in Ecuador giving highly acceptable British style hot-smoked products (Barratt *et al.* 1986) which have attracted considerable interest from local processing companies. Limited semi-commercial trials also indicated that the product could be sold through local supermarkets. As yet no company has committed itself to producing smoked fish, but at least one company has had samples prepared in order to investigate possible export markets. Interestingly, in another ODA supported project in Bolivia, smoked fish have gradually been popularised and have become a profitable activity, albeit small in volume terms.

In Ecuador attempts have been made to apply minced fish technology to the utilisation of small pelagic fish. The Ministry of Education used biscuits enriched with fish minces in a schools' feeding programme. Biscuits were preferred to other products because they can be stored at ambient temperature and need no further preparation; the schools have no facilities for storing or preparing food. Initially shark meat was incorporated (Bostock & Coulter 1985), but experimental batches of similar biscuits containing 20% sardine, mackerel and tuna dark meat (waste from canning) were prepared and an initial trial indicated their acceptability (Wood *et al.* 1988). This product could be compared with alternatives in any future school feeding programmes.

UNITED KINGDOM – MACKEREL (*Scomber scombrus*)

In recent history pelagic fish have been important in British fisheries but, until early this century, herring (*Clupea harengus*) was the principal element (Wood & Hopper 1984). A commercial fishery started to develop in the mid 19th Century; in 1881, some 170 thousand tonnes were landed, mostly from the North Sea, and the fishing and curing industries provided direct but seasonal employment for nearly 70 000 people. It peaked in 1913 when some 577 000 tonnes were landed of which nearly 75% was exported, mostly salt cured in barrels. Thus home demand was substantial, but the sheer size of the industry depended on export markets. Exports then declined progressively as other countries built up fleets to supply their needs and the UK catch halved by the mid 1930s, prompting the Government to set up the Herring Industry Board in 1935.

Immediately after World War II herring was plentiful, cheap and unrationed. Gradual decline of the catch developed from the 1950s, due primarily to the deterioration and eventual collapse of the North Sea fishery (the major herring fisheries were closed in 1977 and 1978). Despite marketing efforts, overall demand both at home and abroad declined. But as total international supplies of herring diminished, the demand for British herring as a food fish increased, resulting in a rapid increase in the proportion klondyked (direct landings or trans-shipments to foreign container or factory vessels for immediate export). This was mainly at the expense of the proportion for petfood, meal and oil. By now, clear consumer preference in the UK was for white fleshed species.

UK pelagic supplies were maintained in the 1970s, as herring declined, by diversion of the fleet to mackerel (Fig. 23.2). Other pelagic species such as sprat (*Sprattus sprattus*), pilchard (*Sardina pilchardus*) and horse mackerel (*Trachurus trachurus*) have been much less significant.

Mackerel provides a particularly good example to consider; it is probably the most studied and well documented pelagic species in terms of its biology and the management and utilisation of the fishery. Whitmarsh & Young (1985) reviewed the economic importance of mackerel to the UK and the historical development of the fisheries. Lockwood (1988) has reviewed the biology and the management of the fishery in detail.

Mackerel have been eaten in Britain for a great

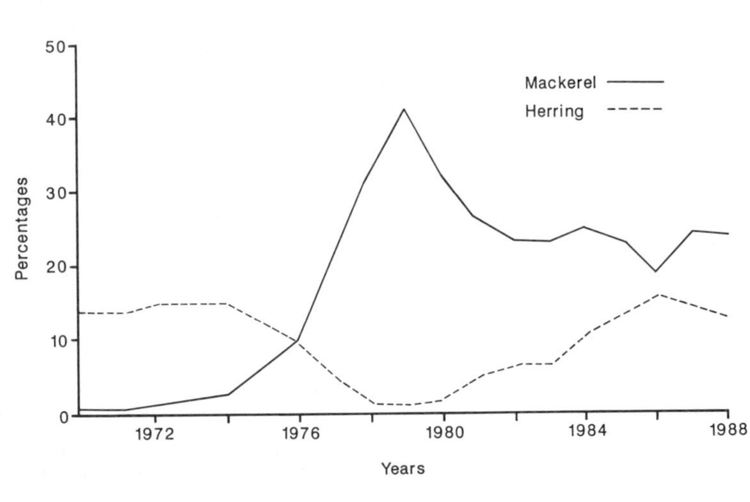

Fig. 23.2 Variability of herring and mackerel catch 1970–88 as percentages of total landings of fish and shellfish by British vessels. Figures for 1986 to 1988 are provisional.

many years but only in small quantities. In the 1850s, although completely overshadowed by herring, mackerel were a more important element of fish supplies than in the 1950s, particularly for the poor. Then Billingsgate Market alone handled roughly 10 000 tonnes of fresh mackerel a year (Mayhew 1851). At the beginning of this century the catch peaked at about 20 000 tonnes, but mackerel landings by British boats steadily declined to less than 3000 tonnes before World War II. After the war they declined further to about 1000 tonnes in 1960, rising slowly again to about 3000 tonnes by 1969. This probably represents the best estimate of the then low scale of home demand or consumption, predominantly in the south of England, for chilled fish.

By the mid 1960s most mackerel for human consumption was supplied by the Cornish summer handline fishery (April–September). Its modest development in the late sixties was based in part on growing demand among the immigrant population (Keay 1970) but for most of the population demand remained weak. This period coincided with the unusual appearance of dense mackerel shoals in the winter months inshore off the south Cornish coast, which later turned out to be a temporary shift in migration patterns probably due to complex environmental changes.

Development of the fishery, supplies and the industry

As the south-west fishery expanded, trawling became more important; the trawlers fished overnight and stowed the catch without ice, but it was often mixed with pilchard (*Sardina pilchardus*) and scad (horse mackerel, *Trachurus trachurus*). The fish were suitable for canning if scad were absent, but most trawl-caught fish were sold for fish meal, and line-caught fish for human consumption. Freezing whole fish extended supplies beyond the catching season and satisfied export markets. Mackerel could be caught relatively easily in quantity by trawl off Cornwall and, initially, were exploited heavily by the Eastern Bloc fleets. The international catch rose to over 300 000 tonnes by 1975, but UK vessels only contributed 32 000 tonnes which, nevertheless, was large by historical standards and far exceeded the demand for human consumption in the UK. Consequently much was processed to fish meal and pet food. However, markets also developed in Europe for supplies of chilled fresh fish transported by refrigerated lorries.

The collapse of the herring fishery and the exclusion from traditional distant water grounds by extensions of fishery limits elsewhere had led fishermen and processors to seek alternatives. Extension of EC fishery limits denied access to the Eastern Bloc fleets and resulted in a spectacular increase in the UK mackerel catch (Fig. 23.2); it became the most important single species by 1978, contributing 322 000 tonnes. Compared with the traditional herring fisheries, this diversion of effort was based on relatively few ports in south-west England and north-west Scotland, areas of above average unemployment. The growth and fluctuations in total landings of mackerel in the UK over the last 20 years, mostly supplied by UK fishing vessels, are shown in Table 23.5.

Mackerel made up 87% of the pelagic landings by 1979 when the south-west fishery peaked, but stock abundance declined rapidly after 1980 and regulatory measures were introduced. Scottish landings assumed increasing importance, contributing 96% of the total mackerel supplies by 1985. Nearly 80% of this was landed at one port, Ullapool on the north-west coast, sparsely populated and with virtually no processing infrastructure. Supplies from north-west Scotland were trucked to processors in the south-west of England, the north-east of Scotland or Humberside. In fact the main mackerel processing and marketing activity had not developed in the south-west where the bulk of the mackerel was landed, but in Humberside and Scotland where some companies had turned to mackerel in the face of herring and cod shortages. The immediate outlook for maintenance of current Scottish landings is not optimistic and the period of seasonal availability is tending to decrease.

In the peak years most of the catch was taken by purse net gear from vessels discharging virtually daily. However, the continuing change in migration pattern to deeper waters further offshore from the north of Scotland has necessitated larger vessels fishing for longer periods, and requiring improved holding facilities onboard to maintain quality. There are also problems catching the fish by purse net in the deeper waters which may necessitate a return to trawling by freezer trawlers (SFIA 1989). Flexibility is essential within the industry to deal with all these variables in the seasonal and annual supplies.

Surprisingly, the UK imports about 20 000 tonnes of mackerel (1987), mostly from Ireland, including vessel landings and frozen fillets. The latter has increased to nearly 1000 tonnes and may reflect trade opinion that the average size of fish available from UK landings, for the domestic smoking industry, had diminished. In addition about 2000 tonnes of canned fish is imported, mostly for home consumption. In total including imports, the equivalent of some 45 000 to 50 000 tonnes of whole fish is immediately poten-

Table 23.5 Mackerel in pelagic landings+ in UK and Scotland 1969–88. (MAFF/DAFS fishery statistics.)

Year	Total pelagic 10^3 t++	Total mackerel 10^3 t++	% pelagic	Scotland mackerel 10^3 t++	% total mackerel	Scotland mackerel £/t
*1989	273	158	65	151	96	118
*1987	308	190	62	170	90	113
1985	299	178	60	171	96	106
1983	276	202	73	120	59	110
1981	268	198	74	115	58	95
1979	415	363	87	108	30	95
1977	340	187	55	54	29	81
1975	234	49	21	17	35	33
1973	288	23	8	8	35	47
1971	215	7	3	1	14	39
1969	187	4	<3	1	—	25

+ Rounded to nearest 1000 tonnes.
++ Includes landings of foreign vessels and trans-shipments to klondykers, ie direct landings to foreign container or factory vessels for immediate export.
* Provisional figures.

Table 23.6 Comparison of the disposal of UK caught mackerel 1981–7. (SFIA 1989.)

	1981 %	1987* %
Klondyked	46	66
Home market	22	13
Landed abroad	19	15
Meal, oil, petfood	13	6

* provisional figures

tially available to the UK market for home consumption, i.e. some 25% of current landings.

Markets, handling and processing

Mackerel successfully provided an almost instant alternative for fishermen; it could simply replace herring or even other species such as cod for the distant water freezer fleet. There were bulk markets (Table 23.6) which could be satisfied fairly readily: the meal, oil and petfood industries; the Eastern Bloc factory vessels; chilled fish for Holland, France, West Germany, etc; frozen fish for West Africa; and there were EC financial incentives for freezing-at-sea in EC waters.

For the processor who had been geared up for herring products for home consumption or export, the problems were both difficult and complex: herring and mackerel products simply were not interchangeable; and there was no immediate substantial market for consumption of mackerel in the UK and it continues to be relatively weak (Table 23.7), being only about 20 000 tonnes. Smoked products, particularly hot

Table 23.7 UK consumption of mackerel (tonnes) 1983–9.

Year	1983	1986	1989
*Household purchases**			
Unsmoked products	1 950	2 219	2 008
fresh: whole fish	1 433	1 584	1 408
fillets	226	458	441
frozen: fillets	291	177	159
Smoked products	2 494	3 198	3 434
fresh: whole fish	192	309	290
fillets	1 864	2 276	2 539
frozen: fillets	438	613	605
Caterers' purchases (estimate)*			
Smoked products	n.a.	1 460	2 696
*Prepared and preserved**			
Canned predominantly	1 660	1 696	1 000
Estimated consumption			
Whole fish equivalents (cf. data above)	12 450	18 400	20 500
UK consumption % of landings	6	14	13

* SFIA, Fishery Economics Research Unit
na: no data available

smoked products, have been the most successful, having shown steady growth in volume and become popular in the pub and catering trade. Indeed, even since 1986 total consumption of smoked products has increased by over 30% in terms of product weight. The home market (Table 23.6) also includes fish processed for export and handles roughly ten times the amount of UK-caught fish handled in 1970. Only about 5% of the catch is now sold directly to the byproducts industry.

The fishery remains heavily dependent on export markets for whole fresh fish, either via the klondykers or directly via landings in foreign ports (Table 23.6). This dependence coupled with the competition for such markets determines the low price achieved for the raw material (Table 23.5), which has failed to keep pace with inflation; the price in 1987 was only 75% in real terms of the price in 1974. However, these outlets together with the byproducts industry can cope with large fluctuations in supply, and provide the flexibility which the current modest requirements of the UK market for human consumption cannot. Nevertheless the proportion of the catch destined for human consumption in all markets at home and abroad has increased to well over 80%, depending on the markets for the fish landed abroad.

Even before any real expansion of the fishery in the south-west took place, there were difficulties in marketing mackerel due to its unfamiliarity as a food fish among consumers in the country as a whole; a firmly and widely held but wrongly based belief that it was a scavenger fish, somehow unhygienic and not suitable for the table; and to the availability of relatively inexpensive, imported, canned mackerel, mostly from Japan (*Scomber japonicus*). Particularly during the period of rapid, one hundred-fold expansion of the fishery, considerable technological effort was devoted towards better utilisation of the resource by means of improved handling, processing, product development and marketing. The emphasis was on the quality and safety of products and hygienic production, and the problems for processing and product quality of the variable nature and composition (Table 23.8) of the fish available, the seasonality of supplies and the poorly developed processing infrastructure in the sourth-west. Apart from the problems of consumer perception of mackerel, manufacturers considered that rancidity was a major impediment to the production and distribution of frozen mackerel products.

As the fishery and onshore handling, processing and distribution expanded, various quality and marketing problems became apparent in the trade and some concerns were expressed about aspects of safety of mackerel products. Some examples are considered below.

Table 23.8 Seasonal variation in % lipid content of SW mackerel.

Month caught	Headed and gutted fish	Fillet flesh Mean	Fillet flesh Range
January	19.1	15.2	10.0–18.2
March	8.7	7.1	3.3–14.1
May	6.3	3.3	1.3–8.0
June	7.6		
July	9.9	12.0	3.7–20.0
August	14.3		
September	11.2	17.2	8.3–23.8
October	17.5		
November		19.0	15.3–25.0
December	19.8		

Scombrotoxin poisoning

Although prized by some consumers, the public perception of mackerel remained poor. There was a widely held view that the fish spoilt rapidly producing noxious substances. Under some conditions, storage of scombroid fish produces histamine and other amines in the flesh, considered as possible sources of the alleged unpleasant effects (Kimata 1961). In mackerel histamine was produced above 0°C as a result of bacterial histidine decarboxylase activity which was highest in liver tissue (Fernandez-Salguero & Mackie 1979). At ambient temperatures it was produced relatively quickly, but all the evidence suggested that mackerel would become putrid, inedible, and bellies would have burst before levels of histamine and the diamines were high enough to cause effects which could be related to histamine itself (Hardy & Smith 1976).

Nevertheless, reported scombrotoxic fish poisoning incidents, although generally not serious and relatively few, became more frequent, particularly in 1979. Spoilage investigations and observations of inadequate handling in catching, processing and distribution (Gilbert *et al.* 1980), led to a concerted effort to advise the industry directly that, since some degree of spoilage seemed necessary to cause poisoning, the problem could be diminished by proper attention to temperature control and hygienic principles throughout production, distribution and storage.

Processors were advised to use good quality raw material, kept chilled below 5°C prior to smoking and after smoking, and to indicate to both retailer and consumer the shelf-life and the storage conditions required for the product. Some loss of business resulted from these outbreaks and improvements in the practices were essential to safeguard further losses. Collaborative, chemical, biochemical, microbiological and medical investigations of the cause of this illness have continued (Chapter 17, Murray 1982, Clifford *et al.* 1989) but, in the absence of a proven aetiology, it is not possible to recommend handling practices to the industry which will ensure that no scombrotoxic incidents occur, only that incidents can be reduced.

Botulism

This neuroparalytic disease resulting from ingestion of toxins produced by the bacterium *Clostridium botulinum* occurs extremely rarely in Britain; fishery products, in particular, have an excellent record of freedom from incrimination in outbreaks of this frequently fatal food poisoning. The organism was found in 4% of Cornish mackerel examined, somewhat higher than the 1% or less usually found in British inshore fish, but still of little significance in most fishery products as any toxin formed during growth of *Clostridium botulinum* during poor storage of the product would be destroyed by normal cooking procedures. If growth and toxin formation does occur in fishery products it is almost always a result of faulty handling and processing (Hobbs 1987).

Hot smoked mackerel were being produced more widely and the fish were more lightly cured. In hot smoked products, eaten without further cooking, there is a potential hazard if hygiene standards are allowed to fall and temperatures are allowed to rise during distribution and storage. Outbreaks of botulism from hot smoked fish had occurred in North America in the 1960s (Lewis and Cassel 1964), and in Germany (Baumgart 1970). A survey in the south-west of production and distribution revealed the need for the following recommendations for quality and safety to be circulated widely:

(1) On receipt, thoroughly eviscerate, de-gill, wash mackerel
(2) Minimal salt concentration 3% (water phase) for smoked fish
(3) All processed products should be stored below 4°C until sold
(4) Fish presented for sale without refrigeration should be discarded if unsold at the end of the day's trading
(5) Smoked mackerel packs should carry a 'sell by . . . date' not more than 7 days from the date of smoking, with a further instruction that the contents be eaten within 24 hours if kept at room temperature, and 3 days if refrigerated.

Detailed studies of samples inoculated with spores and vegetative cells were carried out to establish safe, commercially practicable processing and storage conditions for vacuum packed hot smoked mackerel conpatible with the trends in consumer preferences towards lighter smoke cures. Experimentally a vacuum packed, hot smoked product with a minimal salt concentration of 3% in the water phase had a safe shelf-life of 30 days at 10°C, but only 1 day at 20°C (Cann & Taylor 1979). Thus the reluctance of some companies to market chilled vacuum packed products was unfounded.

Oxidation and rancidity

The oxidation of mackerel during frozen storage was very similar to herring stored under similar conditions (Banks 1952). Fish could be stored glazed at −30°C for 44 weeks before rancid flavours were detected. Frozen in polyethylene bags topped up with water, which filled the voids and surrounded the fish in ice, the fish could be stored for more than 1 year at −30°C. As expected, vacuum packed frozen fillets were protected for much longer than the non-vacuum packed; at −30°C rancid flavours eventually developed by 72 weeks. The detrimental effects of higher storage temperatures and extended chilled storage prior to freezing were elucidated (Hardy & Smith 1976).

'Fluffy' hot smoked mackerel

A number of processors encountered a problem with mackerel fillets frozen after hot smoking. Apparently, perfectly good frozen fillets expanded during cold storage to as much as three times their original thickness, giving a 'fluffy' appearance similar to very severe freezer burn. On thawing, the fillet returned to its normal thickness but was damaged irreversibly, being more friable and unsaleable. Tests showed that this only occurred with hot smoked fillets made from mackerel containing more than 15% oil. Expansion did not occur during freezing or when stored below −17°C. If the susceptible, frozen, hot smoked fillets were raised to a temperature of −5°C or higher and then recooled, expansion took place within 40 minutes; tight packaging reduced expansion. Other factors, such as salt, smoke, fresh *vis* frozen fish, method and rate of freezing, had no influence. Thus the high fat product was obviously susceptible to poor cold storage practice. The advice was to keep temperatures low and avoid refreezing.

'Soft mackerel'

Some 1–2% of the fish landed were 'soft' fish, so soft that the flesh could be squeezed out of the fish like a paste. This was shown to be the result of extensive proteolysis, probably due to infestation by a protozoan parasite, tentatively identified as *Chloromyxum histolyticum*. Such fish were heavily parasitized; others contained small pockets of parasites without exhibiting softness and could be filleted.

Chilled sea water systems

Bulk stowage on board in refrigerated sea water (RSW) or chilled sea water (CSW) systems was required because of the rapid expansion of catching, primarily due to the large catching capacity of purse seine vessels, the large ready market of the klondykers for trans-shipment of quality fish for processing for human consumption, and the need to handle mackerel in bulk. The klondykers could handle 55 to 200 tonnes whole fish per 24 hours. The background research was done initially in Canada and the UK in the 1930s, and the systems first introduced commercially some 15 years later in the Californian sardine industry and subsequently, with further experimentation, by industry in Norway, western North America, Eastern Europe, South Africa and Japan. But problems with stratification, cleansing, hygiene, unloading, poor quality and salt penetration of the product persisted.

In the early 1970s, the White Fish Authority (WFA) and Torry Research Station (Hewitt & McDonald 1972) experimented with various systems on herring vessels. As the purse seiners opted for the advantages of bulk tanks rather than container systems, detailed trials provided the data for advice (Kelman 1977), for the design and installation of RSW, CSW or combination systems (Hewitt *et al.* 1978), and for codes of practice for operation. The type of system installed depended on the economics of capital and operating costs and the availability of ice.

Rapid cooling of fish began immediately after pumping from the net into CSW. Not only was it simple but the quality on landing of large catches was better than best boxing in ice. Mackerel stored in ice developed 'off' flavours 1–2 days earlier than in CSW, in which fish were acceptable for up to 8 days, with no evidence of rancid flavours up to 10 days. Storage in CSW for a few days before canning resulted in an improvement of flavour after processing, although texture was not affected. Salt concentrations increased 2–3 times during 6 days in CSW and were affected by oil content, being least in fish with the highest oil contents in September to November (Smith *et al.* 1980).

Marketing and the consumer

The WFA Market Development Unit initiated studies (Whittle 1976) to build on the proven basic handling and processing techniques established by Torry Research Station, and the devleopment of the modest processing industry sector, to find ways of stimulating home consumption and identifying markets abroad in order to benefit from the potential value added by processing. In general housewives were aware of mackerel but prejudiced against it; the fish was not considered popular.

Canned mackerel was the most familiar product, used as a substitute for other canned fish varieties. Few were familiar with the smoked product, but it was liked by those who tried it. Most were concerned about freshness and the alleged scavenging habits of the fish. Despite the low price most ignored it. Those who ate mackerel regarded it well: it was filling, meaty, strong-tasting, dark fleshed and suitable for a man's main meal, but difficulties were experienced in obtaining it from local fishmongers who in general were thought to be unenthusiastic in promoting it to and preparing it for the customer. In turn, fishmongers believed mackerel was unpopular among customers.

A supermarket promotional exercise (Anon. 1978), backed up by a recipe pamphlet, was set up in the London area for fresh fish, fillets, hot smoked fish, fillets and paté. It was later expanded to the rest of the country. The resulting increased sales for fresh fish, the most popular purchase, appeared to be price-related impulse purchases rather than due to consideration of the convenience or nutritional value of the products. Flavour and taste were more important for smoked products; paté was well regarded.

In the early 1970s canned mackerel consumed in the UK was supplied almost exclusively by Japan. Substitution by home production faced two problems: the keenly competitive price the Japanese suppliers could sustain and the alleged preference of consumers for canned Japanese mackerel. In trials (Anon. 1979), both products were considered good but the flavour and colour of the Japanese one was preferred and likened to salmon, while the dark skin of the British product was disliked. No statistically significant preference between products was indicated on the basis of texture; generally, less oiliness and a firm texture were preferred. This is interesting because the softer texture of the Japanese product had been identified earlier in consultation with British canners as a highly desirable characteristic to try to incorporate in British canned mackerel to make it more attractive to the consumer. Indeed, research successfully softened the texture by controlled use of papain during processing but, in the light of the later consumer information, the work was wrongly directed. The entire problem became academic because Japanese dominance of the UK sales declined rapidly about this time to 38% of market volume, due to scarcity of mackerel in Japan and strong promotion of the British product by one of the brand leaders.

Today the Japanese product is insignificant in the UK market. By 1984 over 80% of the imported canned mackerel (worth £2.5 m but only about 1% of the total value of canned fish imports) came from Denmark (1600 tonnes) in ring-pull flat cans; Denmark was highly competitive because of modern volume production. Some 2700 tonnes canned mackerel were produced in the UK in 1985, over 40% of which (nearly £1.5 m) was exported.

Packaging and shelf-life

Packaging provided an additional dimension to contribute to the convenience, presentation and stability of mackerel products, but it was essential to determine the safe shelf-life of chilled packaged products and to assess the potential dangers of temperature abuse in respect of the scombrotoxic and botulinogenic potential of the packaged products.

Recommendations (Mills & Tiffney 1982) were made for the packaging of fatty fish in modified gas atmospheres (MAP). However, more detailed studies later (Cann *et al.* 1983), found no extension of shelf-life for hot smoked mackerel when stored in either 40/30/30:$CO_2/N_2/O_2$ or 60/40:CO_2/N_2 compared with vacuum packed products. There were advantages in reduced drip loss in MAP, convenience and presentation, but the presence or absence of oxygen had no effect on pack collapse. The flora of the spoilt product was altered but had no effect on sensory assessment of the hot smoked mackerel. Packing in MAP in the absence of oxygen did not in itself significantly increase the food poisoning hazards of the packaged fish. Gas mixtures rich in carbon dioxide are the most effective in inhibiting bacterial spoilage (Statham 1984); such conditions may favour the growth of *C. botulinum* when fish is stored at temperatures above 4°C.

Despite the reassuring conclusions that differences in the rate of toxin production between vacuum packaged (VP), MAP and air packed fish products are marginal (Cann *et al.* 1983), the remaining concern in some quarters about the potential toxicity in VP and MAP fish products which are likely to be consumed without further cooking, is known to have inhibited some large retailers from selling packaged mackerel. Compared with air packs, shelf-life extension was observed for vacuum skin packed, hot smoked mackerel products stored at 5°C which was further enhanced at 0°C.

New developments

The UK mackerel fishery has grown in spite of, not because of, consumer demand in the UK. Nevertheless, significant home production and markets have been built up for canned, hot smoked and fresh products. However, domestic demand has remained modest in comparison with the level of landings, and the bulk of the catch is exported. Landing prices have remained low due to this weak demand even though the exported fish is now used mainly for direct human consumption. The domestic market needs to be further developed by expanding the consumption of established products and introducing new, attractive products.

Basis of consumer choice – product innovation

With the reopening of the major herring fisheries it became very clear that, on the basis of the then current range of products, none of the pelagic species was likely to make much impact on consumption patterns because the preference was for the popular white fleshed fish and shellfish species. Although information on consumer attitudes to fish (and the pelagics) was increasing, it was essential to investigate the real basis for consumer choice of fish compared with other foods, and for choice among the variety of fish products available. The results of some of this recently completed behavioural research related to the pelagic resource were reported to this conference (Chapters 25 and 26).

In the meantime, in response to the widely held views about consumers' attitudes to fatty species, the views of manufactures and the increasingly important changes in food habits towards convenience and snacking, a fresh experimental initiative was taken on product innovation with fatty species following both traditional and alternative approaches. Examples of the former were a fresh look at the characteristics of mackerel marinades, patés and sausage products. In addition, it was clear from earlier market research that the changes occurring in oily fish during chilled storage, and consumer reaction to them, would be of considerable importance in relation to acceptance and possible market expansion (Baird *et al.* 1985). Responses to hot smoked and unsmoked mackerel fillets prepared from whole fish which had been cold stored, since much of the mackerel processed in the UK was now being handled in this way, clearly showed that rate of loss of acceptability during chill storage was much slower for hot smoked than unsmoked fillets; smoking clearly conferred a liked characteristic (Baird *et al.* 1985).

Technically many problems had to be dealt with in

considering alternative approaches: seasonality, differences in composition, mix of species available, dark colour, strong flavour, aftertaste, bones, dark skin, appearance, rancidity. One of the objectives for example, which also related to patés and sausages, was to be able to formulate the composition of the product irrespective of the species or mix of species available, and therefore be independent of variability in supply and composition. It was not considered a constraint that the product need be clearly identifiable as fish.

The substantial retail market for frozen breaded portions was devoid of fatty fish based products and was identified as an opportunity. It is a strongly held view by some that fish fingers are not necessarily identified strongly with fish at all. Some prototype herring products, in the form of fingers of deboned fresh or kippered flesh in oatmeal crumb, were well received in the early 1970s (Bannerman *et al.* 1973) but were not pursued successfully by industry, perhaps because they were fingers but also due to concern about rancidity. Nevertheless, it is worth recalling that commercial decisions at Birds Eye in a previous generation had narrowly opted to launch the cod fish finger (eventually a highly successful product but currently waning), rather than the competing herring finger.

The frozen portion products were based on bone-free, skin-free, mackerel mince produced from the familiar Baader deboning machine. A range of experimental products of different shapes were produced in a typical high capacity, versatile food former, and enrobed with a variety of crumb coatings or pastry. They were formulated from the mince, with a variety of sauces and spices, including tomato, curry, mustard, peppers, pineapple and so on, to confer attractive flavours, and with recovered white fish flesh, or rice or potato, to ameliorate the strong flavours, oil content and colour of mackerel, and to give different eating characteristics. [Mackerel mince could be whitened by treatment with peroxide (Young *et al.* 1980), with some protein insolubility and loss of sulphur amino acids, but the process is not approved.]

The results from taste panel assessments among polytechnic students and from the college restaurant were very encouraging (Keay & Parsons 1984, 1987). Frozen products stored without protective vacuum packaging and simply packed in the usual commercial way for portion products were remarkably stable, as found earlier for frozen herring portion products (Cole & Keay 1976). Six months shelf-life at −30°C was entirely feasible. The minces on which these products were based, when packed properly in conventional block liners avoiding voids, were remarkably stable, but vacuum packaged blocks showed hardly any change in peroxide value over 2 years at −30°C. Thus problems of rancidity did not seem to present any obstacles for these product concepts.

A particular problem with mackerel, identified in further development by SEAFISH (UK Sea Fish Industry Authority), was the control of mince texture in formulated experimental products, particularly those made from fish which had been frozen. The interactions between the ingredients which contribute to texture can be determined experimentally so that differences in fish composition can be overcome by changing the formulation (Chapter 32). The properties of surimi and kamaboko made from mackerel have also been described (Chapter 34).

Major difficulties for the home market remain in terms of prejudice and unfamiliarity with the species and with the lack of success in introducing more acceptable products to overcome the barriers to consumer purchase. This must be the key area to solve for further improvement in the home market. On the one hand it still seems realistic to develop convenient products in the frozen food sector in which the identity of mackerel is hidden and in which the products are not even identified with fish, while on the other hand there also seems to be scope for building on the satisfying attribute of dark flesh, which for some consumers was an advantage but perhaps for the majority was linked to major dislikes such as strong flavours.

Another prospect for maximising the value of the resource must be potential export opportunities for more sophisticated mackerel products in countries where the species is fully accepted. Most of the information required by the fisherman and processor for handling and processing a wide variety of high quality mackerel products is now available.

HOW CAN MORE SMALL PELAGIC FISH BE USED FOR HUMAN CONSUMPTION?

Basically there are two types of problem to be solved: ensuring that good quality fish is available for utilisation, and developing marketable products suitable for the intended consumers. Solutions to both are essential if small pelagic fish are to achieve their full potential.

Improving fish quality

Good quality raw material is a prerequisite to successful product development and the maintenance of markets once captured. Pelagic species spoil rapidly but reduction of the rate of loss of quality begins with good handling and rapid chilling on board the fishing vessel. Well designed and properly operated refrigerated or chilled sea water systems provide an efficient and effective means of rapidly chilling large catches. Even in the developed countries there is plenty of scope for improvement in the design and use of these systems (Chapter 6), while the artisanal fisheries in the LDCs rarely chill the catch in spite of high ambient temperatures, but depend on the brevity of fishing trips to avoid spoilage problems. Similarly at each stage in the production and distribution of products, attention must be paid to proper handling and processing practices, good hygiene and avoidance of temperature abuse to deliver high quality safe products to the consumer. Clearly if fish which is currently used for meal production is to be diverted to direct consumption, improvement of the quality of the fish landed is essential.

Development of marketable products

Poor consumer demand has been the primary reason for the underutilisation of small pelagic fish. In India this problem has eased and domestic demand has expanded. The UK and Ecuador still depend on export markets, with meal production dominating in Ecuador and other countries such as Peru and Chile. In some countries there is strong consumer resistance to the introduction of unfamiliar species previously used for meal production, because fish have an 'only fit for fish meal' image. The greatest potential is undoubtedly for developing local and export markets for products for human consumption in countries such as Ecuador, although both the UK and Indian fisheries industries would greatly benefit from a more attractive and diversified range of products. While there may be political and social pressures to produce inexpensive products for low income consumers in LDCs, economic pressures may indicate that more sophisticated higher value products for export may be preferred. How can these changes in consumption be brought about?

There have been various attempts by Government sponsored institutions and industry to generally popularise the traditionally unwanted species and to develop attractive products which could then be taken up commercially. These products have ranged from inexpensive dehydrated products for the poor in LDCs to the very sophisticated aimed at consumers in industrialised countries. Some of these efforts have had little lasting impact (of which fish protein concentrate is perhaps the most well known disappointment; reviewed by Pariser *et al.* 1978). Our three examples indicate some successes: fresh sardine in India; canned sardine and mackerel in Ecuador; canned, hot smoked and fresh mackerel in the UK. These successes have come about slowly and have occurred because of consumer demand. A strong preference for white fleshed fish in the UK and Ecuador probably explains why consumption of small pelagics in these countries is still relatively low. It is because of this that the experimental development of novel products which disguise the nature of the raw material has received considerable attention.

The conservatism of consumers is reflected by industry's reluctance to adopt new products which use small pelagic species. Various canned mackerel products, fish in sauce products, and frozen enrobed portions have been described (Keay 1970, Keay & Hardy 1974). Fish-in-sauce dishes are now a rapidly growing sector in the UK but popular white fleshed species are mainly used so far. Hot smoked mackerel and other products were introduced to Ecuador (Bostock & Coulter 1985, Barratt *et al.* 1986) but to date, in spite of considerable interest being expressed by several fish processors, the industry has not tried to introduce the products commercially. Only when the products are seen to be potentially successful will companies be prepared to start producing and marketing them.

Small pelagic fish are increasingly being utilised for direct human consumption in many countries. The authors do not doubt that this trend will continue. In countries such as India, where these fish are already popular, expanding use for direct consumption will probably be largely due to an expansion of present methods of usage. Increasing demand will see growing competition for raw material, leading to a decline in meal production. There will probably also be increasing opportunities for more sophisticated processed products as social and economic development follows the patterns which have been seen in industrialised countries. In other countries the pattern of utilisation will have to change considerably if small pelagics are to achieve anything like their full potential in the national diet. Nevertheless, rising populations will increasingly exert pressures on the currently popular fish resources, leading to rising prices and the search for cheaper alternatives. There is a strong case for consumer education campaigns and technical assistance to encourage both the industry and the consumer to try small pelagic species to their mutual benefit.

Technological advances and novel products can help in the process but the introduction of new food products in any country requires some degree of investment and is risky. Fresh and cured fish are inexpensive and seemingly the most suitable products for large scale consumption in LDCs, but little sustained work has been done on promoting them in countries where they are not traditionally consumed.

The successful development and introduction of novel products, and the search for non-traditional raw materials to make products such as surimi, may open up export possibilities for small pelagics. It is noteworthy that in Ecuador canned fish intended primarily for export is an increasingly popular product on the home market. Similar developments could take place with other products. These products will almost certainly be relatively expensive and therefore unlikely to be suitable for lower income consumers, but would be important developments for the industry.

Whether low cost products specifically aimed at poorer consumers will have much impact in the future is debatable. It is a high risk strategy. Experience in Ecuador has shown that the introduction of unsophisticated conventional style, but unfamiliar, products is difficult; these problems could be compounded if more sophisticated, capital intensive technologies are used to produce products which are even less recognisable as fish. The commercial sector would not be interested in risky projects which offer limited returns; the public sector should consider carefully the existing market and consumer preferences in targeting any intervention. Nevertheless, over 20 years the combination of infrastructural and technical improvements coupled with commercial enterprise has greatly improved the utilisation of this resource in many locations, and sustained efforts will undoubtedly lead to continued progress.

PROPOSALS FOR FUTURE STRATEGY

The status of small pelagic fish utilisation varies considerably from country to country and from species to species. Improved utilisation of this resource is desirable for nutritional and economic reasons. In India these changes have taken place to an appreciable extent and in the UK domestic market progress has been limited but significant, while in Ecuador the huge fishery has had very little impact on local fish consumption. Consumer demand has largely controlled the rate of progress in the examples of the three countries discussed.

The following strategy is recommended to improve pelagic fish utilisation:

- evaluate the resource, and local and export markets for products
- identify products acceptable to particular potential markets
- promote products from small pelagic fish
- monitor handling and processing with respect to product safety and quality
- ensure that adequate standards or specifications are met

It must be appreciated that changes will be slow and that the introduction of new products is commercially risky; even after good market research and promotion the product can fail. Expansion in the consumption of existing products, or the substitution of small pelagics for more popular species, may be the most effective way to overcome consumer prejudices against small pelagic fish. Launching unpopular fish as unfamiliar products seems doubly risky.

These considerations are as valid for products intended for low income consumers in LDCs as they are for industrialised countries. Fresh fish is the preferred product in Ecuador (and elsewhere) but sadly there has been little sustained attempt to popularise fresh, small pelagic fish in that country. Food preferences are part of the consumers' culture and are unlikely to change quickly. The authors recommend a cautious, step-by-step approach which considers all the technical, social and marketing aspects mentioned above, but one which is sustained and committed to improving the current situation in target areas.

BIBLIOGRAPHY

Anon. (1978) *Mackerel promotion exercise conducted in six supermarkets in the London area.* Tech. Rep. 161, White Fish Authority.

Anon. (1979). In-home placement trials to compare the acceptability of British canned mackerel with Japanese canned mackerel. Tech. Rep. 170, White Fish Authority.

Baird P.D., Bennett R., Hamilton M., Smith J.G.M. & Whittle K.J. (1985) Consumer acceptability as the basis for determining the chilled shelf-lives of smoked products from thawed herring and mackerel. *Refrig. Sci. and Tech.*, **4**, 243–8. IIR, Commissions C2 & D3, Aberdeen.

Banks A. (1952) *The freezing and cold storage of herrings.* DSIR Food Investigation Special Report 55, HMSO, London.

Bannerman A., Keay J.N., Smith J.G.M. & Hardy R. (1973) Kipper fingers – a new fish product. *Food Manufacture*, **48** (10) 63–4.

Barratt A., Coulter J. & Camba N. (1986) Produccion y mercadeo de pescado ahumado. Una guia para productores. *Boletin Cientifico y Tecnico*, **8** (9) 1–22. Instituto Nacional de Pesca, Guayaquil, Ecuador.

Baumgart J. (1970) Vorkommen und Wachweis von *Clostridium botulinum* Typ E vei Seefischen. *Arch. Lebensmittelhyg*, **23**, 34–9.

Bostock T. & Coulter J. (1985) Low-cost fish products for popular consumption in Ecuador. *Infofish Marketing Digest*, (3) 23–4.

Cann D.C. & Taylor L.Y. (1979) The control of the botulism hazard in hot-smoked trout and mackerel. *J. Fd Technol.* **14**, 123–9.

Cann D.C., Smith G.L. & Houston N.C. (1983) *Further studies on marine fish stored under modified atmosphere packaging.* Occasional publication, Torry Research Station, MAFF.

Clifford M.N., Walker R., Wright J., Hardy R. & Murray C.K., (1989) Studies with volunteers on the role of histamine in suspected scombrotoxicosis. *J. Sci. Fd Agric.*, **47**, 365–75.

Cole B.J. & Keay J.N. (1976) The development of rancidity in minced herring products during cold storage. In *Proceedings of conference on production and utilization of mechanically recovered fish flesh (minced fish)* (Ed. by J.N. Keay), pp. 66–70. Torry Research Station, Aberdeen.

Fernandez-Salguero J. & Mackie I.M. (1979) Histidine metabolism in mackerel, *Scomber scombrus*. *J. Fd Technol.*, **14**, 131–9.

Gilbert R.J., Hobbs G., Murray C.K., Cruickshank J.G. & Young S.E.J. (1980) Scombrotoxic fish poisoning: features of the first 50 incidents to be reported in Britain (1976–79). *British Medical Journal*, **281**, 71–2.

Hardy R. & Smith J.G.M. (1976) The storage of mackerel (*Scomber scombrus*). I, The development of histamine and rancidity. *J. Sci. Fd Agric.*, **27**, 595–9.

Hewitt M.R., Kelman J.H. & McDonald I. (1978) Chilled sea water systems for the preservation of fish. *Proceedings of the Institute of Refrigeration 1977/78*, **74**, 1–8.

Hewitt M.R. & McDonald I. (1972) Containers for herring in chilled sea water. *Fish. News Int.*, **11**, (12) 67.

Hobbs G. (1987) Microbiology of fish. In *Essays in Agricultural and Food Microbiology*, (ed. by J.R. Norris & G.L. Pettipher), pp. 199–226. John Wiley, Chichester.

Keay J.N. (1970) The utilization of mackerel by processing. *Scottish Fisheries Bulletin*, **34**, Dec., 15–17. Dept of Agriculture & Fisheries for Scotland, HMSO, Edinburgh.

Keay J.N. & Hardy R. (1974) Utilisation of fish resources. *J. of Hotel Catering and Institutional Management Ass.*, (25) 9, 11, 13, 16.

Keay J.N. & Parsons E. (1984) *Product innovation with fatty species of fish.* Unpublished paper presented at a meeting of the West European Fish Technologists' Association, Madrid.

Keay J.N. & Parsons E. (1987) The production, acceptability and storage properties of coated fatty fish portions. In *Savoury Coatings* (Ed. by D.B. Fuller & R.T. Parry pp. 37–46. Elsevier Applied Science, London & New York.

Kelman J.H. (1977) *Stowage of fish in chilled sea water.* Torry Advisory Note 73, MAFF, HMSO, Edinburgh.

Kimata M. (1961) In *Fish as Food* (Ed. by G. Borgstrom), **1**, 329. Academic Press, New York & London.

Lewis K.H. & Cassel K. (1964) In *Botulism – Proceedings of a Symposium*, US Dept. Health, Education & Welfare, Public Health Service, Cincinnati.

Lockwood S.J. (1988) *The Mackerel – its Biology, Assessment and the Management of a Fishery*. Fishing News Books, Oxford.

Madhavan P., Unnikrishnan Nair T.S. & Balachandran K.K. (1974) A review of oil sardine. 1, Distribution, preservation and transportation. *Fishery Technology*, **11** (2) 88–92.

Mayhew H. (1851) *London Labour and the London Poor*, **1**, The London Street Folk. Strand Office (1851) London.

Menz A. & French S. (1982) The fishery for small pelagic fishes off the coast of Ecuador: its development and investigation. *Monografias Biologicas*, **2**, 7–17.

Mills A. & Tiffney P. (1982) The use of controlled atmosphere packaging for the storage of fish. *Chilled Foods*, April, 22–3.

Murray, C.K. (1982) New thoughts on the scombrotoxin problem. *J. Sci. Fd Agric.*, **33**, 1245–6.

Nair R.V. (1973) *Indian sardines.* CSIR Zoological Monograph 2, Council of Scientific and Industrial Research, New Delhi.

Pariser E.R., Wallerstein M.B., Corkery C.J. & Brown N.L. (1978) *Fish Protein Concentrate: Panacea for Protein Malnutrition?* The MIT Press, Cambridge, Massachusetts.

SFIA (1989) *Pelagic species review – resources and markets.* Report 2002, January, pp. 165. Market Development Department, Sea Fish Industry Authority, Edinburgh.

Smith J.G.M., Hardy R. & Young K.W. (1980) A study of the storage characteristics of mackerel stored at chill and ambient temperatures. In *Advances in Fish Science and Technology* (Ed. by J.J. Connell), pp. 372–8. Fishing News Books, Oxford.

Statham J.A. (1984) Modified atmosphere storage of fisheries products: the state of the art. *Food Technology in Australia*, **36**, 233–9.

Whitmarsh D.J. & Young J.A. (1985) Management of the UK mackerel fisheries. *Marine Policy*, **9** (3) 220–36.

Whittle J.G. (1976) *A study of mackerel processing and marketing in England and Scotland.* Field Report 592, pp. 19, White Fish Authority, SFIA, Kingston upon Hull.

Wood J.D. & Hopper A.G. (1984) *A report on the UK herring fisheries in the 1980s.* Technical Report 245, August, pp. 38, Sea Fish Industry Authority, Kingston upon Hull.

Wood C.D., Boustead P., Mora Y. & Grijalva M. (1988) Ecuador: a review of the handling, distribution and marketing of fresh and frozen fish from the artisanal sector. *Boletin Cientifico y Tecnico*, **9** (5) 42–81. Instituto Nacional de Pesca, Guayaquil, Ecuador.

Young K.W., Neumann S.L., McGill A.S. & Hardy R. (1980) The use of dilute solutions of hydrogen peroxide to whiten fish flesh. In *Advances in Fish Science and Technology* (Ed. by J.J. Connell) pp. 242–50. Fishing News Books, Oxford.

CHAPTER 24

Some factors affecting the viability of the utilisation of small pelagics

J.A. Young, K. Crean, D.A. Palfreman

INTRODUCTION

The potential contribution of small pelagics to the utilisation of marine fish resources has long been widely recognised as considerable. Yet experience to date indicates that much of this potential has not been fully developed. More specifically, in those situations where the resources have been exploited, there still remains much scope to enhance the benefits accruing to society.

Traditionally the development of the small pelagic fisheries has been essentially supply-led. Generally they have been characterised by heavy investment in technology, especially within the catching sector, and the production of low unit value products often not for human consumption. While this pattern of development is understandable for a variety of reasons, it is less than ideal from a number of different perspectives.

This chapter analyses some of the technological, marketing and economic factors which have contributed to this sub-optimal utilisation of a major sector of fish resources. By highlighting the problems of the traditional approach to exploitation of small pelagic species, the chapter concludes by considering some alternative strategies for future development.

SOME MARKETING CONSIDERATIONS: SUPPLY-LED, BUT WHAT ABOUT THE MARKET?

One of the major problems in fisheries generally has been the tendency only to consider market requirements after the means of supplying the product has been secured; and in some cases not at all. This concentration on catching the fish and, to a lesser extent, the transformation of the raw material into some product thought saleable, is in some contrast to the marketing orientated approach evident in most other market sectors since the 1950s. While the 1980s have witnessed a much greater awareness of the importance of marketing fish rather than simply selling the product landed, the sales orientated approach still persists.

This is especially the case in newly developing fisheries and is evident in many for small pelagics. As has been noted from the investigations of Haywood & Curr (1987) into sardine and mackerel fisheries in Thailand, India, Morocco and Indonesia: 'Fishing operations have received considerable attention and investment but the problems of handling, processing and marketing large quantities of small pelagic species have often been ignored, with the result that the fish meal industry has provided the main marketing outlet . . . There is a great need for investment to be applied throughout the marketing chain.'

Before considering why the fish industry has been slow to adopt a marketing orientation, it is useful to define what is meant by marketing. It has been variously defined: a commonly accepted version is that of Kotler (1984) which states marketing to be 'a social process by which individuals and groups obtain what they need and want through creating and exchanging products and value with others.' Within the context of fisheries, marketing has been described as 'concerned with the broad range of activities of individuals and organisations which relate to the exchange of products and services within a continuum between decisions of fish capture and consumption' (Young 1987). Marketing is thus based around the existence of an exchange which is dependent on participants perceiving satisfaction by entering into such an agreement. In providing such satisfaction marketeers use a 'mix' of product, price, promotion and place – the four Ps. While marketeers can control this mix to advance their own tactics, they must operate within the wider marketing environment where they have little, if any, influence over the numerous forces at play.

In any marketing environment, success is dependent on the ability to respond to ever-changing consumer wants with a suitable mix and, at the same time, to contend with the dynamic constraints and opportunities of the marketing environment. While this applies in any marketing situation, the precise nature of these circumstances in fisheries renders the task all the more difficult. For example, uncertainty is frequently heightened by the absence of property rights over the resource base which, though sustainable, is also destructible. This situation is compounded by the tendency for fisheries to attract excessive fishing effort. Fluctuations in the natural environment can also seriously affect the availability of the raw material on which the marketing rests.

But for marketing to have any long run success, it is vital that consumer wants are identified. In other words, that the consumer is considered first before deciding how to put the fish in front of the consumer. It is in this respect that the traditional supply-led orientation of fisheries fails to realise full market potential; problems of market acceptance are inevitable if consumer acceptability is only considered last.

Numerous reasons can be suggested as to why this supply-led approach exists. Not least is the fact that concentration on the supply side is often initiated by the catching sector. The mobility of their ever-improving technical resources prompts diversification into new fisheries and is encouraged by the declining catch per unit of effort (CPUE). Similarly this tendency has been encouraged by the research and development emphasis of the majority of fishing firms on technical problems of fish capture and product transformation, rather than more explicit marketing considerations (Cunningham *et al.* 1985). Yet from a marketing perspective, such issues are of secondary importance and of little value if they fail to satisfy consumer wants.

The importance of this priority is emphasised by the characteristics of the market in which fish competes. Fish for human consumption is essentially no different from any other food commodity in that it has to compete with a wide range of alternatives. Moreover in many markets, and certainly in those of developed countries which yield the highest financial returns, competition is heightened by the fact that current volumes of food satisfy most consumers. Therefore increased consumption of any product category can be achieved only at the expense of consumption elsewhere. To achieve this the product must offer attributes perceived by the consumer to give greater comparative satisfaction. If it does not, the product is simply relegated to some alternative sector in which returns are considerably reduced.

CHARACTERISTICS OF SMALL PELAGIC RESOURCES

Many intrinsic features of the small pelagic resources affect their exploitation, the stability of commercial catches and the overall industrial base. There are significant conservation problems in managing fisheries for small pelagics; some have experienced the most dramatic declines of all fish stocks (Beddington & Rettig 1983). Similarly, economic and social problems have also resulted from the small pelagic fisheries. Especially notable has been the conflict between using fish for human consumption or for other market outlets. Smaller fish tend to be targeted by vessels producing fish meal, while human consumption markets prefer larger specimens. This creates a problem whereby species which have longer life cycles, e.g. herring, sardinellas and mackerels, are exploited before they have reached their optimum commercial value. Direct competition between different fleets exploiting small pelagics, in the absence of regulation, will tend to favour the fishery for small sized fish (industrial fishery) because it is able to exploit a year class before the fish is suitable for the food fishery.

One cannot foresee optimum fishing effort or sustainable yield for small pelagics until resource exploitation is well under way. The conventional bell-shaped curve showing the relationship between fishing effort and yield is regarded by some to be an unacceptable model from which to derive practical management policy (Hillborn & Sibert 1988). As demonstrated in demersal fisheries, there is never a gradual approach to an optimum exploitation level; in practice it is rapidly exceeded and a period of reduced catches quickly follows.

Resource management is also more complex and unstable with small pelagics, mainly because the stocks show marked seasonal fluctuations resulting from natural environmental factors. For example, the standing crop of small pelagics has been shown to be influenced by the intensity of nutrient-rich ocean currents (CNROP 1984). Moreover it has been established (Ulltang 1975) that for some small pelagic fisheries there is a poor relationship between CPUE and stock size. This is particularly so for purse seining and directed mid-water trawling operations. Thus when heavy fishing operates in concert with unfavourable environmental conditions, the level of uncertainty surrounding the small pelagic resources increases alarmingly and poses the threat of the fishery's collapse. In practice this can lead to 'boom or bust' industry conditions where the element of risk seriously undermines planned exploitation and investment stability.

The dense shoaling behaviour of the small pelagics, combined with their relatively short life cycles and their response to adverse environmental conditions, has precluded the use of indirect techniques of stock assessment. The size of small pelagic resources can be assessed by direct methods, such as acoustic devices, but these are not always reliable. Biological sampling, including assessments of egg and larval production, is preferable but can only be applied with accuracy to those small pelagics with longer life cycles. If the collapse of the small pelagic fishery is to be avoided, the resource management strategy should aim to ensure that the spawning stock does not fall below a sustainable level. This requires monitoring of both catch and stock numbers.

The biological information obtained is combined with commercial catch data to help set limits on exploitation. In recent years it has been established that fixed fishing effort is a reasonable basis for management of small pelagic resources. Such an objective is not, however, without considerable practical difficulties in implementation, such as how to measure the fishing effort, and what variables need to be specified in any resultant licensing scheme. Other views suggest that a system of individual transferable fish quotas might be a more effective means of resource management.

PELAGIC RESOURCES' CHARACTERISTICS: SOME MARKETING IMPLICATIONS

Several marketing problems result from the resource characteristics discussed above. One of the most significant is the precarious stability of the resource base. While the potential for stock depletion is endemic to any fishery, small pelagics are especially vulnerable and therefore critically dependent on appropriate resource management. In the absence of this, firms' propensity to invest and base their marketing infrastructure on such resources is inevitably reduced. Not only are there considerable problems of resource management to begin with, but it ironically becomes all the more important if resource management succeeds in encouraging a greater marketing involvement. *Ceteris paribus*, the higher market prices paid will tend to attract additional fishing effort. A resource management policy perceived to be effective by potential marketing investment is thus necessary for the fisheries' development. This point is returned to in the section on economic considerations.

A management strategy which aims to secure long-run sustainable yields is, however, unlikely to solve all problems. Fluctuations in the natural environment and consequent deviations in fish behaviour, such as changes in migratory routes, are both uncontrollable and a disincentive to invest. Attempts to avoid them, for example with floating processing units, involve high costs, so uncertainty appears to be an inherent and pronounced characteristic of small pelagic activities, which will discourage the risk-averse investor.

The large volume of fish supplies typically encountered also brings its own set of problems. Not least is the need to deal with large quantities of small and highly perishable fish within a short time to avert post harvest losses. While mechanical solutions to this problem may assist greatly, the seasonality of supplies can result in capital investment lying idle for much of the year. Similarly, facilities such as buffer storage to ensure year-round availability diminish marketing margins and, in certain cases, may restrict target markets.

Such constraints may also act as a barrier to entry into the market. Investment demands commensurate with the volume requirements of low unit value raw materials, tend to favour participation by larger firms and institutions. The market power enjoyed by such institutions gives them an advantage on price and so reduces the revenue yielded by the fishery. But against this, the scale and speed of their operations are more suitable to the practical circumstances of many of the world's fisheries for small pelagic fish and so offset any notional loss of income to the fisherman.

Such circumstances can prove frustrating to smaller scale marketing operations which, although willing to pay higher unit prices, cannot compete in respect of volume. Such a situation was experienced in the UK western mackerel fishery where indigenous processors offering a higher price could not always secure supplies in the face of competition from large volume Eastern bloc buyers (Whitmarsh & Young 1985). Clearly this problem of market entry tends to be self-perpetuating and makes it more difficult for smaller operations to develop new target markets.

Variation in the composition of the raw material also creates further marketing difficulties. In different species, and significantly within the same species, different lipid:water ratios and flesh textures are found over the year. Such changes influence the disposition of the catch, notably in relation to the level of post-harvest wastage (Crean 1987), keeping qualities, types of product that can be produced and consumer response. In the past such heterogeneity has restricted certain human consumption markets and so favoured fish meal outlets.

Other product attributes of small pelagics have met with unfavourable consumer reaction: bone content, smell, strong flavour and difficulty in preparation.

However there are some indicators that more favourable reactions are developing. Within the last decade the value of small pelagics in the human diet has been more widely realised. Medical research in a number of countries has demonstrated that the presence of fish oil in the diet may result in a significant decrease in heart disease.

Polyunsaturated fatty acids from fish oil appear to play a special role in reducing the blood fat levels. The omega-3 fatty acids of the fish oil reduce the capacity of blood platelets to aggregate (Barlow 1987). This has an ameliorating effect on the problems associated with heart disease: arteriosclerosis, thrombosis and vascular spasm.

Other health attributes are also evident, notably as a low calorie food and as a rich source of the fat soluble vitamins. Such positive attributes have in particular been noticed in the developed world and have given rise to targeted marketing campaigns with some success. For example in 1987 the Spanish Government agency, FROM, mounted an advertising campaign focusing on the health attributes of the sardine. An outstanding public response was achieved, with demand quickly outstripping supply (FROM, personal communication).

In developing countries small pelagics have been widely promoted for human consumption as a source of protein. Extensive trials have been undertaken in Peru and Ecuador to increase consumption of fresh, frozen and dried products, but in spite of the technological successes in producing new products, no dramatic increase in demand has occurred so far. Out of the 4.1 million tonnes of fish landed (mainly small pelagics) in Peru in 1985, less than 5.5% were destined for direct human consumption (Austlitz & Valencia 1987). Indeed some commentators have argued that the need to boost protein levels in the diet of developing country populations is not so important as increasing the overall level of calorie intake, which might be obtained more cheaply and cost effectively from vegetable sources (Anon 1983, Trimmer *et al.* 1983, Grigg, 1985). This debatable view has not helped support the strategy of using small pelagics for human consumption.

However, more generally there are encouraging signs that small pelagics will receive wider consumer acceptance as a result of product diversification. In recent years the scope for new product development has increased with the advent of surimi-based technologies (Miyake *et al.* 1985) and improved production in the artisanal sector which traditionally has provided fresh, and to a lesser extent frozen, products from small pelagics. The surimi-based technologies counter some of the organoleptic characteristics which have inhibited consumer acceptance of small pelagics. For example, the problem of dark coloration of minced fish flesh, not favoured by consumers, has been reduced (*ibid.*).

The small size of many species, which traditionally has limited the scope for product variation and consumer acceptance, is also much less significant. Consequently innovations should allow the incorporation of product attributes sought by food consumers, and provide a valuable addition to those attributes already inherent in products from small pelagic fish.

FISHERIES DEVELOPMENT AND SMALL PELAGIC FISH: SOME ECONOMIC CONSIDERATIONS

The above considerations, of course, need to be viewed in the wider context of fisheries development. This section examines the role of development planning in small pelagic fisheries and attempts to clarify some of the issues and make suggestions for future research and development.

The case for development planning rests on the view that the allocation of resources resulting from the operation of free market forces can be improved (Thirlwall 1983). This is true of fisheries as much as other sectors. Factors include shortages of capital and foreign exchange, structural surpluses of labour, other market imperfections, external economies and diseconomies, high risks of investment and the need in some countries for structural change. Furthermore the preference of people for present, rather than future, consumption results in a low level of savings and hence investment. These factors are all part of the currency of development planning and can be summarised under the general heading of 'market failure'. In this context the 'market' denotes the economic system, not only the point of sale of fish.

Much of the practical guidance for development planning in fisheries (Engstrom 1974, Campleman 1976) is unclear on the implications and significance of the various distortions in the fisheries sector and on the appropriate remedies. Such guidelines treat project planning as a practical exercise in applied financial analysis, rather than as a complex task which seeks to overcome weaknesses or failures in the operation of markets.

Market failure arises in capture fisheries through open access. The consequences are well known: excessive fishing effort is attracted to the fishery and the potential for the generation of a flow of net benefits over time is diminished. A planned approach to development would deal with the problem of open access.

At project level, the revolution that must occur in fisheries is that the resources must be priced. Deciding on the correct shadow price for use in project plans is likely to be difficult, but this should not prevent the project planner from recognising that fisheries resources are a scarce input to projects. This is equivalent to the current concern in project planning elsewhere over attaching a price to adverse environmental effects. If the fish resource were owned privately, participants would pay a rent as a fee for the privilege of using it. Theoretically this rent (the licence fee) equates to the external costs that each new addition to catching capacity generates. These external costs fall on other vessels engaged in the fishery and manifest themselves as a reduction in the commercial returns from fishing.

The case for including external costs in project plans does not depend on whether or not the resource is open access. Whenever investment is made in fisheries there are repercussions because there is always some impact on other vessels. Any project – including port and fish processing developments – which is intended to improve the financial performance of the industry, generates external costs. Those occasions when benefits from development can be obtained from the expansion of fisheries without generating external repercussions are very much the exception; even the smallest project has its price. Acceptance of this approach to fisheries places them firmly within environmental economics where the central problem of coherent planning is pricing the external effects.

The implications for fisheries development planning are quite clear. Market failure is most likely to occur in the catching sector; indeed, unless the problem has been recognised and dealt with by the authorities, it can be guaranteed. The returns on planning initiatives are therefore likely to be greatest if they tackle the problems of fisheries resource management. Earlier it was noted that small pelagics present particular management problems, but this does not mean that nothing can be done. It need hardly be added that fisheries management measures which address the biological problem and ignore the economic and marketing implications, at best merely succeed in switching from one kind of market failure to another. Moreover, while less likely, it should be appreciated that market failure may also occur in the post-harvest sector, especially in the form of excessive investment. Consequently planning initiatives to ensure that prices reflect resource opportunity costs will be as helpful here as in the catching sector.

A great deal of development effort has been devoted to the creation or improvement of fisheries products from small pelagic fish (see for example Bostock *et al.* 1985). Unfortunately some of the arguments used to justify this now carry less conviction. The economic theory of the consumer depicts him as maximising utility subject to an income or budget constraint, implying that he makes the best use of whatever income he receives according to his own judgement. However rich or poor the consumer may be, a new product must be perceived as providing more satisfaction than the alternative ways of spending his money. Therefore attracting buyers to fish or fish products requires planners to understand how and why consumers currently spend their money as they do.

This understanding obviously has wider implications for the success of a fishery project. In particular it will help predict the possible consequences of alternative product strategies. The benefits of such an approach have been investigated in the sardine fishery of Morocco (Haywood 1982). The model developed highlights the implications of product choice by the private sector, and examines how these decisions might contribute to wider governmental fisheries policy objectives such as employment, taxation and regional development. But without a fundamental appreciation of consumer wants to begin with, results are much less certain and much more likely to lead to failure.

CONCLUSION

The case for a shift in emphasis in the development of small pelagic resources has been clearly established. Essentially it rests on the firm belief that such developments should be led by the market rather than by supplies or technology. While at first sight this may appear to imply only a limited role, if any, for the public sector in fisheries development, fuller examination reveals this is not so.

On the one hand the world demand for attractive fisheries products appears to be strong and responsive to rising incomes. On the other hand we have very large world resources, much of which attract a low valuation as they are used for fish meal production. If there is a means by which consumer perception of small pelagics could be enhanced, it could make a significant difference to the economic welfare of many developing and developed countries. If the public sector bows out of the research and development that is necessary, the private sector will no doubt ultimately produce the required technology. But the innovation is most likely to be from within the largest private sector organisations and, in general, the consequences of this for the distribution of income between rich and poor countries, and within individual fish industries, could be adverse and uneven.

It may therefore be preferable for the public sector, including the development agencies, to retain some involvement but to focus more clearly on the issues of putting the consumer first and then initiating a development programme within the context of a coherent economic framework. This role would beneficially extend in several directions to include, *inter alia*, fish trading, health and hygiene, certification and business practices. The public sector also has a valuable contribution to make, all the more so with forthcoming EC hygiene legislation, to the provision of safe, hygienic and commercially appropriate fish markets at all stages in the marketing chain.

Only by such a reorientation to a more integrated approach, firmly focussed on the satisfaction of consumer wants, can there be any realistic hope of securing improved utilisation of small pelagic resources.

ACKNOWLEDGEMENTS

The authors wish to acknowledge the valuable comments made by R. Blair, Dr I. Cowx, Dr K.H. Haywood, S. Maddock and S. Ridgway on an earlier draft of this paper; but they exclude their responsibility for any remaining errors.

BIBLIOGRAPHY

Anon. (1983) *The potential of fisheries in alleviating undernutrition*. FAO Fisheries Circular 761, Rome.

Austlitz H.J. & Valencia D. (1987) Salted sardines for improving nutritional levels of the Peruvian Andean population. *Infofish Marketing Digest* 6.

Barlow S.M. (1987) Beneficial medical effects of fish oil. *Infofish Marketing Digest* 1.

Beddington J.R. & Rettig R.B. (1983) Approaches to the regulation of fishing effort. *FAO Fish. Tech. Pap.* 243, 26–7. Rome.

Bostock T., Montano R. & Mora-Sanchez Y. (1985) Galletas enriquecidas con proteinas de pescado para alimentacion de ninos en el Ecuador. *Boletin Cientifico y Technico* **VIII**, (7). Instituto Nacional de Pesca, Guayaquil.

Campleman G. (1976) *Manual on the Identification and Preparation of Fishery Investment Projects*. FAO, Rome.

CNROP (1984) Variation de l'intensite de l'upwelling Mauritanien entre 1955 et 1982 – implications ecologiques. *Bulletin du Centre National de Recherches Oceanographiques et des Peches*, **12** (1).

Crean K. (1987) Fishery policy and organisation – influence on post-harvest losses. In *Post-harvest Fishery Losses: Proceedings of an international workshop*, University of Rhode Island.

Cunningham S., Dunn M.R. & Whitmarsh D.J. (1985) *Fisheries Economics: an introduction*. Mansell St Martin's, London.

Engstrom J.E. (1974) *Preparation of Fishery Investment Projects*. FAO, Rome.

Grigg D. (1985) *The World Food Problem 1950–1980*. Blackwell Scientific Publications, Oxford.

Haywood K.H. (1982) *A methodology for investment planning in developing fisheries*. PhD thesis, University of Hull.

Haywood K.H. & Curr C.T.W. (1987) *Project identification in small pelagic fisheries: Indonesia, India, Thailand, Morocco*. FAO Fish. Tech. Pap., (289) 4 Rome.

Hillborn R. & Sibert J. (1988) Adaptive management of developing fisheries. *Marine Policy*, April, p. 112.

Kotler P. (1984) *Marketing Management: Analysis, Planning and Control*. PHI, 5th edn, 4, London.

Miyake Y., Hirasawa Y. & Miyanabe M. (1985) Technology of surimi manufacturing. *Infofish Marketing Digest*, 5.

Thirlwall A.P. (1983) *Growth and Development*, 3rd edn. Macmillan, London.

Trimmer C.P., Falcon W.P. & Pearson S.R. (1983) *Food Policy Analysis*. John Hopkins for the World Bank, Baltimore and London.

Ulltang O. (1975) *Catch per unit effort in the Norwegian purse seine fishery for Atlanto-Scandian (Norwegian spring spawning) herring*. FAO Fish. Tech. Pap., 155, Rome.

Whitmarsh D.J. & Young J.A. (1985) Management of the UK mackerel fisheries: an economic perspective. *Marine Policy*, **9** (3).

Young J.A. (1987) Marketing in a dynamic environment: an overview of the UK fish processing industry. *Food Marketing* **3** (1) 144.

CHAPTER 25

Consumer attitudes towards pelagic fish

D.W. Marshall and J. Currall

INTRODUCTION

Consumers generally do not talk of pelagic fish or even dark fish, for the term is alien to them. Instead they refer to these species as fatty or oily fish. This is more than a description of inherent product characteristics; it is in its own right an evaluation. Any attempt to encourage the consumption of these currently underutilised species needs to consider the implications of such evaluations in relation to their overall acceptability – the sorts of influences that come to bear on the decision to incorporate a particular food into meals. It requires some understanding of the rules of inclusion or exclusion or appropriateness, which influence these decisions.

Qualitative research was conducted and a series of attitudinal statements were drawn up from group discussions, representing the main considerations of the consumer on food choice and meal appropriateness. The next stage in the research attempted to measure the strength of these attitudes and investigate the relationship between them with respect to food acceptability. This chapter reports the results of a national survey to investigate consumers' attitudes towards meat and fish and relates specifically to pelagic (or dark) fish and how it is perceived by consumers.

Why measure attitudes? It is said that attitudes may be our best indicator of behaviour. One problem with attitudes is the diversity of definitions which exist. Early definitions focussed on the affects or liking for or against an object. Allport (1935) proposed a definition which considered the readiness to respond; attitudes were regarded as 'a predisposition to behaviour'. The model which divides attitudes into cognitive (information, beliefs), affective (liking or disliking) and conative (action) elements has received much support (Krech *et al.* 1962, Seibold 1980). However, much of the focus has been on the affective component (Fishbein & Ajzen 1972). Foxall (1981), examining the treatment of attitudes in consumer research, tells us that 'attitudes take explicative priority over other mental constructs and, frequently, are shown to subsume the effects of social and environmental factors. The task of consumer research is to identify and validate more sensitive and powerful means of defining and measuring them'. Most of the focus on attitudes has concerned the relationship between attitudes and their ability to predict behaviour. This research on attitudes towards pelagic fish in relation to white fish and meat is concerned with the knowledge and beliefs which underlie consumer attitudes towards pelagic fish.

The first aim of the research was simply to measure the level of agreement or disagreement with statements relating to food choice across a large sample of the UK population, and to identify those areas where pelagic fish is perceived to be significantly different from white fish and meat. A second aim was to investigate the salience of the attitudinal statements and their ability to distinguish between meat and fish, so characterising each of the four product groups selected (red meat, white meat, dark fish and white fish). The attitudinal statements reflected concerns relating to food choice and appropriateness of foods for inclusion in meals. The evaluation of pelagic fish on these attitudinal statements aids an understanding of the overall acceptability of pelagic fish. It does not give a direct measure of the degree of liking or likelihood of purchase, but by implication suggests possible reasons for the limited acceptability of pelagic fish (supported by the UK's low consumption figures). The focus is on understanding and measuring the relative importance of the beliefs held, rather than with behaviour prediction. Such information could be utilised in the future marketing of these species.

METHODOLOGY

A questionnaire was piloted with the help of students from the department of Agricultural Economics and Food Marketing, Newcastle University, under the supervision of one of the authors. The pilot revealed certain ambiguities and weaknesses in the initial design. It proved lengthy and time consuming with completion times of up to 45 minutes. The pilot questionnaire involved 78 statements extracted from the qualitative research, and each of the four products had to be evaluated across these statements. This not only increased the amount of time involved for completion but made the task of administering the questionnaire extremely tedious both for the interviewer and interviewee. Given the voluntary nature of the students' involvement and the lack of financial recompense for the interviewees, it was important to refine this section of the questionnaire. Sections on preferred foods and current food usage were deleted as these were primarily concerned with liking and behaviour rather than attitude. An attitudinal section specifically designed for analysis using the Fishbein-Ajzen model 1 was deleted.

The remaining 72 attitudinal questions were examined by multi-dimensional scaling for questions with low discriminatory ability and for statements that were eliciting a similar answering pattern. This objective reduction of the statements to a manageable number helped overcome the problem of deciding which statements to retain; the initial problem was that they all seemed relevant.

The final version of the questionnaire contained two sections: a demographic section to record the characteristics of the sample and an attitudinal section using 27 of the original 78 statements. Completion time was reduced to a manageable 20 minutes.

The attitude statements were evaluated on a seven point Likert agree/disagree scale which ran from 1 'very strongly disagree' to 7 'very strongly agree'. Respondents indicated their level of agreement with each of the statements for the four products. The mean scores across the sample were used as the basis for the analysis of attitude statements.

A national survey was carried out in November 1987 with the assistance of seven UK colleges and polytechnics. The survey was administered in Newcastle-upon-Tyne, Liverpool, Birmingham, Bath, London, Aberdeen and Glasgow. Prior to the survey all the colleges were visited and the students briefed on the survey aims and interviewing procedures. Selection of a random sample was not possible due to limited time and funds so an independent quota control was used to ensure representation of particular respondents (Chisnal 1973). A total of nine hundred people were interviewed. To the credit of the interviewers only 2% of the returned questionnaires were unusable due to missing values and incomplete data.

FINDINGS

The findings address the two aims outlined above. The first aim – to measure the level of agreement or disagreement with statements – requires a straightforward analysis of variance to reveal significant differences between product groups across the attitudinal statements. The second aim – the more complex question of identifying the distinguishing features, in perceptual terms, of the different types of fish and meat – involves the use of multi-dimensional scaling. The sample cannot be described as truly representative of the UK population as it is slanted towards a younger sample of respondents with evidence of some bias towards the middle classes, reflected in the classification of students as lower middle class. If, as suspected, the reason for a decline in fish consumption overall is due to the vintage effect (Lesser *et al.* 1982) – i.e. as older generations die they carry food habits with them – it is the attitudes of the young consumer one needs to be concerned with.

Figure 25.1 illustrates the mean scores achieved across the total sample for dark fish and white fish on the 27 attitudinal statements. It appears that fresh dark fish is perceived to taste better than frozen dark fish and specialist knowledge is required when buying this type of fish. Cans of the product are perceived to be good for 'snacks'. It is generally perceived to be healthy although the fat/oil is off-putting and it is 'fiddly to eat'. It is not regarded as a children's food nor believed to have a bland taste, as indicated by the disagreement scores. The attitudinal scores would suggest that the general image of dark fish is characterised by fresh products rather than canned. In certain respects its conferred benefits of convenience in the canned form and of healthy image are offset by the oily associations and the inconvenience of fiddly eating. One suspects that this is related to the presence of fine bones in these species which would also account for its unsuitability for children.

Dark fish are distinctly different from white fish and this is supported by the fact that on the scores for dark and white fish there are statistically significant differences on 23 out of the 27 statements, illustrating the danger of grouping these two together as one product, i.e. fish. Both types are considered equally unpleasant to handle in the raw state, and as boring food which requires special knowledge on both buying

Fig. 25.1 A comparison between attitudes towards dark fish and white fish.

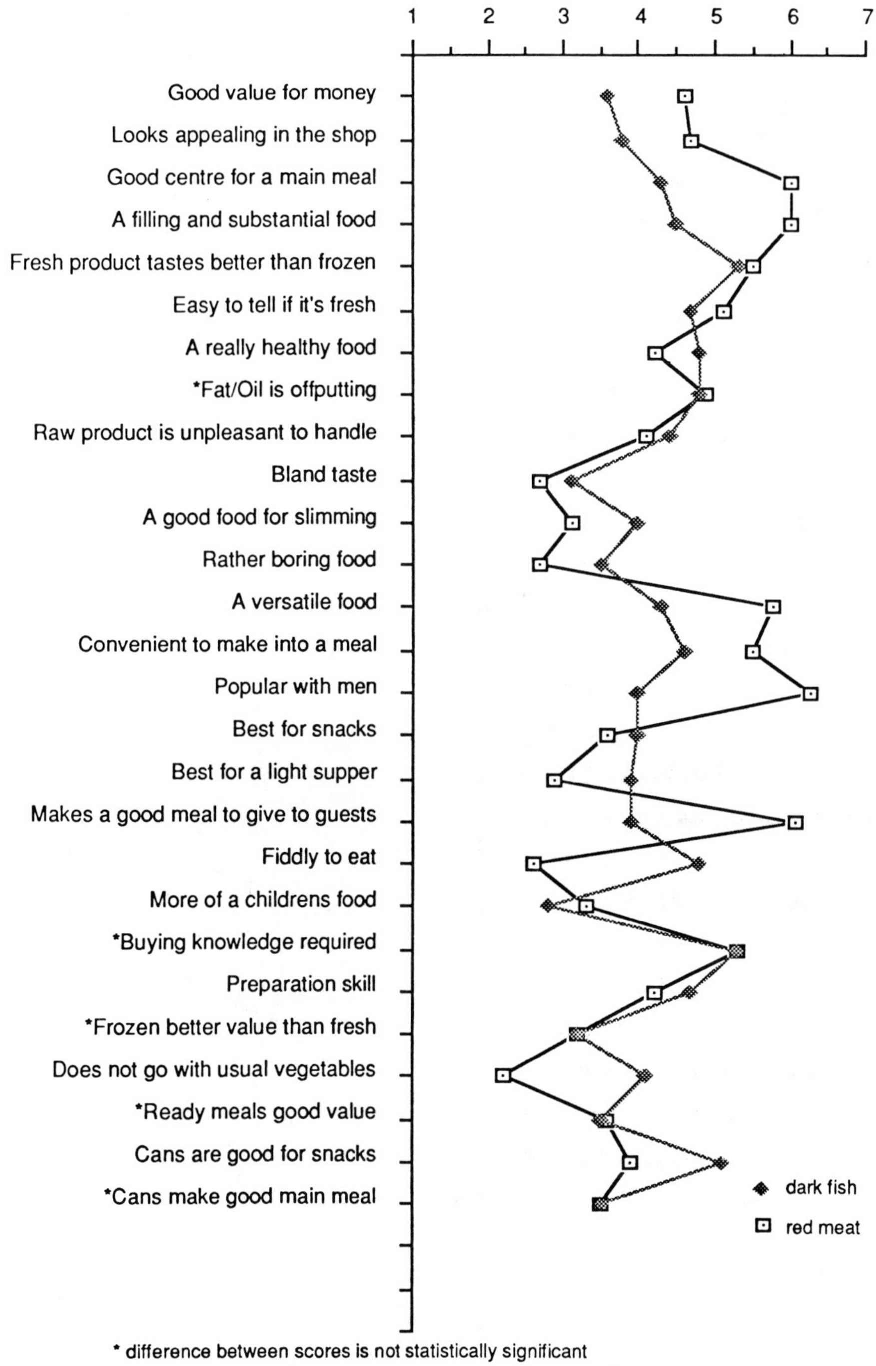

Fig. 25.2 A comparison between attitudes towards dark fish and red meat.

and preparation. One can infer that these are therefore features which are associated with fish per se. Closer examination of the scores does however reveal that the those for being unpleasant to handle and boring lie around the neutral point of the scale. Further, the scores associated with dark and white fish show the former to be perceived in a much poorer light. White fish is generally more appealing, and is for example regarded as better value, healthier, better for slimming and more suited to light supper meals. Although the differences are not so marked on several other aspects, white fish does score better on versatility, convenience and popularity with men, so adding to its overall acceptability. Dark fish scores significantly better on suitability for snacks and the use of canned products for snacks. But it combines less well with other vegetables and is more fiddly to eat. Dark fish is very much the poor companion to white fish.

The importance of (red) meat in our food and its central place in the meal was stressed in the qualitative research which preceded the attitudinal survey. Figure 25.2 compares dark fish with red meat. Again, as in the comparison with white fish, dark fish elicits significantly different scores on most attitudinal statements when compared with red meat. The scores for both products are however similar in several respects. Fat/oil is considered off-putting in both red meat and dark fish. They both require special skill when buying, (although interestingly not when preparing!). There is disagreement on whether frozen products are better value than fresh or whether ready meals containing these products offer good value. This needs to be interpreted with caution given that there are very few frozen dark fish products or ready meals containing this species on the market. The question is, if available on the market would they offer better value than fresh, or good value in ready meal form? The level of disagreement on whether cans of red meat and dark fish make good main meals is not significantly different; again this may reflect a perception of canned foods as much as an evaluation of the two products.

Examining the scores highlights the central role of red meat, which is seen to be filling and substantial and to make a good main meal. Its strength lies in its general acceptability seen by the high levels of agreement with the red meat's versatility, convenience and acceptability among men and guests. One suspects this generally favourable response accounts in part for its higher rating in terms of good value. The main strength of dark fish when compared to red meat again lies in its suitability for snacks from cans and for light suppers. It gets a neutral score on the statement that it does not combine well with the usual vegetables, but in total contrast red meat evokes strong disagreement with this statement. The other area where dark fish evokes stronger agreement is on the statement that it is more fiddly to eat than red meat. In its favour dark fish holds more promise as a healthy food and slimming aid than red meat.

IMAGE OF DARK FISH

The above analysis is extremely illuminating in that it helps highlight and measure consumer perceptions of dark fish. Interpretation is improved by the visual presentation in the figures but therein lies a danger. Presented in this form there is the suggestion that these statements are in some way related. This may be true to a certain extent, given the scores achieved, but who is to say if there is a true relationship between, for example, the healthy image of dark fish and it being fiddly to eat. The following method attempts to obtain an overall impression of dark fish in relation to the other products, examining not only the distinctive and salient characteristics but also the relationships between the statements themselves – something that the previous analysis does not permit. This complements the previous analysis to give a clearer overall picture.

Multidimensional scaling (MDS) is used in this work to show pictorially how the pattern agreement with the 27 statements as a whole relates to the four products (red meat, white meat, white fish and dark fish). A perceptual map is produced that enables one to see the relationships between statements as well as the relationship between statements and products simultaneously. Some statements will largely exhibit the same pattern: for example, 'a really healthy food' and 'good for slimming'; while others will show opposite patterns: 'a good centre for a main meal' and 'good for snacks'. The remainder show either no relationship at all or something between these extremes. The pictorial representation will show those statements in substantial agreement plotted close together, i.e. on the same side of the diagram, those in substantial disagreement plotted far apart on the opposite sides of the diagram, and the remainder in locations appropriate to their relationships. In addition to these 'global positionings', the diagram needs to place those statements that can be particularly positively associated with one of the products at the side of the diagram representing that product, and those particularly negatively associated with the product on the opposite side.

This allows the identification of those statements which discriminate between the products in the perception of those completing the questionnaires. The statements that fail to discriminate and get similar answers for all products will be more centrally placed.

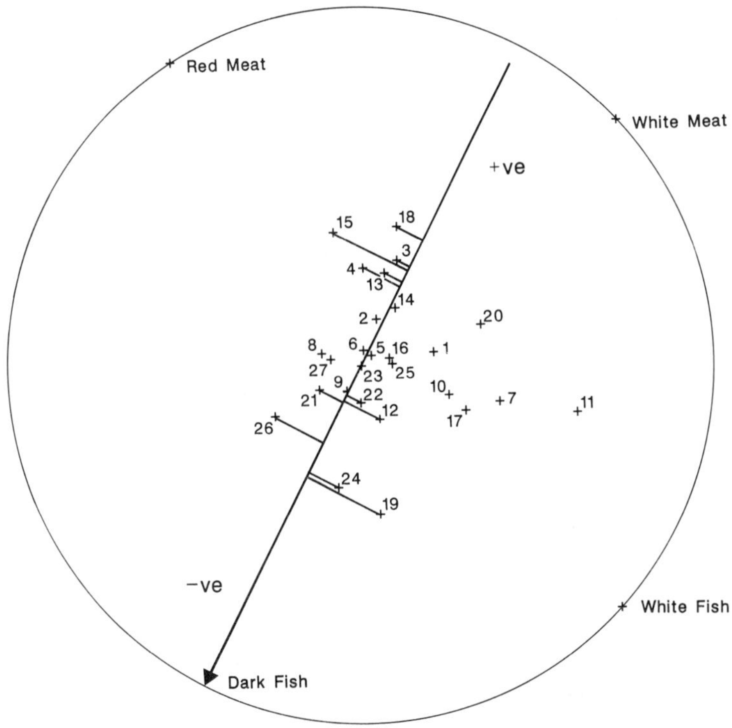

Fig. 25.3 'Perceptual' map of attitudinal statements associated with dark fish.

The 'perceptual' map in Figure 25.3 identifies those attitudinal statements associated with dark fish. The subject points (product groups) are positioned on the periphery of a unit circle and the stimuli (attitudinal statements) are arranged to reflect the relationship between the statements and the four products. A vector has been constructed to identify the statements most strongly associated with dark fish, in both a positive and negative direction. These statements support the previous analysis and are:

Statements positively associated with dark fish

(19) fiddly to eat
(24) not go with usual vegetables
(26) cans are good for snacks
(12) boring
(21) knowledge when buying
(22) preparation skills

Statements negatively associated with dark fish

(18) good to give to guests
(3) good centre for a main meal
(15) popular with men
(13) versatile
(4) filling and substantial

The attitudes associated with dark fish highlight its lack of general acceptability. Despite eliciting the highest levels of agreement that fresh pelagic fish tastes better than frozen, that special skills are required when buying the product and that cans are good for snacks, the most characteristic feature of dark fish relative to the other products is the perception that it is fiddly to eat and does not combine well with the 'usual' vegetables – a characteristic exclusive to dark fish. Being 'fiddly to eat' is probably related to the presence of fine bones, adding inconvenience to eating and reducing acceptability. Previously it was suggested that dark fish was not seen as a children's food. This analysis suggests that of the four products (including white meat), dark fish is least acceptable for guests and as a main meal.

CONCLUSIONS

Attitudes towards dark fish do not present a very promising picture on overall acceptability. The positive attitudes tend to be overshadowed by the negative ones. The advantage of taking the analysis one stage further and examining the interrelation between the statements is that one gets some idea of the extent to which this is true. In other words, it is not enough simply to look at the attitudinal statements in isolation.

ACKNOWLEDGEMENTS

Thanks are due to the staff and students of the colleges which participated in this research study. They included the home economics departments of Newcastle Upon Type, Liverpool and North London Polytechnics, Birmingham College of Food, Bath College of Higher Education, Robert Gordon's Insititute of Technology Aberdeen and the Department of Management Studies, Queen Margaret's College, Glasgow.

BIBLIOGRAPHY

Allport G.W. (1935) Attitudes. In *Handbook of Social Psychology*. (Ed. by C. Murchison), pp. 798–844. Clarke University Press, Worcester, Massachusetts.

Chisnall P.M. (1973) *Marketing Research: Analysis Measurement and Control*, 2nd edn. McGraw-Hill, London.

Fishbein M. & Ajzen I. (1972) Attitudes and opinions. *Annual Review of Psychology*, **23**, 487–544.

Foxall G.R. (1981) *The treatment of 'attitude' in consumer research*. Paper prepared for the marketing theory seminar on buyer behaviour organised by the department of marketing, University of Strathclyde, and supported by the SSRC. Discussion Paper Series B, 77. University of Strathclyde.

Krech D., Crutchfield R.S. & Ballachey E.L. (1962) *The Individual in Society*. McGraw-Hill, New York.

Lesser D., Ritson C. & Gofton L.R. (1982) *Marketing British caught fish*. Report 28, Department of Agricultural and Food Marketing, Newcastle University.

Seibold D.R. (1980) Attitude-verbal report-behaviour relationships as causal processes: formalisation, test, and communication implications. In *Message-Attitude-Behaviour Relationship* (Ed. by D.P. Cushman & R.D. McPhee), pp. 195–244. Academic Press, New York.

CHAPTER 26

Deconstructing sensory preferences; social factors influencing the demand for dark fish

L.R. Gofton and D.W. Marshall

The immense variations in human food preferences across cultures are commonplace, but no less mysterious for that. Most British consumers do not at the moment see dark fish as good food (SFIA 1983). In a recent research project (Gofton and Marshall 1989), we explored some of the reasons for this and found that consumers themselves identified a number of different factors.

Taste is obviously important, and yet there is something of a paradox here since our research (and most other market research into this topic) shows that many present day UK consumers, particularly the young, had never tried these (or indeed, very many other) species (Ross 1983, Gofton & Marshall 1989). Yet these consumers were prepared to express distaste for 'things like that'. Of those who had tried dark fish, many disliked the taste because it was 'too strong', or found the oiliness unattractive. Also, there was a strong and general dislike for the skin and particularly the bones – the distaste for the former and the risk associated with the latter were mentioned more than any other factors. One of the reasons for preferring white fish, especially cod, seems to be that it is perceived to be less boney than dark species. Perhaps more than for other fish, the smell was particularly unappealing, and consumers mentioned both the smell during cooking and the fact that it lingered afterwards.

Appearance is very important, as with other food and drink, and dark flesh colour in fish was generally disliked. The reasons for this seem vague even to consumers themselves – most would simply say that they 'didn't like it' – they seemed to share the general requirement that fish should be 'pure and fresh' (e.g. white), but few were prepared to give clear reasons. Tales were told of darker fish being more risky (likely to be 'off'), of the days when fish fingers were made with 'minced rubbish', of the colour of fish innards, and skin, and so on. Bottom feeding fish, some said, were considered 'dirty', as scavengers. Given the general lack of knowledge about species, their characteristics, and even their names, general dislike led to the conflation of feeding habits and general 'riskiness' to any dark coloured fish.

When pressed, some respondents discussed the importance of making a meal visually appealing on the plate itself, and argued that this was more difficult with fish in general, but particularly with darker species, because neither the flesh and skin of the white species nor the oily dark flesh of pelagic species were regarded as attractive and appetising in themselves and they did not combine with standard vegetables and dressing. To give this kind of meal visual appeal usually involved extra effort in choosing, buying and preparing non-standard vegetable accompaniments to the meal (Schutz 1988. See also Douglas & Isherwood 1979, Murcott 1983, Bourdieu 1984 & Gofton 1989 for ways in which food functions symbolically as well as nutritionally).

Two benchmarks were used by our respondents. First, the general comparison of fish with meat, and secondly, the comparison of the attributes of particular fish products and species with the characteristics of white fish, notably cod and haddock. In some ways, of course, this should be good news for dark fleshed fish, since both colour and flavour were said by many consumers to be more meat-like than those of white fish, although the texture of the flesh, lacking the characteristic 'flakiness' of for instance cod or tuna, was not so well regarded. Historically certain fish were reserved for the monarch or specially saved for meatless days, because of their 'meaty' texture, mouthfeel and flavour (*The Guardian* 1988) but these days few seem to make these connections between meat and fish, at least consciously.

Meat is still, despite recent changes in food habits (NFS 1987), the focal point of the UK diet. Meat was, most agreed, the centre of most major food events in the UK food system. Most important meals featured roast meat or poultry, and the rigid structure of such meals together with the combination of ingredients precluded the substitution of fish. Fish just does not 'fit' alongside the vegetables, gravy, sauces and puddings involved and if it were used most of the other ingredients would have to be altered, while the other courses, and the balance of the whole meal, would be compromised. The function of the main meal – to provide the substance, to fill you up – could not, many people thought, be served by a meal in which fish was used as a substitute for meat.

While we take these to be somehow natural reactions, they are of course cultural conventions. In many other cultures fish is more prized than meat, is accorded higher status and may be preferred for the most important meals. That fish should not 'go' with gravy is not a law of nature, or a natural reaction of the body, but merely something which is part of our folk wisdom about food, along with the idea that flesh and fowl should not be mixed in one meal. This point has been explored by Douglas (1976, 1982, 1984) (see also Nicod 1979) in a justly famous sequence of papers. She argues that the structure of meals in the UK is fairly rigid, and that the main elements cannot be changed without disturbing the social order which that structure both reflects and expresses. As Bourdieu (1984) also argues, the structure of many meals is heavily ritualised and convention-bound, and to disturb the combination of ingredients, the order of serving, the organisation of preparation and consumption and so on, is somehow to subvert the social relationships within the household which serves the meal. When, for instance, some foods are thought unsuitable for men or children, or are especially for invalids or babies, or suitable only for poor people, animals, or social climbers, to challenge such conventions is likely to be seen as a threat to agreed notions of what is 'going on' in the meal, and may thereby upset or deny the participants part of the satisfaction which such an occasion offers. For a meal to be enjoyable we must be served the right sorts of food, within broad classes at least.

The place of fish within the UK meal system then is limited. Fish will generally be used in minor meals rather than major, as part of supplementary courses, as the centre of snacks and for meals eaten alone or informally as a family (most famously, of course, fish and chips).

Note that the reasons for using fish as food depend on how it looks and smells, how it can fit alongside other foods on the plate and in the meal, what is involved in cooking it, and how it compares with meat and other foods as well as how it tastes or feels in the mouth and how far it supplies filling nourishment or provides a feeling of repleteness. A taste for fish, then, is likely to depend on a whole range of criteria, rather than simply sensory impressions from the mouth or the stomach. Food must not, it seems, simply be good to eat but also good to think (Angyal 1941, Levi-Strauss 1969, Tambiah 1969). The criteria do not simply hang in the air by themselves but belong to a whole class of activities and are formed, sustained and subtly modified by the way in which they relate to other areas of life. We might, with the philosopher Wittgenstein, describe taste for foods as having a family resemblance to taste in other areas of life. Thus taste in food might share a good deal with taste in art, music or clothes, i.e. it might be much more to do with social convention than natural inclination: there is likely to be a good deal of learning involved; there may be a strong 'self-expressive' component in the choices made; the activities should be seen alongside other features of work or social life such as courtship or family life; and there may well be systematic relationships between tastes and social institutions such as class structure, educational system etc.

That these types of taste should be closely related to each other can be demonstrated fairly clearly by asking consumers what they do and why. Actual sensory preferences in both food and drink play a fairly minor role (but see Rolls *et al.* 1982). There is typically a limited vocabulary to describe good or bad tastes; brewers, for instance, have pursued the idea of compiling a 'taste vocabulary' used by consumers for many years to no great effect (Gofton 1983, 1986, 1988, University of Newcastle 1983, 1985). Typically tasting is something private, while preferring or rejecting a taste tends to involve descriptions, categories, ideas or motives which have a generally agreed significance; for example, in the case of a distaste for dark fish there are clear generally agreed reasons which can be invoked without qualification. It is simply accepted within the UK culture that strong flavour, dark flesh and oily appearance are undesirable attributes. It is also clear from the work done on the acquisition of taste preferences that there are few, if any, invariant factors in taste preference, and that even pain is not a barrier to the development of a liking (Rozin *et al.* 1986).

As with other language uses, of course, the idea of taste is highly misleading since it suggests that what is involved in choosing, using or preferring various foods is crucially dependent on the way in which the food feels in the mouth or on the tongue, and that somehow this is itself an objective. But in fact, the interpretation of sensory impressions, and arguably a good deal of the character of the sensory impressions themselves, is subject to a whole range of cultural, contextual and psychological factors shaping the cognition of the individual concerned.

The distinction between the fish most often used in meals, which is usually white, and the dark species, rests on criteria which largely again relate to the nature of usage. Fish is required to be safe, bland and quick to prepare, intended for meals in which for one reason or another there is some degree of suspension of normal social rules (eating alone, for convenience, for ease of handling on the plate as with children's meals – much mentioned by our respondents – for 'weight watching' etc.).

The objections consumers raised to dark fish made no sense without taking into account the context. Certainly the taste, smell, appearance, mouthfeel etc. of dark fish were considered and mainly found wanting, but such judgements clearly did not rest simply on objective sensory impressions and were bound up with other sorts of evaluation. To make a judgement on the acceptability of a food means placing it within a food system and relating its characteristics to the criteria of particular food occasions.

Thus, for instance, the colour of dark fish would be unacceptable because, many consumers told us, they 'preferred' white fish (Ross 1983). Some said that dark fish were 'dirty' and older consumers talked of some species 'scouring the ocean floor' or 'scavenging'. Others related the colour to the strong taste – white fish had a 'purer' taste; the taste of the dark fish was believed to be 'too strong' to go with many of the normal accompaniments (such as chips, salad or vegetables; and the 'oily, mushy' mouthfeel of dark fish was compared very unfavourably with the 'flakey, chunky, meaty' texture of cod or haddock.

Almost all judgements about the acceptability or otherwise of fish and fish products, and perhaps all food products, involve this kind of contextualising. It is fatuous to suggest that such judgements should be objective and independent of context, yet most sensory testing, of course, fails to take into account such grounds. At the same time, distinctions between species on the basis of smell, taste or mouthfeel are notoriously inconclusive; many consumers seem unable to make such distinctions at all, while even the keenest palates cannot do so consistently. In the later stages of the project, the testing of new products made from minced kippers showed that many consumers were unable to decide confidently that the products actually contained fish at all, let alone one of the disparaged dark species (Gofton & Marshall 1989).

What, then, is the relationship between the sensory characteristics, in terms of choice, and the usage characteristics which seem to provide the main criteria for choice decisions? We argue that food choices are constructed and justified very much like other types of social actions and the nature of the language involved, and the relations to other social actions, provide the key to 'deconstructing' taste preferences.

As anthropologists have demonstrated convincingly, the variety of human appetites is most likely due to complex interplay between cultural and environmental forces, with tastes functioning both symbolically (to make distinctions between groups, for instance) and to ensure an efficient use of available resources (cow love in India, pig love in New Guinea for instance; Harris 1985, Simoons 1961). At everyday level, these functions fuse into a 'practical philosophy' in which judgements are made and criteria applied seamlessly (Elias 1978, Gofton 1989). Disgust is unproblematic when it arises from characteristics which are obvious and generally agreed – not objective sensory features but socially defined categories in the main. When only sensory evidence is available judgement is often more difficult (Douglas 1976, 1984).

Anyone who has studied food must now agree that its uses by human beings are complex and multifaceted. The study of disgust shows that there is immense variety in the abominations which are held to be natural and self evident by different groups, that these have little to do with natural physical reactions and everything to do with the symbolic forms by which the group marks itself off from outsiders, stratifies itself internally or relates various activities and forms of life within the cultural universe.

Distaste for dark fish within our culture is unlikely to be explicable simply by physical characteristics. But at the same time it seems unsatisfactory simply to assign it to history or culture, which is another common move. To say that dark fish is unpopular because of a range of factors which are partly cultural and partly historical inheritance (for instance, the association of fish with invalid eating, with the poverty of meatless days in the past, with symbolic associations of 'rottenness', danger, wildness, rawness, naturalness etc. (Gofton & Marshall 1989) offers suggestive and plausible theories but the fine details are unprovable. We cannot decide on their priority for consumers as either groups or individuals.

In looking at consumer tastes we must note the duality of the term itself. Taste here carries both the sensory aspect and the aesthetic, but it is by no means clear where the division lies. It may, in fact, be worth thinking about what is involved in consumer reactions to food. Certainly taste, smell, appearance etc. are vital. But what happens is also a symbolic process. Foods are categorised, which involves marshalling the agreed meanings, associations, uses, combinations, history, dangers and conventions – in fact, the state of knowledge about how such food is to be regarded and used within a particular cultural group. These are of course, known by everyone and combine with individual taste, prejudice, specialised knowledge and experience.

In the group interviews we conducted, accounts of these decision processes showed them to be far less idiosyncratic or individual than one would suppose. The main reasons for using or not using fish depended on a range of *practical* issues of which personal or assumed general distaste for fish was only one element. Fish would be rejected because it was not attractive in appearance, was troublesome to prepare, was difficult to handle or messy when cooking – a whole range of features spread across the overall process of planning meals, shopping, choosing food, getting it home, storing it, preparing it along with accompaniments, serving and eating it, and then disposing of the left-overs.

What struck us most forcibly was the wide range of reasons for *rejecting* fish, and the *limited number* of reasons for choosing it. As an example, we asked consumers to imagine that they were planning a meal which had to include fish. Everyone immediately began to plan the accompaniments: what vegetables, sauces, other courses etc. would go with the fish. Other stages of the process were similarly involved: for instance, calculation of how the food would be prepared and cooked took into account the general assumption that both the appearance and smell of fish was offensive to most people, while it was generally assumed that most people liked the smell of meat cooking, if not the appearance of an uncooked joint (Elias 1978); that during the cooking process fish could not be 'left' in the same way as meat, since it had to be cooked 'just so'; that it could not be handled so easily as meat, and so on. While everyone knew how to cook meat, fish was thought much more specialised, varied and difficult; fish was more trouble generally speaking, more 'risky'.

Reasons for using or not using dark fish similarly extended over the overall process. It is pointless to offer general characteristics; these might mislead us when looking at individual products and species, and would reflect 'popular commonsense', general prejudice and belief, folk wisdom etc. which are a bad guide to actual scientific and commercial practice and judgement.

The main point made by our research is that the choice process is precisely that – a process in which foods are evaluated across a range of issues, involving practical criteria, rather than simply based on sensory data. This is certainly important but needs to be set alongside careful analysis of the ways in which dark fish are regarded and used by consumers, describing and analysing the criteria which consumers employ over the whole process of choosing and using it.

BIBLIOGRAPHY

Angyal A. (1941) Disgust and related aversions. *Journal of Abnormal Psychology*, **36**, 393–412.

Bourdieu P. (1984) *Distinction; A Social Critique of the Judgement of Taste.* Routledge & Kegan Paul, London.

Douglas M. (1976) Culture and Food. In *Culture; Essay on the Culture Program.* Annual Report, Russell Sage Foundation, New York.

Douglas M. (1982) *In the Active Voice.* Allen Lane, London.

Douglas M. (Ed) (1984). *Food in the Social Order.* Russell Sage Foundation, New York.

Douglas M. & Isherwood B. (1979) *The World Of Goods.* Allen Lane, London.

Elias N. (1978) *The Civilising Process.* **I**, The History of Manners. **II**, State Formation and Civilisation. Basil Blackwell, Oxford.

Guardian, The (1988) *A Fish That Time Forgot.* May 21.

Gofton L. (1983) Real Men, Real Ale. *New Society*, **66**, 1096.

Gofton L. (1986) Social change, market change. *Food & Foodways*, 253–77.

Gofton L. (1988) Folk devils and the demon drink. In *Drogalkohol, Alkohol und Drogen-Forschung und Praxis der Pravention und Rehabilitation*, 3/88.

Gofton L. (1989) Sociology and Food Consumption. *Brit. Fd J.*, **9** (1) 25–31.

Gofton L. & Marshall D. (1989) *A study of the behavioural variables influencing fish consumption in the UK.* Report prepared for Chief Scientist's Group, Ministry of Agriculture, Fisheries and Food, London.

Harris M. (1985) *Good to Eat: Riddles of Food and Culture.* Allen and Unwin, London.

Levi-Strauss C. (1969) *The Raw and the Cooked.* Harper & Row, New York.

Murcott A. (Ed). (1983) *The Sociology of Food & Eating.* Gower, Aldershot.

NFS (1987) *Household food consumption and expenditure.* Annual report of the National Food Survey Committee, HMSO, London.

Nicod M. (1979) Gastronomically speaking. In *Nutrition and Lifestyle* (Ed. by M.R. Turner). Applied Science Publishers, London.

Rolls B.J., Rolls E.T. & Rowe E.A. (1982) The Influence of variety on human food selection and intake. In *Psychology of Human Food Selection* (Ed. by L.M. Barker), pp. 102–22. AVI, Westport.

Ross Report on Fish (1983) Ross Foods/National Opinion Polls, London.

Rozin P., Fallon A.E. & Pelchat M.L. (1986) Psychological factors influencing food choice. In *The Food Consumer* (Ed. by C. Ritson., L.R. Gofton and J. McKenzie). Wiley, London.

Schutz H. (1988) Beyond preference, appropriateness as

a measure of contextual acceptance of food. In *Food Acceptability* (Ed. by D.M.H. Thomson). Proceedings of the University of Reading SCI Symposium on Food Acceptability.

SFIA (1983) *Marketing British Fish*. Fisheries Economics Research Unit, Sea Fish Industry Authority, Edinburgh.

Simoons F.J.K. (1961) *Eat Not This Flesh*. University of Wisconsin Press, Madison.

Tambiah S.J. (1969) Animals are good to think and good to prohibit. *Ethnology*, 8, (4 October), 424–59.

University of Newcastle (1983) *The North East Beer Drinker*. Restricted circulation report prepared for Newcastle Breweries Ltd.

University of Newcastle (1985) *Pubs and Their Patrons*. Restricted circulation report prepared for Newcastle Breweries Ltd.

CHAPTER 27

Taking new product evaluation beyond the 'sensorial'

D.W. Marshall, N.J. Boyd and L.R. Gofton

INTRODUCTION

New product development is crucial in today's competitive market and for many years the underutilised resources of pelagic fish species have been considered a reservoir from which innovative products may be drawn. One reason for this is the presence in fatty fish such as herring and mackerel of a special type of polyunsaturated fatty acid known as eicosapentaenoic (EPA), which is believed to be beneficial in protecting against heart disease. At present supplies of pelagic fish are abundant in the UK, with increased landings following the lifting of the EC ban on herring fishing over the 1981–3 period offering a relatively cheap source of material. However, the development of new products based on pelagic fish is not without its problems. This chapter reports on an evaluation of a new prototype pelagic fish product developed at Torry Research Station.

BACKGROUND

Preference in the UK is undoubtedly for demersal fish but the reaons are far from clear. It may be a reflection of the proliferation of new products based on demersal fish, combined with the limited range of new products based on pelagic species (Spence 1984). While the demand for fatty fish has been encouraged by the increasing variety of herring marinades available and the appearance of new products like hot-smoked mackerel, the overall consumption of pelagic species remains low. The real promise lies in the frozen and convenience sectors of the market, which have shown the greatest growth in recent years (Young 1982). The acceptance of pelagic products in the market-place has met with some resistance from consumers and manufacturers alike. Associations with fatty fish tend to be negative, such as fine bones and flavours that are considered too rich for the majority of consumers (Chapter 26 of this book and McSween 1973). However, it is possible to remove these negative features, for example by using fish mince as the basis for products (Bannerman *et al.* 1973). Herring and mackerel fish fingers developed in the early 1970s achieved good acceptance ratings in sensory tests but manufacturers remained unconvinced and the products never made it as far as the market-place. One reason suggested for their lack of success was the association of the fish finger shape with demersal fish (Keay & Parsons 1987). Manufacturers seem unwilling to adopt concepts in their entirety substituting pelagic species for demersal species.

New product development (NPD) is a complex and expensive process in which evaluation has an important role. Evaluation exists to enhance the concept or the prototype and move it closer towards the market-place. Despite this, new product evaluation is far from common (Lawton & Parasuraman 1980). Prototype evaluation aims to investigate how well the prototype fulfils the protocol devised from the initial concept, and to suggest ideas for improvement. As well as this it offers an opportunity to examine possible ways in which the use of the product is perceived by respondents. Acceptability needs to consider wider issues, such as appropriateness and situational influences (Meiselman *et al.* 1988, Schutz 1988).

Schutz stresses the fact that preference measurements alone do not explain food acceptance, and much of his work examines food appropriateness by considering the cognitive categorisation of foods. He recognises the fact that 'to consume varies widely, depending on the nature of the consumption situation, who is consuming, occasion, cost, cultural differences and other attributes of the item to be consumed'.

Products tend to fail because:

(1) potential buyers have no real need or desire for the products
(2) they fail to meet existing needs
(3) the marketing has been mishandled.

Previous research showed that there was a market opportunity for a product aimed at the convenience or snack market (University of Newcastle unpublished data). The products developed at Torry drew on this research and, following discussions with the research team at Newcastle, the product concepts were formulated and the snack/convenience prototypes developed.

SENSORY ANALYSIS

The new product was composed of minced herring fillet combined with seasonings and spices to give three basic flavours: Indian, Mexican and Italian.

The three formulations were prepared as 20 g uncoated blocks and cooked by microwave heating before being offered to a taste panel of 84 persons. The object of this preliminary exercise was to gain a 'likeness' or acceptability score for each of the products, using a 9 point hedonic rating system: 1 – dislike extremely, 9 – like extremely.

There was no statistically significant difference between the three flavours. However, the slight bimodal peaking on the histogram for the Italian flavour suggested that assessors form two groups: they either 'like moderately' or 'dislike moderately' the Italian flavour. Scores on the Mexican and Indian flavours tend towards the higher end of the scale, suggesting the products were favourable to the majority of assessors. The general opinion among assessors was, however, that all three flavours were acceptable (Fig. 27.1). For this reason it was decided to use all three flavours for the coating trial, which involved four different coating types so giving 12 combinations.

Balls (15 g) of each formulation were prepared and hand coated in each of the four types of coating: potato flake, Pandora (the Japanese style breadcrumb), Tempura (the traditional style puff batter) and seasoned flaky breader (a crunchy coating of several types of cereal flakes).

The samples were deep fried in vegetable oil and served hot to a panel of 42 assessors who were presented with four coated samples and asked to rank them in order of preference for the coating only, not the contents (each person was presented with all four coating types and all three formulations, one of which was duplicated in the set).

The scores from each assessor were totalled, the

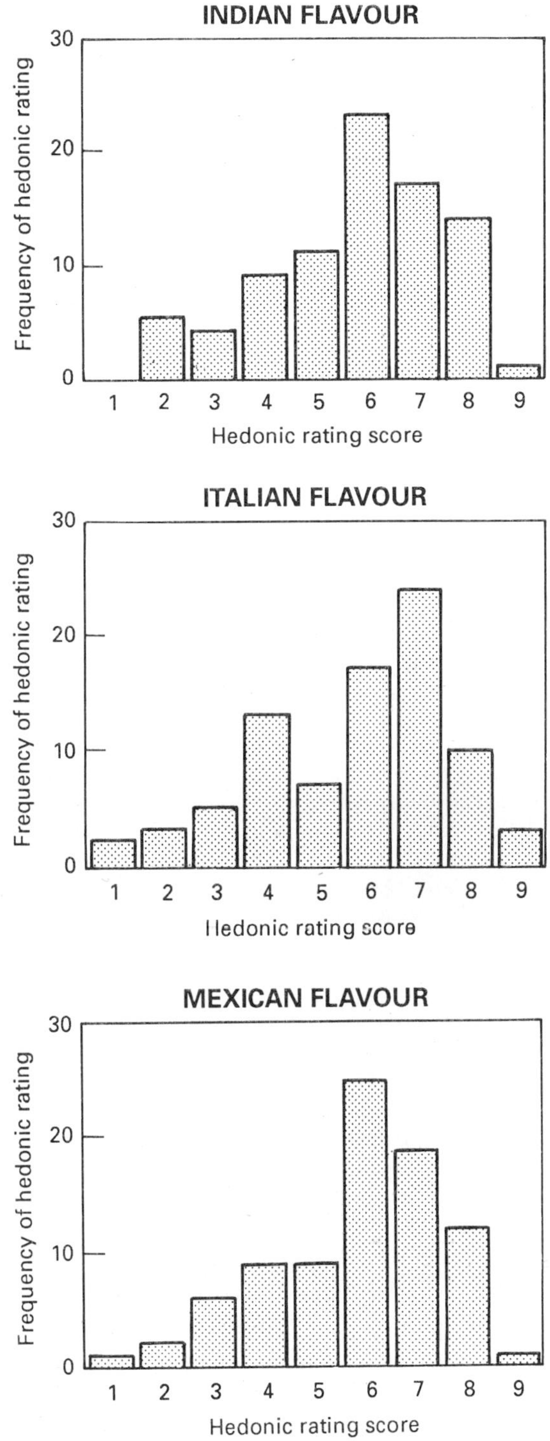

Fig. 27.1 Hedonic scores for Indian, Italian and Mexican flavours.

Table 27.1 Aggregate scores for each combination of flavour and type of coating.

Coatings/flavour	Italian	Mexican	Indian
Potato flake	90	84	90
Pandora	97	91	100
Seasoned flaky	112	116	120
Tempura	121	129	110

Table 27.2 Aggregate scores for differently shaped products.

Sphere	160
Triangle	209
Rectangle	217
Cube	254

lowest number representing the most preferred sample. The results are shown in Table 27.1.

Potato flake proved the most popular for all the formulations, closely followed by Pandora crumb. The rather unusual flaky breading was less popular, many people being put off by the speckled appearance and the very hard nature of some of the flakes. The traditional style puff batter fared slightly worse than expected but was still popular with a few assessors, particularly older people.

The product tested was spherical in shape, primarily because a ball is easy to hand form to give uniform product size. However, on an industrial process line the product would be prepared by a mechanical food former, allowing a variety of shapes to be produced. Four variations of shape were tested at Torry: sphere, cube, rectangle and triangle or wedge shape.

The 84 panel members were presented with the four shapes and asked to rank them in order of preference for appearance, i.e. which shape they considered to look best. All portions were approximately 20 g in weight and coated with potato flake in an attempt to standardise all parameters except shape.

The original shape, the ball or sphere, remained popular, while the cube was considered quite an unsuitable shape for the product. The wedge and rectangle came either second or third with most assessors (see Table 27.2).

The results from the sensory panels suggested that the most acceptable prototype product should be spherical in shape, with a potato flake coating and a savoury centre flavoured with Indian or Mexican seasoning. The Torry panel is not, however, necessarily representative of wider consumer acceptance. On the basis of this sensory work consumer studies were carried out using two product formulations (Indian and Mexican), two product shapes (spheres and wedges) and a potato flake/pandora coating (in order to reduce production costs).

CONSUMER REACTION

The next stage in the research attempted to examine consumer perceptions of the prototype products. The prototype (referred to as the product) was acceptable from a sensory perspective. The next question was how would consumers react to it; would they consider using such a product? In other words, did this product meet an existing need? Product success is likely to depend as much on the positioning strategy adopted as on inherent product features. Investigating consumers' perceptions of the product and ways in which they are likely to use it, aids development of an effective marketing strategy.

The products were tested in a focus group to encourage open discussion. Five groups of eight respondents were professionally recruited, and quota controls were used to ensure a range of age, social grade and sex. The products were presented in a cooked form, as they would be in a sensory test, and respondents received no information regarding product contents or flavours. The two flavours (curry/Indian and chilli/Mexican) were introduced separately to the groups. Marketing issues were left for the consumers to discuss. Initially respondents were simply asked to observe the products and express their reactions.

Findings

Overall the products were well received, although they performed better after tasting. The findings relate to product evaluation before and after tasting, and report on the appearance, taste, shape and perceived usage of the product along with the marketing implications.

Appearance

Initially the reaction was not very encouraging:

'Don't look as if they're going to taste nice . . . not exquisite.'
'Doesn't look very appealing.'

At this stage the products were regarded exclusively as a children's food, certainly not adult, and suitable for a light lunch or snack rather than a main meal. This reaction at least confirmed the appeal of the products to the snack/convenience market. The Indian products were lighter in colour than the Mexican products as a result of different flavourings in the contents. The appearance of the products suggested a high degree of processing and additives, although this was more marked in the case of the Indian products. Some thought the lighter colours of the Indian products 'insipid' but the general reaction was favourable and suggested chicken, vegetable, rice or fish contents.

The main justification for suggesting fish was that several respondents felt the products smelt fishy, but not all were convinced. The darker colours of the Mexican products suggested cheap meat contents, such as spam, corned beef or sausage, and as a result some felt they were poorer quality. The Mexican products were criticised for patchy coating, which reinforced this image. The coating was generally well accepted and although it was perceived to be a higher quality coating than batter, it did not suggest a home-made appearance.

The various associations seem strongly related to product shape; the spheres, or balls as they were referred to, evoked associations with meatballs and cheap, somewhat inferior products. Childlike images were reinforced with references to the products as marbles and footballs. While it was felt that this shape would be preferred by children, the practicalities of tackling them with a knife and fork could present problems. The wedge or triangular shape was preferred. These products were regarded as more substantial and practical. Wedge shaped products presently on the market (e.g. Marks and Spencer chicken bites) seem to have a positive image and look more innovative compared to the balls which are simply seen as a derivative of meatballs.

Taste

There was an interesting change in reaction to the products on tasting, and most were pleasantly surprised:

'They don't taste as bad as they look.'
'. . . some kind of meat with a slightly fishy flavour.'
'Anyone who liked spicy food would find these on the tame side.'
'They taste like meat but the texture isn't meat.'

Tasting permitted respondents to evaluate the flavours, examine the contents and comment on the textural appearance and mouthfeel. The product rating improved following tasting. The perception of the product's appropriateness changed from a children's product to an adult product, suitable for adult buffets and parties. The Indian flavours were more acceptable to the respondents but the taste was described as mild to those who liked Indian food, and was associated with onion bajhis and samosas. Older respondents who were unfamiliar with Indian food felt that the flavour was too spicy. The Indian products were thought to leave an aftertaste and for this reason were felt to be less appropriate for children.

The Mexican flavour was generally less well received and was described as spicy, 'spamy' and 'tomatoey'. They were regarded as too mild for adults but it was felt the flavour might appeal to children. The wedge shapes were felt to have a stronger fish taste, with the balls lacking any distinctive taste.

Texture

Shape played an important role in the evaluation of texture, which for many was the most disappointing aspect of the products. The wedges were preferred because they were regarded as firmer than the balls. This is likely to reflect the surface area to volume ratio of the two products. The balls have a larger ratio of filling to coating and texture could be improved by making smaller balls. It may also result partly from a transference of perceived attributes of the shape itself, in the sense that the wedge was the preferred shape. The texture was likened to fish cakes rather than meat or fish. The 'sogginess' of the inside contrasted with the crispiness of the coating. The internal texture and the fact that the contents were 'mashed up' suggested poor quality ingredients.

Shape

The association of the balls with meatballs and the wedges with chicken bites has already been mentioned. The wedges were felt to be a more practical shape suitable as 'finger food', hence their appropriateness at buffets/parties, although the balls were by no means excluded from such occasions. Alternatively the ball would be suitable for serving on a skewer or stick.

Product usage

The research confirmed the appropriateness of the product as a convenience food. Generally the product was regarded as suitable for:

(1) a children's meal (including teas, snacks or starters)
(2) an adult buffet/party.

The size of the products confirmed their place in children's meals. They were also seen to be more appropriate as a light meal or a snack, ideal for someone who had 'just come in from work', or 'after the pub'. In this situation few regarded the product as a high status food and did not expect it to be highly priced. Such products were likely to be sold in supermarkets and freezer centres. Appropriate accompaniments for the products would be chips, peas and beans, simply because that is what children like.

Salads offered an alternative accompaniment but were not regarded as appealing to children, which could present some problems.

The experience of tasting the products raised their status to that of the adult buffet/party and changed the product expectations. The product size suggested those less formal events where 'finger food' was appropriate. In such a situation they could be served with other spicy foods and possibly savoury dips. Alternatively they could be served hot or cold as starters. Although adults felt that eight of the wedges or balls might achieve satiety, it might also be construed as being greedy. However it was not simply a case of 'how many'; the product was inappropriate for a main meal.

Other suggested uses included picnic meals and packed lunches, barbecues, or retailed through fast food outlets like Wimpy and McDonalds. The products were regarded as very convenient and quick to cook.

When the product contents were specified respondents were again pleasantly surprised. The image of the products was further improved due to the healthy image of fish, although it was thought unusual to find fish in such products. Most felt that the products were based on cod or haddock rather than herring; in fact no one identified herring as the fish used. When they were told, the reaction was one of surprise:

'I don't normally like that fish.'

'... herring doesn't have a good image ... marketing it as a herring product would put people off.'

MARKETING IMPLICATIONS

Marketing implications for product improvements to the prototype relate to the types of market the products are likely to be aimed at and the situations in which the products are to be used. The two are by no means mutually exclusive. Initial reaction to appearance is obviously important as this is likely to be one of the initial ways in which consumers evaluate a product on the shelf. Although we measured reaction only to the cooked products, they were perceived to be relatively cheap and likely to be packaged accordingly. The importance of producing a well-coated product is stressed along with the dangers of vivid colours which are associated with processing or additives. Product shapes and contents are evaluated in the light of previous experiences and of other products currently available.

Tasting dramatically changes the way in which respondents perceive the products and their appropriate uses. The taste needs some further development in the light of these findings and the final product specification needs to carefully consider correct packaging and branding. The Mexican flavours could be improved by increasing their spiciness for the adult market or by producing a blander 'tomatoey' product targeted towards the children's market. Careful evaluation of each alternative is necessary as each requires a different marketing approach. The Indian flavours should be made more spicy, targeted towards the adult market where they are perceived to be more acceptable. The textures need to be improved to give a better quality image and more substantial 'mouthfeel', retaining the crunchiness of the exterior but removing the 'fish cake' texture.

The main distinction in appeal to children and adults may be developed by focusing one the product shapes and targeting the balls at the children's market, and the wedges at the adults' buffet market. The form of packaging and the distribution of the product clearly influences its status. Freezer packs are more likely to be perceived as lower status, in contrast to chilled packaging which creates and image of higher status and consequently commands a higher price. The former may be more appropriate to the children's market, the latter to the adult market. This area needs further research, with the development and evaluation of several package formats.

Fish is currently 'good to think and good to eat', but one needs to proceed with caution before promoting too aggressively on this basis due to the expectations of what is and what is not perceived to be fish. The focus should clearly be on the naturalness, wholesomeness, and additive-free benefits of this speciality 'seafood' product.

On specific product aspects it appears that the flavours need some further development along with the coatings. Indeed, the negative associations with frying as a method for cooking would suggest developments of coatings which permit the products to be grilled or oven-baked. Technically this should present no problems.

The prototype test proved the need for product design to consider the importance of usage characteristics as well as taste, texture, shape and smell. It illustrates the fact that the product attributes are by no means exclusive from the usage characteristics, the occasion and those present. The ultimate success of a new product depends on the marketing programme complementing the product. The basic product itself could be marketed in a variety of ways depending on the mix of marketing elements such as pricing, packaging, distribution and promotion.

The product development is far from complete; the next stage requires commercialisation of the prototype with full market testing to assess the impact of different packaging and promotional mixes for the products. There is still some way to go, but perhaps it is time to overcome our neophobia.

ACKNOWLEDGEMENT

This work was funded by the Ministry of Agriculture, Fisheries and Food.

BIBLIOGRAPHY

Bannerman A., Keay J.N., Smith J.G.M. & Hardy R. (1973) Kipper fingers a new fish product. *Food Manufacture*, October, 63–4.

Keay J.N. & Parsons E.M. (1987) The production, acceptability and storage properties of coated fatty fish portions. In *Savoury Coatings* (Ed. by D.B. Fuller & R.T. Parry), pp. 37–46. Elsevier Applied Science, London.

Lawton L. & Parasuraman A. (1980) So you want your new product planning to be productive. *Business Horizons*, December, 29–34.

McSween I.M. (1973) *Attitudes to fish.* Fisheries Economic Research Unit mimeographed papers. White Fish Authority, Edinburgh.

Meiselman H.L., Hirsch E.S. & Popper R.D. (1988) Sensory, hedonic and situational factors in food acceptance and consumption. In *Food Acceptability* (Ed. by D.M.H. Thompson). Proceedings of The University of Reading/Society of Chemical Industry Symposium on Food Acceptability. Elsevier, London.

Schutz G.H. (1988) Beyond preference: appropriateness as a measure of contextual acceptance of food. In *Food Acceptability* (Ed. by D.M.H. Thompson). Proceedings of The University of Reading/Society of Chemical Industry Symposium on Food Acceptability. Elsevier, London.

Spence L. (1984) Unpublished data, Torry Research Station, Aberdeen.

Young J.A. (1982) *UK fish marketing: a fresh approach.* Research Paper Series 28, Resources Research Unit, Portsmouth Polytechnic, Portsmouth.

CHAPTER 28

Development of a new fishery in Denmark. Catching, handling and utilisation of roundnose grenadier and greater argentine

E.P. Larsen

INTRODUCTION

In the last decade fishermen in Denmark have searched for new resources to exploit, due to declining quotas for traditional fish resources, especially the codfish species. Danish fishermen catch and land only a handful of different species for human consumption (Andersen & Hansen 1979). There is no tradition of introducing new species and the fishing industry is not prepared to accept them or even pay a reasonable price for them.

Because of the decline in catches of the traditional codfish species, the Danish Ministry of Fisheries decided to investigate the possibility of exploiting alternative fish resources, and the fishery in the Skagerrak for deep water species such as grenadier and argentine (Larsen 1987a) was investigated.

Skagen is the most northerly fishing port in Denmark. One of the major fisheries in the Skagerrak is for shrimp. Over the years the fishermen have observed quite a large number of grenadier and argentine as bycatch in the shrimp fishery. Sometimes the fishing fleet has to move away from the fishing grounds because these two species interfere with the normal shrimp fishing. When, however, the two species were brought occasionally to port, they were sold to the fish meal factories and were not registered in the official fishery statistics.

In the summer of 1986, The Fishermen's Organization of Skagen and Skagen Business Council decided to apply for support for an experimental fishery for grenadier and argentine. The project was financed by a grant from the Danish Ministry of Fisheries, and a project group was established.

BIOLOGY

Roundnose grenadier (*Coryphaenoides rupestris* (Gunn.)) belongs to the family Macrouridae which consist of more than 120 deep-water species, almost all bathypelagic, living on the continental slopes and abyssal plains of all oceans.

The body of the roundnose grenadier is compressed and strongly tapered behind a short trunk, to form a long 'tail'. The head is fairly short and compressed, representing about 15% of the body length. Scales on the trunk and tail are large and very deciduous. The habitat of the fish is bathypelagic at between 400 and 1500 m, but it is often caught in midwater from 300 m downwards, where the fish shoal in water of 3–8°C (Madsen 1987). The feed consists of bathypelagic animals, especially shrimps. Distribution is shown in Figure 28.1 (Geistdoerfer 1984).

The greater argentine (*Argentina silus* (Asc.)) belongs to the family Argentinidae, species of which are generally elongate, silvery fishes with parietal bones meeting at the midline. The sub-family Argentininae are deep-water fishes, bathypelagic on the continental shelf and upper slope. The fish probably school close to the bottom, most commonly at 150–550 m, but can be caught down to depths of 1400 m. They feed primarily on epibenthic and pelagic invertebrates and some fishes. Distribution is shown in Figure 28.2 (Cohen 1984).

Both species are characterised by a slow growth rate and it is not unusual to catch individuals which are 30 years old. They seem to be quite stationary and do not migrate like, for example, herring and mackerel (Madsen 1987).

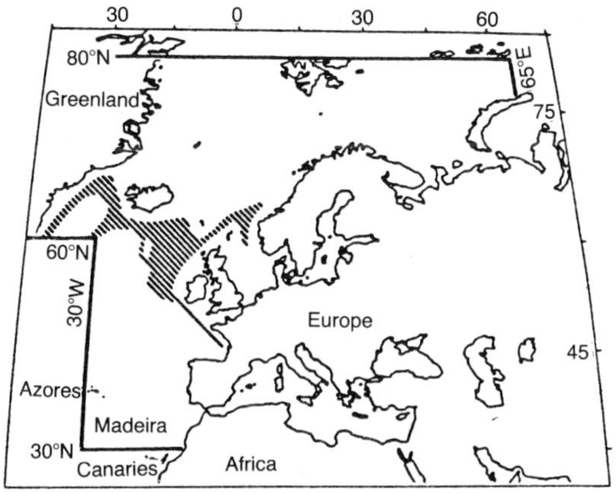

Fig. 28.1 Distribution of roundnose grenadier in the north-east Atlantic (Geistdoerfer 1984).

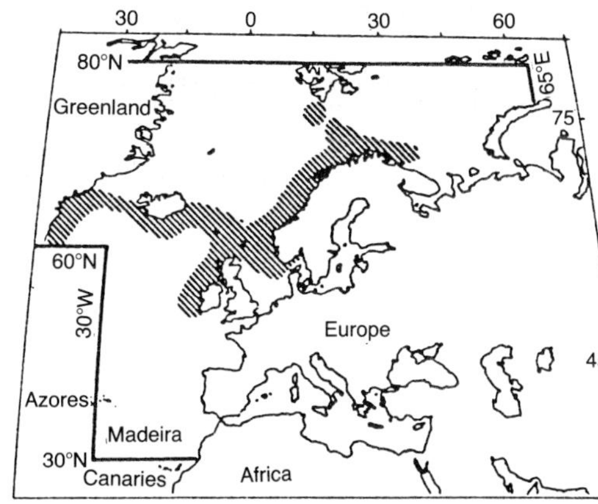

Fig. 28.2 Distribution of greater argentine in the Atlantic (Cohen 1984).

FISHERY POSSIBILITIES

Before experimental fishing with a commercial trawler began, a research cruise was conducted from 30 April to 7 May 1987 by the Danish research vessel *Dana*, from the Danish Institute for Fisheries and Marine Research. The purpose was to make a preliminary estimate of the total resources of grenadier and argentine in the deep part of the Skagerrak. This was done by demersal and pelagic trawling and by echo sounder integration. Length and weight were measured and the total biomass was estimated in different length groups. Determination of the age distribution has not been carried out yet, but material has been collected for this purpose.

In Tables 28.1 and 28.2 the average weight, total weight and the total biomass of each length group are shown, for both grenadier and argentine.

The total biomass of grenadier in the Skagerrak area is estimated to be 85 000 t and for argentine 28 000 t. The estimate for argentine is uncertain and seems to be underestimated compared with the actual catch in the commercial fishery (Madsen 1987).

Two trawl types were tested during the research survey in May 1987: a bottom trawl (GOV-trawl) and a semi-pelagic trawl (EXPO-trawl). The results indicated that the semi-pelagic trawl fished well when used as a bottom trawl, but when used pelagically a pelagic trawl would have performed better. Pelagic fishing, 200 m from the bottom and 450 m from the surface, showed that grenadier was abundant at this level (Yngvesson & Sørensen 1987).

The total annual catch of grenadier and argentine should not exceed 10 000 t (Madsen 1987). This is due to the slow growth rate of both species (Eliassen 1986, Ringdal & Strand 1989). The overall conclusion from the research survey was that commercial exploitation of the deep water fish species in the Skagerrak should include both grenadier and argentine.

CATCH HANDLING

The experimental fishing revealed that catch-handling on board was difficult. Both grenadier and argentine had to be landed ungutted and this leads to the assumption that the catch would have to be handled as for other pelagic fish species, such as herring and mackerel.

The key problem was to sort grenadier from argentine and grade them into different weight classes. The crew on the fishing vessel had to be supplemented with one more fisherman because sorting and grading the catch was so labour intensive. Mechanised grading was tested and a diverting conveyer belt grading machine was purchased but not installed in the fishing vessel during the experimental fishing period from November 1987 to May 1988.

RESULTS FROM THE EXPERIMENTAL FISHING

The Danish Ministry of Fisheries initially granted 700 000 DKr for the experimental fishing. This grant was later increased to more than 1 million DKr. It was decided that the major part of the money should sub-

Table 28.1 Grenadier: composition of the biomass.

Length group* cm	Numbers of fish (× 10^6)	Biomass t**	Average weight g
5.5	0.718	21.50	30.00
6.0	0.059	2.10	35.00
6.5	0.136	6.10	45.00
7.5	0.642	32.10	50.00
8.0	0.229	14.40	63.00
8.5	0.059	4.40	75.00
9.0	0.119	14.30	120.00
9.5	0.059	9.40	160.00
10.0	0.591	112.90	191.00
10.5	0.136	31.30	230.00
11.0	1.521	390.90	257.00
11.5	3.871	1 238.70	320.00
12.0	6.658	2 330.30	350.00
12.5	7.767	2 998.10	386.00
13.0	11.659	4 931.80	423.00
13.5	20.686	9 308.70	450.00
14.0	18.389	8 734.80	475.00
14.5	20.182	10 091.00	500.00
15.0	16.672	9 169.60	550.00
15.5	12.688	7 612.80	600.00
16.0	9.262	6 112.90	660.00
16.5	5.552	4 053.00	730.00
17.0	4.870	3 793.70	779.00
17.5	3.591	3 088.30	860.00
18.0	2.587	2 328.30	900.00
18.5	3.207	3 207.00	1000.00
19.0	1.761	2 025.20	1150.00
19.5	0.220	275.00	1250.00
20.0	0.783	1 057.10	1350.00
20.5	0.246	356.70	1450.00
21.0	0.260	426.40	1640.00
21.5	0.229	407.60	1780.00
23.0	0.059	136.90	2320.00
23.5	0.060	162.30	2704.61
–	155.528	84 485.4	–

* Measured to first anal fin ray.
** From acoustic measurement.

Table 28.2 Argentine: composition of the biomass.

Length group* cm	Numbers of fish (× 10^6)	Biomass t**	Average weight g
14	0.024	0.48	20.00
17	0.024	0.84	35.00
18	0.178	7.57	42.50
19	0.096	4.80	50.00
20	0.048	2.81	58.50
21	0.024	1.62	67.50
22	0.024	1.87	78.00
23	0.054	4.81	89.00
24	0.284	28.83	101.50
25	0.240	27.48	114.50
26	0.806	103.97	129.00
27	0.460	66.70	145.00
28	1.540	248.71	161.50
29	1.956	351.65	179.50
30	2.544	506.26	199.00
31	2.194	482.52	219.75
32	2.434	587.14	242.00
33	3.357	891.69	265.50
34	3.273	950.38	290.50
35	4.307	1 366.19	317.00
36	5.907	2 040.50	345.50
37	7.159	2 688.65	375.50
38	5.482	2 227.38	407.00
39	5.296	2 328.65	440.50
40	6.227	2 958.86	475.00
41	5.324	2 726.16	512.50
42	5.417	2 984.14	551.00
43	2.779	1 645.10	592.00
44	2.173	1 379.82	635.00
45	0.602	408.30	679.50
46	0.497	361.58	726.00
47	0.363	280.77	775.50
49	0.168	147.76	879.50
–	72.261	27 814.00	–

* Measured to first anal fin ray.
** From acoustic measurement.

sidise the fishing vessel, so that the net income could be guaranteed to a certain level (Skagen Fiskeriforening 1989). The fishing vessel was a 167 GRT trawler with a 1014 hp engine. The fishing equipment was suitable for deep water fishing grounds. Nevertheless, considerable repair was required, especially on the trawl winches.

The programme was more successful than expected because a much greater part of the catch was sold to the processing industry than was anticipated initially. The remaining part of the catch was sold as industrial fish. In the period from November 1987 to May 1988, 290 t of grenadier and 794 t of argentine were landed, averaging 4.8 t grenadier and 11.5 t argentine per fishing day.

The variation in Figure 28.3 has to be treated with caution because the usual problems in starting an experimental fishery influenced the catches in November. The gaps in February and the first part of March are due to cod fishing in the Baltic. The peak in April indicates that this period is the spawning period of the argentine; during the other months the fish are more scattered (Ebbesen 1989).

During the experimental fishing period an increased

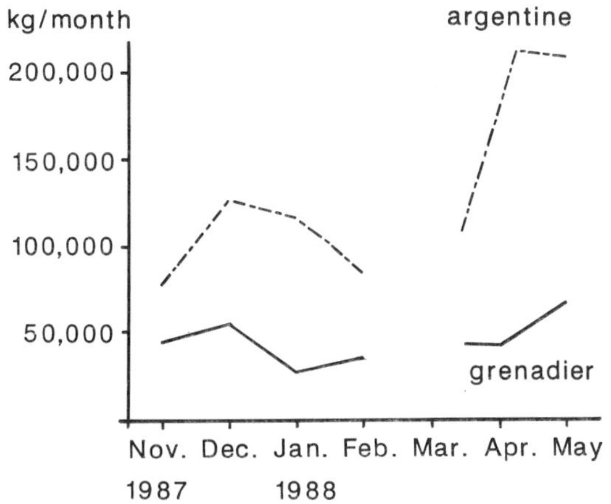

Fig. 28.3 Variation of the catches of grenadier and argentine in kg per month.

demand was observed for argentine, due to an unexpected increase in the market potential. This explains the change in fishing activities from March to May, as shown in Figure 28.3.

Distribution

Grenadier and argentine are not restricted to specific areas of the Skagerrak (Christians 1976). The research survey indicated that the two species were to be found in the larger part of the Skagerrak, but restricted to waters more than 300 m deep. It was observed during the experimental fishery that the argentine migrated north during and after the spawning period and concentrated to a higher density during the spawning period, and also that the grenadier and argentine fishery is best north of 58°N and west of 11°W (Fig. 28.4).

PRODUCTION POSSIBILITIES

Before introducing alternative species of fish to be processed in the industry, it is always necessary to describe the anatomy as well as the flavour of the species in question. The general appearance of the fish is also of significance.

Grenadier

The grenadier has very large scales on the trunk and tail, which give some problems in the skinning process. The scales have to be removed before skinning or even before filleting (Hanussardóttir & Mohr 1984). The skeletal structure does not interfere with the filleting process, and gives a bone-free fillet. The weight of the head of a gutted fish averages 35%. The length of the tail averages 50% of the total length (Sahrage 1986). The fillet yield by machine is 22–23% of the round fish, using the Baader 188. Hand filleting implies access to larger individuals (Ebbesen 1989).

If the grenadier is cooked as a whole fish, the skin is very difficult to remove. This implies that the fish has to be skinned and filleted. The cooked fillet resembles cooked cod fillet, both in consistency and colour. The taste is appealing, with a weak shellfish flavour (FF, Hirtshals 1987).

A broad variety of products is possible for grenadier, the most important being skinned fillets, fresh or frozen, and skin-on fillets as a smoked product. A new variation could be skinned fillets cut into squares for use in pizzas, salads, soups etc. (Larsen 1989).

Argentine

The most characteristic feature of the argentine is its skeletal structure. There are two rows of parietal ribs for 65% of the length of the fillet. These are difficult to detect and remove. The argentine has the same filleting drawbacks as the grenadier.

The yield of machine fillets averages 30% of the round weight after skinning. The highest yield is in the length range 30–35 cm. In filleting bigger fish some of the meat is lost during the head and back cutting (Poulsen & Sigurdsson 1988). The fillet has a light yellow colour and a rich taste. The overall impression is that the taste is acceptable but that the parietal ribs are annoying.

Argentine could be prepared as hot smoked whole fish: the colour is white and the taste is good, reminiscent of smoked mackerel, in spite of a relatively low fat content. The argentine has already proved itself a splendid fish for mince production, due to a high gel forming ability. In Figure 28.5 the gel forming ability of block frozen mince is shown.

ECONOMICS

This section deals with the economical aspects of fillet production. Only grenadier is discussed because fillet production of argentine is considered to be of less interest.

It is always useful to analyse the economics of exploitation of an alternative species of fish. The present estimation is rough but gives a picture of the products assumed to be the most profitable. It concentrates on comparing the cost of production of

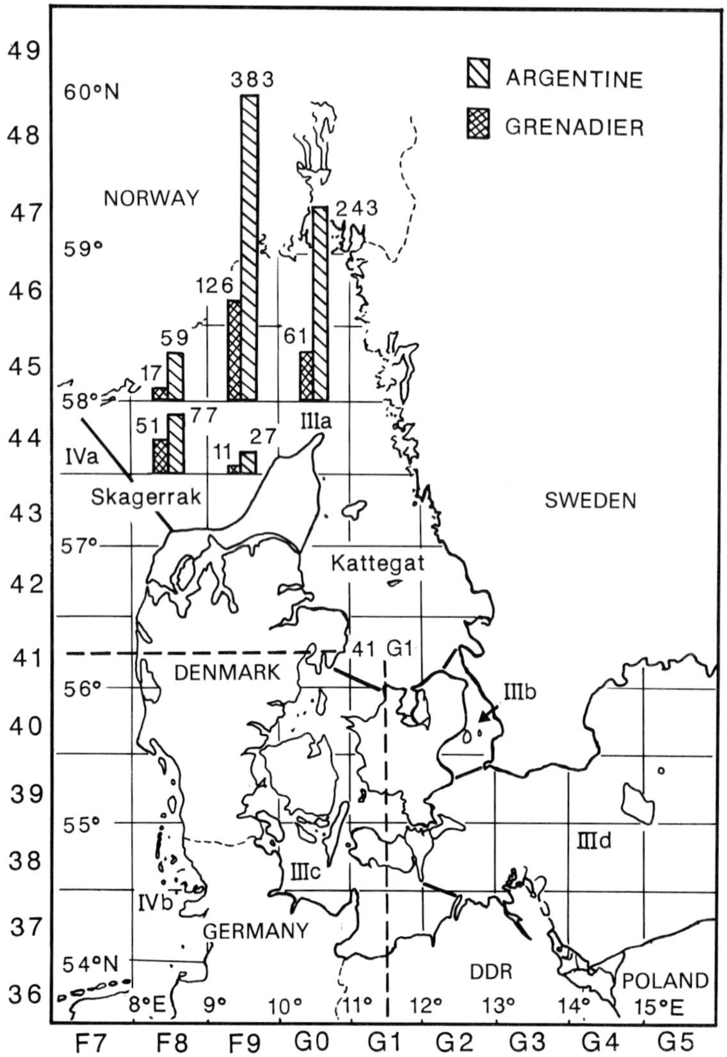

Fig. 28.4 Distribution of the catches in tonnes per ICES statistical rectangle.

ungutted and gutted tailcut grenadier in manual fillet production and machine fillet production.

Assumptions

(1) Estimated total catch of grenadier: 10 000 tons

(2) Weight and length distribution (Matcon 1987):

	average	variation±
total length:	64 cm	6.5 cm
weight	622 g	200 g

(3) Suitable machines for:

headcutting	Norfo, Bornholm or Baader 417
filleting	Baader 188
skinning	Baader 51

(4) Yields:
machine filleting: max. yield 20–25% of round weight (skinned)
hand filleting: max. yield 29% of round weight (skinned)

(5) Price ex fishing vessel:

whole ungutted	3.00 DKr per kg
gutted tailcut	4.00 DKr per kg

(6) Profit: fixed at 10% of the minimum sales price.

(7) Minimum sales price is calculated as variable expenses/(1 − profit).

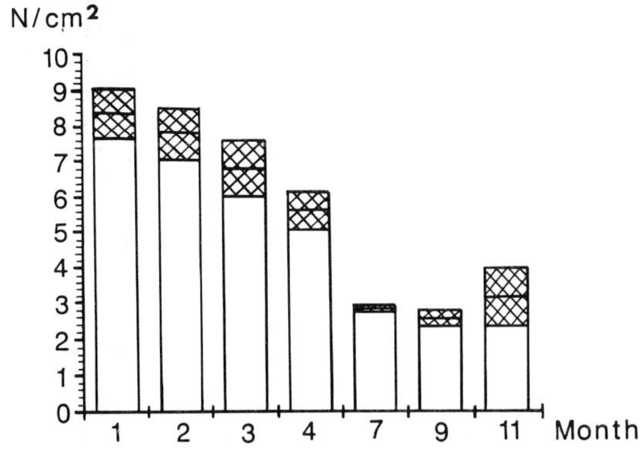

Fig. 28.5 The gel forming ability of block frozen mince after frozen stage. The average values are shown with 95% confidence intervals (Poulsen & Nielsen 1989).

(8) Salary: on average 111.00 DKr per hour
(9) Expenditure on transportation and advertising is not included in the calculations.

The result of the calculation in Table 28.3 shows that it is not profitable to gut and tailcut grenadier before processing. Single frozen fillets give a higher profit because of the better sales price proportional to the variable expenses.

The same calculations have been carried out for machine filleting grenadier. The same sales price has been used to compare hand and machine filleting. The profit per kg raw fish in both ungutted and gutted tailcut are higher when using machine filleting. The results for ungutted grenadier are: 0.62 DKr per kg (block frozen) and 0.63 DKr per kg (single frozen). For gutted tailcut grenadier the results are: 0.92 DKr per kg (block frozen) and 0.94 DKr per kg (single frozen).

The conclusion of this rough calculation is that machine filleting gives the highest profit, and the grenadier should be gutted and tailcut and sold as single frozen fillets. These calculations are meant only as a supporting tool to help decide how to proceed when a new species of fish is exploited commercially. Such considerations as investment in new machinery have not been taken into account.

Table 28.3 Calculation of the profit in a factory with hand cut fillet production. Two parameters are variable. The fillets could be block frozen or single frozen and the fish could be ungutted or gutted and tailcut (Larsen 1989). (All figures in DKr per kg final product.)

	Ungutted		Gutted, tailcutted	
	Block	Single	Block	Single
Fish (raw material)	12.00	12.00	12.50	12.50
Waste (to fish meal)	−2.22	−2.16	−1.57	−1.57
Gross salary	6.22	6.22	5.99	5.99
Packing charges	1.60	1.60	1.60	1.60
Energy for freezing	0.40	0.40	0.40	0.40
Electricity, water	0.15	0.15	0.15	0.15
Variable expenditure	18.15	18.21	19.07	19.07
Sales price	20.17	20.23	20.17	20.23
Profit per kg final product	2.02	2.02	1.10	1.16
Profit %	10.00	10.00	5.50	5.70
Profit per kg raw fish	0.51	0.51	0.35	0.37
Profit per salary hour	36.06	36.06	20.35	21.47
Percentage yield	29	29	32	32

MARKETING POSSIBILITIES

The two processing companies which have bought the grenadier and argentine have used their normal distribution network in Europe to open up a new market for the two species. Both companies estimate that the marketing possibilities can be expanded in both Europe and Japan. If this sales promotion is to be a success the sales price has to be raised from the present price range of just below that of other white fish species, to at least the price of cod products.

It was surprising to find that the prices of grenadier products were more or less independent of the price of cod. This indicates that grenadier has found its own niche in the market (Larsen 1987b).

CONCLUSIONS

Involvement in development of new fisheries, as a fish technologist, demands much more than the traditional laboratory field of activity. The economic aspects of the whole project in particular have to be taken into consideration, from an acceptable income for the fishermen to a reasonable price for the final product.

Several other factors, such as the design of fishing gear, catch handling equipment, storage on board, processing in the factory and different product types and marketing also have to be taken into consideration.

As the project described has clearly shown, whenever a new species is introduced to the processing industry you literally have to start from scratch.

BIBLIOGRAPHY

Andersen E.M. & Hansen U.J. (1979) *Fangst-og anvendelsesmuligheder af nogle utraditionelle fiskearter*. Rap. 2, 53, Nordjyllands Amts Fiskeriprojekt.

Christians O. (1976) *Neu Nutzfische*. Forschungsbericht M 76–03 Bundesforschungsanstalt für Fischerei, Hamburg.

Cohen D.M. (1984) *Fishes of the north-eastern Atlantic and the Mediterranean*, **I**, 387–8. Unesco, Paris.

Ebbesen R. (1989) *Forsøgsprojekt Skolæst. Udvikling af en rentabel konsumproduktion for dybvandsfiskene Skolæst og Guldlaks*. Final report, Danish Ministry of Fishery.

Eliassen J-E. (1986) Undersøkelser av utbredelse, forekomst og bestandsstruktur av skolæst (*Coryphaenoides rupestris Gunn.*) i Trøndelag. *Fisken Hav.*, **1**, 1–19.

FF, Hirtshals, (1987) *Kemisk og organoleptisk vurding af Grenadier som konsumfisk*. Working report, February.

Geistdoerfer P. (1984) *Fishes of the north-eastern Atlantic and the Mediterranean*, **II**, 659–60. Unesco, Paris.

Hanussardóttir M. & Mohr L. (1984) *Forforsøg i forbindelse med råvarebehandling af Guldlaks*. Heilsufrødiliga Starvsstova.

Larsen E. (1987a) *Danish Experimental Fishing in the 1980s*. Paper presented at ICES Fishing Technology and Fish Behaviour Working Group, Hamburg.

Larsen E. (1987b) *Forsøgsfiskeri efter skolæst*. Working paper, Danish Ministery of Fishery.

Larsen E. (1989) *Anvendelse af guldlaks og skolæst som alternativ konsumfisk*. Project report FF, Lyngby, Denmark.

Madsen K.P. (1987) *Togtrapport 3/87*. Survey report. Danmarks Fiskeri-og Havundersøgelser.

Matcon A/S (1987) Forsøgsproduktion. In *Grenadier, Delrapport No. 1*. Forsøgsfiskeri og – produktion. Skagen Erhvervsråd, DK-9990 Skagen, Denmark.

Poulsen B. & Nielsen T. (1989) *Guldlaks til konsum*. Paper presented at FF's Konsumfiskemøde, Aalborg, Denmark.

Poulsen B. & Sigurdsson B. (1988) *Processing af guldlaks*. Final report, pp 109, Aalborg Universitets Center, Denmark.

Ringdal O. & Strand O. (1989) Hvitlaks – en liten – men vigtig ressurs. *Fiskets Gang*, **3**, 6–7, 38.

Sahrhage D. (1986) *Wirtschaftlich wichtige Grenadierfische des Nordatlantiks*. Bundesforschungsanstalt für Fischerei 37, Mitteilungen aus dem Institut für Seefischerei, Hamburg.

Skagen Fiskeriforening (1989) *Statistik over regnskaberne for fiskerflåden i Skagen 1988*.

Yngvesson S. & Sørensen E.F. (1987) *Rapport over skolæst togt med Dana 2–7 maj 1987*. Working report. Dansk Fiskeriteknologisk Institut.

CHAPTER 29

Utilisation of sand eel (*Ammodytes tobianus*), greater sand eel (*Hyperoplus lanceolatus*), sprat (*Sprattus sprattus*) and Norway pout (*Trisopterus esmarkii*) for human consumption

J. Nielsen, B. Rønsholdt, N.C. Jensen and N. Alsted

INTRODUCTION

Small pelagic fish include many species such as capelin, sprat, some sardine species and immature fish of species like herring and mackerel. Many other small fish are underutilised and although they are strictly not pelagic species this chapter includes a description of Danish experiments with sand eel (*Ammodytes tobianus*), greater sand eel (*Hyperoplus lanceolatus*), sprat (*Sprattus sprattus*) and Norway pout (*Trisopterus esmarkii*) for human consumption. These four species are the main raw material used in the fish meal industry in Denmark.

During deterioration of the mentioned species, volatile amines, biogenic amines and toxins are formed. In the past raw material of low quality was often accepted, as some of the amines and toxins formed are destroyed by the normal heat treatment in the fish meal process. As it became obvious that the nutritive value and the yield of the fish meal process is dependent on good quality raw material, i.e. low content of volatile bases and biogenic amines, the Danish fish meal industry introduced a 'pay according to quality' system. In order to obtain good quality, more than 150 Danish trawlers are now equipped with an automated device which divides the catches according to size (thickness) and provides a sufficient dose of ice. The system of size grading, combined with improved chilling in bulk tanks and containers, presented the opportunity of making experiments with sand eel, greater sand eel, sprat and Norway pout for human consumption.

Efficient catch handling, where the fish is chilled immediately after catch, is necessary in order to prevent 'belly burst' and softening of the flesh of small fish.

Many small fish are consumed whole, but in other cases it is necessary to divide them into edible and non edible parts. Machines are available for heading and gutting but they are very expensive and can only be used for fish larger than 15 cm. The fish meat can also be mechanically separated from skin and bone in a meat and bone separator, and then used for minced products. The best quality mince is obtained from filleted or very thoroughly rinsed fish in order to avoid blood and kidney tissue in the meat. In some cases the most valuable part of the small fish is the intestinal fraction which can be used for biotechnological products or for direct consumption, as for example the roe from capelin – an expensive product in Japan.

Small fish are rarely distributed fresh because of the high deterioration rate. Instead they are used for day-to-day consumption in coastal areas. Preservation is needed in cases where increased utilisation of the resource is wanted, and some methods of preservation are examined here.

RESOURCE AVAILABILITY

The supply of raw material in Denmark for the fish meal industry is approximately 1.5 million tonnes per year and 85% of this is sand eel, sprat and Norway pout. Landings are concentrated in the harbours of western and northern Jutland.

Due to the deterioration rate of the small fish, processing for human consumption has to be done near the landing places.

The resource availability changes during the year, with the high season in the summer months. Variations are due to biological conditions, quotas and weather conditions.

INVESTIGATIONS

Sprat

Sprats are herring-like fish. They are pelagic shoaling fish living in fjords and waters near the coast, in summer at depths of 10–50 m, in winter down to 150 m. They reach a length of 10 cm when one year old and can be 16 cm long. Sprat can be 5–6 years old but when caught as industrial fish it is often 2–3 years old. The resource is large: 130 000–150 000 tonnes are caught annually in Denmark.

Sprat is a fatty fish with an oil content varying from 6% in the spring to 14% in the autumn (Jensen 1986).

In the Nordic countries sprats are traditionally used for human consumption. One year old sprats have to be kept in nets after catching so that the gut is emptied, or they have to be caught during the night when the stomach is empty, before they are canned in oil and sterilised (Brisling). If the sprats are smoked before canning, the product is known as, for example, Kieler Sprats. Larger sprats are preserved by pickling before being canned, pasteurised and sold as chilled anchovies.

Sprats can be kept in refrigerated sea water (RSW), chilled sea water (CSW) or iced in boxes for two to three days before processing; otherwise the belly starts bursting (Jensen 1986). Jensen (1986) prepared canned sprat in two new marinades (lemon/oil and Escabeche) and compared the product with the traditional smoked sprat canned in oil. It was not possible to get fresh sprat of good quality so frozen sprats were used as raw material. Sprats caught in August were iced in boxes for two days, vacuum packed and frozen. The products were prepared from thawed, nobbed sprats after three weeks of frozen storage at −30°C. The yield after nobbing was 75%. The sprats were 8–10 cm long and had a total volatile base (TVBN) content of 10 mg N/100 g, an oil content of 12% and a protein content of 17.4%. The flesh was very soft and the fish were difficult to handle.

The products were made in 14 club cans; 100 g of fish were placed in each can and pre-cooked at 80°C for 15 min. Water was decanted and the cans filled with a mixture of lemon/corn oil 4:6 containing 2% salt. In the Escabeche product the fish was battered and fried in oil instead of being pre-cooked. A marinade (fried mixture of onion, carrots, chilli, salt, garlic, bay-leaf and thyme) filled the cans. They were then sealed and autoclaved at 110°C, 0.7 N/cm^2 to an F_0 of 3.5. The smoked sprats were prepared in the traditional way and autoclaved together with the other two products.

The three products were stored for one month and served to a three person expert panel and an eight person laboratory panel with no experience in grading of canned products. All three products were evaluated as excellent in appearance and taste, though it was felt the lemon taste of the lemon/oil product should be adjusted to the market taste. However, both panels preferred the traditionally smoked product.

Sand eel and greater sand eel

Sand eels are caught in the North Sea and along the coasts of the Nordic countries and the UK. In Denmark 600 000–1 000 000 tonnes are caught annually for reduction to fish meal and oil. There are many local families of sand eel, distinguished by the number of vertebrae.

Sand eels normally reach a length of 16 cm with a maximum length of 20 cm. They spawn in the late autumn and live quite near the bottom. Greater sand eels are 20–35 cm in length and spawn from April to August. The two species are caught together, greater sand eel normally making up between 1 and 20% of the catch. Sand eel is a fatty fish with an oil content ranging from 1% after spawning to 10% (Jensen 1986).

Processing of sand eel must start within three days of being caught, and the fish must be chilled immediately after catch. Even then belly burst will be seen in some of the fish (Jensen 1986).

Jensen (1986) did the same canning experiments with sand eel as with sprat. Sand eels in lemon/oil, Escabeche and smoked in oil were accepted to the same extent as sprats. However, the market potential in Denmark for canned products is rather limited and might not exceed 1000 tonnes of sand eel or sprat. In order to start commercial production an export market is needed.

Greater sand eel was hot smoked by Jensen (1986). After capture in June in the North Sea the fish was handled in three different ways:

(1) gutted and frozen immediately after catching
(2) gutted, iced in boxes and frozen after 48 hours.
(3) iced in boxes and gutted before freezing after 48 hours.

After freezing the fish was glazed and packed in polyethylene pouches before storing for 10 days at −30°C. The length of the fish was 24–26 cm. The thawed greater sand eels were salted at 10°C for 15 minutes in a 10% (w/w) brine, (1:1). Before hot smoking the fish were hung on a stick and dried.

The thawed greater sand eel had a TVBN content

of 13 mg N/100 g, an oil content of 4.8% and a protein content of 17.5%. Thiobarbituric acid value (TBA) and free fatty acids (FFA) were measured in the three different types before and after smoking. TBA and FFA were highest in the fish frozen immediately after catch, but after smoking, the flesh of the three different types had the same rather low level of TBA (6–10 μmol malonaldehyde/kg fish) and FFA (2.3–2.6% in oil).

The sensory evaluation was performed by a trained laboratory panel (eight members). After smoking there were no differences between the catch handling methods. However, the fish gutted after 48 hours were very soft and difficult to handle before smoking. Appearance of the three products was very good, the taste was good and the texture was a little dry but acceptable.

The conclusion was that there are technological possibilities of producing a good smoked product of greater sand eel, but in the market it has to compete with smoked herring. Each year approximately 1000 tonnes of smoked herring are produced in Denmark, most of which are sold on the home market.

Nielsen (1984) experimented with the production of roller dried sand eel. Sand eel were nobbed and minced on a meat and bone separator before drying on a roller dryer. Mince from white fish can be dried on a single or double roller dryer at a temperature of 155°C for a few seconds. The mince comes off the roller in a continuous lace-work of dried flattened fibres. It can be mixed with cereals or seasoning before drying. When using a fatty fish like sand eel it is essential to mix the mince with low oil materials prior to roller drying.

Starchy material such as maize, millet, wheat and rice and proteinaceous vegetables such as lentils and chick peas were tested, along with skimmed milk powder, to find the best suited ingredient. Roller drying accelerated development of rancidity and it was necessary to add chemical or natural antioxidant to avoid this. Either rosemary or sage or a mixture of BHA and citric acid appeared to have good antioxidant effects. The product could be stored for seven months by packing it under vacuum in aluminium foil pouches.

Minced sand eel can be used as an ingredient in kroepoek. This is the Malaysian name for an expanded snack, with fish or shellfish, usually shrimp, as the protein ingredient. Sago and tapioca are traditionally used for starch, and the mixture is pregelatinised over an open fire or in an oven. Before serving, kroepoek is expanded in oil. The manufacture is usually a cottage industry, but kroepoek can be pregelatinised on a roller dryer.

Minced, washed sand eel was used in preparing kroepoek. Figure 29.1 gives the sequence of other actions used. The mince was washed in order to remove some of the oil (from 8.1% down to 2.1%). Sand eel was compared with Norway pout (unwashed mince) as raw material. The purpose of the experiment was to replace shrimp with a cheaper protein, but to keep the seafood taste. The pregelatinised mixture was dried to different contents of dry matter and it was found that the most acceptable product should be dried to a water content of between 13% and 15%.

Norway pout

Norway pout is a member of the cod family. It reaches a length of 9–13 cm in the first year, 14–19 cm in the second year and it can be 25 cm long. It matures at two years. It lives at depths of 80–300 m except in Skagerak where it may occur at depths of less than 40 m. This is the most abundant species of the cod family but it is not caught for human consumption because of its small size. In Denmark 200 000–400 000 tonnes of Norway pout are caught each year for reduction to fish meal and oil.

Norway pout is a lean fish and the oil content varies from 3% in spring to 9% in the autumn (Jensen 1986). The oil content in the meat alone is less than 1%.

Experiments with preparation of minced fish and surimi from Norway pout (Jensen 1986, Rønsholdt *et al.* 1984) were carried out using fish chilled in CSW.

Three different ways of removing the entrails were compared:

(1) Cutting the head off just behind the gills and pectoral fin.
(2) Cutting at an angle from the neck to the gut, whereby the entrails are removed along with the head.
(3) Nobbing – most of the entrails are removed when the head is drawn.

The angular cut (Fig. 29.2) was shown to give the best result with regard to yield and removal of intestines.

The cutting was done by hand and a preliminary experiment using an Arenco-CIS machine, normally used for herring, on Norway pout larger than 12 cm showed the need for an extensive rebuilding of the machine. After cutting out the remaining pieces of intestine, the blood and black membranes were removed by washing in cold water and unwashed mince was compared with washed mince.

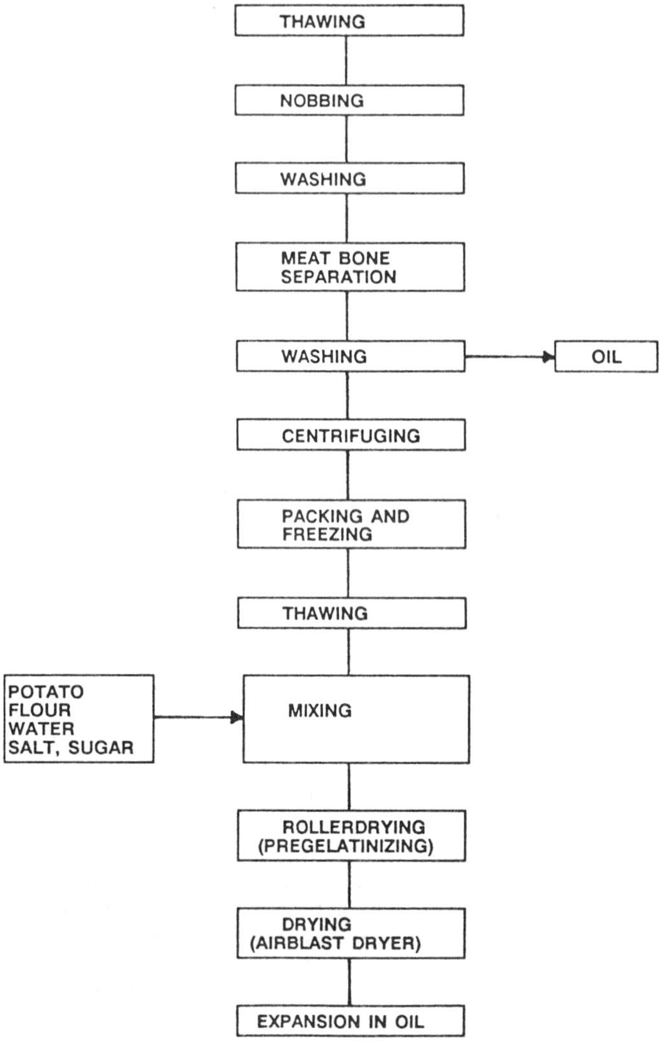

Fig. 29.1 Manufacturing of kroepoek (Rønsholdt & Thomsen 1984).

After draining, the tails were separated in a meat and bone separator equipped with a drum with 5 mm holes covered with a screen of 1.2 mm holes, or a 7 mm drum with a 3 mm screen. Remaining blood and black spots were removed by washing the mince. Draining was performed either in a screw press or in a basket centrifuge. The yield of the mince was approximately 50%, with the highest yield obtained by using the angular cut.

Dewatering by centrifugation gave a minimum loss of dry matter in the wash water. In the experiment with the 1.2 mm screen, part of the washed and unwashed mince was mixed with 8% of skimmed milk powder on a silent cutter to prevent freeze denaturation. It has been shown that skimmed milk powder has a certain cryoprotective effect (Nielsen *et al.* 1985). Yield of mince was 47% and of washed mince 37%, based on whole fish. Figure 29.3 gives the sequence of operations used in preparing Norway pout mince.

After freezing, the minces were stored at −20°C and evaluated by chemical, functional and sensory analyses. Measurement of the functional properties showed a stronger gel of the washed mince (Knudsen *et al.* 1987). Sensory analyses performed on battered fish sticks showed that:

- Unwashed mince was grey with a hard, dry and rubbery texture.

Fig. 29.2 Angular cut of Norway pout (*Trisopterus esmarkii*).

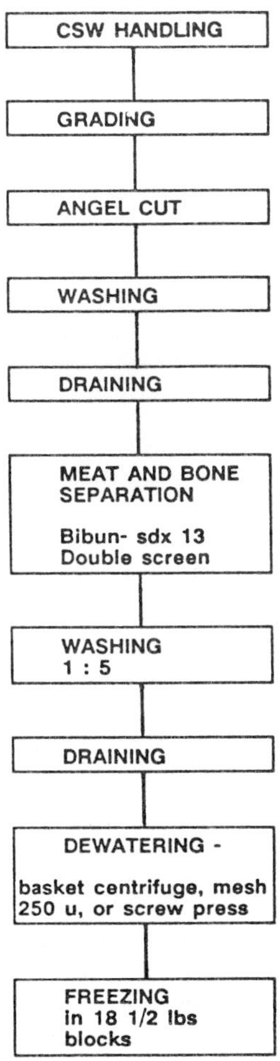

Fig. 29.3 Flow sheet for production of Norway pout mince (Børrensen & Liljemark 1988).

- Washed mince had a lighter colour but a very hard and rubbery texture.
- Mixing with skimmed milk powder did not improve the texture or colour, but added an unpleasant sweet taste to the product.

The sensory analyses were performed by a laboratory panel and a panel from the Danish fishing industry. The fishing industry found the minces unacceptable as a substitute for white fish mince.

In the experiments using the 3 mm screen, two different grades of Norway pout were used: 8–12 cm and 12–20 cm. The mince was not washed and only 2% skimmed milk powder was added in a suspension on a food mixer. Mince from the larger fish gave a yield of 43% and from the smaller 36%, based on whole fish. The minces were block frozen and stored at −30°C.

No cryoprotective effect was found to be attributable to the skimmed milk powder, measured as functional properties (gel strength) or in the sensory evaluation. Mince from the larger fish had a slightly better gel strength than that from the smaller ones.

The 3 mm samples got a better sensory evaluation than the 1.2 mm samples. After three months of storage, however, a yellow/brown colour and an off-taste developed and the minces became hard and rubbery. Again the fish industry found the blocks unacceptable for fish fingers, due to colour and texture.

A strainer (Bibun) was added to the equipment and an experiment with production of surimi (Suzuki 1981) was conducted (Rønsholdt *et al.* 1985).

Mince from the tails was washed three times with water (5 : 1). After final dewatering in a centrifuge, the washed mince was strained and mixed with 0.2% polyphosphate, 4.0% sucrose and 4.0% sorbitol, and frozen as surimi. Dewatering was difficult and it was possible to obtain a final dry matter content of only 14.6% in the dewatered mince, compared with 22.2% in fish meat.

Gel strength in the surimi was doubled to 25 N/cm^2 compared with gel strength of the unwashed tail mince.

The block frozen surimi was stored at −20°C and the functional properties were evaluated after two months and after a year. The surimi kept an acceptable water holding capacity, while the gel strength was reduced immediately after freezing but stayed at the same level after a year of storage. The gel strength was measured as 15 N/cm^2, which compared well with the gel of a good quality Japanese surimi, found to be 12 N/cm^2 using the same method.

An off-taste developed in the surimi immediately after freezing, which might be due to the quality of the water used for washing. The surimi could not be graded as first quality because of a very light grey colour and tiny spots of membrane.

CONCLUSION

Experiments have shown technological possibilities for producing good and acceptable products of small fish such as smoked greater sand eel, different canned products from sand eel and sprat, dried sand eel and kroepoek. Mince of Norway pout might be a future product. However, the mince cannot be used in the same way as mince from larger white fish is used. Further product development is therefore needed.

Surimi stands a good chance of being produced from Norway pout. The tiny membrane spots present in the mince can be removed by use of better technology, but Norway pout has a naturally light grey colour and the mince will obviously have this unless bleached.

Common to the small fish species such as sand eel, greater sand eel, sprat and Norway pout is the fact that they represent a great resource potential, but there are difficulties concerning handling of the fish for human consumption, mainly because of rapid deterioration after catch. Production should therefore preferably be made on board, or the fish must be frozen if being kept for more than two to three days.

In the North Atlantic it might be more obvious to continue the development of products for human consumption with larger species still not fully utilised, such as blue whiting (*Micromesistius poutassou*), round nose grenadier (*Coryphaenoides rupestris*), greater argentine (*Argentina silus*), herring (*Clupea harengus*), mackerel (*Scomber scombrus*) or even horse mackerel (*Trachurus trachurus*). The smaller species would continue to be used for fish meal and oil and silage.

BIBLIOGRAPHY

Børresen T. & Liljemark A. (1988) Fiskmassa. Fiskfärs ur restfisk, småfisk och mindre utnyttjade fiskarter. *Final report from a Nordic co-operative project.* Technological Laboratory, Danish Ministry of Fisheries. Technical University, Lyngby, Denmark.

Jensen N.C. (1986) *Fraktionering og forædling af industrifisk.* Thesis for an industrial PhD.

Knudsen L., Børresen T. & Nielsen J. (1987) Textural parameters in compression testing of a gel made from fresh and frozen mince. *J. of Texture Studies*, **18**, 261–71.

Nielsen J. (1984) Roller-drying of fatty fishes. *Lebensm.-Wiss. u.-Technol.* **17**, 151–4.

Nielsen J., Bruun A. & Madsen A. (1985) Milk powders as cryoprotectants in frozen minced fish. *Refrigeration Science and Technology*, **4**, 169–74. IIR, Commissions C2 & D3, Aberdeen.

Rønsholdt B., Alsted N. & Blak E. (1984) *Sperlings anvendelsmulighed til konsum.* Report 15, minced fish, Technological Laboratory, Danish Ministry of Fisheries.

Rønsholdt B., Alsted N. & Sonnesen P.M. (1985) *Sperlings anvendelse til konsum.* Report 19, minced fish, Technological Laboratory, Danish Ministry of Fisheries.

Rønsholdt B. & Thomsen S.B. (1984) *Mince af tobis og Sperlingmince og tobismince til kroepoek.* Report 16 & 17, minced fish, Technological Laboratory, Danish Ministry of Fisheries.

Suzuki T. (1981) *Fish and Krill Protein: Processing Technology.* Applied Science Publishers, London.

CHAPTER 30

The drying of small pelagic fish: the Icelandic experience

S. Arason, G. Thoroddsson and G. Valdimarsson

INTRODUCTION

In recent years the annual world production of dried unsalted fishery products has been about 350 000 tonnes. Stockfish production is about 10 000 tonnes, the main producers being Iceland and Norway, but the biggest producers of the other dried products are in Asia and Africa (FAO 1989).

A number of fish species have been utilised for the production of fish meal and fish oil in Iceland. In the early 1970s capelin became the main source of industrial fish. The yearly catch was about 1 million tonnes. Blue whiting was also caught from 1976–82, but the highest catch was in 1978 at 35 000 tonnes (Fisheries Association of Iceland 1983).

For the last six years there has been no catch of blue whiting in Icelandic waters. For catching this fish large fishing gear (mid-water trawl) is needed together with ships with powerful engines, but only purse seiners for capelin have the necessary ability and engine power. Obviously this type of fishing is expensive, and because of its low fat content the blue whiting gives a lower yield than fatty fish in fish meal production. It is therefore doubtful that it will be economical to catch blue whiting for fish meal production only.

To make the blue whiting fisheries economical a significant part of the catch must be for human consumption. Much experimental work on this has been done in many countries, by producing minced fish and fillet blocks. (Nordforsk 1981) In Iceland, however, the main emphasis has been on drying blue whiting and other pelagic species such as capelin and Norway pout. For instance blue whiting can be dried artificially, and the product – a 'mini-stockfish' – is considered comparable in quality to traditional products made from cod or haddock. In recent decades much experimental work has been done in many countries for finding efficient means of artificially drying fish. These studies include those of Jason (1958), Waterman (1976) and Kristinsson & Arason (1983).

There has been much interest in Iceland in producing dried fish for human consumption from the various small fish species, but only two have been considered seriously: blue whiting and capelin.

Some of the studies described here were part of a cooperation project on blue whiting named Nordfisk. Participants were from Norway, the Faröe Islands and Iceland, but Denmark and Sweden participated indirectly (Nordforsk 1981).

Blue whiting is one of the species that is considered underutilised in Icelandic waters and can be used for human consumption. Blue whiting is a codfish and its fat is mostly deposited in the liver, in contrast to fatty fishes like herring and capelin. In recent years 600 000 800 000 tonnes of blue whiting have been caught in the North Atlantic. The body fat is constant throughout the year but the fat content of the liver changes significantly from one season to the other.

From 1979 to 1982 about 250 tonnes of dried blue whiting were produced in Iceland for human consumption and about 50 tonnes for the pet food market. Most of this was exported to Nigeria and European countries.

In 1974 a few tonnes of capelin were dried in a conventional klipfisk tunnel dryer for experimental purposes. Mainly brine-salted, male capelin was used as raw material. Although the fat content was as low as 5–6% it was too high for drying unsalted capelin, at least in the conventional type of tunnel dryer. The drying is too slow and capelin, which only contains 5–6% fat, gives an unappealing product with a yellow, rancid layer on the surface.

Between 1978 and 1982 there was a yearly production of 150 tonnes of dried capelin, mainly for human consumption in African countries. But now the yearly production is about 250 tonnes and mainly for the pet food market in Europe. This chapter attempts to summarise the results of various experiments conducted in Iceland for producing dried products from capelin and blue whiting.

MATERIALS AND METHODS

Raw material

Capelin (*Mallotus villosus*) and blue whiting (*Micromesistius poutassou*) were used for the experiments. The fish were transported iced on shore and processed within 3–4 days of catching. Also, fish were frozen in a plate freezer at −25°C and kept for up to 10 weeks prior to thawing and drying. Capelin was also brine salted in 100 litre plastic barrels. The fish was fully salted (flesh containing 18% w/w NaCl) and stored for eight weeks prior to desalting and drying. The capelin was either fully desalted or semi-desalted, containing 2.1% and 5.8% salt respectively in the flesh, prior to drying. Blue whiting was dried either gutted or ungutted. Also ungutted blue whiting was brine salted to a concentration of 6% in the flesh prior to drying.

Drying equipment

The drying experiments were done in commercial tunnel dryers where air speed and temperature could be regulated. Conditions for the drying of fish can be divided into primary (constant rate) and secondary (falling rate) drying. Primary drying is the first step in the drying process, in which the water that evaporates relatively easily is removed and where large volumes of air are necessary to remove the water vapour. During primary drying the fish is placed on frames or conveyor belts inside drying cabinets. In secondary drying lower quantities of the more tightly bound water evaporates with relatively slow air velocities.

The aims of the drying experiments were:

(1) to investigate at which point it was safe with regard to microbial spoilage to take the fish out of the

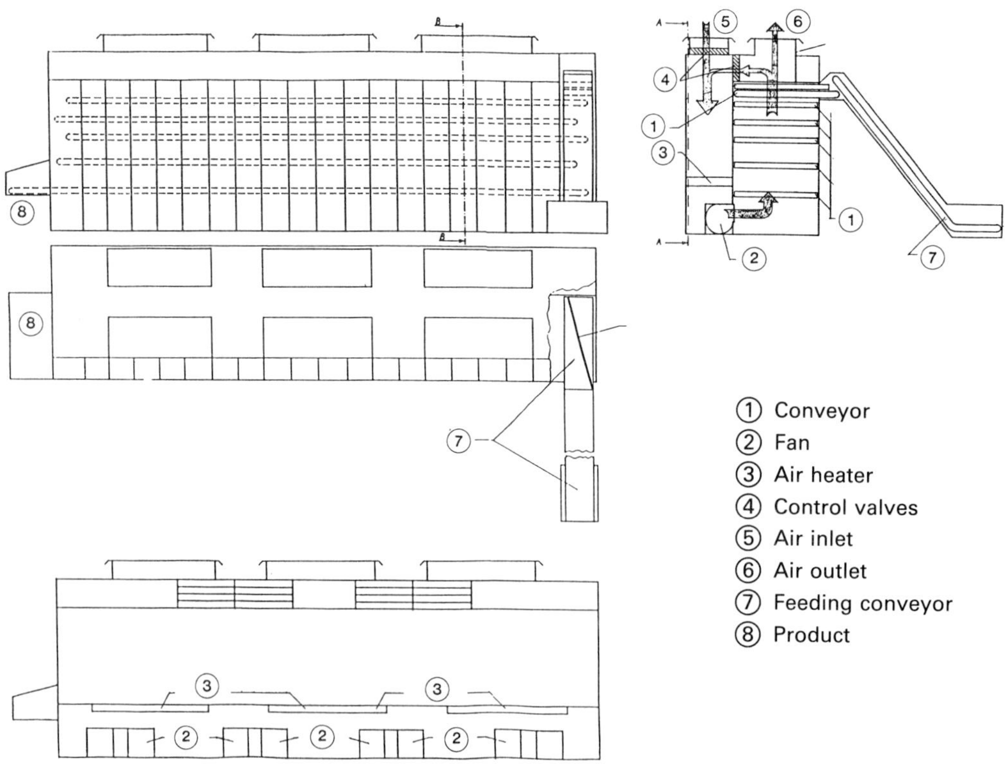

Fig. 30.1 Conveyor type primary dryer.

drying cabinet for secondary drying in a drying room
(2) to optimise temperature and air speed relationships with regard to product quality and process economy.

The production was modified with two markets in mind: the Nigerian and Italian markets for stockfish and the pet food market. Both these can use mini-stockfish, produced from either gutted or ungutted fish. Drying ungutted fish can be very difficult as the fish must not be too fat (Kristinsson & Arason 1983).

Two types of drying cabinets have been developed by the Icelandic Fisheries Laboratories. The first one is a conveyor type dryer (Fig. 30.1), which is suitable for drying small fish species. The raw material enters at the top of the conveyor belt and the belt is then stopped and left idle for about 3 hours. The raw material then moves down to the next belt and new raw material comes in at the top, and so on until the primary drying is over. The three lowest belts move at half the speed of the others. The fish remains for between 20 and 24 hours inside the drying cabinet which is fully automated, has a capacity of 15–20 tonnes of small fish per 24 hours and a maximum loading efficiency of about 90 kg of fish per square metre of conveyer belt. The cabinets have air speeds varying between 0.5 and 4 m/s and the air can be heated to 40°C by heat exchangers using 80°C hot geothermal water, humidity varying between 10 and 80%.

Two fish drying plants in Iceland using this type of primary drying device proved to be very efficient.

Another type of primary drying cabinet is the rack-dryer (Fig. 30.2). Between 10 and 20% of the air passing over the racks is fresh air, the remainder being recirculated air which can be regulated. The drying racks are stacked on pallets, 15 to 22 racks on each. The capacity of the cabinet is 10 tonnes of blue whiting. From 12 to 14 kg of fish can be accommodated per square metre of rack.

When primary drying is over, the semi-dried fish, with 50–60% water content, is put into drying containers each up to 2.5 m^3 capacity, with openings at both ends so that air can be forced through them. The drying conditions are similar to those in the primary

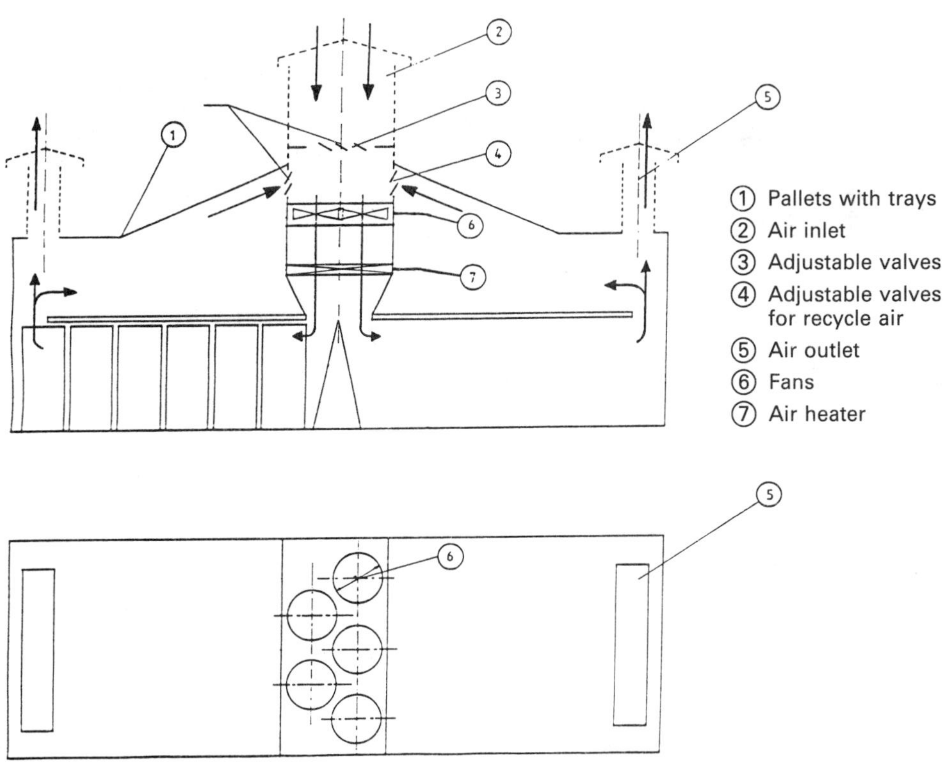

Fig. 30.2 Rack type primary dryer.

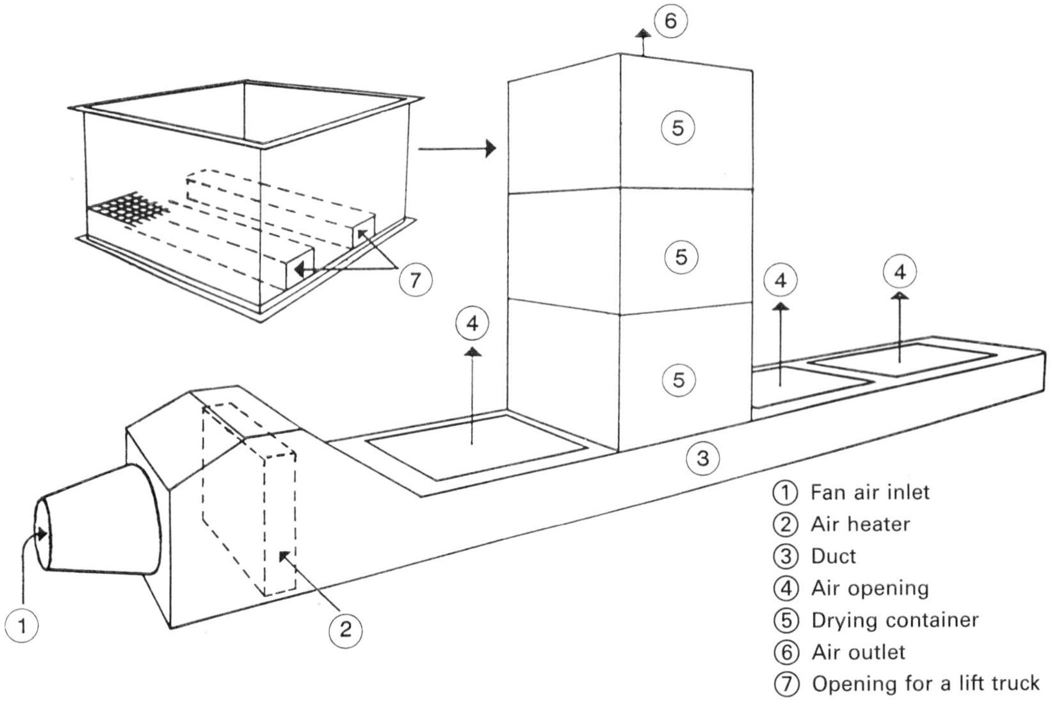

Fig. 30.3 Secondary dryer.

drying except that the air velocity is one third. Figure 30.3 shows such a secondary drying device.

Quality assessment

When the optimum drying parameters had been determined and the drying curves obtained, detailed studies were made of product quality, organoleptic, chemical and microbiological.

Aerobic plate counts were made on Plate Count Agar (PCA, Difco) by the pour plate method. Plates were incubated at 22°C for up to 3 days (Valdimarsson & Guðbjörnsdóttir 1984).

Total volatile bases (TVB) were determined by using a Struer semi-automatic distillation unit, the basic method being that of Antonacopoulos (1971). Water activity measurements were made on a Novasina AG hygrometer with a sensor ZFB-45 (5–100% RH) and ecus-1-0893 (74–100% RH), at 22°C. Saturated salt solutions of known a_w were used as controls (Valdimarsson & Guðbjörnsdóttir 1984).

Organoleptic assessments were made by a panel of trained inspectors.

RESULTS AND DISCUSSION

The most important results regarding both blue whiting and capelin were that the fish should not be taken out of the drying cabinet (primary drying) unless the water ratio (weight of water/weight of dry matter) is down to 1.0 (50% w/w water content). Regarding secondary drying the fish can be taken out of the drying cabinet at a water ratio of 0.18 (15%). The temperature in the drying cabinet must not be higher than 28°C. If it is higher, the fish will be cooked and become brittle and cannot be packed as whole fish.

Regarding blue whiting the maximum total fat content for ungutted fish seems to be around 3.5% if it is to be used for drying. The fat content can vary considerably between individual fish, e.g. when measured in July 1979 it varied from 3–8% in the same catch. The reason for this is probably that blue

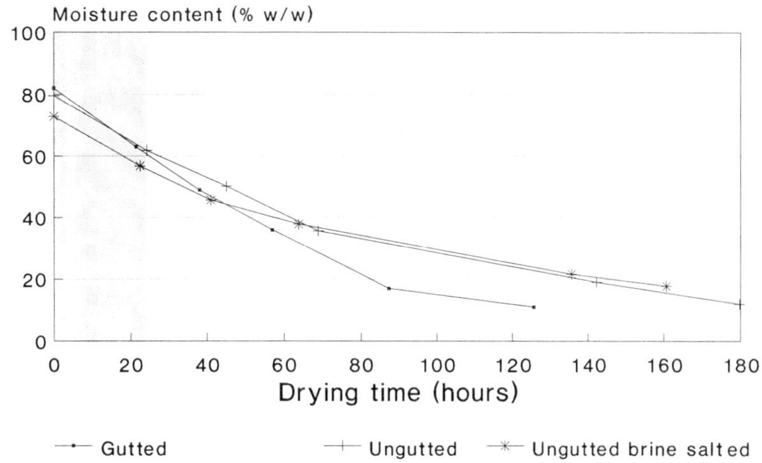

Fig. 30.4 Drying curves for blue whiting.

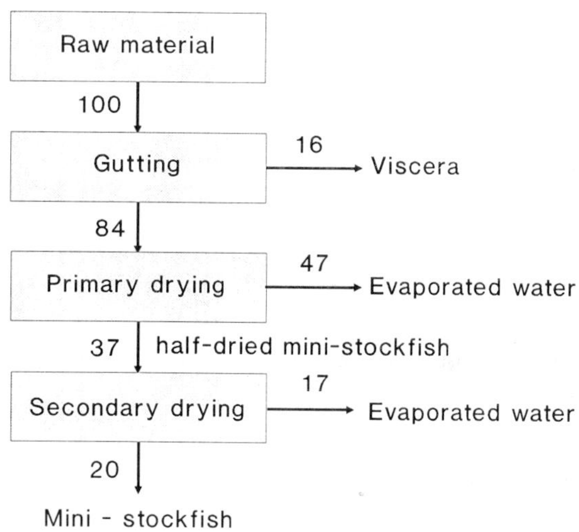

Fig. 30.5 Flow diagram for the drying of blue whiting. Percentage yield figures are given for each step.

whiting do not all migrate at the same time to the fishing grounds off the east coast of Iceland. Because of this, blue whiting must be gutted prior to drying if it is caught later than June.

The optimum conditions in our experience for primary drying of blue whiting and capelin are: air temperature between 20°C and 25°C, airspeed of 1.5–2 m/s and air humidity about 50%. The final water content after the primary dryer is 55–60% (about 50% weight loss) and Figure 30.4 shows that it takes about 24 hours to reach that point.

Figure 30.5 shows a complete flow diagram for the drying of blue whiting, including yield figures. The total drying time for gutted blue whiting is about 80 hours and the yield is 20%. For ungutted blue whiting these figures are 140 and 23% respectively. Figure 30.4 shows the water content as a function of drying time for blue whiting.

In 1978–80 experiments were made with drying capelin in a rackdryer and a conveyor-dryer as primary drying. These experiments showed that the fat content in capelin should not be higher than 4.0% prior to

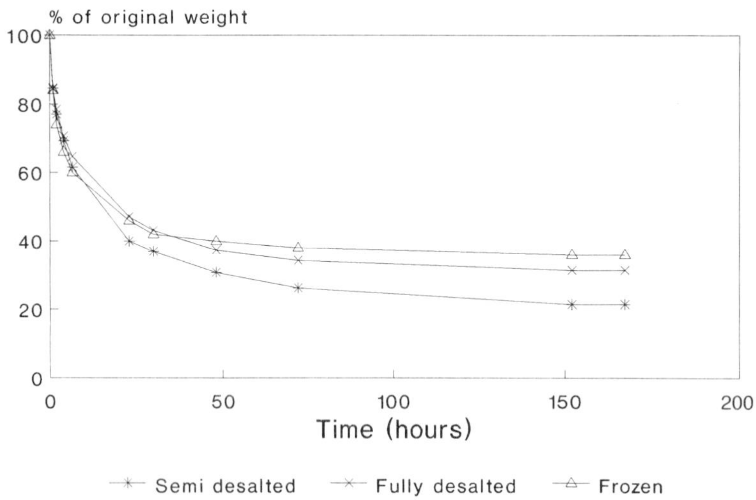

Fig. 30.6 Drying curves for capelin.

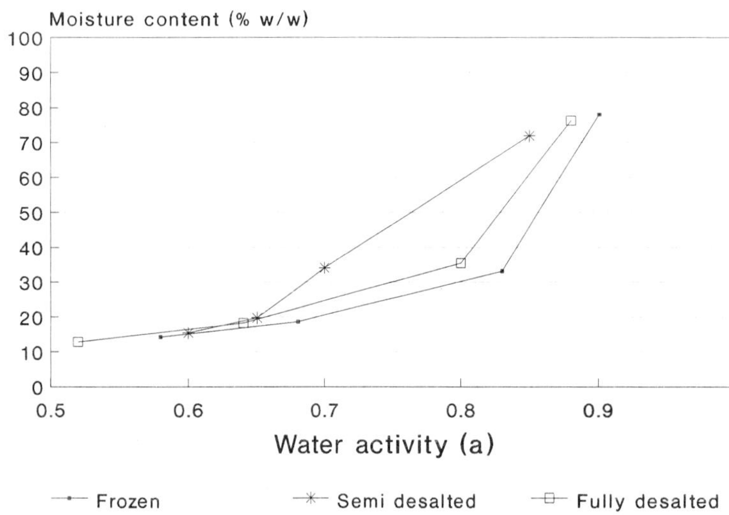

Fig. 30.7 Relationship between moisture and water activity in dried capelin.

drying. Accordingly the economical catching season is about one month in Icelandic waters. Mainly male capelin is used for the drying but female fish can also be used after spawning, at which time they contain only 2% fat. The dried male fish is, however, considered to be of better quality.

Figure 30.6 shows the drying curves. As can be seen the curves and the drying rate for frozen and fully desalted capelin are quite similar, but the curves for the semi-desalted fish are different because part of the water is bound to the remaining salt in the capelin. (Adolfsdóttir & Arason 1988).

Figure 30.7 shows the drying isotherms for capelin. The isotherms for frozen and fully desalted capelin are quite similar. The water activity falls more rapidly when the water content is below 30% in the drying process. Water activity for semi-desalted capelin expectedly dropped more quickly and was about 0.7 with water content 37%.

Results from chemical and microbiological tests are shown in Figures 30.8 and 30.9. In the thawed capelin the aerobic plate counts increased during the whole drying period. This increase was most rapid during the first 24 hours. The drying temperature (25°C) is close to the optimum temperature for growth of the microbial flora of capelin which is psychrotrophic in nature. In salted capelin not many halotolerant bacteria were found. It can be expected that most of

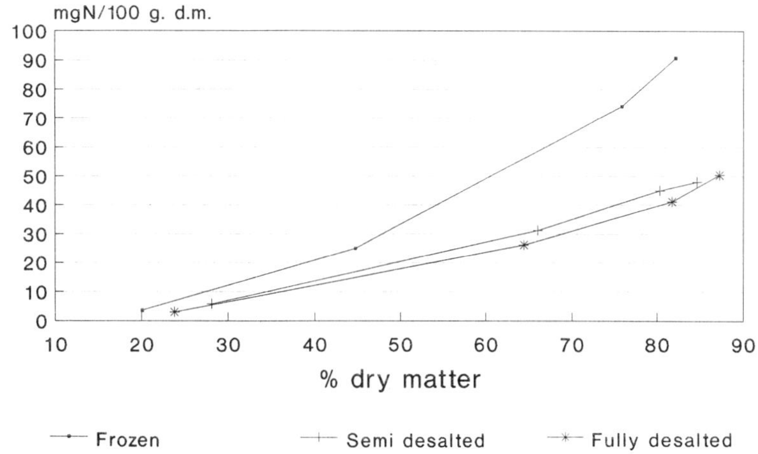

Fig. 30.8 TVB content of capelin.

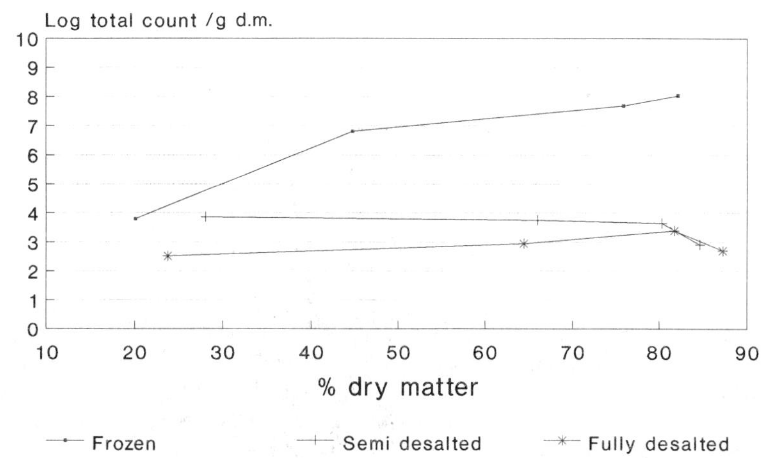

Fig. 30.9 Change in bacterial load during drying of capelin.

the cells have been 'injured' and are thus unable to multiply during the five to six months the capelin was kept in salt. During desalting the bacteria are washed away. Thus at the end of the desalting period the numbers of bacteria had decreased by 85%. During the drying period a further decrease in bacterial numbers was observed. It can be expected that increased 'stress' due to lowered water content decreases the growth rate of bacteria. This must be important from a hygiene point of view. Very similar results were obtained from the TMA and TVB measurements.

The main results from this work were that almost identical drying patterns were obtained for both desalted and thawed capelin. Loss of dry matter (protein and fat) during desalting was 10–15% of total weight. Up to 85% of bacteria were washed away and the same applied to TMA and TVB. Despite the 10–15% weight loss by keeping the capelin salted instead of frozen, the salting process turned out to be considerably more economical for preserving the fish. Furthermore, the reduction of bacteria during this process is important from a hygiene point of view. However, the texture of the dried desalted capelin was 'rougher' than that of the frozen/thawed capelin.

The economy of the drying process is very sensitive to the price of energy. In Iceland geothermal energy is abundant and most of the drying plants are situated in areas where it can be used. Geothermal steam can give trouble by being corrosive, which makes the use of expensive materials necessary. In some cases the exess heat from freezing systems has been used by upgrading it with heat pumps. Fish meal factories usually have excess heat at temperatures that are suitable for the drying process, but such waste heat has not so far been used for this kind of drying.

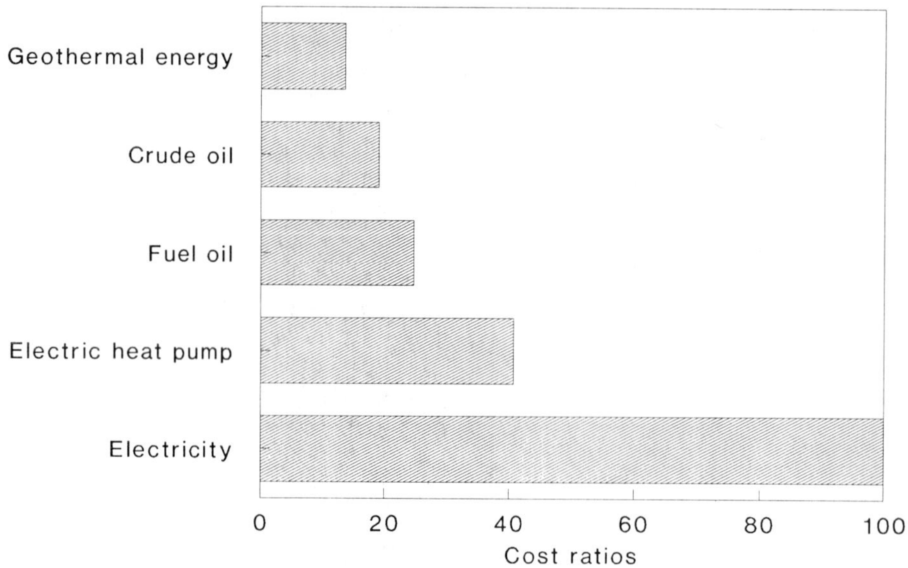

Fig. 30.10 Cost ratios of different energy sources for drying in Iceland.

Various energy sources have been compared and by far the most economical way to heat the air in the dryer is to use geothermal energy (Fig. 30.10) (Arason & Benediktsson 1988).

From the experiments described in this chapter valuable information has been gained on the drying process of blue whiting and capelin. They have been the basis for constructing practical drying equipment which can be used for drying various species of small fish.

BIBLIOGRAPHY

Adolfsdóttir G. & Arason S. (1988) *Drying of capelin.* (In Icelandic.) The Icelandic Fisheries Laboratories, Reykjavík.

Antonacopoulos N. (1971) Comparison of sensory and objective methods for quality evaluation of fresh and frozen saltwater fish. In *Fish Inspection and Quality Control* (Ed. by R. Kreuzer) Fishing News Books, Oxford.

Arason S. & Benediktsson B. (1988) *Utilization of geothermal energy in the fish industry.* A special report. National Energy Authority, Reykjavík.

FAO (1989) *FAO yearbook*, **65**, Fishery Statistics. FAO, Rome.

Fisheries Association of Iceland (1983) *Fishery Statistics in Iceland* (Ed. by J. Blöndal). Fisheries Association of Iceland, Reykjavík.

Jason A.C. (1958) A study of evaporation and diffusion processes in the drying of fish muscle. In *Fundamental Aspects of the Dehydration of Foodstuffs*, proceeding of a conference in Aberdeen. Society of Chemical Industry, London, pp. 103–36.

Kristinsson K. & Arason S. (1983) *Artificial drying of small fish species.* News Summary, **54**, 143–60. International Association of Fish Meal Manufacture, London.

Nordforsk (1981) *Nordfisk, fra industrifisk til konsum.* (Ed. by E. Bornhoft & K. Solvik). Nordforsk, Copenhagen.

Valdimarsson G. & Guðbjörnsdóttir, B. (1984) The microbiology of stockfish during the drying process. *Journal of Applied Bacteriology*, **57**, 413–21.

Waterman J.J. (1976) *The Production of Dried Fish.* FAO, Rome.

CHAPTER 31

Utilisation of roe and milt from herring

R. Richardsen

INTRODUCTION

The Norwegian spring spawning herring stock was heavily overexploited in the 1970s. After recovery in the 1980s several processing units took up production again, but had to face declining prices and low margins for most of the traditional products. This was partly due to increasing supply from the North Sea herring stocks and to shift of demand in major European markets.

The Norwegian Fisheries Research Council financed an industrial programme which originally started within the traditional areas of fish handling and quality. In addition, market studies were carried out on old and potential new marketing areas for Norwegian herring. The latter led to efforts being concentrated on the Japanese market (Osland *et al.* 1985), where there was the possibility of export of whole herring, sex-sorted herring and herring roe. This chapter gives some results of these projects.

SEX-SORTING OF HERRING

Norwegian export of large (>300 g) herring to Japan soon became a success. After a market test in 1982 the quantity exported reached 18 000 tons in three years. A considerable increase in the value of Norwegian exports seemed possible if sex-sorted herring, that is roe-herring, were delivered. Because of the special status of herring roe, the Japanese consumer is willing to pay a significantly higher price for roe-herring than for unsorted packed herring. Manual sorting of whole fish is not possible due to the high labour cost. 'Popping' every fish is labour intensive. A system for automatically sorting female from male herring would make processing and utilisation of roe and/or milt from herring more cost effective (Fig. 31.1).

Sex sorting of herring will be useful in four different processing alternatives:

(1) production of roe-herring for the Japanese market
(2) production of 'kazunoko' for Japan
(3) production of herring eggs for human consumption or as a start-feed in marine fish farming
(4) production of DNA from milt.

Electronic system

The first attempt at sex-sorting Norwegian spring spawning herring was based on earlier developed systems using visible light (Dahl 1957, Haupmann & Green 1977). These tests did not succeed on large herring for two main reasons:

(1) The light source was influenced by other light sources in the processing plant.
(2) Very high intensity light was needed to transmit through the large Norwegian winter herring.

Experiments on the optimum wavelength gave the results shown in Table 31.1.

Table 31.1 Optimum wavelength results for sex-sorting herring.

Wavelength	Intensity		
	Light source uncovered	Through 'red' roe	Through 'light' roe
455 nm	110	—	—
600 nm	28	—	—
740 nm	53	17	40
940 nm	30	10	23
1000 nm	53	17	40

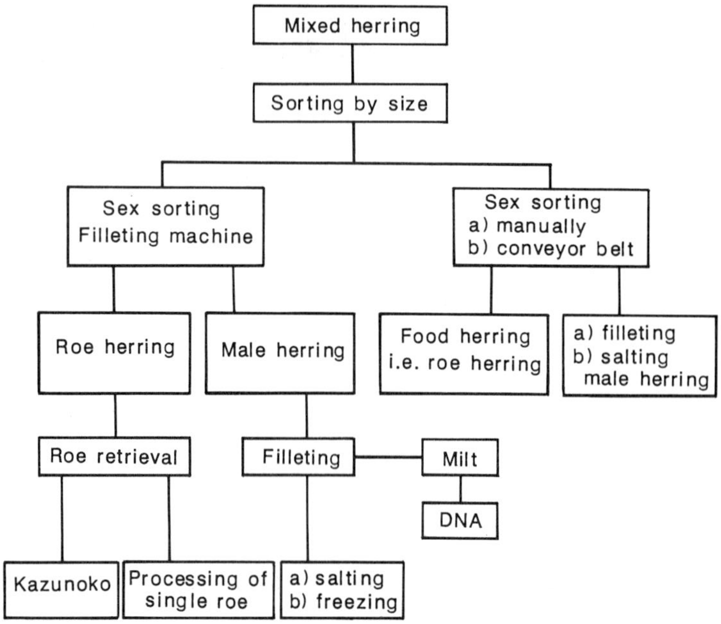

Fig. 31.1 Flow chart of possible production alternatives where sex-sorting machines would be needed.

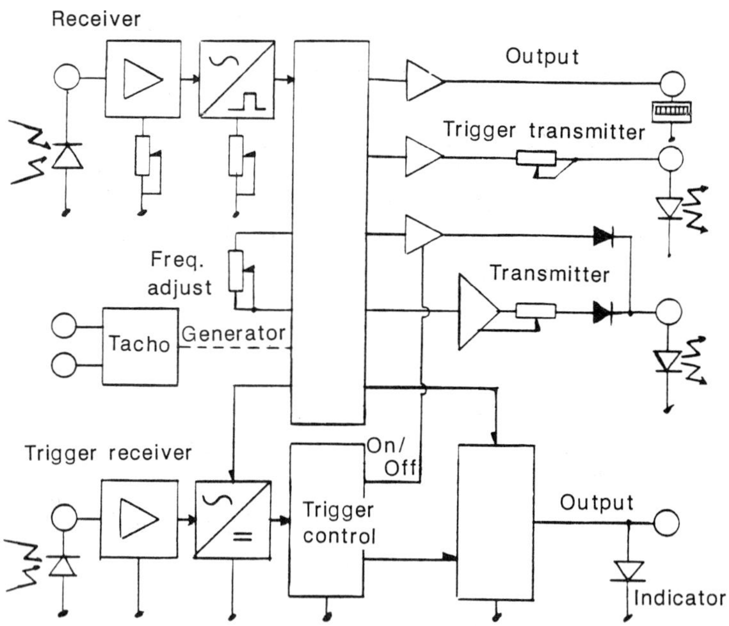

Fig. 31.2 Functional block diagram and typical connections of the electronics.

The amount of transmitted light is best around 940 nm, i.e. infra-red light. We can also see that immature roe ('red' roe) transmits less light than mature 'light' roe.

Due to these findings our system was based on components in the infra-red spectrum (Richardsen & Petterson 1988). The principles of measurement are the same: measuring the light transmitted through the fish. Almost all the light goes straight through a roe berring, but it is completely absorbed by milt in male herring. This gives an easy way of sorting whole fish without costly manual operations.

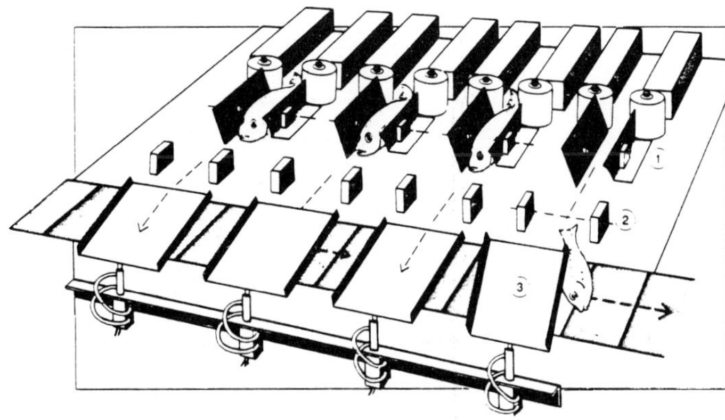

Equipment for differentiating female herring and male herring by means of infrared light. The equipment can be fitted to practically all types of production lines for herring, and in principle, can also be used for other types of fish.

Norsk Forsvarsteknologi A/S
Avd. Narvik
P.O.Box 265 – N-8501 Narvik – Norway

Fig. 31.3 Sex-sorting equipment attached to the VMK-filleting machine.

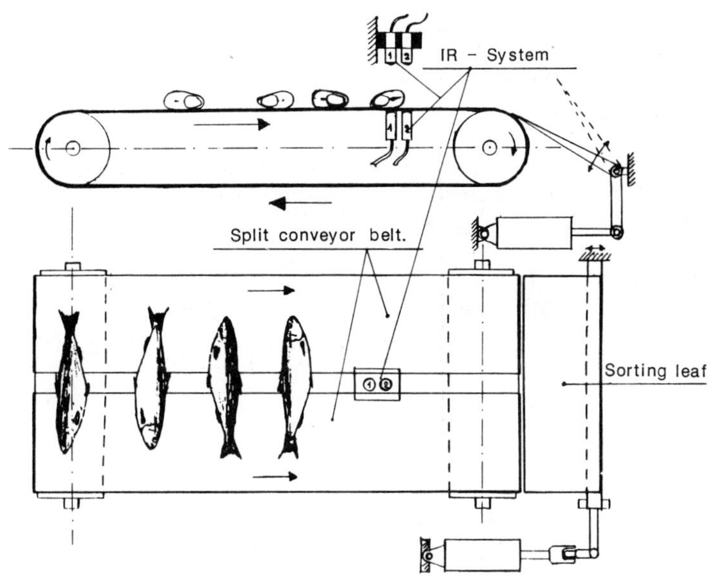

Fig. 31.4 Conveyor belt modified for sex-sorting.

A functional block diagram of the electronics is shown in Figure 31.2.

Figures 31.3 and 31.4 show the sex-sorting equipment mounted in two different types of processing lines:

(1) in a VMK-filleting machine
(2) in a conveyor-belt for production of roe herring.

Sorting results

Depending on how accurately the fish is positioned when passing through the electronic system, the results of sorting can be close to 100%. Tests done with manual feeding gave the results shown in Table 31.2.

Sorting in a filleting machine with automatic feed-

Table 31.2 Results of tests with manual feeding.

	Total number of fish 1800	
	Sorted as female	Sorted as male
	962	838
Control	962	835
% correctly sorted	100	99.6

ing will lower the result to 95–98%, due to problems with the turning operations in the feeding system. Some fish are not turned belly down and this causes incorrect sorting of female fish.

HERRING ROE

Herring roe have been used for human consumption in Japan for a long time. They have been mainly from Pacific herring (*Clupea palasi*) in the form of kazunoko (salted whole roe). We investigated the possibilities of using the large amount of roe from the winter herring fishery; in this fishery the by-products, roe and milt, account for as much as 20% of the total weight landed.

Roe shape, colour and texture are all important for acceptance in the Japanese market. Since texture is one of the most important criteria for good quality, some investigations were made into possible methods for improving roe texture. For producing various products from eggs it is also of interest to identify the nature of the adhesive forces which allow eggs to stick to kelp on the sea bed after spawning.

The histochemical aspects of eggs were studied. Only unpublished information on herring eggs was traced (Senstad 1986), but eggs from other fishes have been studied. Most have a surface layer of mucopolysaccharides and protein which are important for attachment to seaweed and other structures. The inner membrane, which gives the high texture, consists only of proteins. The surface layer is synthesised just prior to spawning and this is the reason why immature eggs do not stick to surfaces.

The described surface layer is very similar to hyaline and very resistant to microbial degradation. Proteolytic degradation is possible only after acid degradation or in the presence of thiol compounds. Hatching enzymes are responsible only for degradation of the inner layer (membrane); they will not degrade the surface layer.

Our experiments (Senstad 1986) have shown that after hydrolysis there is no glucosamine left. Deacetylation of the amino-sugar is not necessary for the identification of glucosamine before hydrolysis; this shows that the amino-sugar is not acetylated.

Hyaluronic acid (hyaluronate), chitin and chitosan were not identified. Histochemical experiments showed that the eggs are composed of protein, glucoprotein, polysaccharides and different mucopolysaccharides (neutral, acidic, carboxylated and sulphonated). It has not been possible to identify a single ionic component in sea-water responsible for attachment of eggs. It seems more likely that the ionic strength is a 'catalyst' of attachment. The surface layer hardens after 20 minutes in sea-water. Eggs cleaned or treated in freshwater will still attach to seaweed and other structures when exposed to sea-water again.

From these results it seems likely that the attachment and hardening of eggs could be due to the formation of strong S–S bonds.

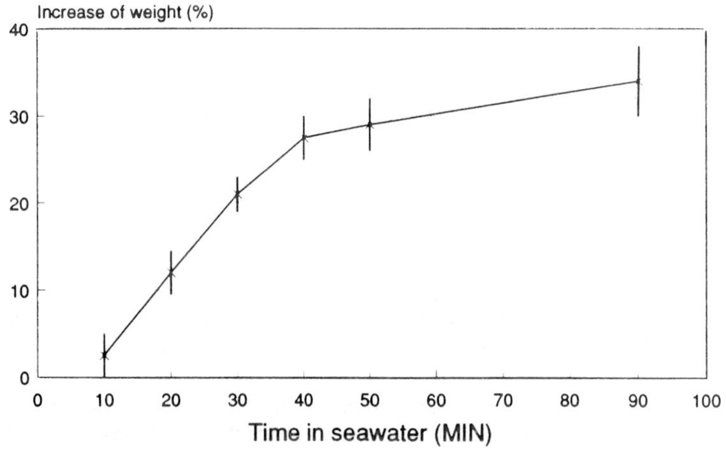

Fig. 31.5 Increase in weight for herring eggs treated in sea-water (Norwegian spring spawning herring). (Source: Inst. of Fish. Techn. Research.)

Table 31.3 Chemical composition of herring roe and milt.

Raw material	% Water	% Ash	% Fat	% FFA	% Protein	Remarks
Whole herring with roe and milt	72.4	2.1	9.5	3.9	16.5	15.3% roe 24.7% milt
Whole herring with roe and milt	72.8	2.0	9.0	3.8	16.7	13.8% roe 24.7% milt
Herring milt	82.5	2.0	2.0	19.8	16.7	Fish with 21% milt
Herring milt	81.8	1.8	2.3	20.4	16.9	Fish with 20.8% milt
Natural herring roe	74.1	1.2	0.7	15.1	19.2	Fish with 12.1% roe
Natural herring roe	75.1	1.1	0.6	16.2	18.3	Fish with 7.8% roe
Separated eggs[1]	81.1	0.2	0.5	12.7	14.5	Fish with 21.1% roe
Separated eggs	78.2	0.7	0.6	12.4	16.1	Fish with 15.7% roe
Separated eggs	75.4	1.2	0.6	15.0	18.3	Fish with 16.4% roe

[1] Eggs separated from the ovarian sac, washed and cleaned in hydrocyclones and flotation tank. See Figure 31.7.

The effect of sea-water on the hardening of the eggs was also found by others (Tsuyuki *et al.* 1980) who added different compounds to salmon eggs but found that only brine had some effect on hardening or texture. We found that herring eggs increase in weight when treated in sea-water (Fig. 31.5).

Glutathione is reported as having a positive effect on the texture of capelin eggs (Bøe 1986). This was tested by us without any significant increase in the texture of herring eggs (Richardsen 1987). Frozen herring eggs have a tendency to become spongy (Bilinski *et al.* 1975, Huynh 1985), and also lose texture (Boyd *et al.* 1972).

Tables 31.3 and 31.4 list the results of some chemical analyses of herring roe and milt and of whole herring.

Table 31.4 Amino acid composition of herring roe (Proportion of protein in the water soluble protein fraction).

	%
Asparagine	1.7
Glutamine	4.0
Serine	4.8
Glycine	7.8
Histidine	2.4
Arginine	5.1
Threonine	8.2
Alanine	11.7
Proline	15.3
NH_3	1.2
Tyrosine	3.1
Valine	9.5
Methionine	1.8
Isoleucine	5.8
Leucine	8.7
Phenylalanine	4.9
Lysine	2.6

Kazunoko

To produce roe of proper quality it is important that fish is captured, and roe processed, at the right stage of maturity. Only mature roe has the right texture for the Japanese market; this means that the gonad index for suitable fish must be around 18:

$$\text{Gonad index (GI)} = \frac{\text{Roe mass}}{\text{Total fish mass}} \times 100$$

Figure 31.6 illustrates the development of the gonad index in Norwegian spring spawning herring.

As seen from Table 31.5, roe from Pacific herring have almost double the texture score of that from Atlantic herring. Norwegian spring spawning herring and herring from Atlantic Canada are at the same level, while roe from North Sea herring have the poorest texture.

This leads to the same conclusion as market opinions indicate: roe from Pacific herring are rated as a premium product over Atlantic herring, due to the higher texture score. In the market Pacific kazunoko are used as gift items during the new year season,

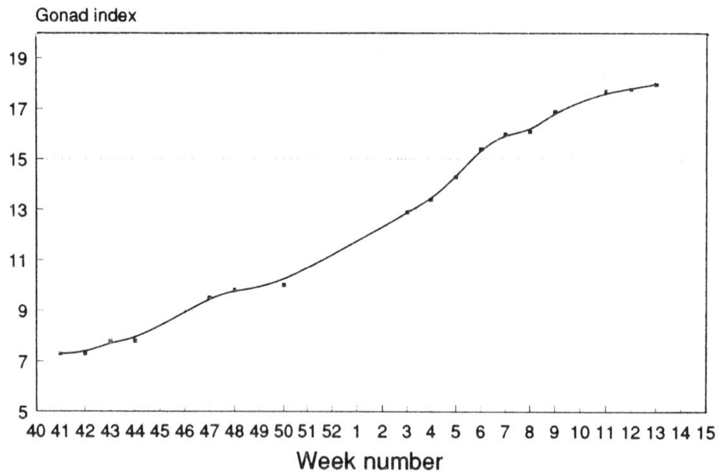

Fig. 31.6 Gonad index development in Norwegian spring spawning herring.

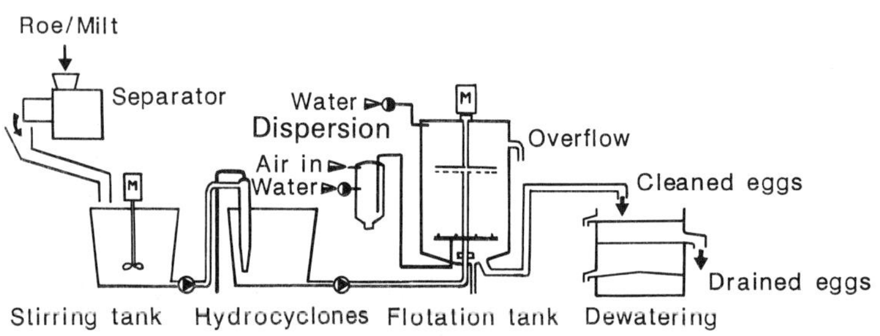

Fig. 31.7 Equipment for processing herring eggs.

while Atlantic roe are used in mixed roe products at lower price levels.

Production of single herring eggs

Since whole sacs of roe (kazunoko) from Atlantic herring have a limited market due to poorer texture, it is important to find other uses for this roe. Eggs, or washed and cleaned herring roe, have potential in various products. For this purpose the Institute of Fisheries Technology Reseach developed equipment for large scale processing of single eggs, useful in utilising roe as a by-product from filleting plants. The equipment used for processing herring eggs is illustrated in Figure 31.7.

In Japan there is a limited market for fish eggs, both as mixed fish egg products and for production of artificial kazunoko. These products must have the highest possible texture, and production must be done from mature roe and processed in sea-water. Since mature roe are only available during a short period of the fishing season, utilisation of immature roe is also of vital interest to the industry.

Table 31.5 Texture of herring roe. Five eggs in each sample were compressed to their breaking points. Force required to break the eggs (g) was measured by a computer based equipment for texture measurements (Sundfær 1985).

Origin	Texture ×(g)	SD (g)	N
Bristol Bay, Alaska (*Clupea pallasi*)	956	111	18
East-Coast, Canada (*Clupea harengus*)	492	123	16
North-Sea, Holland (*Clupea harengus*)	247	92	15
Norwegian spring spawning mature (*Clupea harengus*)	475	105	16

In Scandinavia there is a market for canned fish eggs for human consumption. Normally cod roe or roe from other demersal species are used. Herring roe was tested as an ingredient in this type of product and found to be acceptable.

An interesting use for immature herring roe is as a start-feed for marine fish larvae, which are, with a few exceptions, dependent on a supply of live feed for growth and development. Experiments based on other materials (Garatun-Tjeldstø *et al.* 1984, Opstad *et al.* 1984) concluded that roe-based feeds, fortified with peptone and coated with hard fat or sterol, are capable of initiating development of the gut and filling of the swim bladder, and supporting growth of cod larvae (Garatun-Tjeldstø *et al.* 1987).

Roe-based diets have also been compared with several other experimental diets from various commercial sources. It is concluded that roe-based start diets gave better growth rates and longer survival times than other dry feeds. Roe-based diets were also tested in start feeding of salmon fry, which showed negligible mortality and in some cases a higher growth rate than fry on a commercial salmon diet (Garatun-Tjeldstø *et al.* 1989).

MILT FROM HERRING

Until now nucleic acid has been the only highly purified biochemical in industrial production from fish viscera. The product is used mainly for pharmaceutical purposes and the raw material has been milt from cod and herring and recently also from salmon.

Nucleic acid – structure and isolation

DNA (deoxyribonucleic acid) is a polymer consisting of a sugar (deoxyribose), four different bases and phosphate. In native form it exists as a double helix. Hydrogen bonds between the complementary bases preserve the double-stranded nature of DNA. These hydrogen bonds may be broken either by elevated temperatures (>70°C) or alkaline pH (>11). When the bonding forces between the two strands are disrupted and the strands separated, the DNA is denatured. DNA can be partly renatured either by neutralisation or by lowering the temperature.

Other important bonds for the structural integrity of the molecule are the phosphodiester bond in the sugar-phosphate backbone and the N-glycosidic linkage between the base and the sugar.

Native DNA is a very large molecule with a molecular weight of approximately 8×10^{10} daltons and with a length of about 4 cm. The total length of DNA in a diploid cell is 1–1.5 m. The great length of the DNA molecule makes it very fragile and difficult to isolate in intact form. From higher organisms very careful techniques produce fragmented DNA molecules with molecular weights of 10^7–10^8 daltons.

Commercially two types of DNA are of interest: polymerised DNA (MW $> 6 \times 10^5$ daltons) and standard DNA (oligonucleotides), with the former as the most interesting. In addition to physical breakage of DNA due to shearing (stirring, pouring, pipetting etc.), the molecule may also be fragmented by extreme chemical conditions or enzymes (nucleases) present in the tissue.

The phosphodiester bond in the sugar-phosphate backbone is hydrolysed at pH 1 and approximately half the N-glycosidic bonds are disrupted at pH 4. Generally both these types of bonds in DNA are very resistant to alkaline pH and high temperatures. Breakdown of the DNA by nucleases abundant in the milt may be prevented by having EDTA or citrate present. Usually nuclease activities are dependent on divalent cations, and chelators such as EDTA and citrate will consequently minimise nuclease activity.

Isolation of DNA

(1) Grinding of milt before adding 4 volumes 10 mM potassium citrate and 2M NaCl. Incubation 2 hours, 37°C.
(2) Adjustment of pH; first to 11 with NaOH and then to 4.8 with HCl. Proteins are removed by filtration.
(3) pH adjustment to 7.4 (NaOH) and DNA precipitation by addition of 2.5 volumes cold water. Standing 12 hours.
(4) Water removal. DNA extraction in 2M NaCl (minimum volume).

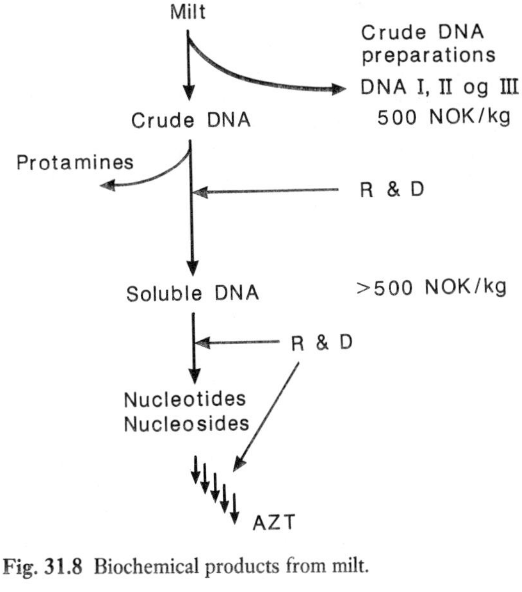

Fig. 31.8 Biochemical products from milt.

(5) DNA precipitation again by 1 part ethanol or 0.2M $CaCl_2$. If potassium chloride is used, the DNA must be washed in water before drying.

This method gives a yield of approximately 1.5–3% dried DNA from frozen milt. Milt from herring gives a better yield than cod milt (Olsen 1987).

No smell or taste of fish should be present in the product and it should be soluble. The average size of the fragmented DNA may be estimated by agarose gel electrophoresis. The purity is determined normally by UV absorption spectroscopy. DNA has a characteristic absorption at 260 nm and for purified, native DNA the ratio between the absorption at 260 nm and 280 nm is usually 1.8–2.0. High molecular weight DNA has a white, threadlike appearance.

DNA from cod and herring milt is now produced in a plant specially set up in Tromsø, Norway, to utilise different waste products from the fishing industry. Health products from milt are also produced from cod milt as shown in Figure 31.8.

BIBLIOGRAPHY

Bilinski E., Borg L., Smith D. & Yamamoto M. (1975) *Quality and quantity of roe obtained from Pacific coast herring under industrial processing conditions.* Technical Report 569, Fisheries and Marine Service, Vancouver.

Boyd J.W., Geiger S.E. & Southcott B.A. (1972) *Herring roe retrieval and processing.* Technical Report 336, Fisheries Research Board of Canada, Vancouver.

Bøe U-B. (1986) Personal communication. NLH, Ås, Norway.

Dahl H. (1957) *Fremgangsmåte til sortering av sild og fisk, og anordning til utførelse av fremgangsmåten.* Norsk patent nr. 90487. Oppfinner: Ing. Helmer Dahl. Rolf Olsen A/S, Bergen.

Garatun-Tjeldstø O., Isaksen E., Raae A., Teigland M., Thomassen J., Walther B.T., Klungsøyr L., Opstad I., Jensen P., Strand B. & Huse I. (1984). Experiments with start-feeding of cod (*Gadus morhua L.*) larvae in the laboratory. In *Acts of the Norwegian-French Workshop on Aquaculture,* 4–8 Dec., pp. 269–86. IFREMER, Brest.

Garatun-Tjeldstø O., Thomassen J., Klungsøyr L., Opstad I., Strand B. & Huse I. (1987) Artificial start-feed for cod larvae (*Gadus morhua L.*) based upon cod roe. *Sarsia,* **72**, 373–4.

Garatun-Tjeldstø O., Opstad I., Hansen T. & Huse I. (1989) Fish roe as a major component in start-feed for marine fish larvae. *Aquaculture,* **79**, 353–62.

Haupmann E.G. & Green J.R. (1977) *Fish Characteristic Detecting and Sorting Apparatus.* United States Patent No. 4, 051, 952. Inventors: E.G. Haupmann & J.R. Green. Assignee: Neptune Dynamics Ltd, North Vancouver.

Huynh M.D. (1985) *Quality changes in herring roe during cold and frozen storage.* Technical Report 16, Fisheries Technology Division, Vancouver.

Olsen R. (1987) *Nukleinsyrer fra melke.* Unpublished data. Institute of Fishery Technology Research, Tromsø.

Osland O., Richardsen R. & Vassdal T. (1985) *Strategi for markedsføring av norsk konsumsild.* Report A39, Institute of Fishery Technology Research, Tromsø.

Opstad I., Strand B., Huse I., Garatun-Tjeldstø O., Klungsøyr L., Thomassen J. & Walther B. (1984) Laboratory studies on the use of dry diets, rotifers (*Brachionus plicatilis* O.F. Müller) for start feeding cod larvae (*Gadus morhua L*). In *Acts of the Norwegian-French Workshop on Aquaculture,* 4–8 Dec., pp. 257–67. IFREMER, Brest.

Richardsen R. (1987) *Tilsetting av glutathione for økning av tekstur i silderogn.* Unpublished data. Institute of Fisheries Technology Research, Tromsø.

Richardsen R. & Petterson K.O. (1988) *Kjønnssortering av sild.* Report U48, Institute of Fishery Technology Research, Tromsø.

Senstad C. (1986) *Klebing av silderogn.* Unpublished data. Institute of Fisheries Technology Research, Tromsø.

Sundfær K. (1985) *Teksturpresse.* Report U30, Institute of Fisheries Technology Research, Tromsø.

Tsuyuki H., Cheng J., Huynh M.D. & Williscroft S.N. (1980) *Pacific salmon roe processing: effects of postmortem age, handling methods and additives on the quality of products prepared from fresh and frozen roe.* Fisheries and Aquatic Science Technical Report 987, Technology Service Branch, Vancouver.

CHAPTER 32

Utilisation of deboned retextured mackerel flesh

A. Mills and H. Teepsoo

INTRODUCTION

In the UK less than 30% of the 250 000 tonnes of home-caught pelagic species are directly consigned for human consumption (DAFS 1988). The remainder is disposed of at relatively low prices for export, for fish meal and oil, or for pet food.

As part of its duty to improve the efficiency of the fishing industry, and also to carry out commissions on behalf of the Ministry of Agriculture, Fisheries and Food, the Sea Fish Industry Authority has undertaken the development of new products based on pelagic species. The aim of the work was to stimulate commercial interest in the potential of pelagic species, so increasing their utilisation and value. The categories of products developed were aimed principally at satisfying the requirements of retail and catering outlets. An essential element of the work was the identification of these requirements and progress of the work was determined by commercial responses and endorsement. Wherever possible existing technology with established processing and production techniques was used, although some innovation was necessary to overcome the technical problems encountered. In addition to the commercially based work, more fundamental studies were made of the properties of the pelagic materials used, to elucidate some of the technical problems encountered.

Table 32.1 Summary of requirements for fish-based products in schools and friers.

Schools
- (1) very low cost
- (2) no bones present
- (3) simple preparation
- (4) quick to cook, by oven baking or deep frying
- (5) uniformity of individual portion sizes
- (6) if frozen, then long storage life
- (7) no competition with existing fish products
- (8) appeal to children and teenagers
- (9) nutritious and low additive content
- (10) capable of withstanding high temperatures during storage between cooking and consumption

Friers
- (1) no bones present
- (2) capable of being frozen and cooked from frozen
- (3) to be cooked by deep frying, baking or microwave
- (4) capable of withstanding high temperatures during storage between cooking and consumption
- (5) be a suitable accompaniment with chips
- (6) if to be deep fried then have a stable coating
- (7) no competition with existing products

INVESTIGATIONS

Identification of product requirements

To identify the criteria perceived as desirable characteristics for fish-based products, over 50 interviews were conducted with school meals' supervisors, fish friers and suppliers of fish products. The product requirements for schools and friers, in approximate order of priority, are summarised in Table 32.1.

One of the major obstacles to the utilisation of pelagic fish for schools and friers was that they required products to be bone-free, a requirement given high priority by both. Bones are readily detectable in pelagic species and for this reason all further work was with deboned material.

Production of deboned material

Deboned material was produced by passing fillets through a Baader bone separator. Fillets were selected as the preferred material since previous work

(Howgate 1976) had clearly shown that raw material containing back bone, kidney tissue etc. would give an inferior mince.

This deboned material was of limited use as there was a significant loss of texture on cooking. In particular the mince was dry and fibrous and acceptability was seriously reduced.

Retexturing of the deboned material

The problem of loss of texture was discussed with a number of food ingredients' suppliers. Trials have been carried out with a number of proposed materials, but only one company could supply a mix which successfully re-created a meaty, succulent texture comparable with that of fresh fillets.

The mix, supplied by Lucas Ingredients Ltd, Bristol, consisted of four commonly used ingredients–salt, polyphosphate, starch and soya flour–none of which has been identified as unsuitable for feeding to children (Hanssen 1987). The mix was added at a level of 7% of dry powder to 13% water and 80% deboned pelagic mince, then stirred for two minutes.

Best results in re-creating a meaty, succulent texture were achieved with fresh mackerel mince. The improvement was not achieved with previously frozen mackerel, nor with other pelagic species such as herring or pilchards. For these reasons all further development work was carried out with deboned, retextured mince from fresh mackerel.

Development of product concepts

The retextured mackerel mince was considered to have potential as a fish base for a wide range of products and dishes. For the purposes of this work potential products were evaluated against the requirements defined above, and after further discussions with end-users four major product types were selected for further work:

(1) Pastry products – pies, pasties, sausage rolls and hot dogs
(2) Extended products – fish cakes and mackerel and potato slice
(3) Burgers – plain and enrobed
(4) Kebabs

Following discussions with manufacturers, batches of products were prepared, wherever possible under normal commercial production conditions.

Evaluation of product concepts

Acceptability of the products was evaluated by setting up demonstrations at exhibitions and to retailers, school meal organisers and fish friers. From the responses the most promising products were selected for further trials. As an example the results for the school trials are given, where the products selected were:

(1) Pastry products – pasties, sausage rolls and hot dogs
(2) Extended products – mackerel and potato slice
(3) Burgers – plain

Schools in three different socio-economic areas were visited and children asked to rate the products on a five-point hedonic scale:

Excellent	5
Good	4
OK	3
Not very good	2
Horrible	1

Products were initially presented as free samples, to introduce the products to the children, and then sold at commerical prices and offered alongside the usual school meal dishes. Results are given in Figures 32.1 and 32.2.

The most frequently selected product was the mackerel and potato slice, which was the product most similar to the familiar fish finger. Pasties were the next most frequently selected, but 20% of the children tasting them did not like them. Only the hot dog did not receive any unfavourable ratings.

Fundamental studies of deboned mackerel mince

To maximise the potential benefits of using retextured deboned pelagic mince, a number of studies on the fundamental properties of the material have been carried out, some of which are still in progress. Part of this work has been carried out as a collaborative exercise with Torry Research Station.

The fundamental studies in progress include:

(1) A study of the frozen storage life of the retextured mince and the use of an antioxidant to extend the life.
(2) Methods for controlling the fat content of the retextured mince
(3) A study of the effects of the ingredients of the

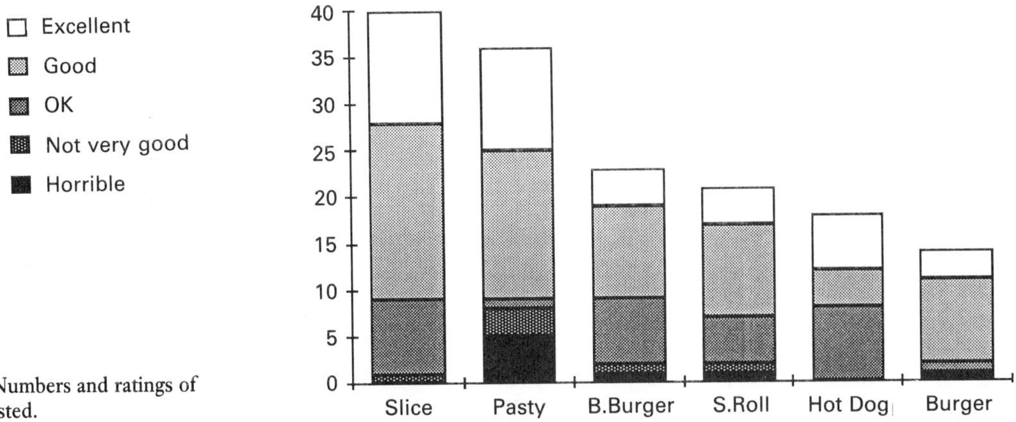

Fig. 32.1 Numbers and ratings of products tested.

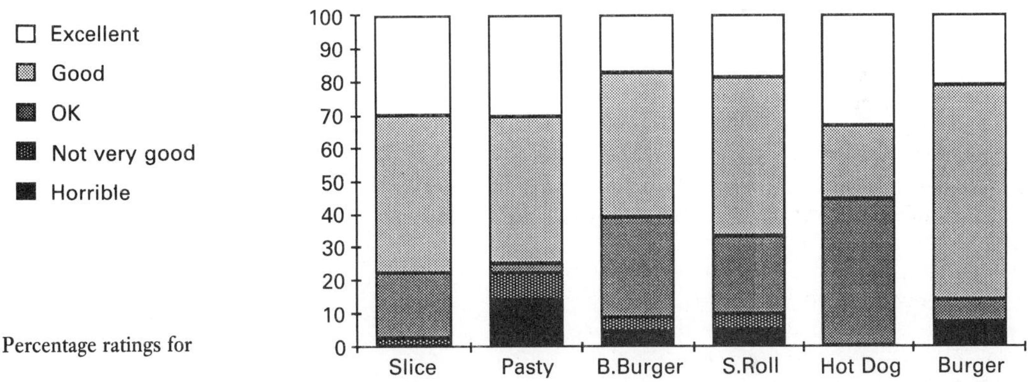

Fig. 32.2 Percentage ratings for products.

Lucas mix on the texture of the mince, both individually and in combination.

Frozen storage trials

Pelagic fish are known to have a significantly shorter frozen storage life than white fish (IIR 1986) and if they are minced that life is reduced. Preliminary trials of blocks of retextured mince produced and stored under commercial conditions have shown that rancidity is detectable after six months' storage. Further trials are in progress with Torry to investigate the use of the antioxidant Ronoxan D20 as a means of extending frozen storage life.

Control of fat content

Control of the fat content, when it was known to vary seasonally from 4% to over 30% (Barnwell 1986), was anticipated to be a problem. For the products taken to the manufacture and evaluation stages a fat content of between 10 and 15% was optimal. The simplest method of controlling fat content was by selecting fish of the appropriate size and fat content. Blending minces from mackerel of different fat contents, or high fat content mackerel mince with white fish mince, were satisfactory alternative methods.

Removal of oil by centrifuging the mince, either with or without preliminary washing, was investigated and although effective was not pursued as a technique for large scale control of fat content. For mince of low fat content the blending of vegetable oils at the time of adding the Lucas mix was satisfactory.

The effects of the Lucas mix

A detailed study of the effects of the ingredients of the Lucas mix on the texture of the mince, as measured by response surface methodology (RSM) and by sensory texture profiling, has identified the separate and combined effects of the ingredients. This study was a collaborative exercise between Seafish, Torry and Lucas Ingredients and for commercial reasons the detailed results are confidential.

CONCLUSIONS

The problems of bones in fresh mackerel can be overcome by passing fillets through a bone separator and re-creating a meaty succulent texture with a mixture of commonly used food additives. This retextured material can be used in a wide range of products, a number of which have been manufactured under typical production conditions. Trials of these products with schools, friers and retailers have shown that the products and the material are well liked. This information is being used to encourage processors and manufacturers to produce pelagic based products, and retail and catering outlets to sell them.

To maximise the potential of the pelagic material, fundamental studies are being made to determine the frozen storage life of the material, ways of controlling the fat content of a highly variable raw material, and the effects of the retexturing mix. Further studies are required to enable prevously frozen material to be retextured and to investigate the use of pelagic species other than mackerel.

BIBLIOGRAPHY

Barnwell E. (1986) *The fat contents of Irish mackerel during 1984/85 and 1985/86.* Fisheries Leaflet 133, Department of Tourism, Fisheries and Forestry, Dublin.

DAFS (1988) *Scottish Sea Fisheries Statistical Tables.* Department of Agriculture and Fisheries for Scotland, Edinburgh.

Hanssen M. (1987) *The New E for Additives.* Thorsons, Wellingborough.

Howgate P. (1976) The sensory properties of minced cod and herring. In *Proceedings of a conference on the production and utilization of mechanically recovered flesh* (Ed. by J.N. Keay). Torry Research Station, Aberdeen.

IIR (1986) *Recommendations for the processing and handling of frozen foods.* International Institute of Refrigeration, Paris.

CHAPTER 33

Response surface methodology in product development

J.E.P. Currall

INTRODUCTION

Many product development cycles involve experiments with the addition of ingredients or additives. These may be used to modify textural and other sensory properties, shelf-life or product yield. The purpose of the experimentation is to find the required quantity of additive to achieve the desired effect. The problem that arises when several additives are used within a product is that the number of combinations of ingredient levels that could be tried, in the search for the required properties, becomes very large.

At Torry Research Station many problems involving multiple ingredient formulations and multiple factor experimentation have arisen. A number of studies have been done involving the modification of the properties of fish minces by ingredient addition. It is not the intention of this chapter to discuss the results, but to outline the methodological framework within which they were carried out. The objective here is to illustrate a particular type of methodology which addresses the problem.

ONE FACTOR AT A TIME

Conceptually the easiest way to examine multiple ingredient formulations is to tackle one factor at a time and to conduct a series of experiments which examine the effect of increasing amounts of a particular additive on the product properties in question.

This approach suffers from a number of important drawbacks:

(1) No attempt is made to examine or account for any interactions between ingredients, such as the formation of an emulsion between water and oil, or chemical binding of water to another of the ingredients.
(2) The experiments involving different ingredients will inevitably involve different batches of the basic product (in this case fish mince), which may well not have identical properties.
(3) In spite of the seemingly simple design of such an experimental layout, it may well involve a large number of treatments overall.

WHAT IS RESPONSE SURFACE METHODOLOGY?

Response surface methodology (RSM) is a system of experimental design and analysis which allows for the simultaneous examination of multiple additives or factors in a single experiment which requires a relatively low number of formulations, even as the number of additives increases.

To achieve these advantages, RSM of the type used at Torry Research Station makes some assumptions. In the context of this chapter, a response is considered to be the result, in terms of a product property, of adding a particular level of ingredient to the product; for example the succulence as a result of adding a certain quantity of tri-polyphosphate to herring mince. A factor is the ingredient itself, for example salt, tri-polyphosphate and oil.

The type of RSM employed in our work assumes that the response as a function of one factor, while holding the others constant, may be characterised by a quadratic polynomial of the following form:

$$\text{response} = \underset{\text{(scale term)}}{\text{constant}_1} + \underset{\text{(linear term)}}{\text{constant}_2 \times \text{factor}} + \underset{\text{(quadratic term)}}{\text{constant}_3 \times \text{factor}^2}$$

Figure 33.1 illustrates the type of relationships between response and factor which can be described by a quadratic polynomial: a straight line relationship is a special case of the equation above, with the quadratic term missing. A very large number of responses to ingredient additions follow this type of relationship and so there is generally little problem with the acceptance of this RSM assumption.

The main problems arise when ingredients have a threshold level up to which there is no response and after which there is a large response. The experimenter should also take care that the range of ingredient addition is realistic in terms of the product. If we make this assumption then we have only to carry out an experiment that enables us to estimate the constants in the equation, rather than examine all the combinations of factor levels. In an experiment involving several factors there would be one scaling term, one linear term for each factor, one quadratic term for each factor and one interaction term for each pair of factors, to represent the interaction between them.

The second assumption concerns the replication of experimental treatments. For non-uniform materials some form of replication is required to gain an estimate of how much the response varies when the factor remains constant. In traditional experimentation this involves making each formulation a number of times, to see how much variability there is. The forms of statistical analysis generally used, however, make the assumption that this variability is the same no matter what level of the factor is used. If this assumption is implicit in our analysis, then we could make the assumption explicit by putting all our replication at one factor combination rather than replicating each and every treatment. This is done in RSM and represents a considerable saving of the number of formulations required for a complete experiment.

If we make the above two assumptions, we can construct a very economical experiment according to a set of statistical principles; tables of designs are readily available (Cochran & Cox 1957).

RSM designs

The factor combinations actually used for the experiments fall into three categories:

(1) A replicated centre point at the midpoint of each factor, which is used to estimate variability and do away with the need for complete experiment replication.
(2) A two to the power n factorial, where n is the number of factors being examined by the experiment.

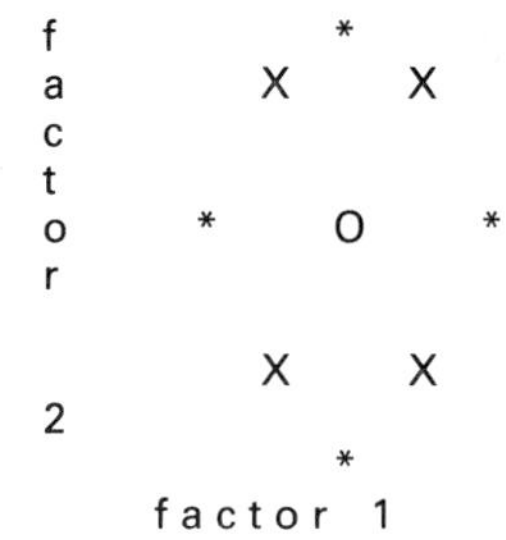

Key
O centre point
X two to the power n factorial
* star points

Fig. 33.1 Layout for two factor experiment.

(3) Points which represent the extremes for each factor at the midpoint of each of the remaining factors – referred to as star points.

The layout for a two factor experiment would be as shown in Figure 33.1.

The number of treatments or formulations required for different numbers of factors are outlined in Table 33.1. The star points are located at the minimum and maximum level of each factor. The centre point is located at the midpoint of the factor space, half-way between the extremes of all the factors. The factorial points are positioned at a distance given by

$$(1 - 1/\text{alpha}) \times (\text{half the range of the factor})$$

above and below the midpoint for each factor. Values for the constant alpha, which are critical to providing essential properties of the design, can be found in Cochran and Cox (1957) and Myers (1976).

For comparison Table 33.1 also shows how many formulations would be required if experiments involving two to six factors were conducted with every combination of three and four levels of each factor. The numbers become very large for any experiment with more than three factors, compared with the RSM designs given in the table. It should also be noted that

Table 33.1 Central composite designs compared with factorial designs.

Number of factors	2	3	4	5	6
2^n factorial	4	8	16	32	64
Star points	4	6	8	10	12
Centre point replications	5	6	7	10	15
Total	13	20	31	52	91
3^n factorial	9	27	81	243	729
4^n factorial	16	64	256	1024	4096

there is no replication at all in these 'every combination' designs, so a doubling or even trebling of these figures may be desired for a realistic experiment.

The type of RSM design illustrated in the table is called a central composite rotatable design and plans for these may be found in Cochran and Cox (1957) and Myers (1976). It is a design of this type that was used by Deng and Tomaszewski (1980), as reported at the Torry Research Station Golden Jubilee Conference in 1979. They examined the effect of a series of ingredients on the quality of fish patties made from minced croaker.

The purpose of RSM

RSM allows us to:

(1) estimate the size and shape of response for each of the ingredients separately
(2) estimate the size of interactions in the response between the ingredients
(3) produce a predictive equation to estimate the response at any ingredient level, whether actually measured in the experiment or not.

The third point is of great importance since it enables us to explore the factor space, as defined in the experiment, for combinations of ingredients that may produce promising product attributes. This activity may be carried out without additional preparation of the product with various levels of ingredients. When promising formulations have been identified, further experiments may be carried out if required to confirm the product attributes.

The importance of the three points above for product development may be summarised as follows:

(1) We can see which ingredients are the most important and what effect each has.
(2) We can see if there are interactions between ingredients and can use this knowledge to reduce quantities of the more expensive ingredients.
(3) We can use the equation to suggest the required ingredient mix to achieve specific properties, without the need for guesswork.

An example

The following example is for illustrative purposes only and is described fully by Hastings and Currall (1989). The experiment involves the addition of four ingredients to surimi: water, oil, albumen and starch. They were added in the ranges given in Table 33.2.

The RSM design involved 27 formulations made up and a range of instrumental and sensory textural tests applied to each. Table 33.3 summarises the results for two of the instrumental tests: the Correx penetrometer and the breaking distance of the gel.

Table 33.2 Ingredient levels used in the experiment.

Ingredient	Percentage (by weight) additions				
Water	0.00	6.25	12.50	18.75	25.00
Oil	0.00	3.75	7.50	11.25	15.00
Albumen	0.00	6.25	12.50	18.75	25.00
Starch	0.00	1.50	3.00	4.50	6.00

Table 33.3 Significance of the factors and the interactions.

	Significance of the coefficient	
Coefficient	Correx	Breaking distance
Scale	***	***
Water	***	***
Oil	***	***
Albumen	***	***
Starch	***	*
$Water^2$		***
Oil^2		**
$Albumen^2$		
$Starch^2$		***
Water* Oil		
Water* Albumen	***	
Water* Starch		
Oil* Albumen	***	***
Oil* Starch	*	
Albumen* Starch		***

*$p < 0.05$; **$p < 0.01$; ***$p < 0.001$

In the case of the Correx penetrometer all the ingredients have an effect, although quadratic effects were not detected. In other words straight lines provide a good approximation for each relationship. There are two main interactions; both involve albumen and indicate that the effect of both water and oil is dependent on the amount of albumen present in the formulation. This is illustrated in the diagrams of Figure 33.2 which show considerable curvature in the contour lines corresponding to the albumen*oil and albumen*water contours. The contour lines in the diagrams are lines of equal response, in the same way that contour lines on a map are lines of equal height. Interactions may be identified in those cases where the pattern of response at the left hand edge of a diagram

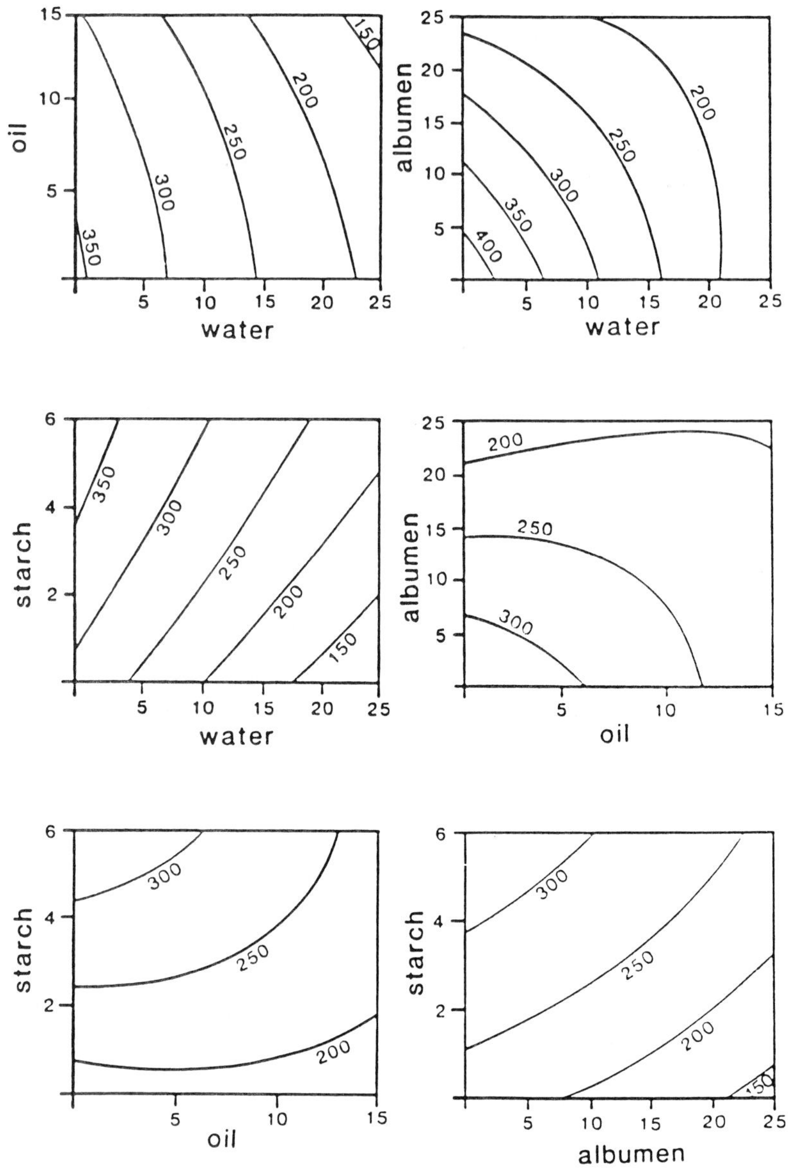

Fig. 33.2 The effect of ingredient levels (parts per 100 parts surimi) on correx reading (contours of gf).

differs from the response at the right hand edge (likewise the top and bottom edges): for example an increasing response along the left hand edge and a decreasing one along the right hand edge. The fact that the contour lines are more or less equally spaced in these diagrams shows that the responses are linear.

In the case of breaking distance albumen has a roughly linear effect while starch, oil and water show curvature. The effects for both water and oil are approximated by a linear effect and a quadratic effect combined. This curvature shows up in the diagrams in Figure 33.3 as unequally spaced contour lines. This is very clear in the diagram representing water* albumen. In addition there two significant interactions, again both involving albumen. They are the interactions of albumen with oil and with starch. In the case of albumen*starch it should be noted that in the absence of starch, increasing the albumen decreases the breaking distance; while at 6% starch, increasing the albumen increases the breaking distance. This

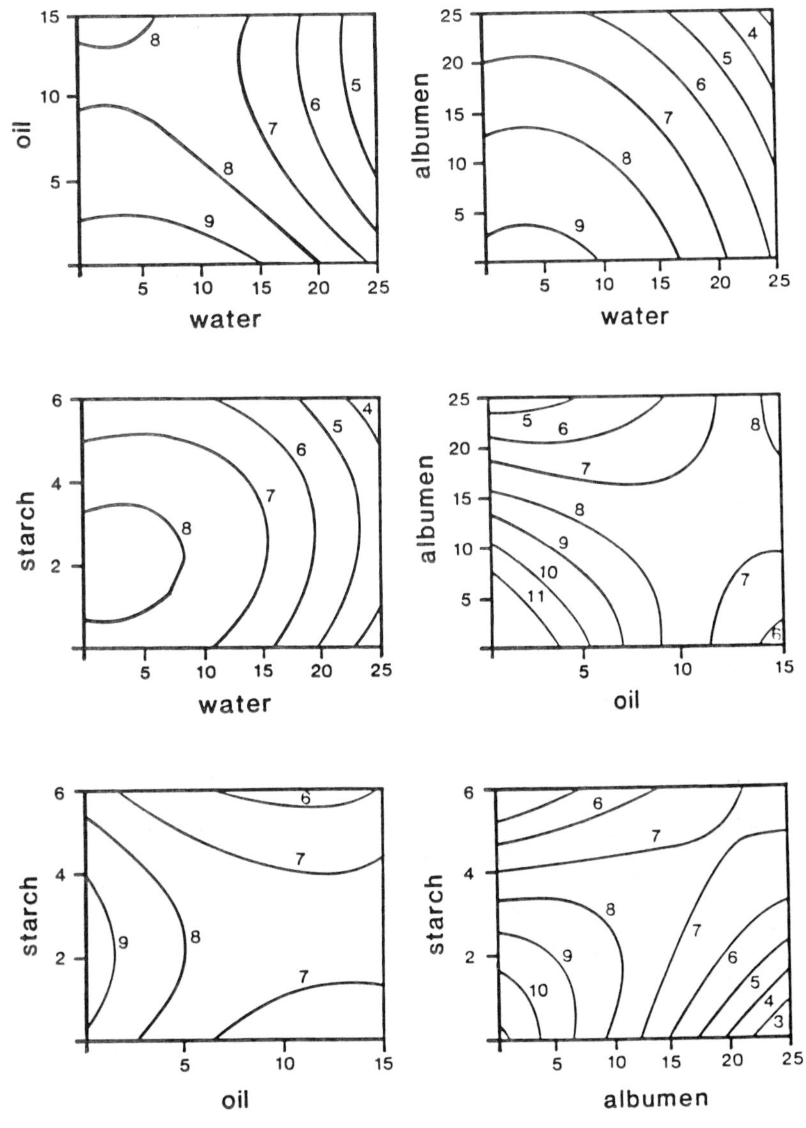

Fig. 33.3 The effect of ingredient levels (parts per 100 parts surimi) on breaking distance by the puncture test (contours of mm).

type of interaction emphasises the need to look for interactions and shows the danger of one-factor-at-a-time experimentation. Investigating one factor at a time we might have determined the effect of adding albumen as decreasing breaking distance; similarly we might have decided that the addition of starch decreased breaking distance. We would then have expected that the lowest breaking distance would be at the highest combination of these two ingredients. This proves false because of the interaction between the two ingredients.

CONCLUSIONS

RSM allows us to carry out product development very economically while at the same time allowing us to predict the result of using different combinations of ingredients. A three phase product development cycle is summarised in the following steps:

(1) Select ingredients believed to be useful or important.
(2) Conduct RSM experiment to identify the useful or

important ones and to suggest ingredient mixes to produce specific product properties.

(3) Try out the suggested mixes and fine tune the ingredient mix to produce the exact properties required.

The third step might include further RSM experiments with restricted ingredient ranges.

ACKNOWLEDGEMENT

The product development work described in the example was carried out by R.J. Hastings.

BIBLIOGRAPHY

Cochran W.G. & Cox G.M. (1957) *Experimental Designs*, 2nd edn. Wiley Int. New York.

Deng, J.C. & Tomaszewski, F.B. (1980) The use of response surface methodology to determine the effects of salt, tripolyphosphate and sodium alginate on the quality of fish patties prepared from minced fish, croaker. In: *Advances in Fish Science and Technology* (Ed. by J.J. Connell) pp. 218–23. Fishing News Books, Oxford.

Hastings R.J. & Currall J.E.P. (1989) Texture of cod surimi gels. *J. Texture Studies*, **19**, 431–51.

Myers, R.H. (1976) *Response Surface Methodology*. Allyn and Bacon, Boston.

CHAPTER 34

Properties of surimi and derived gels from pelagic fish

R.J. Hastings, J.N. Keay and K.W. Young

The preparation of surimi is a method of intensifying the gelling properties of fish protein and preserving them during frozen storage. Fish mince is washed to concentrate the myofibrillar protein, then refined and finely ground with the addition of small percentages of sucrose, sorbitol and phosphates which appear to act as cryoprotectants. A simplified form of the process which can conveniently be used for the laboratory scale preparation of surimi is shown in Figure 34.1. Gels are derived from surimi by adding salt and heat, and white fish such as Alaska pollack give a cohesive elastic texture which is highly prized in Japanese kamaboko products. This textural modification has been used in Japan and more recently in Western countries for the production of up-market analogues of crab, prawn and scallop, and the binding properties of the fish protein may also have potential in sausages, hamburgers, pies etc. A light or white colour is required in kamaboko and the shellfish analogue products, but darker colours may be admissable in other outlets.

With an increasing interest world-wide in the production of surimi as a functional protein, the abundant and cheap fatty fish species have been studied as a potential source in many countries. The main problems have been the removal of fat (with associated flavour) and colour. The method most frequently applied uses sodium bicarbonate in the first wash, and longer washing times (Suzuki 1981, Sonu 1986).

Menhaden (*Brevoortia tyrannus*) was studied in the USA (Bimbo *et al.* 1988). Better results were obtained by refining the mince *before* washing. Batch washing and the Alfa Laval in-line washing system produced surimi of similar quality. A fat content below 2% in the surimi could normally be achieved, independently of the starting amount in the fish (5–12%). High protease activity at 60°C remained in washed menhaden mince (Lanier 1988). Presetting at 40°C or 4°C improved gel strength significantly.

Sardines (*Sardinops melanosticta*) have been studied in Japan (Tokunaga & Nishioka 1988). A new surimi process has been developed which includes sodium pyrophosphate in the first wash and uses a special homogeniser and vacuum washer. The sodium, pyrophosphate is believed to aid dissociation of actomyosin to actin and myosin. The homogeniser breaks down the muscle and the vacuum washer assists the removal of fat, blood and other pigments.

Herring (*Clupea harengus*), mackerel (*Scomber scombrus*) and capelin (*Mallotus villosus*) have been studied in Canada (Spencer *et al.* 1988). They included sodium erythorbate (0.15%) in the wash water to control lipid oxidation during production and reduce the development of unacceptable flavours and odours.

Capelin (*Mallotus villosus*), herring (*Clupea harengus*) and mackerel (*Scomber scombrus*) have been studied in Norway (Langmyhr *et al.* 1988). Various gutting and skinning methods for capelin were investigated. A combination of the FIDECO and surimi processes was found most successful. In the FIDECO process the capelin were cut by a machine with rotary blades, and cleaned and washed to remove skin and black belly membrane. The surimi process used had no special modifications.

Sardines (*Sardina pilchardus*) have been studied in France (Roussel & Cheftel 1988). The method of preparation described by Suzuki (1981) was applied. Fat contents below 1.4% were achieved and presetting at 37°C or 4°C improved gel strength.

At Torry Research Station surimi processes were applied to pelagic species which are plentiful in British waters. The main method was essentially that developed by the Japanese Surimi Association (Sonu 1986) using sodium bicarbonate in the first wash and long washing times (Fig. 34.2). No strainer or refiner was, however, available for use. Surimi was prepared by this method from herring (*Clupea harengus*), horse mackerel (*Trachurus trachurus*) and mackerel (*Scomber*

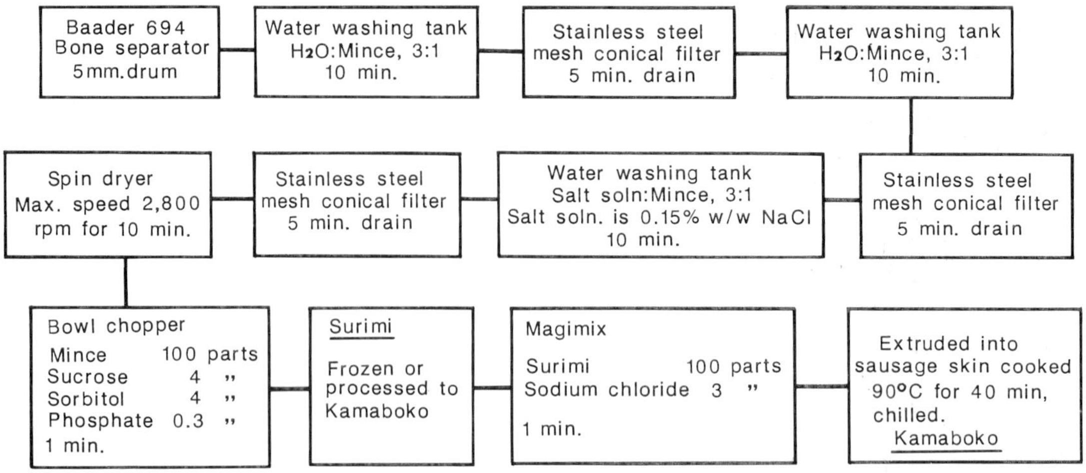

Fig. 34.1 Flow sheet of production process for surimi and kamaboko from white fish.

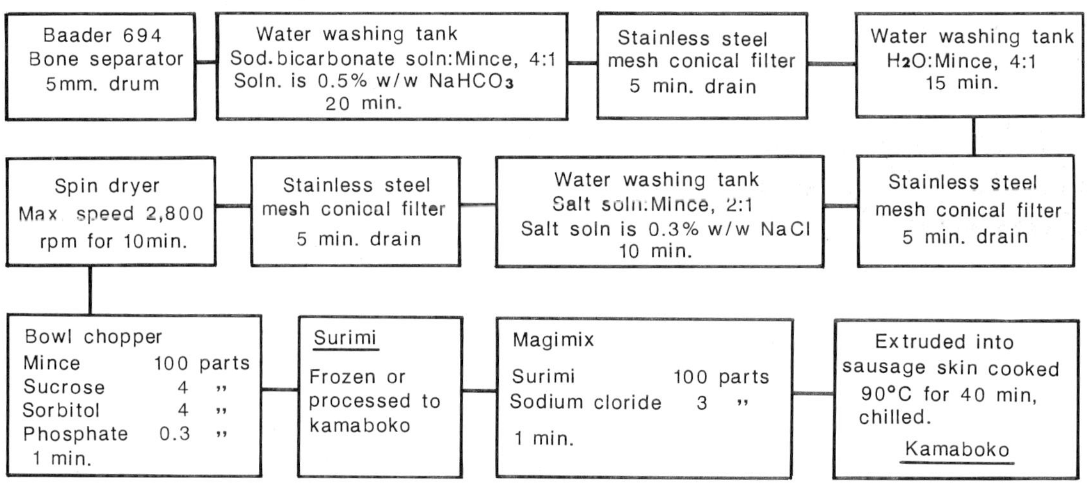

Fig. 34.2 Flow sheet of production process for surimi and kamaboko from fatty fish (JSA method).

Table 34.1 Analyses of pelagic fish mince and surimi (Hastings *et al.* 1990).

	Mince		Surimi			
	Moisture %	Fat %	Moisture %	Fat %	pH	Impurities (number)
Fresh high fat herring	66.0	18.3	69.7	9.0	7.1	62
Fresh low fat herring	76.1	3.8	75.5	2.1	7.2	31
Frozen herring	68.3	15.1	65.3	6.1	7.0	74
Frozen horse mackerel	72.6	10.5	69.8	1.6	7.3	22
Frozen mackerel	66.0	14.7	66.2	5.0	6.9	64
Alaska pollack	—	—	75.97	0.47	6.83	1

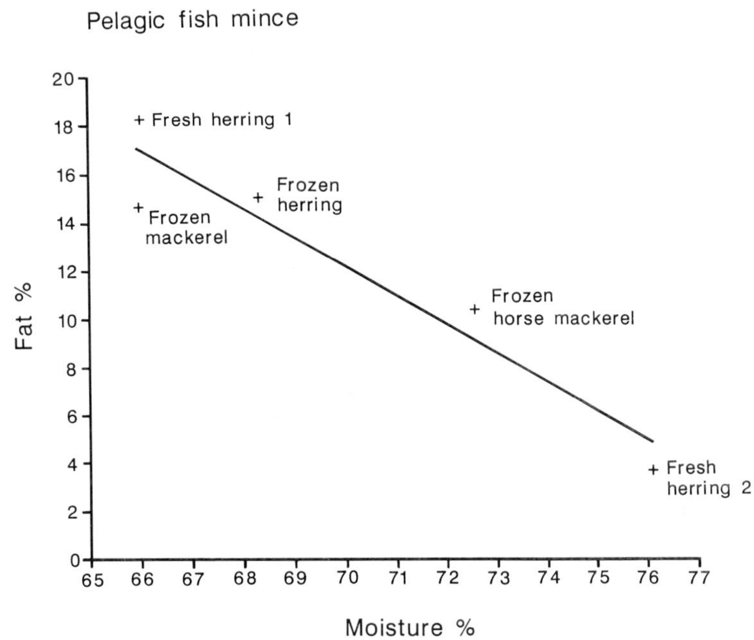

Fig. 34.3 Correlation of fat content with moisture content in pelagic fish mince.

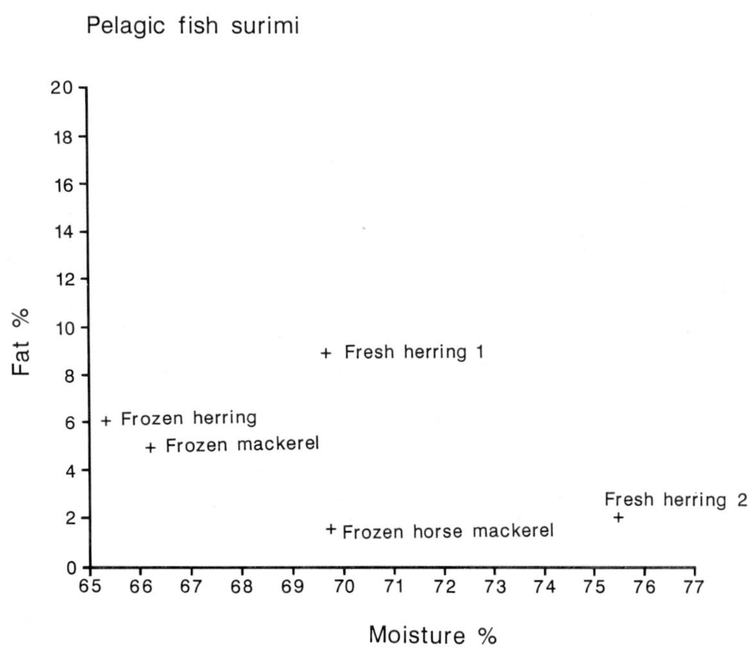

Fig. 34.4 No relationship between moisture content and fat content of pelagic fish surimi.

scombrus) (Table 34.1). Data for Alaska pollack is included as reference standards.

It was found that the fat content of mince from these fish had a negative correlation with moisture content (Fig. 34.3). After processing to surimi there was no relationship between moisture content and fat content (Fig. 34.4 and Table 34.1). This was because the different species lost fat to varying extents, and the small changes in water contents were not consistently in one direction. A particularly high loss of fat was

Table 34.2 Analyses of kamaboko from pelagic fish (Hastings *et al.* 1990).

	Water loss %	Folding test score	Correx score gf
Fresh low fat herring	2.3	6	257
Fresh high fat herring	3.5	3	225
Frozen herring	8.5	2	265
Frozen horse mackerel	1.7	3	409
Frozen mackerel	11.8	2	336
Alaska pollack	2.4	7	651

Table 34.3 Analyses of kamaboko from pelagic fish (Hastings *et al.* 1990).

	Puncture test		
	Breaking force gf	Breaking distance mm	Gel strength gf.cm
Fresh low fat herring	284	8.3	236
Fresh high fat herring	156	3.5	55
Frozen herring	215	3.6	77
Frozen horse mackerel	370	4.6	170
Frozen mackerel	250	4.0	100
Alaska pollack	663	7.0	467

achieved from horse mackerel. The fat floated to the surface during washing. Most of it was removed from the surface and some was lost when draining the wash water from the fish.

The pH of the surimi from all these species was quite high, at 6.9 or above, probably due to the bicarbonate wash (Table 34.1). The impurities were counted in 10 g samples of surimi spread in a thin layer, and were mainly small pieces of skin which passed through the 5 mm holes in the drum of the Baader 694 separator.

Surimi (100 parts) was ground with sodium chloride (3 parts) and cooked at 90°C to form kamaboko. The texture was measured both by instruments and by a trained texture profile panel.

Water binding capacity, expressed as per cent water loss, was the proportion of the initial water content of the kamaboko lost during centrifugation at 1000 rpm for 5 minutes. Low losses were recorded for fresh herring and horse mackerel, indicating good water binding capacity (Table 34.2).

Slices of kamaboko 3 mm thick were folded twice, assessed for cracking, and scored on a Japanese scale of 1 to 7. Higher scores were related to higher elasticity in the samples. The highest score was found for fresh herring of low fat content (Table 34.2).

The Correx instrument was a hand-held strain gauge. A spherical probe, 5 mm in diameter, was pushed into each sample until the surface broke. These results were related to both firmness and elasticity, e.g. a high result was obtained from horse mackerel because of the firmness of this sample (Table 34.2).

A Stevens CR analyser was used for the puncture test and the compression test. In the puncture test a 5 mm diameter probe was used to puncture the flat surface of each sample and force v distance curves were measured on a chart recorder. The breaking point was recorded at the end of each smooth compression curve. Gel strength was the product of breaking force and breaking distance. The highest overall gel strength was given by fresh low fat herring (Table 34.3). This sample also had the highest breaking distance, related to the elasticity of the sample. Frozen horse mackerel had the highest breaking force, which was related to the firmness of this sample.

In the compression test, samples were compressed by 40% of their height with a 50 mm diameter probe, and their firmness at this point was measured. The probe cycled between initial contact point and maximum compression and springiness was the recovery in height of the sample between cycles, expressed as a ratio of its original height. Frozen horse mackerel produced very firm kamaboko (Table 34.4). Samples from fresh low fat herring were the most springy and were less firm.

Sensory assessments of firmness, elasticity and wateriness were made on scales of 0 to 5 (see the Appendix to this chapter). The results are averages from a six member trained panel (Table 34.5). Differences between samples were not so marked as those found with instrumental tests, but a similar pattern for elasticity emerged. Results for firmness were not so consistent. Fresh low fat herring tasted as firm as the frozen horse mackerel gel, though by the compression test it was much less firm.

L a b values were recorded with a Hunterlab Tristimulus Colorimeter. The most pronounced changes occurred in *L* values, which are shown in Table 34.6. The colour was lightened (rise in *L* values) by washing the mince and also during the heating step for the preparation of kamaboko. Lighter colours were achieved with herring than with mackerel or horse mackerel.

Table 34.4 Analyses of kamaboko from pelagic fish (Hastings *et al.* 1990).

	Compression test	
	Firmness gf	Springiness recovery in height
Fresh low fat herring	3590	0.887
Fresh high fat herring	3522	0.834
Frozen herring	5130	0.812
Frozen horse mackerel	7390	0.859
Frozen mackerel	5570	0.815
Alaska pollack	7837	0.875

USE OF BICARBONATE (TABLE 34.7)

Bicarbonate has been recommended for use in the first wash of the surimi process for fatty fish by many workers (Suzuki 1981, Sonu 1986). It is thought to help to remove fat and so improve elasticity.

Our results showed a small improvement in fat removal but a higher moisture content and pH in samples processed with bicarbonate. The rise in moisture content and/or pH had an over-riding effect and produced softer, less elastic gels, both in instrumental tests and sensory tests.

The differences in colour as measured by *L a b* values and as perceived visually were very small.

Table 34.5 Sensory analyses of kamaboko from pelagic fish (Hastings *et al.* 1990).

	Initial impression of texture in the mouth		
	Firmness	Elasticity	Wateriness
Fresh low fat herring	3.9 *a*	3.9 *a*	0.5 *a*
Fresh high fat herring	2.2 *b*	3.0 *b*	0.9 *a*
Frozen herring	2.3 *b*	1.8 *c*	1.0 *a*
Frozen horse mackerel	3.6 *a*	3.3 *ab*	0.5 *a*
Frozen mackerel	2.3 *b*	1.9 *c*	1.0 *a*
Alaska pollack	4.8	4.6	0.7

Different letters in each column indicate samples which were significantly different at the 95% level

Table 34.6 *L* values from tristimulus colour measurement (Hastings *et al.* 1990)

	Mince	Washed mince	Surimi	Kamaboko
Fresh low fat herring	46.7	54.4	55.6	66.2
Fresh high fat herring	43.4	55.4	52.3	64.0
Frozen herring	45.9	52.4	53.5	62.2
Frozen horse mackerel	36.1	48.1	51.4	56.5
Frozen mackerel	48.3	49.4	48.9	54.5
Alaska pollack	—	—	59.49	72.57

Table 34.7 Properties of surimi and kamaboko from fresh herring fillet (18.3% fat) (Hastings *et al.* 1990)

	With bicarbonate	Without bicarbonate	Alaska pollack (without bicarbonate)
Surimi			
Fat %	9.0	10.4	0.5
Moisture %	69.7	66.6	76
pH	7.1	6.6	6.8
Impurities	62	71	1
Kamaboko			
Moisture %	67.6	64.2	73.7
Water loss %	3.5	4.2	2.4
Folding test score	3	3	7
Correx score (gf)	225	398	651
Puncture test			
Breaking force (gf)	156	310	663
Breaking distance (mm)	3.5	4.7	7
Gel strength (gf.cm)	55	146	467
Compression test			
Firmness (gf)	3522	6795	7837
Springiness (dimensionless)	0.834	0.834	0.875
Texture in the mouth			
Firmness	2.2	3.6	4.8
Elasticity	3.0	3.8	4.6
Wateriness	0.9	0.7	0.7

Table 34.8 Properties of surimi and kamaboko from fresh herring fillet (18.3% fat) (Hastings *et al.* 1990).

	Herring fillet	Light flesh only	Alaska pollack (without bicarbonate)
Surimi			
Fat %	10.4	5.0	0.5
Moisture %	66.6	70.4	76
pH	6.6	6.6	6.8
Impurities	71	10	1
Kamaboko			
Moisture%	64.2	68.0	73.7
Water loss %	4.2	3.1	2.4
Folding test score	3	3	7
Correx score (gf)	398	584	651
Puncture test			
Breaking force (gf)	310	367	663
Breaking distance (mm)	4.7	3.8	7
Gel strength (gf.cm)	146	141	467
Compression test			
Firmness (gf)	6795	8095	7837
Springiness (dimensionless)	0.834	0.831	0.875
Texture in the mouth			
Firmness	3.6	4.1	4.8
Elasticity	3.8	3.9	4.6
Wateriness	0.7	0.4	0.7

USE OF LIGHT FLESH ONLY (TABLE 34.8)

The skin and dark flesh immediately beneath it were removed from fresh herring fillets before processing. Surimi was prepared from the light flesh and from the original fillets, without the use of bicarbonate in the wash water. The properties of the preparations were compared.

Surimi prepared from light flesh had a significantly lower fat content (5.0% as compared with 10.4% in fillet surimi); the moisture content was, however, slightly higher. Instrument measurements of firmness (Correx, compression test) were increased, but measurements of elasticity (breaking distance, springiness) were not improved. In sensory tests surimi from light flesh was slightly firmer, but not significantly so.

The colour of mince, washed mince, surimi and kamaboko from light flesh was lighter at all stages, as shown by *L* values (Table 34.9).

Table 34.9 *L* values from tristimulus colour measurement (Hastings *et al.* 1990)

	Herring fillet	Light flesh only	Alaska pollack
Mince	43.4	47.8	—
Washed mince	55.2	58.1	—
Surimi	48.7	56.9	59.5
Kamaboko	64.3	71.8	77.6

USE OF PRE-SETTING (TABLE 34.10)

Surimi which had been prepared from fresh low fat herring (3.8% fat), using bicarbonate in the wash water, and then stored for 3 months at −30°C, was converted to kamaboko by various heating regimes:

(1) 90°C for 40 minutes
(2) 40°C for 40 minutes followed by 90°C for 40 minutes
(3) 2°C for 18 hours followed by 90°C for 40 minutes.

Pre-setting at 40°C or at 2°C produced slightly firmer, more elastic gels. Differences could be detected by instrumental tests, e.g. Correx reading, puncture test, but were not detected by the trained texture profile panel.

CONCLUSIONS

The experimental data presented here shows that in seven cases the functional indices measured for fresh low fat herring most closely match those for Alaska pollack used as standard. Three of the Alaska pollack indices are most closely matched by frozen horse mackerel, and another three by both fresh, low fat herring and frozen mackerel. Nevertheless all the pelagic species give results inferior to Alaska pollack, although in absolute terms potentially useful degrees of functionality have in some cases been obtained, with the prospect of improvement by further experimentation and process development. Flavour was not studied in the present work but reports from other laboratories indicate that this, and its connection with oxidative rancidity, may present the greatest impediment to the use of fatty pelagic species for the production of surimi, and the most severe challenge to workers in the field.

Table 34.10 Properties of herring kamaboko (low fat) heated at (1) 90°C; (2) 40°C and 90°C; (3) 2°C and 90°C.

	1	2	3
Folding test score	7	7	7
Correx score (gf)	235	251	271
Puncture test			
Breaking force (gf)	245	289	285
Breaking distance (mm)	7.5	7.7	7.9
Gel strength (gf.cm)	184	222	227
Compression test (70% compression)			
Firmness (gf)	13 040	15 460	15 150
Springiness (dimensionless)	0.706	0.725	0.703

APPENDIX

Scales used for sensory assessment of kamaboko:

ELASTICITY

The ability of the material to return to its original shape after deformation.

0 completely plastic, retains its deformed state
5 springy, returns to its original shape

FIRMNESS

The resistance to compression when pressed by a fork or between the tongue and palate or between the teeth.

0 very soft, very easily compressed
5 very firm, high resistance to compression

WATERINESS

The release of water on compression. This is the initial response and is to be distinguished from succulence.

0 no water released but not necessarily dry
5 wet, samples release water very readily

BIBLIOGRAPHY

Bimbo A.P., Anderson C.F., Hurley T., Moore R., Means R. & Korhonen R. (1988) US menhaden surimi production: a progress report. In *Fatty Fish Utilization: Upgrading from Feed to Food.* Proceedings of a National Technical Conference in Raleigh, (Ed. by N Davis), pp. 41–76. UNC Sea Grant College Publication 80-04, North Carolina State University.

Hastings R.J., Keay J.N. and Young K.W. (1990) The Properties of Surimi and Kamaboko Gels from Nine British Species of Fish. *International Journal of Food Science and Technology*, **25**, 281–94.

Langmyhr E., Opstvedt J., Ofstad R., Sørensen N.K. & Eide O. (1988) Potential conversion of North Atlantic fatty species into surimi and surimi-derived products. In *Fatty Fish Utilization: Upgrading from Feed to Food.* Proceedings of a National Technical Conference in Raleigh, (Ed. by N Davis), pp. 79–117. UNC Sea Grant College Publication 88-04, North Carolina State University.

Lanier T.C. (1988) Muscle protein functional properties and protease content of surimi prepared from fatty, dark-fleshed fish species. In *Fatty Fish Utilization: Upgrading from Feed to Food.* Proceedings of a National Technical Conference in Raleigh, (Ed. by N Davis), pp. 247–62. UNC Sea Grant College Publication 80-04, North Carolina State University.

Roussel H. & Cheftel J.C. (1988) Characteristics of surimi and kamaboko from sardines. *Int. J. F. Sci. and Technol.*, **23**, 607–23.

Sonu S.C. (1986) *Surimi.* Technical memorandum National Oceanic and Atmospheric Administration, NOAA-TM-NMFS-SWR-013, National Marine Fisheries Service, Southwest Region, US Department of Commerce.

Spencer K.E., Bligh E.G., Chandra C.V. & Lemon D.W. (1988) Canadian testing of fatty species for surimi. In *Fatty Fish Utilization: Upgrading from Feed to Food.* Proceedings of a National Technical Conference in Raleigh, (Ed. by N Davis), pp. 121–39. UNC Sea Grant College Publication 80-04, North Carolina State University.

Suzuki T. (1981) *Fish and Krill Protein: Processing Technology.* Applied Science Publishers, Barking.

Tokunaga T. & Nishioka F. (1988) The improvement of technology for surimi production from fatty Japanese sardines. In *Fatty Fish Utilization: Upgrading from Feed to Food.* Proceedings of a National Technical Conference in Raleigh, (Ed. by N Davis), pp. 143–58. UNC Sea Grant College Publication 80-04, North Carolina State University.

CHAPTER 35

The Java Sea purse seine fishery: resources, production methods and product markets – a case study

J.K. McElroy

INTRODUCTION

Indonesia, the world's largest archipelagic nation, has a population of 180 million people (1989). Most of these people (107 million) live on the island of Java which has a population density of 810 people/km^2, one of the highest in the world. Indonesia's per capita GNP (1988) has been estimated at US$510, although incomes and living standards of the vast majority of the population are low, at hardly more than subsistence level.

Marine fish is seen as a major source of relatively cheap protein, which is considered by the Government of Indonesia to offer substantial further growth potential. Indeed, the 1987 catch accounted for just 30% of the estimated sustainable yield of the country's marine capture resources; over the last 20 years (1968–87), the marine catch has grown at a compound rate of 5.5% per annum, more than double the 2.2% rate of growth of the population.

For much of this period, a substantial part of the growth in the catch has come from the Malacca Strait and Java Sea. Between 1968 and 1980, the main engine of growth in the catch from these waters was the otter trawl, which was first introduced in Riau Province (Sumatera) in 1966 and had spread to the Jave Sea by the early 1970s (Bailey *et al.* 1987). The purse seine was introduced to Indonesia and the Java Sea at Pekalongan in 1968. However, the otter trawl was effectively responsible for the widespread adoption of inboard motorised vessels by the medium-scale sector throughout this area. By the end of the 1970s, the number of trawlers exceeded the licences available by an estimated 40% in the Malacca Straight/Java Sea area, fishing effort was estimated to be some 20% above the level of maximum sustainable yield (MSY), and conflicts with small-scale inshore traditional fishermen, particularly those operating fixed gears, had become a major political issue (Bailey *et al.* 1987). Between 1981 and 1983, a ban on the use of the otter trawl was introduced throughout west Indonesia.

Many of these medium-scale vessels (wooden, 10–30 gross tonnes, with inboard engines generating 100–150 hp) were converted to pelagic fishing, particularly purse seine and gill net. The shift by the owners of the medium-scale vessels to pelagic fishing, and particularly purse seine fishing, also in effect introduced the use of larger, inboard vessels to this fishery, and gave rise to the second spurt in the growth of the catch from the Java Sea during this 20 year period.

While the development of the trawl fishery has received considerable attention (for recent reviews see Bailey *et al.* 1987 and Martosubroto & Chong 1987), little has yet been written on the recent development of the purse seine fishery of the Java Sea its success to date and opportunities for its development in the future (Atmaja & Sadhotomo 1985, Atmaja *et al.* 1986, Sadhotomo *et al.* 1986, Wahyuono *et al.* 1986, Widianto & Linting 1986, Suwarso *et al.* 1987).

This chapter reviews available information on the state of the Java Sea fish resources, the recent development of the Java Sea purse seine fishery in terms of its production and productivity, the processing and marketing of its catch and its effect on fish supplies, prices, and individual fishermen's incomes. Information on the development of purse seine fishing technology in the Java Sea is also presented.

DATA SOURCES USED

Both primary and secondary data sources were used.

Primary data sources comprised:

(1) The National Fishing Port Monitoring Programme which reports basic data on fishing port activity for 24 ports on a monthly basis, e.g. fishing

vessels by gear type, landings by gear type (volume and average value by major species), data on the use of port facilities (ice, water, fuel, salt sales), disposition of the catch, etc.

(2) Field survey data on the catching, processing and marketing of fish obtained as part of the preparation of feasibility studies for expansion and rehabilitation of fishing port and fish landing centre facilities. This information was collected between September 1986 and October 1988 as part of the Fisheries Infrastructure Sector Project (FISP), within the Directorate of Infrastructure, Directorate General of Fisheries (DGF).

(3) Field survey data collected in August 1989 by DGF students as part of a training course in Project Benefit Monitoring and Evaluation, also within the FISP project.

Secondary data sources comprised provincial and national statistics plus relevant internal reports and published papers (see bibliography).

FISH RESOURCES OF THE JAVA SEA

The level and quality of published information to date on the pelagic resources of the Java Sea is generally comparatively poor, with wide confidence limits required around any figures used; the figures are based on primary productivity calculations, supplemented by limited acoustic survey data. While this is true for the 'global' Java Sea estimates, more recent estimates have been made based on catch and effort data for some specific local areas or fisheries, and these should be considerably more robust; for example, Bailey *et al.* (1987) give a value of 290 000–391 000 tonnes for the pelagic resources of the Java Sea, the lower figure being derived from catch effort data for the period 1975–9 (i.e. before the large fleet expansion of the 1980s) from the north Java purse seine fleet.

At present no estimates for particular pelagic species or stocks in the Java Sea have been published. A revision of Indonesia's marine fisheries resources potential, however, is reportedly underway (Martosubroto, personal communication, 1989). By contrast, the estimates for the demersal and penaeid shrimp resources are reasonably sound, being derived from a minimum of 5 years catch effort data immediately prior to the 1981 trawl ban, using Schaefer and Gulland-Fox models (Bailey *et al.* 1987).

According to the most recent DGF marine fisheries resource estimates (DAF/MFRI 1983, and DFR-DGF 1987 unpublished), both pelagic and demersal fish species in the Java Sea (area 381 000 km^2) were relatively abundant and still underexploited, particularly in the northern half of the Java Sea.

The DGF estimates the maximum sustainable yield (MSY) in the Java Sea at about 1 100 000 tonnes annually (i.e. half the standing stock), of which about 52% (568 100 tonnes) is demersal, 47% (519 500 tonnes) pelagic and 1% (15 200 tonnes) penaeid shrimp (Table 35.1). None of the larger pelagics (e.g. tuna and billfish) are found in these waters, although they are taken seasonally at the margins. By contrast, the small coastal tuna-like fish or 'tongkol' (mostly consisting of *Euthynnus affinis*) are quite abundant, these species being grouped, for statistical purposes, with the small pelagic fish stocks.

Table 35.1 Estimated maximum sustainable yields with 1984 and 1987 landings from the Java Sea.

Species	Resource Potential[1] (tonnes/year)	1984[2] Landings (tonnes)	%	1987[3] Landings (tonnes)	%
Pelagic*	519 500**	363 909	70	401 353	77
Skipjack, tuna	—	2 886		3 675	
Demersal	568 100	86 448	15	96 862	17
Shrimp	15 200	11 053	73	15 629	103
Total	1 102 800 (100%)	464 292	42	517 519	47

[1] DGF/MFRI (1983).
[2] DGF (1986) Fisheries statistics of Indonesia, No. 14, 1984, Directorate General of Fisheries, Jakarta.
[3] DFG (1989) Fisheries statistics of Indonesia, No. 17, 1987, Directorate General of Fisheries, Jakarta.

* In the Java Sea context, pelagic is equivalent to small pelagic.
** This DGF value differs substantially from those given in Bailey *et al.* (1987). They give a range of values for the MSY for small pelagics in the Java Sea of 290 000–391 000 tonnes/y, respectively 56–75 per cent of the DGF value.

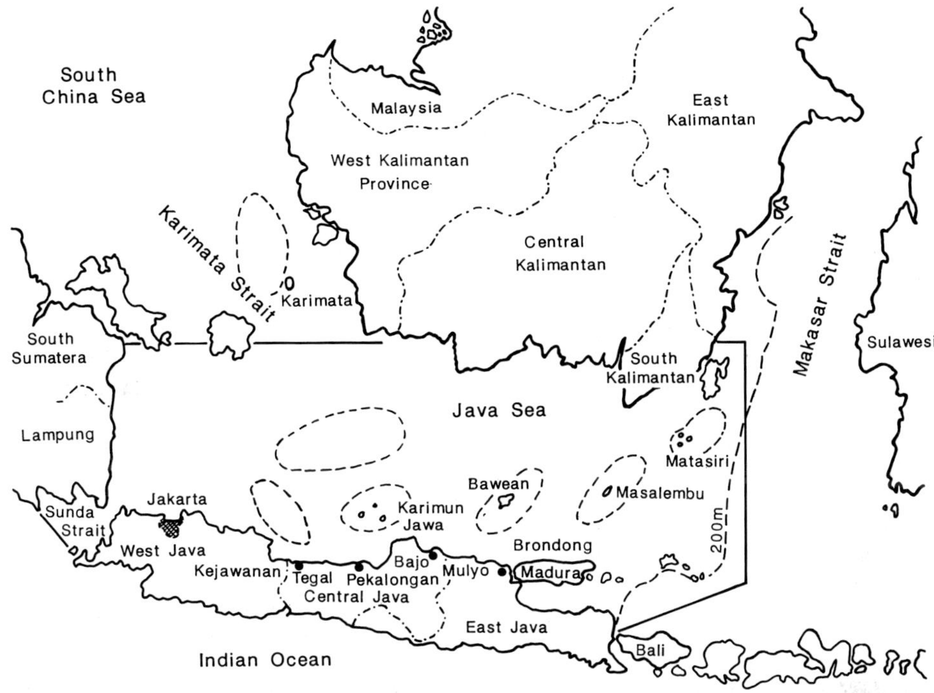

Fig. 35.1 Java Sea and the main pelagic fishing grounds for purse seiners from Pekalongan and Tegal (adapted from Atmaja & Sadhotomo 1985).

While these official figures are used here, earlier DGF and FAO sources (discussed in ADB/IBRD 1983), extrapolating data from the Gulf of Thailand, gave somewhat lower MSY estimates. A point on which all recent sources agreed, however, was the heavy exploitation to overexploitation of the small pelagic fish resources in the coastal waters off north Java.

The evidence presented here suggests that the small pelagic fish species which form the bulk of the purse seine catch show distinct signs of heavy to over-exploitation, not just off north Java but now throughout the Java Sea.

PRODUCTION

Seven provinces border the Java Sea and fish there (Fig. 35.1). They are the north coast fishermen of East, Central and West Java and Metropolitan Jakarta, the east coast fishermen of Lampung (Sumatera) and the south coast fishermen of South and Central Kalimantan (formerly Borneo). In 1987 total marine finfish landings by these seven provincial coastal areas reached 502 000 tonnes, three quarters of which was landed on the north coast of Java. Assuming this catch approximates to the total catch from the Java Sea, according to the latest DGF statistics, 1987 landings of finfish amounted to approximately half (46%) of the potential annual yield from the Java Sea (derived from Table 35.1).

This gross figure obscures four important points:

(1) The main growth in catches since 1981 has come from the pelagic stocks and by 1984 these resources were already approaching their estimated MSY; indeed since 1984 four out of every five tonnes of finfish caught in the Java Sea have been pelagic, with just three genera and six main species accounting for approximately half of this pelagic catch (Table 35.2): scads ('layang', *Decapterus macrosoma* and *D. russelli*, 18–22%), Indo-Pacific mackerel ('kembung', *Rastrelliger brachysoma*, *R. kanagutta*, 10–14%), and sardine, both Fringescale sardinella ('tembang', *Sardinella fimbriata*, 8–12%) and Indian oil sardine ('lemuru', *Sardinella longiceps*, 5–6%).

Table 35.2 Main pelagic species landed by province from the Java Sea in 1984 (tonnes).

Species	Lampung	West Java (+Jkt)	Central Java	East Java	Central K'mantan	South K'mantan	Total	%
Small pelagic	29 260 (8.0)	66 286 (18.2)	115 756 (31.8)	110 469 (30.4)	18 423 (5.1)	23 715 (6.5)	363 909	100
of which:								
scads (layang)	2	2 339	49 269	21 206	—	468	73 284	20.1
mackerel (kembung)	1 328	11 252	14 380	9 252	2 616	8 826	47 654	13.1
sardine (tembang)	554	8 839	12 394	17 027	609	1 438	40 861	11.2
oil sardine (lemuru)	128	1 997	4 787	13 875	39	169	20 995	5.8
Subtotal	2 012	24 427	80 830	61 360	3 264	10 901	182 794	50.2

Source: DGF (1986). Fisheries statistics of Indonesia, 1984, Directorate General of Fisheries, Jakarta.

(2) The level of exploitation of the offshore demersal stocks is likely to remain moderate so long as all trawling (demersal and pelagic, with and without selective mesh sizes) is banned in these waters.
(3) The present yield from the shrimp stocks, even without trawling, has again reached the level of their long-term sustainable yield.
(4) The MSY estimates in Table 35.1, particularly for the pelagic group, fail to take proper account of food chain interactions and ignore the fact that not all the different fish stocks which go to make up the total MSY are commercially exploitable.

Late in 1986, and on the basis of information up to 1985, the author summarised the prospects thus: 'As in the recent past, pelagic fish will continue to comprise the dominant component of the incremental catch . . . (while) increased exploitation of the demersal resource, which offers the greatest potential for growth in catch, is likely to develop only slowly so long as trawling is banned and the economics of pelagic fishing retain their comparative advantage over this and other demersal fishing methods' (PMO/RMC 1987).

PURSE SEINE FLEET

The Java Sea purse seine fleet is based entirely along the north Java coast. Table 35.3 gives the number of vessels by province, with their total landings and average productivity (tonnes/vessel/year) from 1984–7. Some average figures for the period 1975–9 are also included. Table 35.4 gives the breakdown by vessel size category (gross tonnes) for this fleet in 1984 and 1987.

The largest fleet is based in East Java province, where the seasonal fishery for the Indian oil sardine (*Sardinella longiceps*) in the Bali Strait is of particular importance. These vessels are largely open-deck day boats powered by small outboard or inboard engines, although larger boats of up to 30 gross tonnes also take part. By comparison the fleet in Central Java consists of generally much larger vessels (5–150 gross tonnes), with the coastal fleet (3–10 gross tonnes) mostly powered by longshaft outboard engines fishing for between 4–7 days in 1986, depending on season. The larger vessels (20–150 gross tonnes) are powered by diesel engines (25–250 hp) and normally fished in 1986 for 15–21 days (longer when in the eastern part of the Java Sea around Masalembu out to Matasiri Islands). Over the last 2–3 years, the largest vessels increasingly range outside the Java Sea (Makassar Strait and southern part of the South China Sea) in search of less exploited stocks. Such trips may last up to 40 days. A small number of ex-trawler purse seine vessels also operate from bases in West Java.

With the introduction of the trawl ban at the start of 1981, the Government of Indonesia made a large amount of credit available in an effort to encourage expansion into the pelagic fisheries, particularly purse seining. As a result a large number of trawlers were converted, but there was also a marked increase in the number of new purse seine vessels designed to fish further afield. With the growth in the size of vessels, there was a need for deeper harbours with better infrastructure and marketing facilities ashore. At the time, Central Java province was best equipped to handle such vessels, particularly the ports of

Table 35.3 Recent production data for Java Sea purse seine fleet, purse seine (PS) vessels only – excludes payang vessels.

	1984	1985	1986	1987	Average 1984–7	Average 1975–9
Purse seine vessels						
Jakarta	59	52	40	46		
W. Java	11	11	8	6		
C. Java	724	708	680	732		
E. Java	1 366	1 280	1 195	1 387		
Total	2 160	2 051	1 923	2 171	2 076	814
Purse seine landings tonnes						
Jakarta	4 152	1 765	821	242		
W. Java	2 341	2 938	1 001	752		
C. Java	92 238	128 017	124 784	92 995		
E. Java	35 482	28 891	27 911	31 710		
Total	134 213	161 611	154 517	125 699	144 010	48 794
Production/vessel tonnes/vessel/year						
Jakarta	70.4	33.9	20.5	5.3		
W. Java	212.8	267.1	125.1	125.3		
C. Java	127.4	180.8	183.5	127.0		
E. Java	26.0	22.6	23.4	22.9		
Total	62.1	78.8	80.4	57.9	69.4	59.9
PS as % of small pelagic landings	36.9%	42.2%	40.1%	31.3%	37.6%	28%
Total small pelagic landings	363 909	383 050	385 578	401 353	383 472	164 000
PS as % of fin fish landings	29.6%	33.0%	32.0%	25.0%		
Total fin fish landings	453 243	489 828	482 262	501 890		

Source: DGF (1986, 1987, 1988, 1989). Fisheries Statistics of Indonesia, Directorate General for Fisheries, Jakarta. Data for 1975–9 derived from Bailey *et al.* (1987).

Table 35.4a Number and size (gross tonnes) of fishing vessels on the north coast of Java in 1984.

		Inboard					
Province	Outboard	5–10	10–20	20–30	30–50	>50	subtotal
Jakarta	470	150	281	81	16	—	528
West Java	6 448	56	7	13	—	—	76
East Java	9 962	163	11	4	—	—	178
Central Java[1]	8 025	12	90	215	142	4	463
of which Pekalongan[2]	78	2	6	130	101	4	243
	(<1%)	(<1%)	(2%)	(42%)	(64%)	(100%)	(20%)
Total	24 905	381	389	313	158	4	1 245

Sources: DGF (1986). Fisheries statistics of Indonesia 1984, Directorate General of Fisheries, Jakarta. Anon (1985). Annual Report for 1984. Pekalongan Fisheries Service.

[1] Central Java includes the figures for Pekalongan.
[2] Given as % of the total number of vessels in the respective category on the north coast of Java.

Table 35.4b Number and size (gross tonnes) of fishing vessels on the north coast of Java in 1987.

		Inboard					
Province	Outboard	5–10	10–20	20–30	30–50	>50	subtotal
Jakarta	704	233	335	140	10	2	720
West Java	8 479	29	11	6	—	—	46
East Java	11 742	25	7	31	—	2	65
Central	11 713	176	114	112	303	8	713
Total	32 638	463	467	289	313	12	1 519

Sources: DGF (1989). Fisheries statistics of Indonesia 1987. Directorate General of Fisheries, Jakarta.

Pekalongan and Tegal (and Bajomulyo since 1984).

Figure 35.2 graphically illustrates this shift in the concentration of landings, which developed with the growth in the purse seine fleet, away from a large number of small coastal landing places towards a few major landing centres, predominant among which is the National Fishing Port of Pekalongan.

METHOD OF FISHING

Purse seiners have largely taken over from the 'traditional' method of seine net fishing of north Java known as 'payang'. They are considerably more efficient as, unlike payang, they do not have to manoeuvre to aim the net at a school of fish like a pelagic trawl (the 'payang' method does not use pelagic trawl boards). Purse seiners simply attempt to encircle a school of fish with a wall of netting which, when the circle is closed, is pursed in at the bottom, hopefully before the fish can escape. In 1985 the typical purse seine vessel in the Java Sea was 20–30 gross tonnes, with a purse seine net of 450–600 m × 25–70 m, with a mesh size of 2.5 cm.

Even the largest vessels employ the same fishing technique. A fish aggregating device called a 'rumpon', usually consisting of a buoy or bamboo raft with a flag

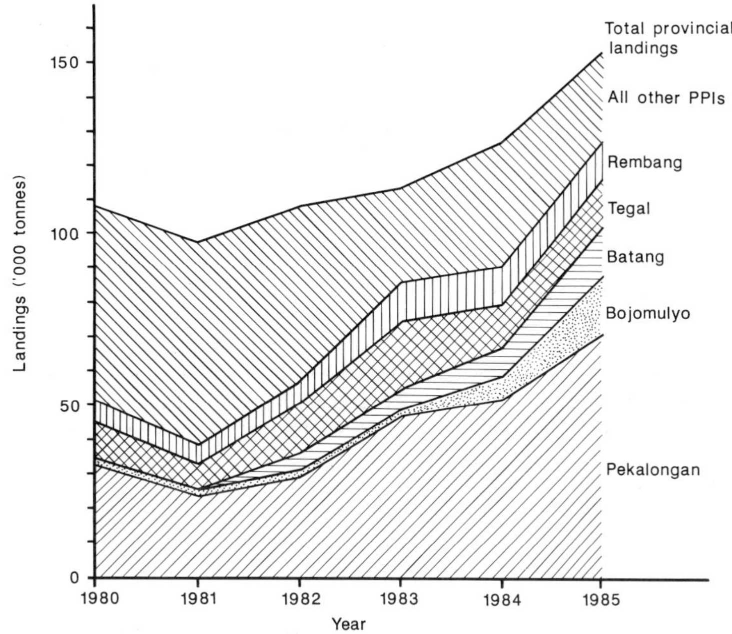

Fig. 35.2 Development of landings at the NFP Pekalongan and the main principal port landings (PPIs) on the north coast of Central Java Province 1980–85.

pole and trailing coconut leaves, is left in the water in a known fishing area. Lamps are generally used at night. The vessel stands-by (with engine off) until a reasonable-sized school accumulates below the rumpon. This process may be repeated at several rumpon sites in the same day. Typically the purse line is hauled in by engine-driven winch, but the net is pulled aboard manually with the aid of a boom and pulley block, hence the large crews of 30–35 men. Currently no other mechanical or electronic fishing aid – with the possible exception of radio – is employed (i.e. no powerblock, no sonar). Indeed, few if any vessels employ rollers to help haul the net over the side, so the number of hauls is physically limited to 2–4 a day, and it is generally felt that the only way to increase the catch rate is to operate in less congested, more distant fishing grounds – i.e. where more successful hauls can be made per day – but this requires more days at sea per trip (Widianto & Linting 1986). Meanwhile active searching for fish schools is considered expensive in fuel, with considerably less certain results, and is no longer normal practice. Fishing normally ceases for about a week over the full moon period.

This 'passive' method of purse seine fishing is probably a sensible strategy for medium-scale vessel owners, given the prevailing techno-economic conditions during the 1980s. First, in changing fishing method to purse seine, the ex-trawlers were overpowered (i.e. too high operating costs), especially as diesel fuel, the largest operational expense, increased 4.5 fold over a two year period (from Rp.52/litre in 1982 to Rp.225/litre in 1984). Second, the cost of imported nets and diesel engines (the latter often more expensive than the boat itself) more than doubled as the Indonesian Rupiah suffered sharp devaluations in 1983 and 1986. Thus, while PT Tirta Raya Mina's small purse seine fleet (two vessels active in 1987), also based at Pekalongan, has used hydraulic powerblock and sonar since about 1976, so far neither technology has been adopted by the medium-scale sector because of the lack of exposure to them and the high cost of such imported equipment.

This situation is likely to change quickly once this gear is proven by one or two medium-scale vessel owner-innovators. Interestingly, in 1987, an Indonesian medium-scale vessel owner in Belawan, North Sumatera, was having good results using a powerblock and sidescan sonar, this technology being widely used by Malaysian fishermen in the Malacca Strait. In the case of the demersal trawl fishery, the technology, once adopted by owner-innovators in Sumatera, was quickly transferred to the larger area of the Java Sea. A similar pattern of adoption is considered likely to occur for the powerblock and sidescan sonar in the Java Sea purse seine fishery in the early 1990s.

PRODUCTION AND PRODUCTIVITY OF THE PURSE SEINE FLEET

This section considers the catch of the purse seine fleet and the changes in its productivity both in total and for the pre-eminent part of this fleet which is based at Pekalongan.

Just five years after the trawl ban was introduced, landings by purse seine vessels operating in the Java Sea peaked in 1985 at 161 600 tonnes, with about 80% of this catch being landed by a little over 700 vessels in Central Java province. Average productivity reached about 80 tonnes/vessel/year, while that in Central Java province was 180 tonnes/vessel/year (Table 35.5). Indeed in 1985 both production and productivity were even more concentrated, with some 251 purse seine vessels making an average of 1.0 trip/month landing 67 880 tonnes at Pekalongan port alone. Thus 42% by weight of purse seine landings from the Java Sea were landed by a mere 12% of such vessels at just one port. In that year landings by the Pekalongan purse seine fleet averaged 270.5 tonnes/ vessel or 22.5 tonnes/trip (average 24 days).

Between 1984 and 1987 in the Java Sea purse seine fishery, while the number of purse seine vessels remained more or less stable at about 2075, there was a dramatic increase in the average (modal) vessel size from 20–30 gross tonnes to 30–50 gross tonnes. Again the Pekalongan fleet was at the forefront of this increase, with the typical vessel at Pekalongan already in the 30–50 gross tonnes size category by 1985, and with the largest vessels over 100 gross tonnes by 1987. This trend towards larger and larger vessels was still evident in 1989, although some of the largest vessels (in the 50–100 gross tonnes and 100–150 gross tonnes categories) were by then using the port of Bajomulyo, because of its deeper access channel and closer proximity to the fishing grounds in the main fishing season.

In all probability, total effective fishing effort (perhaps most sensibly measured in this fishery at present by the single parameter 'days spent at sea per year') has been at a level above the MSY-equivalent level of fishing effort, at least since 1985. Almost certainly this measure of fishing effort has continued to increase since that time. This can be inferred from:

(1) the decline in the total catch by the Java Sea purse seine fleet despite a more or less stable number of (increasingly larger) fishing vessels
(2) the increase in the average length of trip (to 32 days by 1988) and in the average catch per trip by vessels landing at Pekalongan, while their total catch has fallen from the peak of 270 tonnes/ vessel/year in 1985 to 193 tonnes/vessel/year in 1988
(3) the increase in the average consumption/vessel/ year of the major variable inputs: diesel fuel, salt (used in greater quantities/tonne of fish caught in place of more ice the longer the trip), and, at least up to 1986, ice (Table 35.5).

Consequently, real average vessel costs have risen while average vessel productivity (catch/vessel/year) has fallen every year since 1985.

Fortunately from an analytical viewpoint, this fishery has such a mjaor impact on total fish supplies that any substantial fall in landings produces an upward shift in prices. The question of how this affects both vessel owners' and fishermen's incomes is discussed later in the chapter.

CATCH RATES, SPECIES COMPOSITION AND STOCK CONDITION

From the above analysis, the following simple hypothesis can be put forward to cover the recent development and current situation in the Java Sea purse seine fishery: as each separate small pelagic fish 'stock' in the Java Sea has come under continuous fishing pressure, the yield of the furthest 'stock' has fallen in line with the economic 'open access' level of fishing effort. From this, two predictions can be made:

(1) As the catch rate falls the largest vessels will tend to extend their area of operation towards their own economic 'normal profit rate' position.
(2) As fishing pressure comes to bear on the small pelagic resources throughout their geographic range (and possibly life-cycle also), changes (oscillations) in the relative abundance and/or geographic distribution of some species will take place, resulting in fluctuations in the composition of the catch and probably in the size of the total catch in different years. The total volume of landings will also tend to decline (Laevastu & Favorite 1988).

These two predictions would seem to describe accurately the recent developments in the Java Sea pelagic fishery, at least from 1985 to the present. As noted in the previous section, fishing pressure on the small

Table 35.5 Changes in total landings and landings per trip by type of vessel and other indicators of vessel activity at NFP Pekalongan from 1985–88.

		Purse seine		Mini PS		Gill net		Others	
Year	Total	tonnes	%	tonnes	%	tonnes	%	tonnes	%
Landings (tonnes)									
1985	70 475	67 649	96.0	233	0.3	2 502	3.5	92	0.1
1986	67 302	64 045	95.2	700	1.0	2 542	3.8	34	0.1
1987	48 438	43 647	90.1	2 122	4.4	2 625	5.4	44	0.1
1988	45 461	37 519	82.5	4 883	10.7	2 755	6.1	305	0.7
No. of vessel entrances									
1985	4 243	1 921	45	510	12	1 334	31	477	11
1986	3 892	1 670		622		1 191		408	
1987	3 274	1 064		986		1 072		152	
1988	2 533	580	23	840	33	719	28	395	15
Average landings per trip (tonnes)									
1985	16.6	35.2		0.5		1.9			
1986	17.3	38.3		1.1		2.1			
1987	14.8	41.0		2.2		2.5			
1988	17.9	64.7		5.8		3.8			
Average income per trip (in constant Indonesian Rupiah millions)									
1985	5.4	11.4		0.15		0.62			
1986	5.9	13.2		0.39		0.73			
1987	6.1	16.9		0.89		1.01			
1988	7.7	27.7		2.49		1.64			

	Number of vessels by gear type			Average landings/vessel/year (tonnes)		
	Total	Purse seine	Gill net	Total	Purse seine	Gill net
1985	326	251	75	216	270.4	33
1986	318	241	77	211	268.6	33
1987	366	258	108	132	177.4	24
1988	336	249	116	135	192.7	24

	Landings		Average consumption of inputs per trip for all vessels					
	(tonnes)	(Rp million)	Diesel (tonnes)	Petrol (litre)	Oil (litre)	Ice (tonnes)	Salt (tonnes)	Total (Rp million)
1985	16.6	5.41	3.3	73	11	17	0.04	1.1
1986	17.3	5.95	3.6	310	46	31	0.56	1.9
1987	14.8	6.12	4.2	355	42	25	2.92	2.1
1988	17.9	7.69	4.9	354	46	22	5.57	2.4

Source: Derived from Bandjono *et al.* (1989). Field study report on NFP Pekalongan, Project Monitoring Evaluation/Appraisal. Directorate General of Fisheries – Institute of Management, Faculty of Economics, University of Indonesia, Jakarta.

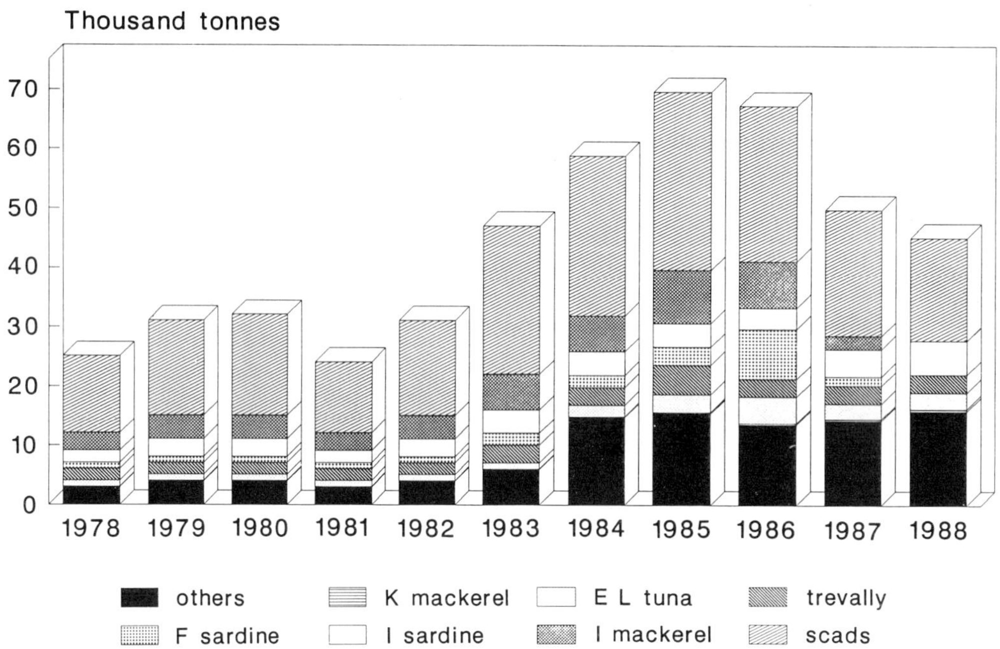

Fig. 35.3 Species composition of fish landed at Pekalongan 1978–88 (Source: Dir. of Infrastructure, DGF).

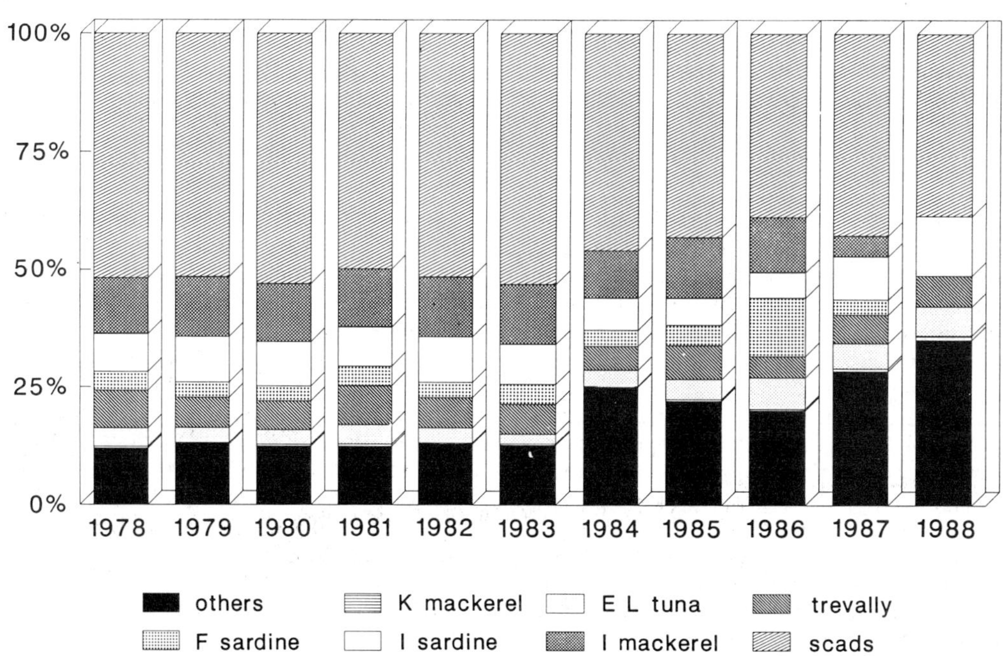

Fig. 35.4 Composition of main species landed at Pekalongan 1978–88 (Source: Dir. of Infrastructure, DGF).

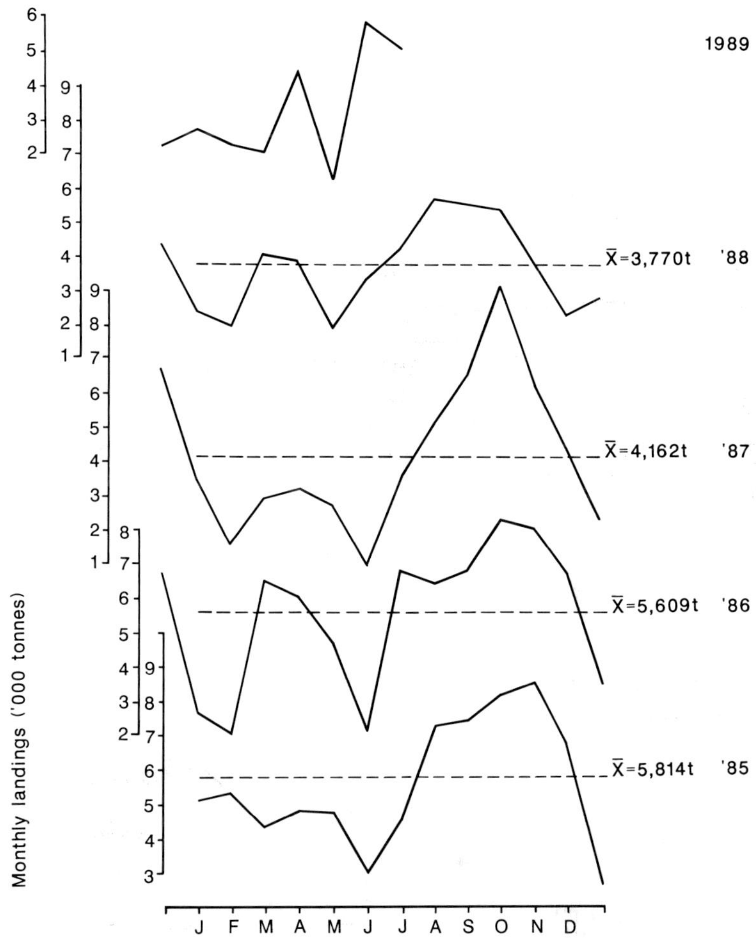

Fig. 35.5 Monthly landings at NFP Pekalongan 1985–89 (Source: Dir. of Infrastructure, DGF).

pelagic stocks of the Java Sea has increased in its geographic range and probably also in its intensity throughout the 1980s. So while total landings by the purse seine fleet continued to grow until 1985, they have declined every year since.

At Pekalongan also the volume of landings has fallen consistently since the 1985 peak, with the species composition, which had been remarkably constant up to 1983–5, also showing marked changes since then. Of particular note since 1985 is the decline (both in absolute and relative terms) in the scad and Indo-Pacific mackerel catch (Figs. 35.3 and 35.4), normally the dominant target species of the larger, wider-ranging (i.e. non-coastal fishing) purse seine vessels. Significantly also, the coastal mini-purse seine vessels have seen their main target species, which also includes a high proportion of sprat and anchovy, show large annual fluctuations since 1983.

Figure 35.5 depicts the changing pattern of monthly landings at Pekalongan from 1985 to mid-1989. While total landings dropped to 65% of their 1985 peak by 1988, there is an increasingly pronounced bi-modality in the pattern of the monthly landings. The relative abundance of the two scad species, *D. macrosoma* (which makes up the bulk of the total scad catch is mainly taken in the East of the region in the period September to November) and *D. russelli* (which is mainly taken in the West of the region) accounts for the major part of the variation in the strength of these two peaks from year to year (consistent with data in Atmaja *et al.* 1986 and Suwarso *et al.* 1987).

Finally, as Table 35.2 and Figure 35.6 graphically illustrate, this pattern of a fall in total pelagic landings has been repeated along the whole north Java coast, but particularly towards the East which is where, in terms of the relative contribution to total landings,

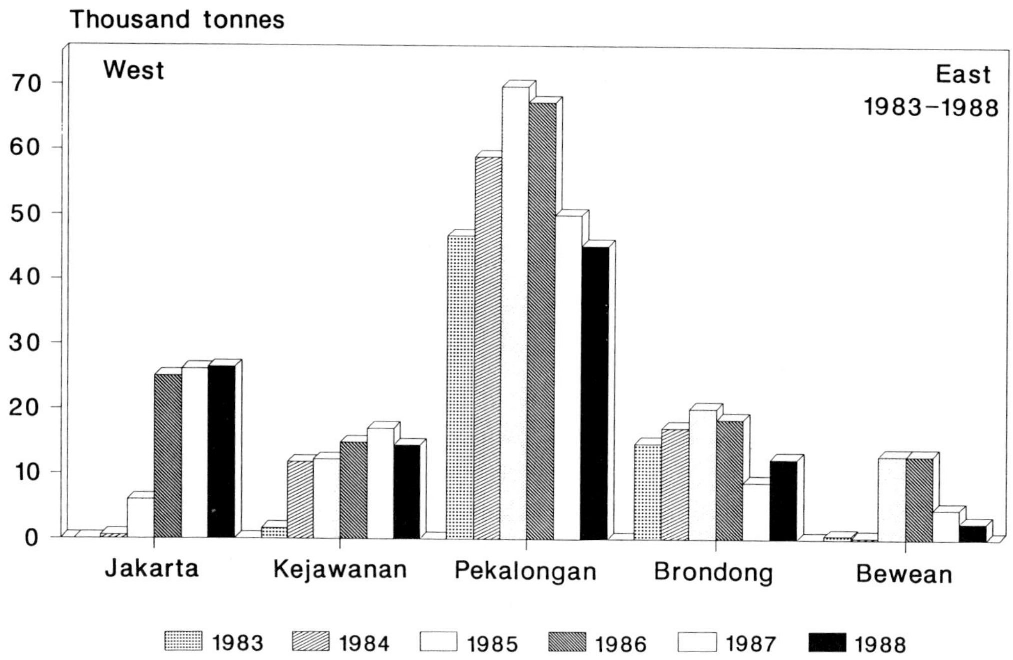

Fig. 35.6 Fish production by national fishing port, north coast of Java. Source: Dir. of Infrastructure, DGF

pelagic fishing methods (and especially purse seining) and the catch of scad predominate.

STORAGE, PROCESSING, DISTRIBUTION AND MARKETING

Storage on board and auctioning

Some improvement of the fish hold has taken place as vessels have grown in size. Separate fish holds are constructed of wood (5–10 cm thick), with fish boards up to 1–1.5 m high to separate the catch. Galvanized iron is increasingly used around the sides, roof and cover. Drainage is provided but is not used while at sea, so after several days at sea the fish 'swim' in an ice and ice-water mix. Sea salt is added to the new fish entering the hold, with more salt being used for the early part of the trip. An initial ratio of fish to ice (taken on board in 25–50 kg blocks) of 0.5 : 1 to 1 : 1 is used, and chunks of ice are usually still visible in vessels returning after 15 days. On longer trips the vessels may take additional ice on board, but this is by no means standard practice. The condition of the catch obviously varies with the time since capture but, in general terms, the fish is in a fair to poor condition and, for fish more than 15 days old, may even be breaking up when it is landed. However up to this point the fish is usually still safe to eat.

All fish landed at Pekalongan goes through the auction (officially 09.00–14.00 hours, extended to 06.00–16.00 hours in periods of heavy landings) in plastic boxes (capacity 40 kg). The fish are washed on board in water taken from the polluted Pekalongan River, and a little chip ice may be added after the fish are weighed. The fish spend on average 2–3 hours in the open air (average temperature 30°C). There are about 35 buyers in the auction.

Table 35.6 gives landed volumes and Table 35.7 gives the fish auction prices by major species at Pekalongan in recent years. Figure 35.7 maps out the relationship between the total quantity landed and the annual average auction price (in constant 1986 Rupiah) over the 10 years 1979–88. The relationship approximates to a downward sloping 'partial demand curve for fish in the whole of Java', and by tracing

Table 35.6 Fish landings at Pekalongan by major species (1978–89) ('000 tonnes)

Species[1]	1978	1979	1980	1981	1982	1983	1984	1985	1986	1987	1988	1989[2]
Scad (layang)	13	16	17	12	16	25	27	30	26.1	21.3	17.4	7.9
Ind. mackerel (kembung)	3	4	4	3	4	6	6	9	7.9	2.2	0.01	0.01
I. oil sardine (lemuru)	2	3	3	2	3	4	6	4	3.6	4.6	5.7	1.82
Trevally (bentong)	2	2	2	2	2	3	3	5	3	3	3	—
F. sardine (tembang)	1	1	1	1	1	2	2	3	8.4	1.6	—	—
E. little tuna (tongkol)	1	1	1	1	1	1	2	3	4.5	2.7	2.8	1.44
King mackerel (tenggiri)	0.1	0.1	0.1	0.1	0.1	0.1	0.1	0.2	0.2	0.25	0.4	0.19
Others	3	4	4	4	4	6	12.8	15.5	13.6	14.2	15.9	12.24
Total	25.1	31.1	32.1	24.1	31.1	47.1	58.9	69.7	67.3	49.9	45.2	23.6

Sources: NFP Pekalongan and Fisherman's Co-operative. Data for 1978–85 period is published in RMC/PMO (1987). Feasibility study on NFP Pekalongan, Fisheries Infrastructure Sector Project, Directorate of Infrastructure, DGF, May 1987. Data for 1986–9 period is unpublished data from Directorate of Infrastructure, DGF. Although the original source is the same Fish Auction Hall records at NFP Pekalongan, it should be pointed out that the DoI-DGF data used here differs substantially from that presented in Bandjono *et al.* (1989). Field study report on NFP Pekalongan, Project Monitoring Evaluation/Appraisal, Directorate General of Fisheries – Institute of Management, Faculty of Economics, University of Indonesia, Jakarta. Nevertheless, the comments made in the text apply irrespective of which data set is used.

[1] *Common name*	*Scientific name*
Scad	*Decapterus macrosoma* & *D. russeli*
Indo-Pacific mackerel	*Rastrelliger brachysoma* & *R. kanagutta*
King mackerel	*Scomberomorus commersoni* & *S. guttatus*
Trevally	*Selar* spp.
Eastern little tuna	*Euthynnus* spp.
Indian oil sardine	*Sardinella longiceps*
Fringescale sardinella	*Sardinella fimbriata*

[2] Covers January–July only.

Table 35.7 Indicative fish prices at NFP Pekalongan (1984–89) (Rp/kg).

Species	1984	1985	1986	1987[1]	1988[1]	1989[3]
F. sardinella	260	245		—	—	—
scads (layang)	289	263		330	780	380
Indian oil sardine	296	309		220	650	470
Trevally (Selar sp.)	420	—		590	830	730
Eastern little tuna	505	525		550	1020	1070
Indian mackerel	627	406		370	1160	1110
King mackerel	1005	1095		1300	1510	1750
Weighted average[2]	321	291	344	419	541	492

Sources: RMC/PMO (1987). Feasibility study on NFP Pekalongan, Fisheries Infrastructure Sector Project, Directorate of Infrastructure, DGF, for data covering period 1984–1985. NFP Pekalongan Fisheries Service monthly port monitoring reports to Directorate of Infrastructure, DGF, for data covering period 1986–89, (unpublished data).

[1] Species prices during 1987–9 are based on indicative monthly prices.
[2] The weighted average prices are based on the total annual auction values divided by the annual sales volume.
[3] 1989 indicative prices cover January to March, while the weighted average price covers January to July period.

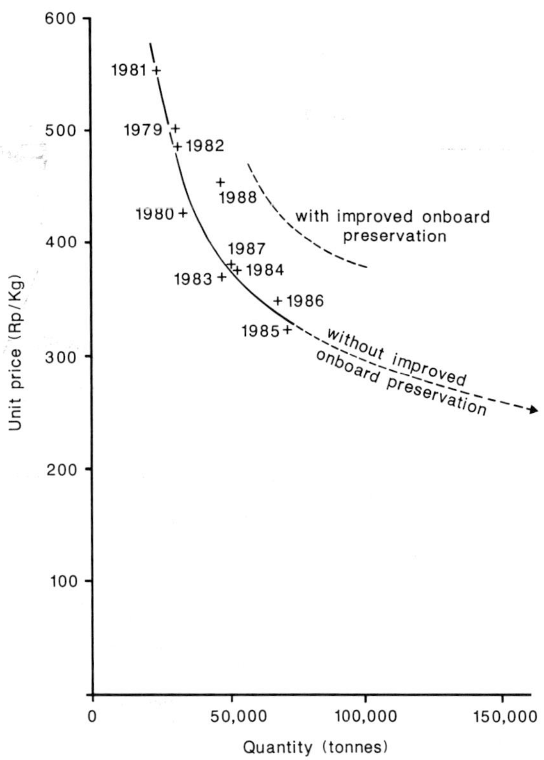

Pekalongan retail price index

1979	1980	1981	1982	1983	1984	1985	1986	1987	1988
40.9	50.5	57.3	62.7	69.4	84.1	89.3	100	108	119

Fig. 35.7 Demand curve for all fish landed at Pekalongan, at 1986 prices.

out each successive year's points it is clear that the partial demand curve has shifted outward (as might be expected) over this period.

Processing, distribution and marketing

The average volume of fish passing through the auction at Pekalongan in 1985 was 193 tonnes/day, while peak landings were about twice this volume. Of this, fresh fish traders and dried fish processors purchased about 42 tonnes/d or 22% each, with salt boiled fish (pindang) processors purchasing the bulk of the catch (56%). In 1988, when the average volume of fish passing through the auction had dropped to 125 tonnes/d, fresh fish traders and dried fish processors purchased the same average volume of 42.5 tonne/d, but it now represented 34% each of the total catch, with the pindang producers taking up the slack of 32%.

The shelf-life of fresh fish depends on time since capture and storage conditions. Typically, fresh fish has a shelf-life of 2–5 days after landing if kept on ice, but it is normally consumed within 48 hours of landing. Only fresh fish is brought to market packed in ice.

Pindang is placed in 0.5 kg bamboo pouches, packed in wire cages, dressed with sea salt before being boiled in saline open baths for about 15 minutes. Pindang has a shelf-life of between 1 and 2 days. Essentially this product form serves to extend the shelf-life of fresh fish without using ice.

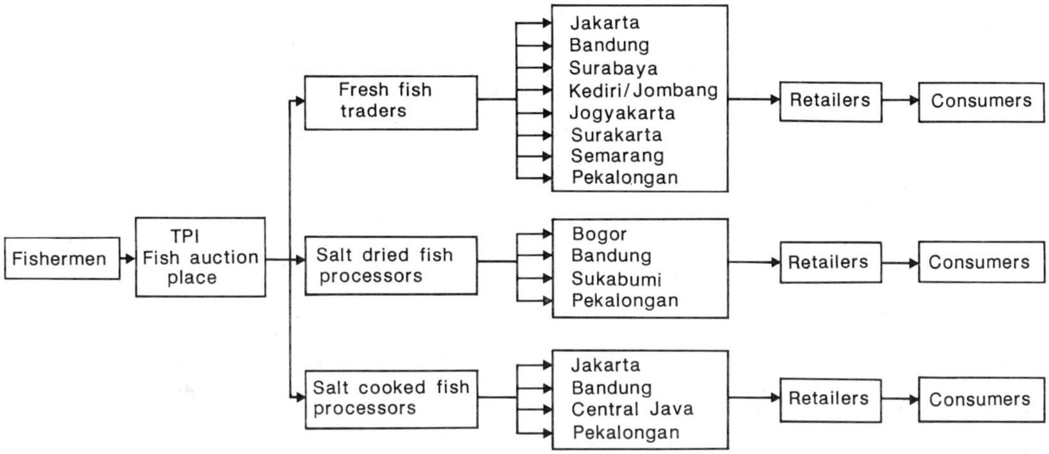

Fig. 35.8 Pekalongan fish processing and marketing chain.

By contrast, salted sun-dried fish is usually split, then soaked in a saline bath for twenty minutes to several hours (depending on the final water content required) before being laid out on racks to dry. Drying takes between 2 and 5 days (depending on season), and dried fish typically has a shelf-life of about 1 month.

Both fresh fish traders and pindang processors dispatch their products to market in wooden boxes, cylindrical plastic bins or bamboo baskets by tarpaulin-covered truck to the night markets in the major cities on Java. Figure 35.8 shows the distribution of fish by major market from Pekalongan in 1985. The pattern of distribution was similar in 1989 (Bandjono *et al.* 1989). Fresh fish and pindang (a product peculiar to Java within Indonesia) are sold in a number of agricultural retail markets in the major cities, normally the day after the fish is landed. Dried fish also finds its way on to these retail markets.

In general terms the price mark-up by the fresh fish wholesaler averages 25% (range 20–100% depending on quality, species and market, with Jakarta giving the highest prices), while the retailer adds a further 25%. In the hawker economy the total mark-up can be higher, but the fish then passes through several hands.

The best quality fresh fish is sold to the hotel, restaurant and supermarket trade, with premium prices paid for certain species. Iced small pelagic fish however, tended to sell for between Rp.800–2000/kg (US$0.45–1.10/kg) in the main supermarkets of Jakarta in 1989. This is between two and four times the average auction price. The average price paid to the wholesaler from Pekalongan was about Rp.1050/kg. Meanwhile the wholesale price of salted dried fish and pindang from Pekalongan in 1989 was Rp.1120/kg and Rp.915/kg (Bandjono *et al.* 1989).

Figure 35.9 illustrates changes in the relative disposition of the total marine fish catch in Indonesia between the time of the trawl ban in 1981, and 1987, while Figure 35.10 presents information on changes in the disposition of marine fish landings between 1984 and 1987 for the north coast of Java alone.

In Indonesia as a whole since the trawl ban in 1981, disposal of the catch as fresh and smoked fish has increased proportionately by about 4% and 1%, while the proportion going towards dried and preserved fish (mostly fermented) has declined by 2% and 1% respectively. It is noteworthy that the proportion going into frozen and canned fish has remained unchanged (at 3% and 1% respectively). These shifts are believed to reflect consumer preference for fresh and smoked fish, while the high price of frozen and canned fish restricts more rapid growth in their consumption.

The disposition of the catch landed in north Java shows considerable differences with that for the country as a whole, reflecting the problems of supply of fresh fish to this large market, although the direction of recent trends (between 1984 and 1987) is generally similar. Most notably, fresh fish consumption is considerably lower in north Java, with dried fish making up most of the difference. These two product forms had an almost identical share of the north Java market in 1987 (37% each), with the share of each growing since 1984. The other outstanding feature is the decline in boiled fish, from 22% to 19%, of the total supply. This pattern reflects that seen at

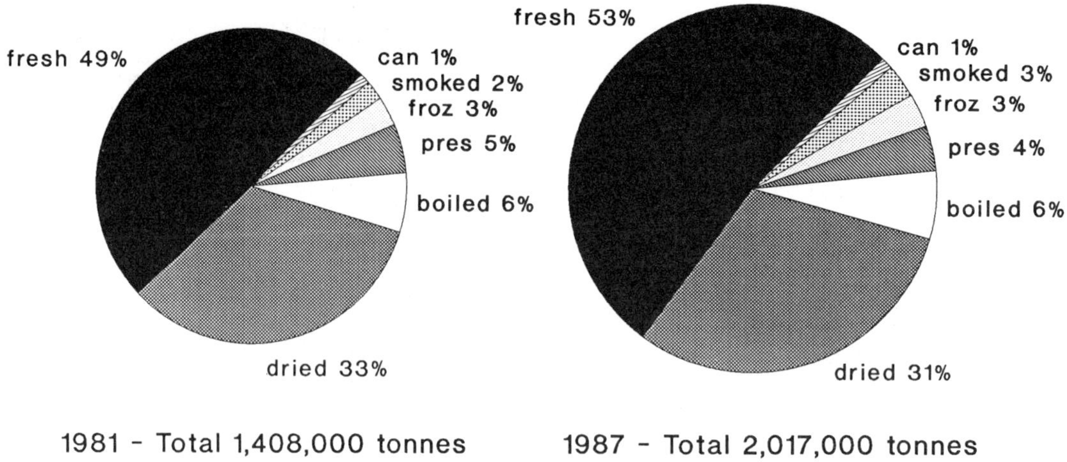

Fig. 35.9 Disposition of marine fish landings in Indonesia 1981 and 1987 (Source: Fish stats. of Indonesia, DGF).

Pekalongan (see above). The other uses of the catch show little change, except that preserved and canned fish have also declined recently in relative terms.

RECENT DEVELOPMENT OF SUPPLIES, PRICES, COSTS AND FISHERMEN'S INCOMES FOR THE PURSE SEINE FLEET WITH SPECIAL REFERENCE TO PEKALONGAN

There have been few meaningful studies on the costs and earnings of vessel owners and their crews, and no attempt has been made previously to monitor changes in their incomes or in the profitability of fishing vessels over time.

The Fisheries Infrastructure Sector Project (FISP) attempted to establish baseline socio-economic data for a sample of the main fishing methods in operation at four ports in Java (including Pekalongan) in August 1989. One explicit purpose was to monitor how the costs and earnings of the main fishing methods (using a stratified sample, in the case of Pekalongan, of purse seine, mini-purse seine and gill net vessels) are affected over time by the particular port expansion/rehabilitation project about to be undertaken. While the objective of the study was largely met, unfortunately the student's reports failed to present the baseline data collected in full (e.g. Bandjono *et al.* 1989). Nevertheless, the standardised questionnaires and relational database format developed for this purpose, when analysed in more detail, should provide a reasonable baseline for comparative purposes with future cost and income surveys.

This objective could best be met with 12 locations being monitored, as was originally intended, on an annual basis to collect cost and earnings data from the same sampling frame of vessel owners. And this could best be done if each vessel is provided with a vessel trip log book.

Previous studies on the economic performance of both small- and medium-scale fishing vessels have recently been reviewed (Bailey *et al.* 1987). Purse seine crews at Pekalongan in 1978 earned an average of Rp.220 000–260 000/crewman/year (i.e. equivalent to Rp.590 000–700 000 in constant 1986 Rupiah, the basis of the rest of the figures quoted here). The capital cost of the purse seine vessel was Rp.32 million (about 20 gross tonnes size), and the owner earned between Rp.5.7–27 million (depending on the study).

Table 35.8 gives a very generalised view of the changes in gross receipts earned by purse seiners, and of average annual income to both the owner and fishermen (crew member) for the period 1985–88 at Pekalongan. Both total receipts and average incomes peaked in 1986, when a 1% share of total receipts was worth Rp.912 000. This measure is a reasonable proxy for a year to year comparison of average crew income. It also allows comparison with earlier studies (refer Bailey *et al.* 1987). Average crew income on this measure dropped back to the Rp.700 000/crewman/year mark in 1987 and 1988, about where it was for purse seine crewmen 10 years earlier, when this fishery had plenty of potential for expansion of the total catch, and vessel sizes were considerably smaller.

As landings from the purse seine fleet have declined in the last 3 years, so the proportion of the catch marketed fresh has tended to increase, as noted above.

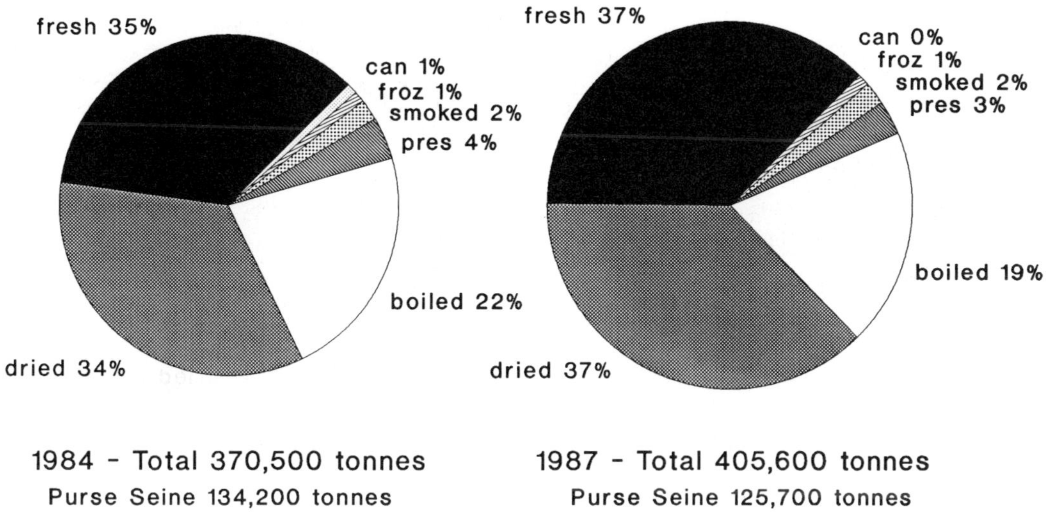

Fig. 35.10 Disposition of marine fish landings, north coast of Java 1984 and 1987 (Source: Fish stats. of Indonesia, DGF).

Table 35.8 The development of gross income to the purse seine fleet at Pekalongan in recent years (1985–8). Derived from PMO/RMC (1987) and Bandjono *et al.* (1989).

	1985	1986	1987	1988
Gross receipts to PS vessels (Rp M)	22 934	23 143	20 048	19 476
Average catch/vessel/year (tonnes)	269.5	265.7	169.2	170.5
Average fish price (in constant 1986 Rp) (Rp/kg)	326	344	414	428
Gross receipts/vessel/year (Rp M)	87.9	91.2	70.0	72.7
Crewman share (Rp 1000) (1% share/crew member)	879	912	700	727

This is, at least in part, a result of the increased weighting of landings by other gears which fish for a shorter period, particularly the mini-purse seine and gill net (trip lengths by these two methods generally lasted for between 9–12 days in 1989).

For the Pekalongan purse seine fleet, the real value of fish landings has probably been declining in recent years, even further than Table 35.8 indicates. This is largely a result of the reduction in average landings per vessel as accumulated year classes are fished down in previously moderately fished stocks, with the ensuing prospect that catches will now vary with the relative strength of just one or two year classes of the main small pelagic fish stocks. Increased fishing effort on small pelagic stocks in the Java Sea will therefore no longer bring forth concomitant rises in catches and may indeed cause the stocks to go into further decline.

With the large annual increase in total landings of pelagic species since 1981, overall, fishermen's real incomes have on average tended to fall (Fig. 35.7), with the only compensating factor for the purse seine crews being their increased productivity per man as the average size of vessels and of their catches has increased. If total supplies do not grow, the future development of fishermen's incomes will depend on (a) increasing the value of their catch (e.g. by landing fresher fish), and (b) rising demand pushing up average fish prices.

DEVELOPMENT SCENARIO

The development scenario outlined below for the Java Sea purse seine fishery is driven by a series of constraints, biological, social and economic.

(1) The Java Sea small pelagics resource base is being heavily exploited throughout its range and for many stocks throughout much of their life-cycle by a

variety of fishing gears, including bagan (lift net), gill net, payang as well as purse seine. In part, this is because a large subsistence and artisanal economy depends on these resources for their own nutrition and as a source of income. Conventional fisheries management techniques involving limiting total effort do not feature as a viable option in this fishery because, given the 'open access' nature of this fishery, politically such measures cannot be enforced. For example, bagan (lift net) as a method of fishing is banned in Java, but this method accounts for several thousand tonnes per year of juvenile small pelagics in Central Java Province alone (Wright *et al.* 1990).

Many of the fishermen dependent on the small pelagic resources have no alternative employment options open to them. For them, fishing represents their 'employment opportunity of last resort'. The only feasible enforceable management measures relate to items such as minimum mesh sizes and gear size restrictions. Area and seasonal closures are considered to be unenforceable.

(2) The introduction of active fishing techniques, involving searching for schools with sidescan sonar and powerblock to increase vessel productivity, should make larger hauls and shorter trips possible. However, such active fishing methods represent a considerable jump in fishing technology and investment by vessel owners. Also, at the moment, it is not possible to predict whether such new fishing techniques will be cost effective, nor to what extent they are likely to be adopted. However, if successful, they will substantially reduce employment opportunities in this fishery.

Furthermore, the introduction of such sophisticated fishing technology is likely to create additional conflict between the small-scale and medium-scale fishermen, similar to the conflict between the shrimp trawlers and the inshore fixed-gear fishermen in the late 1970s. In that instance the small-scale fishermen 'won' because of their political importance. However, provided the medium-scale offshore fishery remains economically viable, the small-scale inshore fishermen are more likely to lose out, because the medium-scale purse seiners fish throughout the area, on a number of different stocks and at different stages in the life-cycle of these fish stocks.

(3) Increasing the value of the catch by improving its quality (freshness) is very cost effective (PMO/RMC 1987). This can be achieved through:

- better insulation of the hold, using thicker, e.g. 10 cm, polystyrene foam
- draining the hold of meltwater
- sorting the catch and icing the fish in boxes on board
- shorter trips (maximum 15–20 days), perhaps requiring smaller vessels.

(4) In the longer term fishermen will only leave this fishery if they have better opportunities elsewhere (either in fishing, e.g. the demersal stocks, or outside fishing altogether).

(5) Given current fish supply constraints, the outlook for fish prices on Java is that they will continue to increase in line with growth in income and population.

(6) The prospects for this fishery achieving a rational level of exploitation (either maximum economic yield or MSY) are not encouraging in the short to medium term. Thus the small pelagic fish stocks of the Java Sea are likely to remain heavily fished to overfished for the foreseeable future.

BIBLIOGRAPHY

ADB/IBRD (1983) *Indonesian Fisheries sector study*. Prepared by the Asian Development Bank and the International Bank for Reconstruction and Development, Manila.

Atmaja S.B. & Sadhotomo B. (1985) *Operational aspects of purse seine fishing in the Java Sea*. Marine Fisheries Research Report 32, Marine Fisheries Research Institute, Jakarta. (In Indonesian.)

Atmaja S.B., Suwarso & Nurhakim S. (1986) Purse seine catch according to season and fishing ground in the Java Sea. *J. Marine Fisheries Research*, **36**, 57–65. Department of Agriculture, Jakarta. (In Indonesian.)

Bailey C., Dwiponggo A. & Marahudin F. (1987) *Indonesian marine capture fisheries*. ICLARM Studies and Reviews 10. International Centre for Living Aquatic Resources Management, Manila, Philippines; Directorate General of Fisheries, and Marine Fisheries Research Institute, Department of Agriculture, Jakarta, Indonesia.

Bandjono S., Soepardjo A., Rasyid H.A., Bangun M., Cahya Indra S. & Endroyono A. (1989) Field study report on National Fishing Port Pekalongan, Project Monitoring Evaluation/Appraisal. Directorate General of Fisheries, Institute of Management, Faculty of Economics, University of Indonesia, Jakarta.

Chullasorn S. & Martosubroto P. (1986) *Distribution and important biological features of coastal fish resources in Southeast Asia*. Fish. Tech. Pap. 279, FAO, Rome.

DFR-DGF (1987) *Marine resources potential and 1985 landings by sea area*. Directorate of Fisheries Resources, Directorate General of Fisheries, Jakarta. (Unpublished data).

DGF/MFRI (1983) Evaluation report on the marine fisheries resources potential in Indonesian and EEZ waters. Directorate of Fisheries Resources Development, Directorate General of Fisheries and Marine Fisheries Research Institute, Department of Agriculture, Jakarta, Indonesia. (In Indonesian.)

Laevastu T. & Favorite F. (1988) *Fishing and stock fluctuations*. Fishing News Books, Oxford.

Martosubroto P. & Chong K.C. (1987) An economic analysis

of the 1980 Indonesian trawl ban. *Indonesian Agricultural Research and Development Journal*, **9**, (1–2), 1–12.

PMO/RMC (1987) *Feasibility study for expansion/rehabilitation of Pekalongan fishing port*. Prepared by Project Management Office and Rogge Marine Consultants. Fisheries Infrastructure Sector Project, Directorate of Infrastructure, DGF, Jakarta.

Sadhotomo B., Nurhakim S. & Atmaja S.B. (1986) Development of catch composition and catch rate of purse seine in the Java Sea. *J. Marine Fisheries Research*, **35**, 111–120. Department of Agriculture, Jakarta. (In Indonesian.)

Suwarso, Atmaja S.B. & Wahyono M. (1987) The catch composition of scad (*Decapterus* spp.) caught by purse seine according to fishing ground in the Java Sea. *J. Marine Fisheries Research*, **38**, 55–8. Department of Agriculture, Jakarta. (In Indonesian.)

Wahyuono H., Soselisa J. & Badrudin M. (1986) Catch rate differences during the 1977/8, 1978/9 and 1985/6 northwest monsoon in Semarang Kendal water, Central Java. *J. Marine Fisheries Research*, **36**, 15–21. Department of Agriculture, Jakarta. (In Indonesian.)

Widianto & Linting M.L. (1986) Notes on the purse seine fishing in the region of Tegal. *J. Marine Fisheries Research*, **34**, 57–8. Department of Agriculture, Jakarta.

Wright P.J., Willoughby N.G. & Edwards A.J. (1990) Growth, size and age composition of *Stolephorus heterolobus* in north Central Java. In *Proceedings of Workshop, Tuna Baitfish in the Indo-Pacific Region* (Ed. S.J.M. Blaber & J.W. Copland), Honiara, Solomon Islands, ACIAR Proceedings No. 30, pp 141–6.

Index

acceptability, 249, 272
acetic acid with RSW, 43
acids with RSW, 43
activation energy, 220
additives, 311
  in sardine mince, 148–156
adenosine triphosphate *see* ATP
aerobic microorganisms, 128
Aeromonas, 128
alanine, 224
albacore, 142, 200
  spoilage in ice, 170–77
Alteromonas, 128
amberjack, 117, 120–24
  landings, 2
amine oxidases, 182
amino acids, 224, 233–6, 303
Ammodytes tobianus see sand eel
anaerobic microorganisms, 128
anchovy, 117, 120–24, 142, 224, 227, 233
  chilled, 286
  landings, 2, 4
Antarctic krill *see* krill
antibiotics, 229
antioxidants, 132, 232, 287, 309
  in sardine mince, 148–56
aquaculture feed *see* fish feed
Arenco-CIS gutting machine, 287
Argentina silus see greater argentine
arginine, 224
Arrhenius, 219, 220
arteriosclerosis *see* heart disease and fish oil
ash content, 119, 223, 225, 232, 233
aspartic acid, 224
Atlantic herring *see* herring, Atlantic
ATP, 126, 127, 128
attracting fish, 22
autolysis, 232
autoxidation, 129
  control, 132
  effect of environment, 130
  effect of fish size, 130
available lysine, 235

bacterial spoilage, 128
  control, 131
bait, 22
behavioural research, 249
belly burst, 127, 285
bigeye tuna, 200
biogenic amines, 178–86, 285
  quantitative analysis, 178
biological value of fish meal, 232
blood clotting, effect of fish oil, 233
blue whiting, 118, 120–24, 214, 291
  dried, 291–8
  landings, 2, 4
  nematodes in, 49
  processing on board, 49
  surimi, 82
bluefin tuna, 142, 200
boiling point of fish oil, 226
Bolivian smoked fish, 243
bonga shad, landings, 2, 4
bonito, 200, 201
bottom trawl, 279
boxing, 58
Brevoortia spp. *see* menhaden
brisling, 286
browning, non-enzymic, 130
burst bellies, 127, 285
by-products, 222–31

cadaverine, 179
calcium, 225
calorific value of fish oil, 226
canned
  fish, 339
  sand eel, 286, 290
  sprat, 286, 290
  tuna, species identification, 200–207
canning, 39, 286
capelin, 117, 120–24, 224, 227, 234, 285
  dried, 291–8
  landings, 2, 4
  roe processing, 46, 285
  sorting, 45

carbon dioxide, 131
  with RSW, 43
catalases, 132
catching methods, 21–37, 330, 342
  effect on fish quality 40, 170–77
chemical deterioration, 129
chilled sea water *see* CSW
chilling, 39, 131
  tanks, 285
chinook salmon, 233
cholesterol level, effect of fish oil, 232
citrates, 132
citric acid, 149
cleaning of RSW tanks, 44
Clostridium spp., 128, 132
Clostridium botulinum, 247
Clupea harengus see herring, Atlantic
Clupea pallasi see herring, Pacific
CO_2 *see* carbon dioxide,
cod, 267, 269, 278, 283
  milt, 306
composition, 119–26
  amino acids, 224, 233–6, 303
  edible flesh, 120–24
  effect on autoxidation, 130
  fatty acids *see* fatty acids
  protein *see* proteins
  proximate, 232
  roe, 303
  variation, 256
  whole fish, 120–24
  see also fish meal, composition; fish oil, composition
consumer
  preference, 249, 339
  reaction, 249, 256, 257, 260–66, 267–71, 274
  studies, 274
  tastes, 269
containers, 66
conversion ratio of fish meal, 238
cooking, 131
copper, 225
coronary disease *see* heart disease and fish oil
Correx penetrometer, 313
Coryphaena hippurus see dorado
Coryphaenoides rupestris see roundnose grenadier
CSW, 39, 65, 248, 287
  containers, 66
  see also RSW
cystine, 224

damage
  by pumping, 68
  to net-caught albacore, 172
dark muscle, 115
Decapterus macrosoma see scad
Decapterus russelli see scad
denaturation, 130
design of krill factory vessels, 94–114
deterioration, 126
  chemical, 129
  effect of fish size, 129
  enzymic, 127
  factors affecting, 126
  microbiological, 128
  physical, 126
  prevention and control, 130
differential scanning calorimetry, 126
dimethylamine, 128
discoloration, 130
docosahexaenoic acid, 142, 233
dolphin fish *see* dorado
dorado, skin colour changes and histamine in, 187–93
draining fish on board, 41
dried-salted fish, 165–9, 339
  chemical evaluation, 166
  nutritional evaluation, 166
drift
  lines, 34
  netting, 32
dryers, 223, 291
drying of fish, 291–8
  drying curves, 295, 296
  energy costs, 297, 298
  final moisture content, 294, 295
  product quality, 294, 296, 297
  stockfish, 291, 293

eastern bloc fleets, 244
EC grading scheme, 215
economic considerations in fisheries development, 257
economics of
  krill factory vessel operation, 94–114
  stern trawlers, 70
Ecuador
  fish handling, 187
  fish industry, 240
  fish mince, 240
  new developments, 241
  processing and marketing, 240
eggs, capelin, 46
eicosapentaenoic acid, 142, 233
elasticity, 208
electrophoresis, 200–207
  agarose gel, 306
  SDS-PAGE, 208
El Niño 14, 241
energy yield from herring, 230
Engraulidae see anchovy
environment, effect on microorganisms, 128
enzyme electrode, 181
enzymes, proteolytic, 70, 72, 127, 228, 229, 235
enzymic
  deterioration, 127
    control, 130
  lipolysis, 127
escabeche marinades, 286
ethoxyquin, 233
ethylene diamine tetracetic acid, 132
Euphausia superba see krill

Euthynnus alleteratus see little tunny
evaporation plant, 223

factory vessels, 94–114
farming, 225
fat, 225, 232
  and water in herring, 125, 157, 158, 163
  cells, 130
  content, 120–24, 224, 233
  measurement, 157–64
    accuracy, 163, 164
  meter, 157–64
  see also fish oil; lipids
fatty acids, 125, 127, 226, 232
  analysis, 143
  see also free fatty acids
feed, 225–8
fermented fish, 339
fertiliser, 222
filleting on board, 48
fish catching methods *see* catching methods
fish composition *see* composition
fish drying *see* drying of fish
fish farming, fish meal for, 40, 225, 235
fish feed, 222, 225–8, 235
fish handling in Ecuador, 187, 240
fish hydrolysate, 228, 229
fish meal, 222–5
  antioxidants in, 132
  ash content, 223, 225, 233
  biological value, 232
  composition, 223–5, 232–4
  conversion ratio in cattle, chicken, pigs, 238
  energy yield, 230
  fat content, 224, 232
  full, 223
  handling of fish for, 222
  manufacture, 222, 223
  minerals, 225
  moisture content, 223, 224, 233
  nutritional value, 234
  oil content, 224, 233
  pelletised, 223
  presscake, 222
  protein content, 223, 224, 232
  quality, 232
    measurement, 235
  salt content, 223
  world production, 225
fish mince *see* minced fish
fish offal, 222, 223, 227, 230
fish oil
  composition, 125, 225
  energy yield, 230
  fatty acids, 225–7
    monoenoic, 225
    polyunsaturated, 225, 232, 257
  heart disease, 142, 233, 257
  manufacture, 222, 223
  physical properties, 226
  triglycerides, 225
  uses, 227
  world production, 226
fish pastes, 128
fish protein concentrate, 222
fish pumping, 48, 56
fish quality, 38–54
fish sauces, 128
fish silage, 47, 227
fish size
  effect on deterioration, 129
  effect on shelf life, 131
fish slime, 129
fish solubles, 222, 223
fishing methods *see* catching methods
  effect on fish quality, 40
  effect on quality of albacore, 170–77
flash point of fish oil, 226
flavour, 127, 211
floating long lines, 33
flow charts for krill products, 108
fluctuation in stocks, 11–16
folding test, 215, 216, 220
food
  choice, 260, 269
  habits, 249
  poisoning, 129, 132, 178, 194–9, 246, 247
formaldehyde, 128
FPC, 222
free fatty acids, 149, 287
freezing, 131
freshness, 214
freshness index, 215, 219
fringescale sardinella, 327
frying
  effect on Indian mackerel, 165–9
  effect on nutritional quality, 165–9

Gadus luscus see pout
gel
  elasticity, 208
  folding test, 215, 216, 220
  setting, 214
  strength, 208, 214–20
  water content, 215
gelling properties, 214, 216, 281
gizzard erosion, 234
gizzerosine, 234
glazing, 130
glucose oxidase, 131
glutamic acid, 224
glutathione, 303
glycine, 224
gonad index, 303
grading, 45, 49, 58, 279, 285
  EC scheme, 215
greater argentine, 278–83, 290
  anatomy, 281
  biology, 278
  catch handling, 279

fishery, 279
gel forming ability, 281
marketing, 283
parietal ribs, 281
spawning period, 280, 281
greater sand eel, 285–90
free fatty acids, 287
hot smoked, 286
thiobarbituric acid value in products, 287
total volatile base nitrogen, 286
gutting machine, 287

haddock, 267, 269
hand lines, 34
handling
in food fisheries, 39
on board, 55–69
sprat for canning, 39
heart disease and fish oil, 142, 233, 257
heat damage to fish meal, 200, 235
heat of fusion of fish oils, 226
heat processing, 214
heat treatment, 131
heavy metals, 233
herring, Atlantic, 117, 120–24, 224, 227, 230, 233, 278, 285, 290, 299–306
biological change, 18
chemical analyses, 77
eggs, 302, 304
energy yield, 230
female, 299
filleting machine, 301
gonad index, 303
history of UK industry, 243
kamaboko, 320
landings, 2
male, 299
matjes, 70
microwave properties, 157, 158
milt *see* herring milt
nematodes, 72
North Sea, 10
Norwegian spring spawning, 9, 303
pickled, 230
roe *see* herring roe
RSW treatment, 73
sensory assessment, 77
sex sorting, 299
stocks, 9
surimi, 317
utilisation, 243, 299–306
water and fat relationship, 125, 158, 163
water content, 157, 158
worms, 72
herring milt, 299–306
composition, 303
DNA, 305
herring, Pacific, 302
herring roe, 299–306
amino acid composition, 303
chemical composition, 303
immature (red rose), 300
kazunoko (salted whole roe), 299, 302, 304
mature light roe, 300
texture, 302
high pressure liquid chromatography, 179
histamine, 126, 129, 178–86, 187–93, 194–9, 246
effect of temperature on formation, 192
enzymatic determination, 181, 185
by enzyme electrode, 181
by fast plate reader, 183
by test strip, 183
formation of, 234, 242, 247
HPLC determination, 179
in iced albacore, 174
in tuna, 185
maximum permitted concentration, 178
qualitative analysis, 179–85
quantitative analysis, 178
histidine, 126, 187, 194, 224, 234
decarboxylase activity, 246
hoki, 232, 234
Hoplostethus atlanticus see orange roughy
hormones, 229
horse mackerel, 117, 121, 142, 214, 227, 243, 290, 317
landings, 2, 4, 5
HPLC, 179
human consumption of pelagic fish, 238–53
Humberside, 244
hydrolysate(s), 128, 222, 228, 229
hydrolysis, 127
hydroperoxides, 129
Hyperoplus lanceolatus see greater sand eel
hypoxanthine, 128
in iced albacore, 174

ice, handling, 41
icing, 41
India, Central Institute of Fisheries Technology (CIFT), 240
south-western, 239
utilisation of pelagic fish, 239–40
Indian mackerel, 117, 120–24, 165–9, 327
landings, 7
Indian oil sardine *see* sardine, Indian oil
Indonesia *see* Java Sea fishery
Indo-Pacific mackerel *see* Indian mackerel
industrial fishing, 279, 291
infra red light in sex sorting, 300
insulation of RSW tanks, 42
iodine, 225
value, 227
Ireland, 244
iron, 225
irradiation, 132
isoleucine, 224

jack mackerel, 117, 120–24
Japanese sardine, 214
Java Sea fishery, 325–43
developments, 340

fishing methods, 330, 342
processing, distribution and marketing, 338
production, 327, 332
purse seine fleet, 328–36
resources, 326
species, 327, 334, 338
storage, 336

kamaboko, 208, 214–21, 250
colour, 317, 321, 323
production, 317–24
texture measurement, 320–23
Katsuwonus pelamis see skipjack tuna
kazunoko, 299
Kieler sprats, 286
krill, 118, 120–24
landings, 9
products, 94
krill-finfish fishery, techno-economic model, 94–114
kroepoek, 287, 288, 290

lactic-acid-producing bacteria, 131, 227, 229
lampara net, 28
landings *see* pelagic fish
LDCs (less developed countries), 238, 240, 251, 250
malnutrition, 238
leucine, 224
lift nets, 34
Likert agree/disagree scale, 261
lipase, 127
lipases, thermal stability, 131
lipid
autoxidation, 129
oxidation, control, 132
lipids, 229
analysis, 143
in pelagic fish, 142–7
seasonal variation, 143
size effect, 143
see also fat; fish oil
lipolysis, 127, 232
lipoxidase, 129
liquor loss, 127
little tunny, 200
locating fish, 21
long lines, 33
lysine, 224, 235
lysozyme, 129

maasbanker, 224
machines
filleting, 49, 301
grading, 49
gutting, 287, 120–24
mackerel, 117, 120–24, 142, 227, 233, 234, 240, 243–50, 255, 307–310
botulism, 247
canned, 242, 248
chemical analyses, 77
consumption in UK, 245
CSW systems, 248
deboned flesh *see* mackerel mince
fat content, 245
'fluffy' hot smoked, 247
handling and processing, 245
history of UK industry, 243
hot smoked, 247, 249
kamaboko, 320
landings, 2, 4, 5, 243–5
markets, 245
minced *see* mackerel mince
modified atmosphere packaging, 249
product innovation, 249
rancidity, 247
RSW treatment, 73
salted, 242
scombrotoxin, 246
sensory assessment, 77
smoked, 242
soft condition, 247
stock, 16
surimi, 317
texture modification by papain, 248
see also Indian mackerel
mackerel mince, 307–310
control of fat content, 309
frozen storage, 309
product concepts, 308
texture modification, 307, 308, 309
Macruronus novaezelandiae see hoki
magnesium, 225
mahi-mahi *see* dorado
Malacca Strait, 325, 331
Mallotus villosus see capelin
malnutrition in less developed countries, 238
manganese, 225, 233
marinades, 131, 286
marketing of small pelagic fish, 254
matjes herring, 70
meal *see* fish meal
meat bone separator, 285
menhaden, 117, 120–24, 224, 227
landings, 2, 3, 4
mesophilic microorganisms, 128
Metal Box Company, 240
methionine, 224
methods of catching, 21–37
microbiological deterioration, 128
control, 131
Micromesistius poutassou see blue whiting
microorganisms, 128
microwave properties of herring, 157, 158
mid water
trawling, 255
trawl net, 31
minced fish, 208–213, 214, 285, 287, 308
dewatering, 215
washing, 208–213
see also mackerel mince; surimi
minced fish product, 272

minerals, 224, 233
model for krill vessel design, 94–114
modified atmosphere packaging, 249
modori effect, 214
moisture content *see* water content of fish
monoenoic fatty acids, 225
multidimensional scaling, 261, 264
muscle, dark and white, 115

nematodes in herring, 72
net shaker, 57
Netherlands, 70–72
new product evaluation, 272–7
nitrogen
  non-protein, 209, 213, 215
  protein, 209
non-enzymic browning, 130
North Sea herring, stocks, 10
Norway pout, 227, 285–90
Norwegian Fisheries Research Council, 299
Norwegian herring, stocks, 9, 299–306
nucleic acid, 229
nucleotides, 229
nutritional
  quality, changes in, 165–9
  value of fish meal, 234

offal, 222, 223, 227, 230
oil *see* fat; fish oil; lipids
omega-3 fatty acids, 142, 233, 257
  health effects, 142, 233, 257
organe roughy, 232
orange roughy, 232
organic acids with RSW, 43
Overseas Development Administration (ODA), 240
oxidation
  chemical evaluation, 149
  control, 132
  of lipids, 129, 233
  of sardine mince, 148–156
oxidative rancidity, 129

Pacific herring *see* herring, Pacific
Pacific saury *see* saury
packaging, 131, 276
  drip loss, 249
  hot smoked mackerel, 249
  in modified atmosphere, 249
  pack collapse, 249
  toxin production, 249
  vacuum, 247, 249
pair trawling, 31
palm oil, 226
papain, texture modification by, 248
pasteurisation, 132
pelagic fish
  definition, 1
  geographical distribution, 1, 3, 226
  human consumption, 238–53
  landings, 38, 325–43
  migration, 16
  problems in utilisation, 238
  resources, 1–20, 255, 325
  stock management, 18
  utilisation, 238–53
  world landings, 2, 9
pelagic trawl, 279
penetration test, 215
peptides, 128, 200
perosis, 233
peroxide value, 149
Peruvian anchoveta, stock fluctuation, 9, 12, 13
pH of sardine muscle, 220
pharmaceutical products from fish, 229
phenylalanine, 224
phosphates, 132
phospholipids, 129
phosphorus, 225, 233
physical deterioration, 126–7
physical stress, effect on fish quality, 40
pickled herring, 230
pilchard, 117, 120–24, 142–7, 208–13, 215–21, 227, 233, 243, 257
  landings, 2, 3, 4
pilchard meal and oil, 224
pilchard, South American, 240
pindang (salt boiled fish), 338
poisoning, scombroid *see* scombrotoxin
pole and line fishing, 36
polyphosphates, 149
polyunsaturated fatty acids *see* PUFA
potassium, 225
pout, 214
power block, 25
preservation of quality, 38–54
preservation with RSW, 43
presscake, 222
press liquor, 222
pressure/vacuum pump, 56
prevention of deterioration, 130
processing on board, 38–54
processing pre-rigor fish, 127
product development, 272
product evaluation, 272–7
proline, 224
protein hydrolysates, 128
proteins, 125
  actin, 210
  alkali soluble, 208–213
  content, 223, 232
  denaturation, 130, 211
  digestibility, 234
  myofibrillar, 208–213
  myosin, 210
  salt extractable, 215
  salt insoluble, 215, 220
  salt soluble, 215, 220
  sarcoplasmic, 126, 200, 208–213
  solubility, 215, 220
  stroma, 208–213

thermal stability, 126
tropomyosin, 210
troponin, 210
utilisable, 233, 234
water soluble, 211, 215, 220, 222
proteolysis, 127, 211
proteolytic enzymes, 70, 72, 127, 228, 235
Pseudocyttus maculatus see oreo dory
Pseudomonas, 128
psychrophilic microorganisms, 128
PUFA, 125, 129, 142–57, 225, 232, 257
puncture test for surimi, 315
purse seining, 23–7, 240, 248, 325–43
putrescine, 179

quality
assessment, 216
fish meal, 232
index, 215

rainbow trout, 234
rancidity, 129
Rastrelliger brachysoma see Indian mackerel
Rastrelliger kanagutta see Indian mackerel
rate constant, 219
recruitment to stocks, 12
refrigerated sea water *see* RSW
resources, 1–20, 255, 325
response surface methodology (RSM), 309, 311–16
designs, 312, 313
product development, 311–16
rigor mortis, 127, 220
extension, 127
ring net, 28
roe
capelin, 46
herring, 299–306
Ronoxan D20 (antioxidant), 309
roundnose grenadier, 278–84, 290
anatomy, 281
biology, 278
catch handling, 279
economics of filleting, 281
fillets, 281
fishery, 279
hand and machine filleting, 282, 283
marketing, 283
yield, 281
RSW, 41–5, 64, 73–81, 233, 241, 248
see also CSW
herring
chemical analyses, 77
sensory assessment, 77
mackerel
chemical analyses, 77
sensory assessment, 77
tanks
design, 80
temperature in, 73

Salmo gairdneri see rainbow trout
salmon, 233, 248
salmonella, elimination by irradiation, 132
salt, 131
salt content of fish meal, 223, 224
salted dried fish, 339
salt uptake in RSW, 41, 223
sand eel, 227, 285–90
canned products, 286, 290
dried, 285–90
fish meal, 285–90
oil, 285–90
oil content, 286
spawning, 286
sarcoplasmic proteins, 126, 200, 208–213
Sarda sarda see bonito
Sardina pilchardus see pilchard
sardine *see* pilchard
sardine, Indian oil (*Sardinella longiceps*), 239–40, 327
curing, 240
development of fishery, 239
handling and processing, 239
in flexible retortable pouches, 240
markets, 239
new developments, 240
salted, 240
sardine mince, oxidation, 148–56
sardinella, 118, 327
landings, 2, 3, 4
Sardinella fimbriata see fringescale sardinella
Sardinella longiceps see sardine, Indian oil
Sardinops melanosticta, 214, 220
Sardinops sagax see South American pilchard
saury, 118, 120–24
landings, 2, 4
scad, 118, 120–24, 327
scombrotoxin, 126, 129, 178, 194–9, 246
seasonal variation of lipids, 143
Selar spp., 337
selenium, 225, 233
semi pelagic trawl, 279
sensory
analysis, 215, 269, 273, 287–9
evaluation, 219, 269, 286
preference, 267–71
serine, 224
serum triglyceride, 232
sex sorting, 45, 299
shelf life, 309, 338
effect of fish size, 131
effect of packaging, 132
shellfish, 287
shrimp, 241, 287
bycatch, 278
silage, 47, 227
size effect on lipids, 143
size grading, 58, 279, 285
skimmed milk powder, 288
skin colour change in dorado, 187–93

skipjack tuna, 120–24, 200–207
  landings, 2, 4, 8
slime, 129
slip melting point of fish oil, 226
smoke, 131
smoked fish, 242, 247
  greater sand eel, 286, 290
  sand eel, 286
sodium, 225
solubles, 222, 223
sorbitol, 149, 215
sorting, 45, 58, 279. 285, 299
South American pilchard, 240
soybean oil, 226
species identification, 126, 200–207
specific gravity of fish oil, 226
specific heat of fish oil, 226
spermidine, 179
spoilage of albacore
  chemical analyses, 171
  effect of fishing method, 170–77
  an ice, 170–77
  microbiological methods, 171
  sensory assessment, 171
sprat, 118, 120–24, 227, 243, 285–90
  canned, 286, 290
  in RSW/CSW, 286
  landings, 2, 4
Sprattus sprattus see sprat
squid, landings, 9
starch, added to kamaboko, 215
sterilisation, 132
stick nets, 34
stickwater, 222, 223, 232
stockfish drying, 291, 293
stocks
  fluctuation, 11–16
  management, 18
  recruitment, 12
storage
  chilled, 214
  frozen, 214–19
  iced, 208–213, 220
stress, effect on fish quality, 40
stroke *see* heart disease and fish oil
structure of fish, 115
submersible pump, 56
sucrose, added to kamaboko, 215
sulphite in enzymic deterioration, 131
sulphur dioxide, 132
superchilling, 131
surimi, 49, 82, 211, 214–21, 257, 287, 289, 317–24
  chemical analyses, 84
  colour, 84, 321, 323
  effect of initial pH, 214
  effect of washing, 89
  effects of processing variables, 84
  fat content, 317
  fat removal, 317, 319, 321
  from blue whiting, 50
  from capelin, 51
  from sardine, 214–21
  manufacture on board, 50, 82–93
  moisture content, 318
  pH, 318
  physical properties, 84, 214
  production, 317–24
  properties, 317–24
  see also kamaboko; minced fish

techno-economic model for krill vessel design, 94–114
temperature effect on
  autoxidation, 130
  bacterial spoilage, 131
  microorganisms, 128
texture, 214, 248, 275, 307
thermal stability of proteins, 126
thiobarbituric acid value, 149, 287
threonine, 224
thrombosis *see* heart disease
Thunnus alalunga see albacore
Thunnus albacares see yellowfin tuna
Thunnus obesus see bigeye tuna
Thunnus thynnus see bluefin tuna
TMA
  in blue whiting, 84
  in iced albacore, 172
tocopherols, 132
total volatile base nitrogen *see* TVN
total volatile bases *see* TVN
total volatile nitrogen *see* TVN
toxins, 285
  see also scombrotoxin
Trachurus spp. *see* horse mackerel
trawling, 30, 279
trevally, 337
triglycerides, 129, 225, 232
trimethylamine oxide, 128
trimethylamine *see* TMA
Trisopterus esmarkii see Norway pout
trolling, 34
tropical fish, deterioration, 128
tryptophan, 224
tuna, 118, 123, 142, 224
  dark meat, 243
  species identification, 200–207
TVN, 285, 286
  in blue whiting, 84
  in fish for fish meal, 40
  in iced albacore, 172
two boat purse seining, 26
tyramine, 179
tyrosine, 224

ubiquinones, 132
ultra-violet radiation in control of bacteria, 44
United Kingdom, utilisation of pelagic fish in, 243–50
unloading, 67

valine, 224
vibrio, 128
viscosity of fish oil, 226
vitamins, 227, 257
VMK filleting machine, 301
volatile amines *see* TVN

washing of sardine mince, 148–56
water and fat in herring, 125, 158, 163
water content of fish, 119, 157, 158, 215, 223, 232
White Fish Authority (WFA), 248
white muscle, 115
wood smoke, 131
world catch, use of, 239
world production of fish meal and oil, 225, 227
worms in herring, 72

yellow colour change in dorado, 187–93
yellowfin tuna, 200
yield, 117, 281

zinc, 225